INFORMATION THEORY AND STATISTICS

Solomon Kullback

DOVER PUBLICATIONS, INC.
Mineola, New York

Published in Canada by General Publishing Company, Ltd., 30 Lesmill Road, Don Mills, Toronto, Ontario.
Published in the United Kingdom by Constable and Company, Ltd., 3 The Lanchesters, 162–164 Fulham Palace Road, London W6 9ER.

Bibliographical Note

This Dover edition, first published in 1997, is an unabridged republication of the Dover 1968 edition which was an unabridged republication of the work originally published in 1959 by John Wiley & Sons, New York, with a new preface and corrections and additions by the author. The brief note about the author which appears on p. xvi of this edition was reprinted from the program of the Kullback Memorial Research Conference held in Washington, D. C., May 23–25, 1996.

Library of Congress Cataloging-in-Publication Data

Kullback, Solomon.
Information theory and statistics / Solomon Kullback.
p. cm.
"An unabridged republication of the Dover 1968 edition which was an unabridged republication of the work originally published in 1959 by John Wiley & Sons, New York, with a new preface and corrections and additions by the author"—T.p. verso.
Includes bibliographical references (p. –) and index.
ISBN 0-486-69684-7 (pbk.)
1. Mathematical statistics. 2. Information theory. I. Title.
QA276.K8 1997 97–14382
519.5—dc21 CIP

Manufactured in the United States of America
Dover Publications, Inc. 31 East 2nd Street, Mineola, N.Y. 11501

To Minna

26 September 1908—16 November 1966

Preface to the Dover Edition

This Dover edition differs from the original edition in certain details. A number of misprints have been corrected, some tables have been revised because of computer recalculation and certain erroneous statements have been corrected.

Major additions include theorem 2.2 in chapter 3, theorem 2.1a and lemma 2.2 in chapter 4, and table 8.6 in chapter 12. Section 5.4 of chapter 12 has been greatly revised, as have been examples 5.4 and 5.5 of that section. Some of these changes have been incorporated into the body of the text; others are to be found in the appendix, which begins on page 389.

I am grateful to colleagues and correspondents who expressed interest in the work and raised pertinent questions about certain statements or numerical values in the original edition. I must also express my appreciation for the partial support by the National Science Foundation under Grant GP-3223 and the Air Force Office of Scientific Research, Office of Aerospace Research, United States Air Force under Grant AF-AFOSR 932-65 in the revision of the original edition and further research. Finally, I am grateful to Dover Publications, Inc. for their interest and preparation of this edition.

SOLOMON KULLBACK

The George Washington University
November 1967

Preface to the First Edition

Information in a technically defined sense was first introduced in statistics by R. A. Fisher in 1925 in his work on the theory of estimation. Fisher's definition of information is well known to statisticians. Its properties are a fundamental part of the statistical theory of estimation.

Shannon and Wiener, independently, published in 1948 works describing logarithmic measures of information for use in communication theory. These stimulated a tremendous amount of study in engineering circles on the subject of information theory. In fact, some erroneously consider information theory as synonymous with communication theory.

Information theory is a branch of the mathematical theory of probability and mathematical statistics. As such, it can be and is applied in a wide variety of fields. Information theory is relevant to statistical inference and should be of basic interest to statisticians. Information theory provides a unification of known results, and leads to natural generalizations and the derivation of new results.

The subject of this book is the study of logarithmic measures of information and their application to the testing of statistical hypotheses. There is currently a heterogeneous development of statistical procedures scattered through the literature. In this book a unification is attained by a consistent application of the concepts and properties of information theory. Some new results are also included.

The reader is assumed to have some familiarity with mathematical probability and mathematical statistics. Since background material is available in a number of published books, it has been possible here to deal almost exclusively with the main subject. That this also covers classical results and procedures is not surprising. The fundamentals of information theory have been known and available for some time and have crystallized in the last decade. That these fundamentals should furnish new approaches to known results is both useful and necessary.

The applications in this book are limited to the analysis of samples of fixed size. Applications to more general stochastic processes, including sequential analysis, will make a natural sequel, but are outside the scope of this book.

In some measure this book is a product of questions asked by students and the need for a presentation avoiding special approaches for problems that are essentially related. It is my hope that the experienced statistician will see in this book familiar things in a unified, if unfamiliar, way, and that the student will find this approach instructive.

In chapter 1, the measures of information are introduced and defined. In chapter 2, I develop the properties of the information measures and examine their relationship with Fisher's information measure and sufficiency. In chapter 3, certain fundamental inequalities of information theory are derived, and the relation with the now classic inequality associated with the names of Fréchet, Darmois, Cramér, and Rao is examined. In chapter 4, some limiting properties are derived following the weak law of large numbers. In chapter 5, the asymptotic distribution theory of estimates of the information measures is examined.

The developments in these first five chapters use measure theory. The reader unfamiliar with measure theory should nevertheless be able to appreciate the theorems and follow the argument in terms of the integration theory familiar to him by considering the integrals as though in the common calculus notation.

The rest of the book consists of applications. In chapters 6, 7, and 8, the analysis of multinomial samples and samples from Poisson populations is studied. The analysis of contingency tables in chapter 8 depends on the basic results developed in chapter 6. Chapter 9 is essentially an introduction to various ideas associated with multivariate normal populations. In chapter 10, the analysis of samples from univariate normal populations under the linear hypothesis is studied and provides the transition to the generalizations in chapter 11 to the multivariate linear hypothesis. In chapter 12, the analysis of samples from multivariate normal populations for hypotheses other than the linear hypothesis is developed. The familiar results of the single-variate normal theory are contained in the multivariate analyses as special cases. In chapter 13, some general questions on linear discriminant functions are examined and raised for further investigation.

The book contains numerous worked examples. I hope that these will help clarify the discussion and provide simple illustrations. Problems at the end of each chapter and in the text provide a means for the reader to expand and apply the theory and to anticipate and develop some of the needed background.

The relevance of information theory to statistical inference is the unifying influence in the book. This is made clear by the generalizations that information theory very naturally provides. Chapters 8, 11, and 12 demonstrate this. In section 4 of chapter 11, it is concluded that the test statistic for the multivariate generalization of the analysis of variance is a form of Hotelling's generalized Student ratio (Hotelling's T^2). The basic facts on which this conclusion rests have been known for some time. Information theory brings them together in the proper light.

Sections are numbered serially within each chapter, with a decimal notation for subsections and sub-subsections; thus, section 4.5.1 means section 4, subsection 5, sub-subsection 1. Equations, tables, figures, examples, theorems, and lemmas are numbered serially within each section with a decimal notation. The digits to the left of the decimal point represent the section and the digits to the right of the decimal point the serial number within the section; for example, (9.7) is the seventh equation in section 9. When reference is made to a section, equation, table, figure, example, theorem, or lemma within the *same* chapter, only the section number or equation, etc., number is given. When the reference is to a section, equation, etc., in a *different* chapter, then in addition to the section or equation, etc., number, the chapter number is also given.

References to the bibliography are by the author's name followed by the year of publication in parentheses.

Matrices are in boldface type. Upper case letters are used for square or rectangular matrices and lower case letters for one-column matrices (vectors). The transpose of a matrix is denoted by a prime; thus one-row matrices are denoted by primes. A subscript to a matrix implies that the subscript also precedes the subscripts used to identify the elements within a matrix, for example, $\mathbf{A} = (a_{ij})$, $\mathbf{A}_2 = (a_{2ij})$, $\mathbf{x}' = (x_1, x_2, \cdots, x_k)$. There are some exceptions to these general rules, but the context will be clear.

An abbreviated notation is generally used, in the sense that multiple integrals are expressed with only a single integral sign, and single letters stand for multidimensional variables or parameters. When it is considered particularly important to stress this fact, explicit mention is made in the text.

A glossary is included and is intended to supplement the reader's background and, with the index, to provide easy access to definitions, symbols, etc.

SOLOMON KULLBACK

The George Washington University
February 1958

Acknowledgment

Critical comments and questions from friends, colleagues, and referees have improved the exposition in this work. Its shortcomings are mine alone. Thanks are due my students and colleagues at The George Washington University for their interest, understanding, and encouragement. The full and hearty support of Professor Frank M. Weida is gratefully acknowledged. Mr. Harry M. Rosenblatt practically prepared most of sections 2 through 8 of chapter 10. Mr. Samuel W. Greenhouse has practically prepared chapter 13. Dr. Morton Kupperman and Mr. Austin J. Bonis have read versions of the manuscript most carefully and critically, and their remarks, although not always adopted, were invariably helpful. Other contributions are acknowledged in the text. The careful and uniformly excellent typing of Mrs. Frances S. Brown must be mentioned, since it lightened the problems of preparing the manuscript.

S. K.

Contents

About the Author

Solomon Kullback received his Ph.D. from the George Washington University in 1934. Dr. Kullback was employed in the Department of the Army and subsequently in the Department of Defense until his retirement in 1962. He was one of the small group of scholars responsible for breaking the Japanese Code in World War II. During the latter part of his career, he was Director of Research in the National Security Agency.

Dr. Kullback taught part-time at GW from 1938 until his retirement from the Department of Defense. He then became a Professor of Statistics at GW, chairing the Department from 1964 to 1972. Dr. Kullback made major contributions in statistical theory and methodology. He introduced the concept of information theory to statistics—a topic on which he wrote extensively.

Adapted from the May 7, 1977 citation used when Dr. Kullback received the George Washington University Distinguished Alumni Award.

CHAPTER 1

Definition of Information

1. INTRODUCTION

Information theory, as we shall be concerned with it, is a branch of the mathematical theory of probability and statistics. As such, its abstract formulations are applicable to any probabilistic or statistical system of observations. Consequently, we find information theory applied in a variety of fields, as are probability and statistics. It plays an important role in modern communication theory, which formulates a communication system as a stochastic or random process. Tuller (1950) remarks that the statistical theory of communications is often called information theory. Rothstein (1951) has defined information theory as "abstract mathematics dealing with measurable sets, with choices from alternatives of an unspecified nature." Pierce (1956, p. 243) considers communication theory and information theory as synonyms. Gilbert (1958, p. 14) says, "Information will be a measure of time or cost of a sort which is of particular use to the engineer in his role of designer of an experiment." The essential mathematical and statistical nature of information theory has been reemphasized by three men largely responsible for its development and stimulation, Fisher (1956), Shannon (1956), Wiener (1956).

In spirit and concepts, information theory has its mathematical roots in the concept of disorder or entropy in thermodynamics and statistical mechanics. [See Fisher (1935, p. 47) and footnote 1 on p. 95 of Shannon and Weaver (1949).] An extensive literature exists devoted to studies of the relation between the notions and mathematical form of entropy and information. Stumpers (1953) devotes pp. 8–11 of his bibliography to such references, and some others are added here: Bartlett (1955, pp. 208–220), Brillouin (1956), Cherry (1957, pp. 49–51; 212–216), Fisher (1935, p. 47), Grell (1957, pp. 117–134), Joshi (1957), Khinchin (1953, 1956, 1957), Kolmogorov (1956), McMillan (1953), Mandelbrot (1953, 1956), Powers (1956), Quastler (1953, pp. 14–40).

R. A. Fisher's (1925b) measure of the amount of information supplied by data about an unknown parameter is well known to statisticians. This measure is the first use of "information" in mathematical statistics, and was introduced especially for the theory of statistical estimation. Hartley (1928) defined a measure of information, the logarithm of the number of possible symbol sequences, for use in communication engineering. Interest in, and various applications of, information theory by communication engineers, psychologists, biologists, physicists, and others, were stimulated by the work of Shannon (1948) and Wiener (1948), particularly by Wiener's (1948, p. 76) statement that his definition of information could be used to replace Fisher's definition in the technique of statistics. However, note that Savage (1954, p. 50) remarks: "The ideas of Shannon and Wiener, though concerned with probability, seem rather far from statistics. It is, therefore, something of an accident that the term 'information' coined by them should be not altogether inappropriate in statistics." Powers (1956, pp. 36–42) reviews the fundamental contributions of Wiener, Shannon, and Woodward as an introduction to his development of a unified theory of the information associated with a stochastic process. Indeed, Stumpers (1953) lists 979 items in his bibliography and only 104 of these were published prior to 1948. Although Wald (1945a, 1945b, 1947) did not explicitly mention information in his treatment of sequential analysis, it should be noted that his work must be considered a major contribution to the statistical applications of information theory. [See Good (1950, pp. 64–66), Schützenberger (1954, pp. 57–61).]

For extensive historical reviews see Cherry (1950, 1951, 1952, 1957). A most informative survey of information theory in the U.S.S.R. is given by Green (1956, 1957), who takes information theory to mean "the application of statistical notions to problems of transmitting information." The current literature on information theory is voluminous. Some references are listed that will give the reader who scans through them an idea of the wide variety of interest and application: Ashby (1956), Bell (1953), Bradt and Karlin (1956), Brillouin (1956), de Broglie (1951), Castañs Camargo (1955), Cherry (1955, 1957), Davis (1954), Elias (1956), Fano (1954), Feinstein (1958), Gilbert (1958), Goldman (1953), Good (1952, 1956), Jackson (1950, 1952), Jaynes (1957), Kelly (1956), Lindley (1956, 1957), McCarthy (1956), McMillan et al., (1953), Mandelbrot (1953), Quastler (1953, 1955), Schützenberger (1954), Shannon and Weaver (1949), Wiener (1948, 1950), Woodward (1953).

We shall use information in the technical sense to be defined, and it should not be confused with our semantic concept, though it is true that the properties of the measure of information following from the technical definition are such as to be reasonable according to our intuitive notion of information. For a discussion of "semantic information" see Bar-Hillel (1955), Bar-Hillel and Carnap (1953).

Speaking broadly, whenever we make statistical observations, or design and conduct statistical experiments, we seek information. How much can we infer from a particular set of statistical observations or experiments about the sampled populations? [Cf. Cherry (1957, p. 61).] We propose to consider possible answers to this question in terms of a technical definition of a measure of information and its properties. We shall define and derive the properties of the measure of information at a mathematical level of generality that includes both the continuous and discrete statistical populations and thereby avoid the necessity for parallel considerations of these two common practical situations [Fraser (1957, pp. 1–16), Powers (1956)].

2. DEFINITION

Consider the *probability spaces* $(\mathcal{X}, \mathcal{S}, \mu_i)$, $i = 1, 2$, that is, a basic set of elements $x \in \mathcal{X}$ and a collection $\mathcal{S}$ of all possible events (sets) made up of elements of the *sample space* $\mathcal{X}$ for which a probability measure μ_i, $i = 1, 2$, has been defined. $\mathcal{S}$ is a σ-algebra of subsets of $\mathcal{X}$, a Borel field, or an additive class of measurable subsets of $\mathcal{X}$. The pair $(\mathcal{X}, \mathcal{S})$, that is, the combination of the sample space $\mathcal{X}$ and the σ-algebra $\mathcal{S}$ of subsets of $\mathcal{X}$, is called a *measurable space* [Fraser (1957, p. 2)]. The elements of $\mathcal{X}$ may be univariate or multivariate, discrete or continuous, qualitative or quantitative [Fraser (1957, pp. 1–2)]. For an engineer, the elements of $\mathcal{X}$ may be the occurrence or nonoccurrence of a signal pulse, $\mathcal{S}$ may be a collection of possible sequences of a certain length of pulse and no pulse, and μ_1 and μ_2 may define the probabilities for the occurrence of these different sequences under two different hypotheses. For a statistician, the elements of $\mathcal{X}$ may be the possible samples from a univariate normal population, $\mathcal{S}$ may be the class of Borel sets of R^n, n-dimensional Euclidean space (if we are concerned with samples of n independent observations), and μ_1 and μ_2 may define the probabilities of the different samples for different values of the parameters of the populations.

We assume that the probability measures μ_1 and μ_2 are *absolutely continuous* with respect to one another, or in symbols, $\mu_1 \equiv \mu_2$; that is, there exists no set (event) $E \in \mathcal{S}$ for which $\mu_1(E) = 0$ and $\mu_2(E) \neq 0$, or $\mu_1(E) \neq 0$ and $\mu_2(E) = 0$. [μ_1 is absolutely continuous with respect to μ_2, $\mu_1 \ll \mu_2$, if $\mu_1(E) = 0$ for all $E \in \mathcal{S}$ for which $\mu_2(E) = 0$; μ_2 is absolutely continuous with respect to μ_1, $\mu_2 \ll \mu_1$, if $\mu_2(E) = 0$ for all $E \in \mathcal{S}$ for which $\mu_1(E) = 0$.] Since there is no essential problem in the rejection of statistical hypotheses that may have been possible prior to the observations but are impossible after the observations, our mathematical assumption is such as to exclude this contingency. According to Savage

(1954, p. 127), "· · · definitive observations do not play an important part in statistical theory, precisely because statistics is mainly concerned with uncertainty, and there is no uncertainty once an observation definitive for the context at hand has been made." For further study of absolute continuity see Fraser (1957, p. 12), Halmos (1950, pp. 124–128), Halmos and Savage (1949), Loève (1955, pp. 129–132). Let λ be a probability measure such that $\lambda \equiv \mu_1$, $\lambda \equiv \mu_2$; for example, λ may be μ_1, or μ_2, or $(\mu_1 + \mu_2)/2$. By the Radon–Nikodym theorem [Fraser (1957, p. 13), Halmos (1950, pp. 128–132), Loève (1955, pp. 132–134)], there exist functions $f_i(x)$, $i = 1, 2$, called *generalized probability densities*, unique up to sets of measure (probability) zero in λ, measurable λ, $0 < f_i(x) < \infty$ $[\lambda]$, $i = 1, 2$, such that

$$\mu_i(E) = \int_E f_i(x)\, d\lambda(x), \qquad i = 1, 2, \tag{2.1}$$

for all $E \in \mathscr{S}$. The symbol $[\lambda]$, pronounced "modulo λ," following an assertion concerning the elements of $\mathscr{X}$, means that the assertion is true except for a set E such that $E \in \mathscr{S}$ and $\lambda(E) = 0$ [Halmos and Savage (1949)]. The function $f_i(x)$ is also called the Radon-Nikodym derivative, and we write $d\mu_i(x) = f_i(x)\, d\lambda(x)$ and also $f_i(x) = d\mu_i/d\lambda$. In example 7.1 of chapter 2 is an illustration of a probability measure μ_1 absolutely continuous with respect to a probability measure μ_2, but not conversely. If the probability measure μ is absolutely continuous with respect to the probability measure λ, and the probability measure ν is absolutely continuous with respect to the probability measure μ, then the probability measure ν is also absolutely continuous with respect to the probability measure λ, and the Radon–Nikodym derivatives satisfy $\frac{d\nu}{d\lambda} = \frac{d\nu}{d\mu} \cdot \frac{d\mu}{d\lambda}\ [\lambda]$ [Halmos (1950, p. 133), Halmos and Savage (1949)].

If H_i, $i = 1, 2$, is the hypothesis that X (we use X for the generic variable and x for a specific value of X) is from the statistical population with probability measure μ_i, then it follows from Bayes' theorem, or the theorems on conditional probability [Feller (1950), Fraser (1957, pp. 13–16), Good (1950), Kolmogorov (1950), Loève (1955)], that

$$P(H_i|x) = \frac{P(H_i)f_i(x)}{P(H_1)f_1(x) + P(H_2)f_2(x)}\ [\lambda], \qquad i = 1, 2, \tag{2.2}$$

from which we obtain

$$\log \frac{f_1(x)}{f_2(x)} = \log \frac{P(H_1|x)}{P(H_2|x)} - \log \frac{P(H_1)}{P(H_2)}\ [\lambda], \tag{2.3}$$

where $P(H_i)$, $i = 1, 2$, is the prior probability of H_i and $P(H_i|x)$ is the posterior probability of H_i, or the conditional probability of H_i given $X = x$. See Good (1956, p. 62), Savage (1954, pp. 46–50). The base of the logarithms in (2.3) is immaterial, providing essentially a unit of measure; unless otherwise indicated we shall use natural or Naperian logarithms (base e). (See the end of example 4.2.)

The right-hand side of (2.3) is a measure of the difference between the logarithm of the odds in favor of H_1 after the observation of $X = x$ and before the observation. This difference, which can be positive or negative, may be considered as the information resulting from the observation $X = x$, and we define the logarithm of the likelihood ratio, $\log [f_1(x)/f_2(x)]$, as *the information in* $X = x$ *for discrimination in favor of* H_1 *against* H_2. [Cf. Good (1950, p. 63), who describes it also as the *weight of evidence* for H_1 given x.] The mean information for discrimination in favor of H_1 against H_2 given $x \in E \in \mathcal{S}$, for μ_1, is

$$(2.4)\qquad \begin{aligned} I(1:2; E) &= \frac{1}{\mu_1(E)} \int_E \log \frac{f_1(x)}{f_2(x)}\, d\mu_1(x) \\ &= \frac{1}{\mu_1(E)} \int_E f_1(x) \log \frac{f_1(x)}{f_2(x)}\, d\lambda(x), \qquad \mu_1(E) > 0, \\ &= 0, \qquad \mu_1(E) = 0, \end{aligned}$$

with

$$d\mu_1(x) = f_1(x)\, d\lambda(x).$$

When E is the entire sample space $\mathcal{X}$, we denote by $I(1:2)$, rather than by $I(1:2; \mathcal{X})$, the mean information for discrimination in favor of H_1 against H_2 per observation from μ_1, that is, omitting the region of integration when it is the entire sample space,

$$(2.5)\qquad \begin{aligned} I(1:2) &= \int \log \frac{f_1(x)}{f_2(x)}\, d\mu_1(x) = \int f_1(x) \log \frac{f_1(x)}{f_2(x)}\, d\lambda(x) \\ &= \int \log \frac{P(H_1|x)}{P(H_2|x)}\, d\mu_1(x) - \log \frac{P(H_1)}{P(H_2)}. \end{aligned}$$

Note that the last member in (2.5) is the difference between the mean value, with respect to μ_1, of the logarithm of the posterior odds of the hypotheses and the logarithm of the prior odds of the hypotheses. Following Savage (1954, p. 50) we could also call $I(1:2)$ the information of μ_1 with respect to μ_2. Note that the integrals in (2.4) or (2.5) always exist, even though they may be $+\infty$, since the minimum value of the integrand for its negative values is $-1/e$. A necessary condition (but not sufficient) that $I(1:2)$ be finite is $\mu_1 \equiv \mu_2$. As an example in which the mean information $I(1:2)$ is infinite, take $\mathscr{X} = (0, 1)$, μ_1 = Lebesgue measure, $f_2(x)/f_1(x) = ke^{-1/x}$, $k^{-1} = \int_0^1 e^{-1/t}\, dt$. It may be verified that $I(1:2)$ is infinite [Hardy, Littlewood, and Pólya (1934, p. 137)]. See problem 5.7.

3. DIVERGENCE

Following section 2, we may define

$$I(2:1) = \int f_2(x) \log \frac{f_2(x)}{f_1(x)}\, d\lambda(x) \tag{3.1}$$

as the mean information per observation from μ_2 for discrimination in favor of H_2 against H_1, or

$$-I(2:1) = \int f_2(x) \log \frac{f_1(x)}{f_2(x)}\, d\lambda(x)$$

as the mean information per observation from μ_2 for discrimination in favor of H_1 against H_2. Our previous assumption about the mutual absolute continuity of μ_1 and μ_2 ensures the existence of the integral in the definition of $I(2:1)$, even though it may be $+\infty$.

We now define the *divergence* $J(1, 2)$ by

$$\begin{aligned} J(1,2) &= I(1:2) + I(2:1) \\ &= \int (f_1(x) - f_2(x)) \log \frac{f_1(x)}{f_2(x)}\, d\lambda(x) \\ &= \int \log \frac{P(H_1|x)}{P(H_2|x)}\, d\mu_1(x) - \int \log \frac{P(H_1|x)}{P(H_2|x)}\, d\mu_2(x). \end{aligned} \tag{3.2}$$

The middle version of the above expressions for $J(1, 2)$ was introduced by Jeffreys (1946, 1948, p. 158), but he was mainly concerned with its use, because of invariance under transformation of parameters, as providing a prior probability density for parameters. $J(1, 2)$ is a measure of the divergence between the hypotheses H_1 and H_2, or between μ_1 and μ_2, and is a measure of the difficulty of discriminating between them. [Cf. Chernoff (1952), Huzurbazar (1955), Jeffreys (1948, p. 158), Kullback (1953), Sakaguchi (1955), Suzuki (1957).] Note that $J(1, 2)$ is symmetric with respect to μ_1 and μ_2, and the prior probabilities $P(H_i)$, $i = 1, 2$, do not appear. The divergence $J(1, 2)$ (as will be seen) has all the properties of a distance (or metric) as defined in topology except the triangle inequality property and is therefore not termed a distance. The information measures $I(1:2)$ and $I(2:1)$ may in this respect be considered as *directed divergences.* (See problem 5.9.) For other measures of distances between probability distributions see Adhikari and Joshi (1956), Bhattacharyya (1943, 1946a), Bulmer (1957), Fraser (1957, p. 127), Rao (1945, 1952, pp. 351–352).

4. EXAMPLES

Before we consider properties resulting from the definition of information and divergence, and supporting the use of "information" as a name, it may be useful to examine some instances of (2.3), (2.5), and (3.2) for illustration and background.

Example 4.1. As an extreme case, suppose that H_2 represents a set of hypotheses, one of which must be true, and that H_1 is a member of the set of hypotheses H_2; then $P(H_2) = 1$, $P(H_2|x) = 1$, and the right-hand side of (2.3) yields as the information in x in favor of H_1 the value $\log P(H_1|x) - \log P(H_1) = \log [P(H_1|x)/P(H_1)]$. When is this value zero? If the observation x proves that H_1 is true, that is, $P(H_1|x) = 1$, then the information in x about H_1 is $-\log P(H_1)$ [Good (1956)]. Note that when H_1 is initially of small probability the information resulting from its verification is large, whereas if its probability initially is large the information is small. Is this intuitively reasonable?

Example 4.2. To carry this notion somewhat further, suppose a set of mutually exclusive and exhaustive hypotheses $H_1, H_2, \cdots, H_n$ exists and that from an observation we can infer which of the hypotheses is true. For example, we may have a communication system in which the hypotheses are possible messages, there is no garbling of the transmitted message, and there is no uncertainty about the inference after receiving the message. Or we may be dealing with an experiment for which the outcome may be one of n categories, there are no errors of observation, and there is no uncertainty about the inference of the category after making the observation. Here, the mean information in an observation about the hypotheses is the mean value of $-\log P(H_i)$, $i = 1, 2, \cdots, n$, that is,

$$(4.1) \qquad -P(H_1)\log P(H_1) - P(H_2)\log P(H_2) - \cdots - P(H_n)\log P(H_n).$$

The expression in (4.1) is also called the entropy of the H_i's. See Bell (1953), Brillouin (1956), Goldman (1953), Good (1950, 1956), Grell (1957), Joshi (1957), Khinchin (1953, 1956, 1957), McMillan (1953), Quastler (1956), Shannon (1948), Woodward (1953). When logarithms to base 2 are used, the unit of the (selective) information in (4.1) is called a "bit" (binary digit), and it turns out that one bit of information is the capability of resolving the uncertainty in a situation with two equally probable hypotheses or alternatives. Thus, in a "yes" or "no" selection with a probability of $\frac{1}{2}$ for each alternative, $-\frac{1}{2}\log_2\frac{1}{2} - \frac{1}{2}\log_2\frac{1}{2} = \log_2 2 = 1$ "bit." When the n hypotheses are equally probable, so that $P(H_i) = 1/n$, $i = 1, \cdots, n$, we find that $-\sum_{i=1}^{n} P(H_i)\log P(H_i) = \log n$, Hartley's information measure.

It has been suggested that when logarithms to base 10 are used, the unit of information in (4.1) be called a "Hartley" [Tuller (1950)], and when natural logarithms to base e are used, the unit of information be called a "nit" [MacDonald (1952)].

Example 4.3. As another area of illustration, suppose that the sample space $\mathscr{X}$ is the Euclidean space R^2 of two dimensions with elements $X = (x, y)$, and that under H_1 the variables x and y are dependent with probability density $f(x, y)$, but that under H_2, x and y are independent, with respective probability densities $g(x)$ and $h(y)$. Now (2.5) may be written as

$$I(1:2) = \int\int f(x, y) \log \frac{f(x, y)}{g(x)h(y)}\, dx\, dy, \tag{4.2}$$

which has also been defined as the mean information in x about y, or in y about x. See Gel'fand, Kolmogorov, and Iaglom (1956), Good (1956), Kolmogorov (1956), Lindley (1956), Shannon (1948), Woodward (1953, pp. 53–54). Since, as will be shown in theorem 3.1 of chapter 2, $I(1:2)$ in (4.2) is nonnegative, and is zero if and only if $f(x, y) = g(x)h(y)$ $[\lambda]$, the mean information in (4.2) may also serve as a measure of the relation between x and y. [Cf. Castañs Camargo and Medina e Isabel (1956), Féron (1952a, p. 1343), Linfoot (1957).] In particular, if H_1 implies the bivariate normal density

$$f(x, y) = \frac{1}{2\pi\sigma_x\sigma_y(1 - \rho^2)^{1/2}} \exp\left[-\frac{1}{2(1 - \rho^2)}\left(\frac{x^2}{\sigma_x^2} - 2\rho\frac{xy}{\sigma_x\sigma_y} + \frac{y^2}{\sigma_y^2}\right)\right],$$

and H_2 the product of the marginal normal densities

$$g(x) = \frac{1}{\sigma_x\sqrt{2\pi}} \exp\left(-\frac{x^2}{2\sigma_x^2}\right), \qquad h(y) = \frac{1}{\sigma_y\sqrt{2\pi}} \exp\left(-\frac{y^2}{2\sigma_y^2}\right),$$

we find that

$$I(1:2) = \int\int f(x, y) \log \frac{f(x, y)}{g(x)h(y)}\, dx\, dy = -\tfrac{1}{2}\log(1 - \rho^2), \tag{4.3}$$

so that $I(1:2)$ is a function of the correlation coefficient ρ only, and ranges from 0 to ∞ as $|\rho|$ ranges from 0 to 1. Corresponding multivariate values are given in (6.12) and (7.4) of chapter 9.

Example 4.4. As a specific illustration of $J(1, 2)$ let f_1 and f_2 be the normal densities used in (4.3). We find that

$$J(1, 2) = \int\int (f(x, y) - g(x)h(y)) \log \frac{f(x, y)}{g(x)h(y)}\, dx\, dy = \rho^2/(1 - \rho^2), \tag{4.4}$$

so that $J(1, 2)$ is a function of the correlation coefficient ρ only, and ranges from 0 to ∞ as $|\rho|$ ranges from 0 to 1.

Note that Pearson (1904) showed that if a bivariate normal distribution is classified in a two-way table, the contingency and the correlation are related by the expression $\phi^2 = \chi^2/N = \rho^2/(1 - \rho^2)$, when it is assumed that the number of observations N is large and the class intervals are very narrow [Lancaster (1957)]. The corresponding k-variate value is given in (6.13) of chapter 9, but differs from the value of ϕ^2 as given by Pearson (1904). See also (7.5) of chapter 9.

Example 4.5. To illustrate a result in communication theory, suppose that in (4.2) x is a transmitted signal voltage and y the received signal voltage which is taken to be the transmitted signal with noise added, that is, $y = x + n$, where n is the noise voltage. The noise and transmitted signal may be taken as independent, so that

$$f(x, y) = g(x)h(y|x) = g(x)h(y - x). \tag{4.5}$$

$I(1:2)$ in (4.2), a measure of the relation between the received and transmitted signals, is then a characteristic property of the transmission channel. If we assume normal distributions, since the bivariate normal density $f(x, y)$ in example 4.3 may be written as

$$\frac{1}{\sigma_x\sqrt{2\pi}} \exp\left(-\frac{x^2}{2\sigma_x^2}\right) \cdot \frac{1}{\sigma_y\sqrt{2\pi(1-\rho^2)}} \times \exp\left[-\frac{1}{2\sigma_y^2(1-\rho^2)}\left(y - \frac{\rho\sigma_y}{\sigma_x}x\right)^2\right], \tag{4.6}$$

we see from a comparison of (4.6) and (4.5) that $h(y|x) = h(y - x)$ if

$$\rho\frac{\sigma_y}{\sigma_x} = 1, \qquad \rho^2 = \frac{\sigma_x^2}{\sigma_y^2} = \frac{S}{S+N}, \tag{4.7}$$

where $S = E(x^2)$ is the mean transmitted signal power and $N = E(n^2)$ the noise power [Lawson and Uhlenbeck (1950, p. 55), Woodward and Davies (1952)]. With the value of ρ^2 from (4.7) substituted in (4.3) and (4.4), we find that the mean information in the received signal about the transmitted signal and the divergence between dependence and independence of the signals are respectively

$$I(1:2) = -\tfrac{1}{2}\log\left(1 - \frac{S}{S+N}\right) = \tfrac{1}{2}\log\left(1 + \frac{S}{N}\right), \tag{4.8}$$

$$J(1, 2) = \frac{S/(S+N)}{1 - S/(S+N)} = \frac{S}{N}. \tag{4.9}$$

We shall show in chapter 2 that $I(1:2)$ and $J(1, 2)$ are additive for independent observations. The sampling theorem [Shannon (1949), Whittaker (1915)] states that $2WT$ independent sample values are required to specify a function of duration T and bandwidth W. We thus have

$$I(1:2;\, W, T) = 2WT\,I(1:2) = WT\log\left(1 + \frac{S}{N}\right), \tag{4.10}$$

$$J(1, 2;\, W, T) = 2WT\,J(1, 2) = 2WT\,S/N = 2T\,S/N_0 = 2E/N_0, \tag{4.11}$$

where $N = WN_0$, with N_0 the mean noise power per unit bandwidth, and E the total transmitted signal energy. The interpretation of (4.10) as channel capacity is well known in communication theory [Bell (1953), Goldman (1953), Shannon (1948), Woodward (1953)]. The signal-to-noise ratio has long been used by engineers to specify the performance of communication channels.

Example 4.6. To illustrate a less general form of Lindley's (1956) definition of the information provided by an experiment, take y in (4.2) as a parameter θ ranging over a space Θ, so that $f(x, \theta)$ is the joint probability density of x and θ, $h(\theta)$ is the prior probability density of θ, $g_1(x|\theta)$ is the conditional probability density of x given θ, and the marginal probability density of x is

$g(x) = \int_\Theta g_1(x|\theta)h(\theta)\,d\theta$. An experiment $\mathscr{E}$ is defined as the ordered quadruple $\mathscr{E} = (\mathscr{X}, \mathscr{S}, \Theta, g_1(x|\theta))$, and the information provided by the experiment $\mathscr{E}$, with prior knowledge $h(\theta)$, is

$$I(1:2) = \int\int f(x, \theta) \log \frac{f(x, \theta)}{g(x)h(\theta)}\,dx\,d\theta.$$

These illustrations will suffice for the present. In chapter 2 we consider the properties of $I(1:2)$ and $J(1, 2)$.

5. PROBLEMS

5.1. How many "bits" of information (in the mean) are there in a dichotomous choice (*a*) with probabilities $p = 0.99$, $q = 1 - p = 0.01$; (*b*) $p = 1$, $q = 1 - p = 0$?

5.2. Compute the value of $I(1:2)$ and $J(1, 2)$ for:

(*a*) Prob $(x = 0|H_i) = q_i$, Prob $(x = 1|H_i) = p_i$, $p_i + q_i = 1$, $i = 1, 2$.

(*b*) The binomial distributions $B(p_i, q_i, n)$, $p_i + q_i = 1$, $i = 1, 2$.

(*c*) The Poisson distributions with parameters m_i, $i = 1, 2$.

(*d*) The normal distributions $N(\mu_i, \sigma^2)$, $i = 1, 2$, that is, the normal distributions with mean μ_i and variance σ^2.

(*e*) The normal distributions $N(\mu, \sigma_i^2)$, $i = 1, 2$.

(*f*) The normal distributions $N(\mu_i, \sigma_i^2)$, $i = 1, 2$.

5.3. Derive the result given in (4.3).

5.4. Derive the result given in (4.4).

5.5. Let $1 + x$ be the number of independent trials needed to get a success, when the probability of a success is constant for each trial. If

$$P_i(x) = \text{Prob}\,(X = x|H_i) = p_i q_i^x, \qquad x = 0, 1, 2, \cdots; \; q_i = 1 - p_i, \; i = 1, 2,$$

then

$$I(1:2) = E(1 + x|H_1)\left(p_1 \log \frac{p_1}{p_2} + q_1 \log \frac{q_1}{q_2}\right),$$

that is, the mean information for discrimination is the product of the expected number of trials and the mean information per trial.

5.6. Let $f_i(x) = \exp\,(u(\theta_i)v(x) + a(x) + b(\theta_i))$, $i = 1, 2$, where u and b are functions of θ_i, $i = 1, 2$, and v and a are functions of x, with $\int f_i(x)\,dx = 1$. Show that $J(1, 2) = (u(\theta_1) - u(\theta_2))(E_1(v(x)) - E_2(v(x)))$, where $E_i(v(x))$ is the expected value of $v(x)$ in the distribution with $f_i(x)$, $i = 1, 2$. [See Huzurbazar (1955) for the multivariate, multiparameter, distributions admitting sufficient statistics.]

5.7. Let $k_1 = \sum_{n=2}^{\infty} \frac{1}{n(\log n)^2} < \infty$, $k_2 = \sum_{n=2}^{\infty} \frac{1}{n^2(\log n)^2} < \infty$, $p_1(x = n) = \frac{1}{k_1 n(\log n)^2}$, $p_2(x = n) = \frac{1}{k_2 n^2(\log n)^2}$, $n = 2, 3, \cdots$.

Show that $I(1:2) = \sum_{n=2}^{\infty} p_1(x = n) \log \frac{p_1(x = n)}{p_2(x = n)} = \infty$, and that $I(2:1) = \sum_{n=2}^{\infty} p_2(x = n) \log \frac{p_2(x = n)}{p_1(x = n)} < \infty$. [See Joshi (1957), who credits this to Schützenberger.]

5.8. Compute the value of $I(1:2)$ and $J(1, 2)$ for the discrete bivariate distributions defined by Prob $(x = 0,\ y = 0|H_1) =$ Prob $(x = 1,\ y = 1|H_1) = q/2$, Prob $(x = 0,\ y = 1|H_1) =$ Prob $(x = 1,\ y = 0|H_1) = p/2,\ p + q = 1$, Prob $(x = 0,\ y = 0|H_2) =$ Prob $(x = 0,\ y = 1|H_2) =$ Prob $(x = 1,\ y = 0|H_2) =$ Prob $(x = 1, y = 1|H_2) = \frac{1}{4}$.

5.9. Show that $\int\int f_1(x)f_2(y) \log \frac{f_1(x)f_2(y)}{f_2(x)f_1(y)}\, dx\, dy$ may be written as $\int (f_1(x) - f_2(x)) \log \frac{f_1(x)}{f_2(x)}\, dx$, where f_1 and f_2 are probability densities, and x, y are random variables over the same range. [Cf. Barnard (1949), Girshick (1946, pp. 123–127).]

5.10. Let $N = \frac{n!}{n_1!n_2! \cdots n_k!}$, $n = n_1 + n_2 + \cdots + n_k$. Use Stirling's approximation to show that when n_i, $i = 1, 2, \cdots, k$, is large, approximately,

$$\log N = -n \sum_{i=1}^{k} \hat{p}_i \log \hat{p}_i,$$

where $\hat{p}_i = n_i/n$. [Cf. Brillouin (1956, pp. 7–8).]

5.11. Consider sequences of k different symbols. Show that the observation that a sequence of n symbols contains respectively $n_1, n_2, \cdots, n_k$, of the k symbols is approximately an information value of $n \sum_{i=1}^{k} \hat{p}_i \log \hat{p}_i + n \log k$, where $\hat{p}_i$ is defined in problem 5.10.

5.12. Let $P(n_1, n_2, \cdots, n_k) = \frac{n!}{n_1!n_2! \cdots n_k!} p_1^{n_1} p_2^{n_2} \cdots p_k^{n_k}$, $n = n_1 + n_2 + \cdots + n_k$, $p_1 + p_2 + \cdots + p_k = 1$, $p_i > 0$, $i = 1, 2, \cdots, k$.

(*a*) Show that, as in problem 5.10, approximately $\log \frac{1}{P(n_1, n_2, \cdots, n_k)} = n \sum_{i=1}^{k} \hat{p}_i \log \frac{\hat{p}_i}{p_i}$.

(*b*) Show that $\log \frac{1}{P(n_1, n_2, \cdots, n_k)}$, for $p_1 = p_2 = \cdots = p_k = 1/k$, is the information value in problem 5.11.
[Cf. Chernoff (1952, p. 497), Sanov (1957, p. 13).]

5.13. Compute the value of $I(1:2)$ for the discrete bivariate distributions defined by Prob $(x = x_i, y = y_i|H_1) = p_i > 0$, $i = 1, 2, \cdots, n$, Prob $(x = x_i, y = y_j|H_1) = 0$, $i \neq j$, Prob $(x = x_i,\ y = y_j|H_2) =$ Prob $(x = x_i|H_2) \cdot$ Prob $(y = y_j|H_2) = p_i p_j$, $i, j = 1, 2, \cdots, n$ (0 log 0 is defined as 0).

CHAPTER 2

Properties of Information

1. INTRODUCTION

We shall now study the properties of the measure of information that we have defined and examine the implications of these properties [cf. Kullback and Leibler (1951)]. We use the notation $I(1:2; E)$, $I(2:1; \mathscr{X})$, $J(1, 2; X, Y)$, etc., when it is deemed necessary to indicate explicitly sets, spaces, variables, etc., that are concerned. Where necessary for clarity, we shall use X, Y, etc., for generic variables and x, y, etc., for observed values of the generic variables. We shall also generally use only one integral sign even when there is more than one variable.

2. ADDITIVITY

THEOREM 2.1. *$I(1:2)$ is additive for independent random events; that is, for X and Y independent under H_i, $i = 1, 2$*

$$I(1:2; X, Y) = I(1:2; X) + I(1:2; Y).$$

Proof.

$$\begin{aligned} I(1:2; X, Y) &= \int f_1(x, y) \log \frac{f_1(x, y)}{f_2(x, y)}\, d\lambda(x, y) \\ &= \int g_1(x)h_1(y) \log \frac{g_1(x)h_1(y)}{g_2(x)h_2(y)}\, d\mu(x)\, d\nu(y) \\ &= \int g_1(x) \log \frac{g_1(x)}{g_2(x)}\, d\mu(x) + \int h_1(y) \log \frac{h_1(y)}{h_2(y)}\, d\nu(y) \\ &= I(1:2; X) + I(1:2; Y), \end{aligned}$$

where, because of the independence, $f_i(x, y) = g_i(x)h_i(y)$, $i = 1, 2$, $d\lambda(x, y) = d\mu(x)\, d\nu(y)$, $\int g_i(x)\, d\mu(x) = 1$, $\int h_i(y)\, d\nu(y) = 1$, $i = 1, 2$.

Additivity of information for independent events is intuitively a fundamental requirement, and is indeed postulated as a requisite property in most axiomatic developments of information theory [Barnard (1951),

Fisher (1935, p. 47), Good (1950, p. 75), Lindley (1956), MacKay (1950), Reich (1951), Schützenberger (1954), Shannon (1948), Wiener (1950, pp. 18–22)]. Additivity is the basis for the logarithmic form of information. A sample of n independent observations from the same population provides n times the mean information in a single observation. Fisher's measure of the amount of information for the estimation of a parameter also has this additive property [Fisher (1925b, 1956, pp. 148–150), Savage (1954, pp. 235–237)]. In section 6 we shall study the relation between Fisher's measure and the discrimination information measure in (2.5) of chapter 1.

If X and Y are not independent, an additive property still exists, but in terms of a conditional information defined below. To simplify the argument and avoid the measure theoretic problems of conditional probabilities [see, for example, Fraser (1957, p. 16)], we shall deal with probability density functions and Lebesgue measure. We leave it to the reader to carry out the corresponding development for discrete variables. With this understanding, we then have,

$$\begin{aligned} I(1:2;\,X,\,Y) &= \int f_1(x,y)\log\frac{f_1(x,y)}{f_2(x,y)}\,dx\,dy \\ &= \int g_1(x)\log\frac{g_1(x)}{g_2(x)}\,dx + \int g_1(x)\left[\int h_1(y|x)\log\frac{h_1(y|x)}{h_2(y|x)}\,dy\right]dx, \end{aligned}$$

where $g_i(x) = \int f_i(x,y)\,dy$, $h_i(y|x) = f_i(x,y)/g_i(x)$, $i = 1, 2$.

We now set

$$I(1:2;\,Y|X=x) = \int h_1(y|x)\log\frac{h_1(y|x)}{h_2(y|x)}\,dy,$$

and

$$I(1:2;\,Y|X) = E_1(I(1:2;\,Y|X=x)) = \int g_1(x)I(1:2;\,Y|X=x)\,dx,$$

where $I(1:2;\,Y|X=x)$ may be defined as the conditional information in Y for discrimination in favor of H_1 against H_2 when $X = x$, under H_1, and $I(1:2;\,Y|X)$ is the mean value of the conditional discrimination information $I(1:2;\,Y|X=x)$ under H_1. [Cf. Barnard (1951), Feinstein (1958, p. 12), Féron (1952a), Féron and Fourgeaud (1951), Good (1950), Lindley (1956), Powers (1956, pp. 54–62), Shannon (1948).]

We may obtain similar results by an interchange of the procedure with respect to X and Y, so that we state:

THEOREM 2.2.

$$\begin{aligned} I(1:2;\,X,\,Y) &= I(1:2;\,X) + I(1:2;\,Y|X) \\ &= I(1:2;\,Y) + I(1:2;\,X|Y). \end{aligned}$$

Example 2.1. Consider the bivariate normal densities

$$f_i(x, y) = \frac{1}{2\pi\sigma_x\sigma_y\sqrt{1-\rho_i^2}} \exp\left[-\frac{1}{2(1-\rho_i^2)}\left(\frac{(x-\mu_{ix})^2}{\sigma_x^2} - 2\rho_i\frac{(x-\mu_{ix})(y-\mu_{iy})}{\sigma_x\sigma_y} + \frac{(y-\mu_{iy})^2}{\sigma_y^2}\right)\right],$$

so that

$$g_i(x) = \frac{1}{\sigma_x\sqrt{2\pi}} \exp\left(-\frac{(x-\mu_{ix})^2}{2\sigma_x^2}\right)$$

and

$$h_i(y|x) = \frac{1}{\sigma_y\sqrt{2\pi}(1-\rho_i^2)^{1/2}} \exp\left[-\frac{[y-\mu_{iy}-\beta_i(x-\mu_{ix})]^2}{2\sigma_y^2(1-\rho_i^2)}\right],$$

where $\beta_i = \rho_i\sigma_y/\sigma_x$. Note that the variances are the same for $i = 1, 2$.
With these densities we find [or by substitution in (1.2) of chapter 9],

$$I(1:2;\ X) = \frac{(\mu_{2x}-\mu_{1x})^2}{2\sigma_x^2},$$

$$I(1:2;\ Y|X=x) = \frac{1}{2}\log\frac{1-\rho_2^2}{1-\rho_1^2} - \frac{1}{2} + \frac{1}{2}\frac{1-\rho_1^2}{1-\rho_2^2} + \frac{1}{2}\frac{[\mu_{2y}+\beta_2(x-\mu_{2x})-\mu_{1y}-\beta_1(x-\mu_{1x})]^2}{\sigma_y^2(1-\rho_2^2)},$$

$$I(1:2;\ Y|X) = \frac{1}{2}\log\frac{1-\rho_2^2}{1-\rho_1^2} - \frac{\rho_1^2-\rho_2^2}{2(1-\rho_2^2)} + \frac{[(\mu_{2y}-\mu_{1y})-\beta_2(\mu_{2x}-\mu_{1x})]^2}{2\sigma_y^2(1-\rho_2^2)} + \frac{(\rho_2-\rho_1)^2}{2(1-\rho_2^2)},$$

$$I(1:2;\ X, Y) = \frac{1}{2}\log\frac{1-\rho_2^2}{1-\rho_1^2} + \frac{\rho_2(\rho_2-\rho_1)}{1-\rho_2^2} + \frac{1}{2(1-\rho_2^2)}\left[\frac{(\mu_{2x}-\mu_{1x})^2}{\sigma_x^2} - 2\rho_2\frac{(\mu_{2x}-\mu_{1x})(\mu_{2y}-\mu_{1y})}{\sigma_x\sigma_y} + \frac{(\mu_{2y}-\mu_{1y})^2}{\sigma_y^2}\right].$$

Note that $I(1:2;\ X, Y) = I(1:2;\ X) + I(1:2;\ Y|X)$. If $\rho_1 = \rho_2 = 0$, so that X and Y are independent under H_1 and H_2, $I(1:2;\ Y|X) = (\mu_{2y}-\mu_{1y})^2/2\sigma_y^2 = I(1:2;\ Y)$ and $I(1:2;\ X, Y) = \frac{(\mu_{2x}-\mu_{1x})^2}{2\sigma_x^2} + \frac{(\mu_{2y}-\mu_{1y})^2}{2\sigma_y^2} = I(1:2;\ X) + I(1:2;\ Y)$.

3. CONVEXITY

THEOREM 3.1. *$I(1:2)$ is almost positive definite; that is, $I(1:2) \geqq 0$, with equality if and only if $f_1(x) = f_2(x)$ $[\lambda]$.*

Proof. Let $g(x) = f_1(x)/f_2(x)$. Then

$$I(1:2) = \int f_2(x)g(x) \log g(x)\, d\lambda(x) \tag{3.1}$$

$$= \int g(x) \log g(x)\, d\mu_2(x),$$

with $d\mu_2(x) = f_2(x)\, d\lambda(x)$.

Setting $\phi(t) = t \log t$, since $0 < g(x) < \infty\ [\lambda]$, we may write [cf. Hardy, Littlewood, and Pólya (1934, p. 151)],

$$\phi(g(x)) = \phi(1) + [g(x) - 1]\phi'(1) + \tfrac{1}{2}[g(x) - 1]^2\phi''(h(x))\ [\lambda], \tag{3.2}$$

where $h(x)$ lies between $g(x)$ and 1, so that $0 < h(x) < \infty\ [\lambda]$. Since $\phi(1) = 0$, $\phi'(1) = 1$, and

$$\int g(x)\, d\mu_2(x) = \int f_1(x)\, d\lambda(x) = 1, \tag{3.3}$$

we find

$$\int \phi(g(x))\, d\mu_2(x) = \tfrac{1}{2}\int [g(x) - 1]^2\phi''(h(x))\, d\mu_2(x), \tag{3.4}$$

where $\phi''(t) = 1/t > 0$ for $t > 0$. We see from (3.4) that

$$\int g(x) \log g(x)\, d\mu_2(x) = \int f_1(x) \log \frac{f_1(x)}{f_2(x)}\, d\lambda(x) \geqq 0, \tag{3.5}$$

with equality if and only if $g(x) = f_1(x)/f_2(x) = 1[\lambda]$.

Theorem 3.1 tells us that, in the mean, discrimination information obtained from statistical observations is positive [cf. Fisher (1925b)]. There is no discrimination information if the distributions of the observations are the same $[\lambda]$ under both hypotheses. Theorem 3.1 may be verified with the values of $I(1:2)$ computed in example 2.1.

COROLLARY 3.1.

$$\int_E f_1(x) \log \frac{f_1(x)}{f_2(x)}\, d\lambda(x) \geqq \left(\int_E f_1(x)\, d\lambda(x)\right) \log \frac{\int_E f_1(x)\, d\lambda(x)}{\int_E f_2(x)\, d\lambda(x)}$$

$$= \mu_1(E) \log \frac{\mu_1(E)}{\mu_2(E)},$$

for $\lambda(E) > 0$, *with equality if and only if* $\dfrac{f_1(x)}{f_2(x)} = \dfrac{\mu_1(E)}{\mu_2(E)}\ [\lambda]$ *for* $x \in E$.

Proof. If the left-hand member of the inequality in the corollary is ∞, the result is trivial. Otherwise truncate the distributions to the set E

and write $g_1(x) = f_1(x)/\mu_1(E)$, $g_2(x) = f_2(x)/\mu_2(E)$. From theorem 3.1 we now have

$$\int_E g_1(x) \log \frac{g_1(x)}{g_2(x)} \, d\lambda(x) \geqq 0,$$

with equality if and only if $g_1(x) = g_2(x)$ $[\lambda]$, and the corollary follows.

Defining $t \log t = 0$ for $t = 0$, the equality in corollary 3.1 is trivially satisfied for $\lambda(E) = 0$.

COROLLARY 3.2. *If $E_i \in \mathcal{S}$, $i = 1, 2, \cdots$, $E_i \cap E_j = 0$, $i \neq j$, and $\mathcal{X} = \cup_i E_i$, that is, for the partitioning of $\mathcal{X}$ into pairwise disjoint $E_1, E_2, \cdots$,*

$$I(1:2) \geqq \sum_i \mu_1(E_i) \log \frac{\mu_1(E_i)}{\mu_2(E_i)},$$

with equality if and only if $\dfrac{f_1(x)}{f_2(x)} = \dfrac{\mu_1(E_i)}{\mu_2(E_i)}$ $[\lambda]$, *for* $x \in E_i$, $i = 1, 2, \cdots$.

Proof. Use corollary 3.1 and (see problem 8.37)

$$I(1:2) = \int f_1(x) \log \frac{f_1(x)}{f_2(x)} \, d\lambda(x) = \sum_i \int_{E_i} f_1(x) \log \frac{f_1(x)}{f_2(x)} \, d\lambda(x).$$

The properties in theorem 3.1 and corollaries 3.1 and 3.2 [cf. Lindley (1956), Savage (1954, p. 235)] are convexity properties related to the fact that $t \log t$ is a convex function and are in essence Jensen's inequality [Jensen (1906)]. [See problem 8.31. For details on convex functions the reader may see Blackwell and Girshick (1954, pp. 30–42), Fraser (1957, pp. 52–55), Hardy, Littlewood, and Pólya (1934)]. We also see from corollary 3.1 that the grouping of observations generally causes a loss of information [cf. Fisher (1925b), Wiener (1948, p. 79)]; the left-hand side of the inequality of corollary 3.1 is the discrimination information *in the elements of the set E*, whereas the right-hand member of the inequality is the discrimination information *in the set E*. The necessary and sufficient condition of corollary 3.1 that the information not be diminished by the grouping may also be written as $\dfrac{f_1(x)}{\mu_1(E)} = \dfrac{f_2(x)}{\mu_2(E)}$ $[\lambda]$ for $x \in E$, which states that the conditional density of x given E, is the same under both hypotheses. We may treat all $x \in E$ for which the condition for equality in corollary 3.1 is satisfied as equivalent for the discrimination.

As illustrations of theorem 3.1 and corollaries 3.1 and 3.2, we have the following:

Example 3.1. (See example 4.2 in chapter 1 and theorem 3.1.)

$$p_1 \log \frac{p_1}{1/n} + p_2 \log \frac{p_2}{1/n} + \cdots + p_n \log \frac{p_n}{1/n} \geq 0, \tag{3.6}$$

where $p_i > 0$, $i = 1, 2, \cdots, n$, $p_1 + p_2 + \cdots + p_n = 1$. It follows that $\log n \geq -\Sigma p_i \log p_i$, with equality if and only if $p_i = 1/n$, $i = 1, 2, \cdots, n$, corresponding to the fact that the greatest uncertainty in a situation with n alternatives occurs when all the alternatives are equally probable [Shannon (1948)].

Example 3.2. (See corollary 3.1.)

$$(3.7) \quad p_{11} \log \frac{p_{11}}{p_{21}} + p_{12} \log \frac{p_{12}}{p_{22}} + \cdots + p_{1n} \log \frac{p_{1n}}{p_{2n}} \geq (p_{11} + p_{12} + \cdots + p_{1n}) \log \frac{p_{11} + p_{12} + \cdots + p_{1n}}{p_{21} + p_{22} + \cdots + p_{2n}},$$

for $p_{ij} > 0$, $i = 1, 2, j = 1, 2, \cdots, n$, with equality if and only if

$$\frac{p_{11}}{p_{21}} = \frac{p_{12}}{p_{22}} = \cdots = \frac{p_{1n}}{p_{2n}} = \frac{p_{11} + p_{12} + \cdots + p_{1n}}{p_{21} + p_{22} + \cdots + p_{2n}}.$$

Example 3.3. (See corollary 3.2.) For Poisson populations with parameters λ_1 and λ_2 we have,

$$\sum_{x=0}^{\infty} e^{-\lambda_1} \frac{\lambda_1^x}{x!} \log \frac{e^{-\lambda_1}\lambda_1^x/x!}{e^{-\lambda_2}\lambda_2^x/x!} = e^{-\lambda_1} \log \frac{e^{-\lambda_1}}{e^{-\lambda_2}} + e^{-\lambda_1}\lambda_1 \log \frac{e^{-\lambda_1}\lambda_1}{e^{-\lambda_2}\lambda_2} + \sum_{x=2}^{\infty} \frac{e^{-\lambda_1}\lambda_1^x}{x!} \log \frac{e^{-\lambda_1}\lambda_1^x/x!}{e^{-\lambda_2}\lambda_2^x/x!}$$

$$\geq e^{-\lambda_1}(\lambda_2 - \lambda_1) + e^{-\lambda_1}\lambda_1(\lambda_2 - \lambda_1) + e^{-\lambda_1}\lambda_1 \log \frac{\lambda_1}{\lambda_2} + (1 - e^{-\lambda_1} - \lambda_1 e^{-\lambda_1}) \log \frac{1 - e^{-\lambda_1} - \lambda_1 e^{-\lambda_1}}{1 - e^{-\lambda_2} - \lambda_2 e^{-\lambda_2}},$$

with equality if and only if

$$\frac{e^{-\lambda_1}\lambda_1^x}{e^{-\lambda_2}\lambda_2^x} = \frac{1 - e^{-\lambda_1} - \lambda_1 e^{-\lambda_1}}{1 - e^{-\lambda_2} - \lambda_2 e^{-\lambda_2}}, \qquad x = 2, 3, \cdots.$$

A numerical illustration, grouping values $x \geq 4$, is in table 2.1 of example 2.2 of chapter 4.

Example 3.4. (See corollary 3.1.)

$$(3.8) \quad \iint f_1(x, y) \log \frac{f_1(x, y)}{f_2(x, y)}\, dx\, dy \geq \int dx \int f_1(x, y)\, dy \log \frac{\int f_1(x, y)\, dy}{\int f_2(x, y)\, dy} = \int g_1(x) \log \frac{g_1(x)}{g_2(x)}\, dx,$$

with equality if and only if $\dfrac{f_1(x, y)}{f_2(x, y)} = \dfrac{g_1(x)}{g_2(x)}$, where $g_i(x)$, $i = 1, 2$, are the marginal densities of x. The necessary and sufficient condition for equality may also be written as $f_1(x, y)/g_1(x) = f_2(x, y)/g_2(x)$ or $h_1(y|x) = h_2(y|x)$, with $h_i(y|x)$, $i = 1, 2$, a conditional density of y given x.

As a matter of fact, (3.8) is also an illustration of:

COROLLARY 3.3. *(a)* $I(1:2; X, Y) \geqq I(1:2; X)$ *with equality if and only if* $I(1:2; Y|X) = 0$; *(b)* $I(1:2; X, Y) \geqq I(1:2; Y)$ *with equality if and only if* $I(1:2; X|Y) = 0$; *(c)* $I(1:2; X, Y) \geqq I(1:2; Y|X)$ *with equality if and only if* $I(1:2; X) = 0$; *(d)* $I(1:2; X, Y) \geqq I(1:2; X|Y)$ *with equality if and only if* $I(1:2; Y) = 0$ [cf. Lindley (1956)].

Proof. Use theorem 3.1 in conjunction with theorem 2.2.

4. INVARIANCE

If the partitioning of the space $\mathscr{X}$ in corollary 3.2 is such that the necessary and sufficient condition for equality is satisfied, that is, if the conditional density of x given E_i is the same under both hypotheses for all E_i of the partitioning, we may designate the partitioning $\mathscr{X} = \cup_i E_i$ as a *sufficient partitioning* for the discrimination. Note that the coarser grouping of a sufficient partitioning is as informative for discrimination as the finer grouping of the space $\mathscr{X}$. In terms of the concept that a statistic is a partitioning of $\mathscr{X}$ into sets of equivalent x's [Lehmann (1950b, pp. 6–7)], we may say that the statistic defined by the partitioning $\mathscr{X} = \cup_i E_i$ is sufficient for the discrimination if the necessary and sufficient condition for the equality to hold in corollary 3.2 is satisfied. This is consistent with the original criterion of sufficiency introduced by R. A. Fisher (1922b, p. 316): "the statistic chosen should summarise the whole of the relevant information supplied by the sample," and further developments, for example, by Fisher (1925a, b), Neyman (1935), Dugué (1936a, b), Koopman (1936), Pitman (1936), Darmois (1945), Halmos and Savage (1949), Lehmann and Scheffé (1950), Savage (1954), Blackwell and Girshick (1954, pp. 208–223), Bahadur (1954). [Cf. Fraser (1957, pp. 16–22).]

To continue the study of the relation between "information" and "sufficiency," let $Y = T(x)$ be a statistic, that is, $T(x)$ is a function with domain $\mathscr{X}$ and range $\mathscr{Y}$, and let $\mathscr{T}$ be an additive class of subsets of $\mathscr{Y}$. We assume that $T(x)$ is measurable, that is, for every set $G \in \mathscr{T}$, the inverse image set $T^{-1}(G) = \{x : T(x) \in G\}$ [$T^{-1}(G)$ is the set of elements x such that $T(x) \in G$] is a member of the class $\mathscr{S}$ of measurable subsets of $\mathscr{X}$ (see section 2 of chapter 1). The class of all such sets of the form $T^{-1}(G)$ is denoted by $T^{-1}(\mathscr{T})$. We thus have a measurable transformation T of the probability spaces $(\mathscr{X}, \mathscr{S}, \mu_i)$ onto the probability spaces

$(\mathscr{Y}, \mathscr{T}, \nu_i)$, where by definition $\nu_i(G) = \mu_i(T^{-1}(G))$, $i = 1, 2$ [Fraser (1957, pp. 1–16), Halmos and Savage (1949), Kolmogorov (1950, pp. 21–22), Loève (1955, p. 166)]. If we define $\gamma(G) = \lambda(T^{-1}(G))$, then $\nu_1 \equiv \nu_2 \equiv \gamma$ (the measures are absolutely continuous with respect to one another), and as in section 2 of chapter 1, the Radon–Nikodym theorem permits us to assert the existence of the generalized probability density $g_i(y)$, $i = 1, 2$, where

$$\text{(4.1)} \qquad \nu_i(G) = \int_G g_i(y)\, d\gamma(y), \qquad i = 1, 2, \qquad G \in \mathscr{T},$$

for all $G \in \mathscr{T}$. The function value $g_i(y)$ is the conditional expectation of $f_i(x)$ given that $T(x) = y$ and is denoted by $E_\lambda(f_i|y)$ [Fraser (1957, p. 15), Halmos and Savage (1949), Kolmogorov (1950, pp. 47–50), Loève (1955, pp. 337–344)].

In terms of the probability spaces $(\mathscr{Y}, \mathscr{T}, \nu_i)$, $i = 1, 2$, the discrimination information is [cf. (2.4) of chapter 1]

$$\text{(4.2)} \qquad I(1:2; G) = \frac{1}{\nu_1(G)} \int_G g_1(y) \log \frac{g_1(y)}{g_2(y)}\, d\gamma(y), \qquad \nu_1(G) > 0,$$

$$= 0, \qquad \nu_1(G) = 0,$$

and [cf. (2.5) of chapter 1]

$$\text{(4.3)} \qquad I(1:2; \mathscr{Y}) = \int g_1(y) \log \frac{g_1(y)}{g_2(y)}\, d\gamma(y).$$

We shall need the following lemma for the proof of theorem 4.1.

Following the notation of Halmos and Savage (1949), if g is a point function on $\mathscr{Y}$, then gT is the point function on $\mathscr{X}$ defined by $gT(x) = g(T(x))$.

LEMMA 4.1. *If g is a real-valued function on $\mathscr{Y}$, then*

$$\int_G g(y)\, d\nu_i(y) = \int_{T^{-1}(G)} gT(x)\, d\mu_i(x), \qquad i = 1, 2,$$

for every $G \in \mathscr{T}$, in the sense that if either integral exists, then so does the other and the two are equal.

Proof. See Halmos (1950, p. 163), lemma 3 of Halmos and Savage (1949), Loève (1955, p. 342).

THEOREM 4.1. *$I(1:2; \mathscr{X}) \geqq I(1:2; \mathscr{Y})$, with equality if and only if $f_1(x)/f_2(x) = g_1(T(x))/g_2(T(x))$ $[\lambda]$.*

Proof. If $I(1:2; \mathscr{X}) = \infty$, the result is trivial. Using lemma 4.1 above,

$$I(1:2; \mathscr{Y}) = \int d\nu_1(y) \log \frac{g_1(y)}{g_2(y)} = \int d\mu_1(x) \log \frac{g_1T(x)}{g_2T(x)},$$

and therefore

$$I(1:2;\mathscr{X}) - I(1:2;\mathscr{Y}) = \int d\mu_1(x)\left[\log\frac{f_1(x)}{f_2(x)} - \log\frac{g_1T(x)}{g_2T(x)}\right]$$
$$= \int f_1(x)\log\frac{f_1(x)g_2T(x)}{f_2(x)g_1T(x)}\,d\lambda(x).$$

Setting $g(x) = \dfrac{f_1(x)g_2T(x)}{f_2(x)g_1T(x)}$,

(4.4) $$I(1:2;\mathscr{X}) - I(1:2;\mathscr{Y}) = \int \frac{f_2(x)g_1T(x)}{g_2T(x)}\,g(x)\log g(x)\,d\lambda(x)$$
$$= \int g(x)\log g(x)\,d\mu_{12}(x),$$

where $\mu_{12}(E) = \displaystyle\int_E \frac{f_2(x)g_1T(x)}{g_2T(x)}\,d\lambda(x)$, for all $E \in \mathscr{S}$. Since

$$\int g(x)\,d\mu_{12}(x) = \int \frac{f_1(x)g_2T(x)}{f_2(x)g_1T(x)}\cdot\frac{f_2(x)g_1T(x)}{g_2T(x)}\,d\lambda(x) = 1,$$

the method of theorem 3.1 leads to the conclusion that $I(1:2;\mathscr{X}) - I(1:2;\mathscr{Y}) \geqq 0$, with equality if and only if

(4.5) $$\frac{f_1(x)}{f_2(x)} = \frac{g_1T(x)}{g_2T(x)} = \frac{g_1(T(x))}{g_2(T(x))}\,[\lambda].$$

The necessary and sufficient condition for the equality to hold in theorem 4.1 may also be written as [see (4.1)]

$$\frac{f_1(x)}{g_1(y)} = \frac{f_2(x)}{g_2(y)}\,[\lambda], \qquad \text{or} \qquad \frac{f_1(x)}{E_\lambda(f_1|y)} = \frac{f_2(x)}{E_\lambda(f_2|y)}\,[\lambda],$$

that is, the conditional density of x, given $T(x) = y$, is the same under both hypotheses. A statistic satisfying the condition for equality in theorem 4.1 is called a sufficient statistic for the discrimination. [Cf. Mourier (1951).]

Suppose now that the two probability measures μ_1 and μ_2 are members of a set m of measures, for example, a set with all members of the same functional form, but differing values of one or more parameters. We assume that the set m of measures is *homogeneous*, that is, any two members of the set are absolutely continuous with respect to each other. By means of the Radon–Nikodym theorem, we may represent each member of the homogeneous set by a generalized probability density with respect to a common measure [Fraser (1957, p. 21), Halmos and Savage (1949)].

THEOREM 4.2. *If μ_1 and μ_2 are any two members of a homogeneous set m of measures, then $I(1:2;\mathscr{X}) \geqq I(1:2;\mathscr{Y})$, with equality if and only if the statistic $Y = T(x)$ is sufficient for the homogeneous set m.*

Proof. The necessary and sufficient condition given by (4.5) is now equivalent to the condition that the generalized conditional density of x, given $T(x) = y$, is the same $[\lambda]$ for all measures of the homogeneous set m or the defining condition for $T(x)$ to be a sufficient statistic [Fisher (1922b), Neyman (1935), Darmois (1936), Doob (1936), Halmos and Savage (1949), Lehmann and Scheffé (1950), Rao (1952), Savage (1954), Blackwell and Girshick (1954), Bahadur (1954), Loève (1955, p. 346), Fraser (1957, p. 17)].

LEMMA 4.2. *If f is a real-valued function on $\mathscr{X}$, then a necessary and sufficient condition that there exist a measurable function g on $\mathscr{Y}$ such that $f = gT$ is that f be measurable $T^{-1}(\mathscr{T})$; if such a function g exists, then it is unique.*

Proof. See lemma 2 of Halmos and Savage (1949).

COROLLARY 4.1. *$I(1:2;\mathscr{X}) = I(1:2;\mathscr{Y})$ if $Y = T(x)$ is a nonsingular transformation.*

Proof. If T is nonsingular, $T^{-1}(\mathscr{T})$ is $\mathscr{S}$ and therefore $f_i(x)$, $i = 1, 2$, is measurable $T^{-1}(\mathscr{T})$, and the conclusion follows from lemma 4.2 and theorem 4.2. Note that an alternative proof follows from the successive application of theorem 4.1 for the transformation from $\mathscr{X}$ to $\mathscr{Y}$ and the inverse transformation from $\mathscr{Y}$ to $\mathscr{X}$.

COROLLARY 4.2. *$I(1:2;T^{-1}(G)) = I(1:2;G)$ for all $G \in \mathscr{T}$ if and only if $I(1:2;\mathscr{X}) = I(1:2;\mathscr{Y})$; that is, if and only if $Y = T(x)$ is a sufficient statistic.*

Proof. Let $\chi_E(x)$ be the characteristic function of the set E, that is, $\chi_E(x) = 1$ if $x \in E$, and $\chi_E(x) = 0$ if $x \notin E$. We have

$$I(1:2;G) = \int_G \frac{d\nu_1(y)}{\nu_1(G)} \log \frac{g_1(y)}{g_2(y)} = \int \chi_G(y) \frac{d\nu_1(y)}{\nu_1(G)} \log \frac{g_1(y)}{g_2(y)}$$

$$= \int \chi_{T^{-1}(G)}(x) \frac{d\mu_1(x)}{\mu_1(T^{-1}(G))} \log \frac{g_1T(x)}{g_2T(x)}$$

$$= \int_{T^{-1}(G)} \frac{d\mu_1(x)}{\mu_1(T^{-1}(G))} \log \frac{g_1T(x)}{g_2T(x)},$$

$$I(1:2;T^{-1}(G)) = \int_{T^{-1}(G)} \frac{d\mu_1(x)}{\mu_1(T^{-1}(G))} \log \frac{f_1(x)}{f_2(x)}.$$

An application of the method of theorem 4.1 to $I(1:2;T^{-1}(G)) - I(1:2;G)$ and use of theorem 4.2 completes the proof.

We may "randomize" corollary 4.2 by introducing the function $\psi(y)$ such that $0 \leq \psi(y) \leq 1$, for example, $\psi(y)$ may represent the probability for a certain action if y is observed. From the definition of conditional expectation [Fraser (1957, p. 15), Halmos (1950, p. 209), Halmos and Savage (1949), Kolmogorov (1950, p. 53), Loève (1955, p. 340)] we have

$$(4.6) \quad \int \phi(x)\, d\lambda(x) = \int \psi(y)\, d\gamma(y),$$

$$\int \phi(x) f_i(x)\, d\lambda(x) = \int \psi(y) g_i(y)\, d\gamma(y), \qquad i = 1, 2,$$

where $\phi(x) = \psi T(x) = \psi(T(x))$, $\psi(y) = E_\lambda(\phi(x)|T(x) = y)$, that is, $\psi(y)$ is the conditional expectation (using the measure λ) of $\phi(x)$ given $T(x) = y$. (See lemmas 3.1 and 3.2 in chapter 3.)

COROLLARY 4.3.

$$\int \phi(x) f_1(x) \log \frac{f_1(x)}{f_2(x)}\, d\lambda(x) \geqq \int \psi(y) g_1(y) \log \frac{g_1(y)}{g_2(y)}\, d\gamma(y),$$

with equality if and only if $Y = T(x)$ *is a sufficient statistic.*

Proof. An application of the method of proof of theorems 4.1 and 4.2 yields the result.

The preceding theorems and corollaries show that the grouping, condensation, or transformation of observations by a statistic will in general result in a loss of information. If the statistic is sufficient, there is no loss of information [cf. Fisher (1925b, 1935, 1956, pp. 150–152)]. There can be no gain of information by statistical processing of data. A numerical illustration of this loss of information is in section 2 of chapter 4. [Cf. Feinstein (1958, pp. 70–71).]

Corollaries 4.2 and 4.3 show that the sufficiency of a statistic for a set of distributions is not affected by truncation or by selection according to the function $\phi(x) = \psi(T(x))$ [cf. Bartlett (1936), Pitman (1936), Tukey (1949)]. Averaging, on the other hand, is a statistical procedure or transformation that will generally result in a loss of information. A transformation that considers only a marginal distribution in a multivariate situation (ignores some of the variates) also is one that will generally result in a loss of information. (See corollary 3.3; also section 8 of chapter 9.)

5. DIVERGENCE

In view of our assumption in section 2 of chapter 1 that the probability measures μ_1 and μ_2 are absolutely continuous with respect to one another,

$I(2:1)$, defined in (3.1) of chapter 1, satisfies theorems and corollaries similar to those thus far developed for $I(1:2)$. Since $J(1, 2) = I(1:2) + I(2:1)$, we also have a similar set of results for $J(1, 2)$ that we shall state, using the notation and symbols of sections 2, 3, and 4, leaving the proofs to the reader.

THEOREM 5.1. *$J(1, 2)$ is additive for independent random events; that is, for X and Y independent* $J(1, 2; X, Y) = J(1, 2; X) + J(1, 2; Y)$.

THEOREM 5.2.

$$\begin{aligned} J(1, 2; X, Y) &= J(1, 2; X) + J(1, 2; Y|X) \\ &= J(1, 2; Y) + J(1, 2; X|Y). \end{aligned}$$

THEOREM 5.3. *$J(1, 2)$ is almost positive definite; that is,* $J(1, 2) \geqq 0$, *with equality if and only if* $f_1(x) = f_2(x)$ $[\lambda]$.

COROLLARY 5.1.

$$\begin{aligned} \int_E (f_1(x) - f_2(x)) \log \frac{f_1(x)}{f_2(x)} \, d\lambda(x) & \\ &\geqq \left(\int_E f_1(x)\, d\lambda(x) - \int_E f_2(x)\, d\lambda(x)\right) \log \frac{\int_E f_1(x)\, d\lambda(x)}{\int_E f_2(x)\, d\lambda(x)} \\ &= (\mu_1(E) - \mu_2(E)) \log \frac{\mu_1(E)}{\mu_2(E)}, \end{aligned}$$

for $\lambda(E) > 0$, *with equality if and only if* $f_1(x)/f_2(x) = \mu_1(E)/\mu_2(E)$ $[\lambda]$ *for* $x \in E$.

COROLLARY 5.2. *If* $E_i \in \mathscr{S}$, $i = 1, 2, \cdots$, $E_i \cap E_j = 0$, $i \neq j$, *and* $\mathscr{X} = \cup_i E_i$,

$$J(1, 2) \geqq \sum_i (\mu_1(E_i) - \mu_2(E_i)) \log \frac{\mu_1(E_i)}{\mu_2(E_i)},$$

with equality if and only if $f_1(x)/f_2(x) = \mu_1(E_i)/\mu_2(E_i)$ $[\lambda]$ *for* $x \in E_i$, $i = 1, 2, \cdots$.

COROLLARY 5.3. (*a*) $J(1, 2; X, Y) \geqq J(1, 2; X)$, *with equality if and only if* $J(1, 2; Y|X) = 0$; (*b*) $J(1, 2; X, Y) \geqq J(1, 2; Y)$, *with equality if and only if* $J(1, 2; X|Y) = 0$; (*c*) $J(1, 2; X, Y) \geqq J(1, 2; Y|X)$, *with equality if and only if* $J(1, 2; X) = 0$; (*d*) $J(1, 2; X, Y) \geqq J(1, 2; X|Y)$, *with equality if and only if* $J(1, 2; Y) = 0$.

THEOREM 5.4. $J(1, 2; \mathscr{X}) \geqq J(1, 2; \mathscr{Y})$, *with equality if and only if* $f_1(x)/f_2(x) = g_1(T(x))/g_2(T(x))$ $[\lambda]$.

THEOREM 5.5. *If* μ_1 *and* μ_2 *are any two members of a homogeneous set* m *of measures, then* $J(1, 2; \mathscr{X}) \geqq J(1, 2; \mathscr{Y})$, *with equality if and only if the statistic* $Y = T(x)$ *is sufficient for the homogeneous set* m.

COROLLARY 5.4. $J(1, 2; \mathscr{X}) = J(1, 2; \mathscr{Y})$ *if* $Y = T(x)$ *is a nonsingular transformation.*

COROLLARY 5.5. $J(1, 2; T^{-1}(G)) = J(1, 2; G)$ *for all* $G \in \mathscr{T}$ *if and only if* $J(1, 2; \mathscr{X}) = J(1, 2; \mathscr{Y})$; *that is, if and only if* $Y = T(x)$ *is a sufficient statistic.*

COROLLARY 5.6.

$$\int \phi(x)(f_1(x) - f_2(x)) \log \frac{f_1(x)}{f_2(x)} \, d\lambda(x) \geqq \int \psi(y)(g_1(y) - g_2(y)) \log \frac{g_1(y)}{g_2(y)} \, d\gamma(y),$$

with equality if and only if $Y = T(x)$ *is a sufficient statistic.*

At this point it may be appropriate to describe the problem of discrimination between two hypotheses H_1 and H_2 in terms of the language of communication theory, and to derive a result that may throw further light on the meaning of $J(1, 2)$. We shall consider a model consisting of a source that generates symbols, a channel that transmits the symbols imperfectly (a noisy channel), and a receiver which will ultimately act on the basis of the message it has received (or thinks it has received). For general models of the communication problem and the basis for the terms used see Shannon (1948), Shannon and Weaver (1949), McMillan (1953), Khinchin (1957), Joshi (1957), Feinstein (1958).

Suppose the source, or input space, is a state of nature characterized by the hypotheses H_1 and H_2, with $P(H_1) = p$ and $P(H_2) = q = 1 - p$. The input space then consists of only two symbols H_θ, $\theta = 1, 2$. These symbols are transmitted by some discrete random process with probabilities p and q, successive symbols being independently selected. The receiver, or output space, is the sample space $\mathscr{X}$ of elements x in section 2 of chapter 1. The noisy channel is the observation procedure described by the generalized probability densities $f_\theta(x)$, $\theta = 1, 2$, of section 2 of chapter 1, such that $\mu_\theta(E)$ is the conditional probability that the transmitted symbol H_θ is received as $x \in E \in \mathscr{S}$. This communication system may be denoted by $(p; f_1, f_2)$, and the channel by (f_1, f_2). The rate $R(p; f_1, f_2)$ of transmission of information by the communication system $(p; f_1, f_2)$ is defined by Shannon (1948) as the difference between the entropy (see section 4 of chapter 1) of the source, or input entropy (the prior uncertainty), and the mean conditional entropy of the input symbols at the output (the posterior uncertainty), that is,

$$R(p; f_1, f_2) = \mathscr{H}(\theta) - \mathscr{H}(\theta | X),$$

where $\mathscr{H}(\theta)$, the prior uncertainty, and $\mathscr{H}(\theta|X)$, the posterior uncertainty, are given by

$$\begin{aligned}(5.1)\qquad \mathscr{H}(\theta) &= -P(H_1)\log P(H_1) - P(H_2)\log P(H_2)\\ &= -p\log p - q\log q,\end{aligned}$$

$$\begin{aligned}(5.2)\ \mathscr{H}(\theta|X) &= E(-P(H_1|x)\log P(H_1|x) - P(H_2|x)\log P(H_2|x))\\ &= -\int (P(H_1|x)\log P(H_1|x) + P(H_2|x)\log P(H_2|x))f(x)\,d\lambda(x),\end{aligned}$$

with $f(x) = pf_1(x) + qf_2(x)$. [Cf. Lindley (1956, pp. 986–990).] The rate of transmission of information by the communication system is also a measure of the relation between the input and output symbols. Using the values for $\mathscr{H}(\theta)$ and $\mathscr{H}(\theta|X)$ in (5.1) and (5.2) above gives

$$\begin{aligned}(5.3),\qquad R(p;f_1,f_2) &= \sum_{\theta=1}^{2}\int P(H_\theta,x)\log\frac{P(H_\theta,x)}{P(H_\theta)f(x)}\,d\lambda(x)\\ &= \int\left(pf_1(x)\log\frac{f_1(x)}{f(x)} + qf_2(x)\log\frac{f_2(x)}{f(x)}\right)d\lambda(x)\\ &\geqq 0,\end{aligned}$$

where $P(H_\theta, x)\,d\lambda(x) = P(H_\theta|x)f(x)\,d\lambda(x)$ is the joint probability of H_θ and x. Note that $\sum_{\theta=1}^{2}\int P(H_\theta,x)\log\frac{P(H_\theta,x)}{P(H_\theta)f(x)}\,d\lambda(x)$ may be defined [cf. (4.2) of chapter 1] as the mean information in X about H_θ.

The capacity $C(f_1,f_2)$ of the channel (f_1,f_2) is defined by Shannon (1948) as $\max_{0\leq p\leq 1} R(p;f_1,f_2)$, that is, the maximum rate of transmission for all choices of the source. Denoting the maximum of $C(f_1,f_2)/J(1,2)$ over all f_1 and f_2 that are generalized densities with respect to a common measure by $\max_{(f_1,f_2)}\frac{C(f_1,f_2)}{J(1,2)}$, we can state [cf. Sakaguchi (1955, 1957a)]:

THEOREM 5.6.

$$\max_{(f_1,f_2)}\frac{C(f_1,f_2)}{J(1,2)} \leqq \frac{1}{4}.$$

Proof. Note that as a function of p, $0\leqq p\leqq 1$, $R(p;f_1,f_2)$ in (5.3) is concave (the second derivative is never positive); $R(0;f_1,f_2) = R(1;f_1,f_2) = 0$; and R' denoting the derivative with respect to p, $R'(0;f_1,f_2) = I(1:2)$ defined in (2.5) of chapter 1, $R'(1;f_1,f_2) = -I(2:1)$ defined in (3.1) of chapter 1; $R(p;f_1,f_2)$ is a maximum for p such that

$$\int f_1(x)\log\frac{f_1(x)}{f(x)}\,d\lambda(x) = \int f_2(x)\log\frac{f_2(x)}{f(x)}\,d\lambda(x).$$

Next, by writing $f_i(x) = pf_i(x) + qf_i(x)$, $i = 1, 2$, and using the convexity property as in example 3.2, we have

$$\int f_1(x) \log \frac{f_1(x)}{f(x)}\, d\lambda(x) = \int (pf_1(x) + qf_1(x)) \log \frac{pf_1(x) + qf_1(x)}{pf_1(x) + qf_2(x)}\, d\lambda(x)$$

$$\leqq \int pf_1(x) \log \frac{pf_1(x)}{pf_1(x)}\, d\lambda(x) + \int qf_1(x) \log \frac{qf_1(x)}{qf_2(x)}\, d\lambda(x)$$

$$= qI(1:2).$$

Similarly, $\int f_2(x) \log \frac{f_2(x)}{f(x)}\, d\lambda(x) \leqq pI(2:1)$, so that $R(p; f_1, f_2) \leqq pq(I(1:2) + I(2:1))$, or $C(f_1, f_2) = \max_{0 \leq p \leq 1} R(p; f_1, f_2) \leqq \frac{1}{4}J(1, 2)$, from which we finally get the inequality in the theorem.

6. FISHER'S INFORMATION

The information measures that we have been studying are related to Fisher's information measure. Consider the parametric case where the members of the set m of theorem 4.2 are of the same functional form but differ according to the values of the k-dimensional parameter $\theta = (\theta_1, \theta_2, \cdots, \theta_k)$. Suppose that θ and $\theta + \Delta\theta$ are neighboring points in the k-dimensional parameter space which is assumed to be an open convex set in a k-dimensional Euclidean space, and $f_1(x) = f(x, \theta)$, $f_2(x) = f(x, \theta + \Delta\theta)$. We shall show in this section that $I(\theta:\theta + \Delta\theta)$ and $J(\theta, \theta + \Delta\theta)$ can be expressed as quadratic forms with coefficients defined by Fisher's information matrix. [Cf. Savage (1954, pp. 235–237).] We may write

$$J(\theta, \theta + \Delta\theta) = \int (f(x, \theta) - f(x, \theta + \Delta\theta)) \log \frac{f(x, \theta)}{f(x, \theta + \Delta\theta)}\, d\lambda(x)$$

$$= \int f(x, \theta) \frac{\Delta f(x, \theta)}{f(x, \theta)} \Delta \log f(x, \theta)\, d\lambda(x),$$

and

$$I(\theta:\theta + \Delta\theta) = -\int f(x, \theta) \Delta \log f(x, \theta)\, d\lambda(x),$$

where $\Delta f(x, \theta) = f(x, \theta + \Delta\theta) - f(x, \theta)$ and $\Delta \log f(x, \theta) = \log f(x, \theta + \Delta\theta) - \log f(x, \theta)$.

Suppose that the generalized density $f(x, \theta)$ satisfies the following regularity conditions [cf. Cramér (1946a, pp. 500–501), Gurland (1954)]:

1. For all $x[\lambda]$, the partial derivatives $\frac{\partial \log f}{\partial \theta_\alpha}$, $\frac{\partial^2 \log f}{\partial \theta_\alpha \partial \theta_\beta}$, $\frac{\partial^3 \log f}{\partial \theta_\alpha \partial \theta_\beta \partial \theta_\gamma}$

exist, for all $\alpha, \beta, \gamma = 1, 2, \cdots, k$, for every $\theta' = (\theta_1', \theta_2', \cdots, \theta_k')$ belonging to the nondegenerate interval $A = (\theta_\alpha < \theta_\alpha' < \theta_\alpha + \Delta\theta_\alpha)$, $\alpha = 1, 2, \cdots, k$.

2. For every $\theta' \in A$, $\left|\frac{\partial f}{\partial \theta_\alpha}\right| < F(x)$, $\left|\frac{\partial^2 f}{\partial \theta_\alpha \partial \theta_\beta}\right| < G(x)$, $\left|\frac{\partial^3 \log f}{\partial \theta_\alpha \partial \theta_\beta \partial \theta_\gamma}\right| < H(x)$, for all $\alpha, \beta, \gamma = 1, 2, \cdots, k$, where $F(x)$ and $G(x)$ are integrable $[\lambda]$ over the whole space and $\int f(x, \theta) H(x)\, d\lambda(x) < M < \infty$, where M is independent of $\theta = (\theta_1, \theta_2, \cdots, \theta_k)$.

3. $\int \frac{\partial f}{\partial \theta_\alpha} d\lambda(x) = 0, \int \frac{\partial^2 f}{\partial \theta_\alpha \partial \theta_\beta} d\lambda(x) = 0$, for all $\alpha, \beta = 1, 2, \cdots, k$.

We may now use the Taylor expansion about θ and obtain

$$(6.1)\quad \log f(x, \theta + \Delta\theta) - \log f(x, \theta) = \sum_{\alpha=1}^{k} \Delta\theta_\alpha \frac{\partial \log f}{\partial \theta_\alpha} + \frac{1}{2!} \sum_{\alpha=1}^{k} \sum_{\beta=1}^{k} \Delta\theta_\alpha \Delta\theta_\beta \frac{\partial^2 \log f}{\partial \theta_\alpha \partial \theta_\beta} + \frac{1}{3!} \sum_{\alpha=1}^{k} \sum_{\beta=1}^{k} \sum_{\gamma=1}^{k} \Delta\theta_\alpha \Delta\theta_\beta \Delta\theta_\gamma \left(\frac{\partial^3 \log f}{\partial \theta_\alpha \partial \theta_\beta \partial \theta_\gamma}\right)_{\theta + t\Delta\theta},$$

where in the last term θ is replaced by $\theta + t\,\Delta\theta = (\theta_1 + t_1\,\Delta\theta_1, \theta_2 + t_2\,\Delta\theta_2, \cdots, \theta_k + t_k\,\Delta\theta_k)$, $0 < t_\alpha < 1$, $\alpha = 1, 2, \cdots, k$. We also have

$$(6.2)\quad \frac{\partial \log f}{\partial \theta_\alpha} = \frac{1}{f} \frac{\partial f}{\partial \theta_\alpha}; \quad \frac{\partial^2 \log f}{\partial \theta_\alpha \partial \theta_\beta} = \frac{1}{f} \frac{\partial^2 f}{\partial \theta_\alpha \partial \theta_\beta} - \frac{1}{f^2} \frac{\partial f}{\partial \theta_\alpha} \frac{\partial f}{\partial \theta_\beta}.$$

We may therefore write

$$(6.3)\quad I(\theta:\theta + \Delta\theta) = \int f(x, \theta) \log \frac{f(x, \theta)}{f(x, \theta + \Delta\theta)}\, d\lambda(x)$$

$$= -\int \left(\sum_{\alpha=1}^{k} \Delta\theta_\alpha f \cdot \frac{\partial \log f}{\partial \theta_\alpha}\right) d\lambda(x) - \frac{1}{2!} \int \left(\sum_{\alpha=1}^{k} \sum_{\beta=1}^{k} \Delta\theta_\alpha \Delta\theta_\beta f \cdot \frac{\partial^2 \log f}{\partial \theta_\alpha \partial \theta_\beta}\right) d\lambda(x) - \frac{1}{3!} \int \left[\sum_{\alpha=1}^{k} \sum_{\beta=1}^{k} \sum_{\gamma=1}^{k} \Delta\theta_\alpha \Delta\theta_\beta \Delta\theta_\gamma f \cdot \left(\frac{\partial^3 \log f}{\partial \theta_\alpha \partial \theta_\beta \partial \theta_\gamma}\right)_{\theta + t\Delta\theta}\right] d\lambda(x)$$

$$= -\sum_{\alpha=1}^{k} \Delta\theta_\alpha \int \frac{\partial f}{\partial \theta_\alpha} d\lambda(x) - \frac{1}{2!} \sum_{\alpha=1}^{k} \sum_{\beta=1}^{k} \Delta\theta_\alpha \Delta\theta_\beta \int \left(\frac{\partial^2 f}{\partial \theta_\alpha \partial \theta_\beta} - \frac{1}{f} \frac{\partial f}{\partial \theta_\alpha} \frac{\partial f}{\partial \theta_\beta}\right) d\lambda(x) - \frac{1}{3!} \sum_{\alpha=1}^{k} \sum_{\beta=1}^{k} \sum_{\gamma=1}^{k} \Delta\theta_\alpha \Delta\theta_\beta \Delta\theta_\gamma \int f \cdot \left(\frac{\partial^3 \log f}{\partial \theta_\alpha \partial \theta_\beta \partial \theta_\gamma}\right)_{\theta + t\Delta\theta} d\lambda(x).$$

Accordingly, because of the regularity conditions, to within second-order terms, we have

$$I(\theta:\theta + \Delta\theta) = \frac{1}{2}\sum_{\alpha=1}^{k}\sum_{\beta=1}^{k} g_{\alpha\beta}\,\Delta\theta_\alpha\,\Delta\theta_\beta, \tag{6.4}$$

with

$$g_{\alpha\beta} = \int f(x,\theta)\left(\frac{1}{f(x,\theta)}\frac{\partial f(x,\theta)}{\partial\theta_\alpha}\right)\left(\frac{1}{f(x,\theta)}\frac{\partial f(x,\theta)}{\partial\theta_\beta}\right)d\lambda(x),$$

and $\mathbf{G} = (g_{\alpha\beta})$ the positive definite Fisher information matrix [Bartlett (1955, p. 222), Doob (1934), Fisher (1956, p. 153), Huzurbazar (1949), Jeffreys (1948, p. 158), Mandelbrot (1953, pp. 34–35), Rao (1952, p. 144), Savage (1954, pp. 235–238), Schützenberger (1954, p. 54)].

We shall sketch the proof of the related result for $J(\theta, \theta + \Delta\theta)$:

$$\Delta \log f(x,\theta) = \log\left(1 + \frac{\Delta f(x,\theta)}{f(x,\theta)}\right) \approx \frac{\Delta f(x,\theta)}{f(x,\theta)},$$

$$\begin{aligned} J(\theta, \theta+\Delta\theta) &\approx \int f(x,\theta)\left(\frac{\Delta f(x,\theta)}{f(x,\theta)}\right)^2 d\lambda(x) \\ &\approx \int f(x,\theta)\left(\frac{1}{f}\frac{\partial f}{\partial\theta_1}\Delta\theta_1 + \cdots + \frac{1}{f}\frac{\partial f}{\partial\theta_k}\Delta\theta_k\right)^2 d\lambda(x) \\ &= \sum_{\alpha=1}^{k}\sum_{\beta=1}^{k} g_{\alpha\beta}\,\Delta\theta_\alpha\,\Delta\theta_\beta. \end{aligned}$$

7. INFORMATION AND SUFFICIENCY

In the definition of $I(1:2)$ in section 2 of chapter 1 we assumed that the probability measures μ_1 and μ_2 were absolutely continuous with respect to each other. The essential reason for this was the desire that the integrals in $I(1:2)$ and $I(2:1)$ be well defined, so that $J(1, 2)$ could exist. If we do not concern ourselves with $J(1, 2)$, but limit our attention only to $I(1:2)$, we may modify somewhat the initial assumptions, as well as the assumption about the homogeneous set of measures in theorem 4.2. If we re-examine the integrals in (2.4) and (2.5) of chapter 1, we see that they are still well defined if $f_1(x) = 0, x \in E$, but $f_2(x) \neq 0, x \in E, \lambda(E) \neq 0$, since $0 \log 0$ is defined as zero. Thus, limiting ourselves only to $I(1:2)$, we need assume simply that the probability measure μ_1 is absolutely continuous with respect to the probability measure μ_2; that is, $\mu_1(E) = 0$ for every measurable set E for which $\mu_2(E) = 0$. According to the Radon–Nikodym theorem (see section 2 of chapter 1, and references there):

A necessary and sufficient condition that the probability measure μ_1 *be absolutely continuous with respect to the probability measure* μ_2 *is that there exist a nonnegative function* $f(x)$ *on* $\mathscr{X}$ *such that*

$$\mu_1(E) = \int_E f(x)\, d\mu_2(x), \tag{7.1}$$

for every E *in* $\mathscr{S}$. *The function* $f(x)$*, the Radon–Nikodym derivative, is unique in the sense that if*

$$\mu_1(E) = \int_E g(x)\, d\mu_2(x), \tag{7.2}$$

for every E *in* $\mathscr{S}$*, then* $f(x) = g(x)\ [\mu_2]$. *We write* $d\mu_1(x) = f(x)\, d\mu_2(x)$ *and also* $f(x) = d\mu_1/d\mu_2$.

The properties in sections 2 and 3 are valid if the probability measure μ_1 is absolutely continuous with respect to the probability measure μ_2, since with $f(x)$ defined as in (7.1) we have [compare with (3.1), noting that $f_1(x) = f(x), f_2(x) = 1$, since $\mu_2(E) = \int_E d\mu_2(x)$]

$$I(1:2) = \int \log f(x)\, d\mu_1(x) = \int f(x) \log f(x)\, d\mu_2(x). \tag{7.3}$$

Note that according to corollary 3.1 a set E provides no information for discrimination in favor of H_1 if $\mu_1(E) = 0$ but $\mu_2(E) \neq 0$. Theorem 4.2 also holds if the requirement that the probability measures μ_1 and μ_2 are members of a homogeneous set of probability measures is modified so that they are members of a *dominated set* of probability measures; a set M of measures on $\mathscr{S}$ is called dominated if there exists a measure λ on $\mathscr{S}$, λ not necessarily a member of M, such that every member of the set M is absolutely continuous with respect to λ. [See Fraser (1957, p. 19), Halmos and Savage (1949).] The Radon–Nikodym theorem can then be applied in the form where for every μ_i of the set of dominated measures, we have

$$\mu_i(E) = \int_E f_i(x)\, d\lambda(x), \qquad \text{for all } E \in \mathscr{S}.$$

Example 7.1. Suppose that the populations under H_1 and H_2 are respectively rectangular populations with $0 \leq x \leq \theta_1$, $0 \leq x \leq \theta_2$, $\theta_1 < \theta_2$, and

$$f_1(x) = \frac{1}{\theta_1},\ 0 \leq x \leq \theta_1, \qquad f_2(x) = \frac{1}{\theta_2},\ 0 \leq x \leq \theta_2,$$

$$= 0, \text{ elsewhere}, \qquad\qquad = 0, \text{ elsewhere},$$

$$\mu_1(E) = \int_E \frac{dx}{\theta_1}, \qquad \mu_2(E) = \int_E \frac{dx}{\theta_2}.$$

Note that $\mu_1(E) = \int_{\theta_1}^{\theta_2} f_1(x)\,dx = 0$, but that $\mu_2(E) = \int_{\theta_1}^{\theta_2} f_2(x)\,dx = (\theta_2 - \theta_1)/\theta_2 \neq 0$, when $E = \{x: \theta_1 \leq x \leq \theta_2\}$. We see that μ_2 is not absolutely continuous with respect to μ_1, but that μ_1 is absolutely continuous with respect to μ_2, since $\mu_1 = 0$ whenever $\mu_2 = 0$. Both μ_1 and μ_2 are absolutely continuous with respect to Lebesgue measure.

Now

$$I(1:2) = \int_0^{\theta_1} \frac{1}{\theta_1} \log \frac{1/\theta_1}{1/\theta_2}\,dx + \int_{\theta_1}^{\theta_2} 0 \log \frac{0}{1/\theta_2}\,dx, \tag{7.4}$$

or in the notation of (7.3),

$$I(1:2) = \int_0^{\theta_2} f(x) \log f(x) \cdot \frac{dx}{\theta_2},$$

with $f(x) = \theta_2/\theta_1$ for $0 \leq x \leq \theta_1$, and $f(x) = 0$ for $\theta_1 < x \leq \theta_2$, so that

$$I(1:2) = \left(\frac{\theta_2}{\theta_1} \log \frac{\theta_2}{\theta_1}\right) \frac{\theta_1}{\theta_2} = \log \frac{\theta_2}{\theta_1}, \tag{7.5}$$

and therefore for a random sample O_n of n independent observations $I(1:2;\, O_n) = n \log(\theta_2/\theta_1)$. If $\mathscr{X}$ is the space of n independent observations, and $Y = T(x) = \max(x_1, x_2, \cdots, x_n)$, it is known that $g_i(y) = ny^{n-1}/\theta_i^{\,n}$, $0 \leq y \leq \theta_i$, and zero elsewhere, $i = 1, 2$ [Wilks (1943, p. 91)].

We thus have

$$I(1:2;\, \mathscr{Y}) = \int_0^{\theta_1} \frac{ny^{n-1}}{\theta_1^{\,n}} \log \frac{\theta_2^{\,n}}{\theta_1^{\,n}}\,dy = n \log \frac{\theta_2}{\theta_1}. \tag{7.6}$$

Since $n \log(\theta_2/\theta_1) = I(1:2;\, \mathscr{X}) = I(1:2;\, \mathscr{Y})$, we conclude from theorem 4.2 that the largest value in a sample from a rectangular population, with lower value of the range at zero, is a sufficient statistic. [Cf. Lehmann (1950a, p. 3-3).]

Example 7.2. Consider the exponential populations defined by $f_i(x) = e^{-(x-\theta_i)}$, $\theta_i \leq x < \infty$, $f_i(x) = 0$, $-\infty < x < \theta_i$, $i = 1, 2$, $\theta_1 > \theta_2$. We find that

$$I(1:2) = \int_{\theta_1}^{\infty} e^{-(x-\theta_1)}(\theta_1 - \theta_2)\,dx = \theta_1 - \theta_2, \tag{7.7}$$

and for a random sample O_n of n independent observations $I(1:2;\, O_n) = nI(1:2;\, O_1) = n(\theta_1 - \theta_2)$. If $\mathscr{X}$ is the space of n independent observations and $Y = T(x) = \min(x_1, x_2, \cdots, x_n)$, then it is known that $g_i(y) = ne^{-n(y-\theta_i)}$, $\theta_i \leq y < \infty$, and zero elsewhere, $i = 1, 2$ [Wilks (1943, p. 91)].

We thus have

$$I(1:2;\, \mathscr{Y}) = \int_{\theta_1}^{\infty} ne^{-n(y-\theta_1)}(n\theta_1 - n\theta_2)\,dy = n(\theta_1 - \theta_2). \tag{7.8}$$

Since $n(\theta_1 - \theta_2) = I(1:2;\, \mathscr{X}) = I(1:2;\, \mathscr{Y})$, we conclude from theorem 4.2 that the smallest value in a sample from populations of the type $e^{-(x-\theta)}$, $\theta \leq x < \infty$, zero elsewhere, is a sufficient statistic.

Example 7.3. Consider the Poisson populations with parameters λ_1, λ_2. We find that

$$(7.9)\qquad I(1:2) = \sum_{x=0}^{\infty} \frac{e^{-\lambda_1}\lambda_1^{\,x}}{x!} \log \frac{e^{-\lambda_1}\lambda_1^{\,x}}{e^{-\lambda_2}\lambda_2^{\,x}} = \lambda_1 \log \frac{\lambda_1}{\lambda_2} + (\lambda_2 - \lambda_1),$$

and for a random sample O_n of n independent observations $I(1:2; O_n) = nI(1:2; O_1) = n\lambda_1 \log (\lambda_1/\lambda_2) + n(\lambda_2 - \lambda_1)$. If $\mathscr{X}$ is the space of n independent observations, and $Y = T(x) = \sum_{i=1}^{n} x_i$, then it is known that

$$g_i(y) = e^{-n\lambda_i}(n\lambda_i)^y/y!,\ y = 0, 1, 2, \cdots; i = 1, 2$$

[Cramér (1946a, p. 205)]. We thus have

$$(7.10)\quad I(1:2; \mathscr{Y}) = \sum_{y=0}^{\infty} \frac{e^{-n\lambda_1}(n\lambda_1)^y}{y!} \log \frac{e^{-n\lambda_1}(n\lambda_1)^y}{e^{-n\lambda_2}(n\lambda_2)^y} = n\lambda_1 \log \frac{\lambda_1}{\lambda_2} + n(\lambda_2 - \lambda_1).$$

Since $I(1:2; \mathscr{X}) = I(1:2; \mathscr{Y})$, we conclude from theorem 4.2 that $\sum_{i=1}^{n} x_i$ is a sufficient statistic for Poisson populations. [Cf. Lehmann (1950a, p. 3-3).]

Example 7.4. Consider the Poisson populations in example 7.3 so that $I(1:2)$ is given by (7.9). Suppose $\mathscr{X}$ is the space of nonnegative integers and $Y = T(x)$ is 0, 1, 2, according as x is 0, 1, or ≥ 2. In example 3.3 we saw that $I(1:2; \mathscr{X}) > I(1:2; \mathscr{Y})$ and therefore Y is not a sufficient statistic for the Poisson populations. [Cf. Lehmann (1950a, p. 3-4).]

8. PROBLEMS

8.1. Compute $I(1:2; X)$, $I(1:2; Y|X = x)$, $I(1:2; Y|X)$, $I(1:2; X, Y)$ in example 2.1: (*a*) when $\rho_1^2 = \rho_2^2 = \rho^2$; (*b*) when $\mu_{1x} = \mu_{2x}$; (*c*) when $\mu_{1x} = \mu_{2x}$, $\mu_{1y} = \mu_{2y}$.

8.2. Verify corollary 3.3, using appropriate cases of example 2.1.

8.3. Show that the equality holds in example 3.4 if x is a sufficient statistic.

8.4. If $f_1(x)$, $f_2(x)$, $f(x)$ are generalized densities of a homogeneous set of measures, then

$$\int f_1(x) \log \frac{f_1(x)}{f_2(x)}\, d\lambda(x) \geq \int f_1(x) \log \frac{f(x)}{f_2(x)}\, d\lambda(x).$$

When does the equality hold?

8.5. Prove the theorems and corollaries in section 5.

8.6. What is the maximum value of $R(p; f_1, f_2)$ in (5.3) for variations of p?

8.7. In the notation of section 6, what is the value of $I(\theta + \Delta\theta:\theta)$ as a quadratic form?

8.8. Show that for the populations and statistics in examples 7.1 and 7.2, the conditions for equality in theorem 4.1 are satisfied.

8.9. In (7.5) take $\theta_2 = \theta + \Delta\theta$, $\theta_1 = \theta$. Compare with the results according to section 6.

8.10. In (7.7) take $\theta_1 = \theta + \Delta\theta, \theta_2 = \theta$. Compare with the results according to section 6.

8.11. Derive the values for $g_i(y)$ given in examples 7.1, 7.2, and 7.3.

8.12. Show that the number of "successes" observed in a sample of n independent observations is a sufficient statistic for binomial populations.

8.13. Show that the sample average is a sufficient statistic for normal populations with a common variance.

8.14. Let $f(x)$ be a probability density with mean μ and finite variance σ^2, and such that $f(x) \log f(x)$ is summable ($t \log t$ is defined to be zero if $t = 0$). Show that

$$\int_{-\infty}^{\infty} f(x) \log f(x)\, dx \geq \log (1/\sigma\sqrt{2\pi e})$$

with equality if and only if $f(x)$ is equal almost everywhere to the normal probability density $\frac{1}{\sigma\sqrt{2\pi}} \exp\left(-\frac{(x-\mu)^2}{2\sigma^2}\right)$. [Cf. Shannon (1948, pp. 629–630), Woodward (1953, p. 25), *Am. Math. Monthly*, Vol. 64 (1957, pp. 511–512).]

8.15. Generalize the result in problem 8.14 to multivariate probability densities.

8.16. Compute $J(1, 2; X)$, $J(1, 2; Y|X = x)$, $J(1, 2; Y|X)$, and $J(1, 2; X, Y)$ for the populations in example 2.1.

8.17. Compute the value of $I(1:2; X, Y)$ in example 2.1 for $\rho_1 = \rho_2 = \rho$, $\mu_{1x} = \mu_{1y}$, $\mu_{2x} = \mu_{2y}$, $\sigma_x^2 = \sigma_y^2$. Compare the value you get with $2I(1:2; X)$, as ρ varies from -1 to $+1$.

8.18. Compute $J(1, 2)$, $J(1, 2; O_n)$, $J(1, 2; \mathscr{Y})$ for the populations in example 7.3. What are the corresponding values for the populations in examples 7.1 and 7.2?

8.19. In (5.3) take $f_\theta(x) = \mu_\theta(E_i)$, $x \in E_i$, $i = 1, 2$, $\theta = 1, 2$, where $\mathscr{X} = E_1 \cup E_2$, $E_1 \cap E_2 = 0$, $\mu_1(E_1) = \mu_2(E_2)$, and $p = q = \frac{1}{2}$. Show that with these values $R(p; f_1, f_2) = \mu_1(E_1) \log 2\mu_1(E_1) + \mu_1(E_2) \log 2\mu_1(E_2)$, which is the same as the value of $I(1:2)$ for the binomial distributions with $N = 1$, $p_1 = \mu_1(E_1)$, $q_1 = 1 - p_1 = \mu_1(E_2)$, $p_2 = q_2 = \frac{1}{2}$. [See your answer to problems 5.2(*a*) and 5.8 in chapter 1.] What is the value of $R(p; f_1, f_2)$ if $\mu_1(E_1) = \mu_2(E_2) = 1$?

8.20. Compute the values of $I(1:2; X)$, $I(1:2; Y|x = 0)$, $I(1:2; Y|X)$ for the distributions in problem 5.8 of chapter 1. Do your values confirm corollary 3.3(*c*)?

8.21. Suppose that in $I(1:2) = \int f_1(x) \log \frac{f_1(x)}{f_2(x)}\, d\lambda(x)$, $f_2(x) = \chi_A(x)/\lambda(A)$, $f_1(x) = \chi_{A\cap B}(x)/\lambda(A \cap B)$, where $A \in \mathscr{S}$, $B \in \mathscr{S}$, $\lambda(A) \neq 0$, $\lambda(A \cap B) \neq 0$, and $\chi_A(x)$ is the characteristic function of the set A. Show that, for any set $E \in \mathscr{S}$,

(*a*) $\mu_1(E) = \lambda(E \cap A \cap B)/\lambda(A \cap B)$, $\mu_1(B) = 1$.
(*b*) $\mu_2(E) = \lambda(E \cap A)/\lambda(A)$, $\mu_2(B) = \lambda(B \cap A)/\lambda(A)$.

(*c*) If $\lambda(E) = 0$, then $\mu_1(E) = \mu_2(E) = 0$.
(*d*) If $\mu_2(E) = 0$, then $\mu_1(E) = 0$.
(*e*) If $f_2(x) = 0$, then $f_1(x) = 0$.
(*f*) $I(1:2) = \log \dfrac{\lambda(A)}{\lambda(A \cap B)} = -\log \mu_2(B)$.

Note that (*f*) yields Wiener's definition of the information resulting from the additional knowledge that $x \in B$ when it is known that $x \in A$ [Powers (1956, pp. 44–45)].

8.22. With the data in problem 5.8 of chapter 1, show that the partitioning $\mathscr{X} = E_1 \cup E_2$, where $E_1 = (x = 0, y = 0) \cup (x = 1, y = 1)$ and $E_2 = (x = 0, y = 1) \cup (x = 1, y = 0)$ is a sufficient partitioning, or the statistic $T(x, y) = 0$ for $x = y$, $T(x, y) = 1$ for $x \neq y$, is a sufficient statistic.

8.23. With the data in problem 5.8 of chapter 1, show that the statistic $T(x, y) = x$ is not a sufficient statistic.

8.24. Let $f_i(x)$, $i = 1, 2, \cdots, n$, be generalized densities of a homogeneous set of probability measures, and $p_i \geq 0$, $i = 1, 2, \cdots, n$, such that $p_1 + p_2 + \cdots + p_n = 1$. If $f(x) = p_1 f_1(x) + p_2 f_2(x) + \cdots + p_n f_n(x)$, show that the maximum value of $R(p_i; f_i) = \int \left(\sum_{i=1}^{n} p_i f_i(x) \log \dfrac{f_i(x)}{f(x)} \right) d\lambda(x)$ for variations of the p_i occurs when the p_i are such that

$$\int f_1(x) \log \frac{f_1(x)}{f(x)}\, d\lambda(x) = \int f_2(x) \log \frac{f_2(x)}{f(x)}\, d\lambda(x) = \cdots = \int f_n(x) \log \frac{f_n(x)}{f(x)}\, d\lambda(x)$$

and that $\max\limits_{0 \leq p_i \leq 1} R(p_i; f_i)$ is then this common value. Show that $R(p_i; f_i) \leq \sum\limits_{i<j} p_i p_j J(i, j)$. Describe the related communication model as in the last part of section 5.

8.25. Let $f_i(x)$, p_i, $i = 1, 2, \cdots, n$, and $f(x)$ be defined as in problem 8.24, and suppose that $g(x)$ is also a generalized density of the same homogeneous set of probability measures. Show that

$$\sum_{i=1}^{n} p_i \int f_i(x) \log \frac{f_i(x)}{g(x)}\, d\lambda(x) \geq \int f(x) \log \frac{f(x)}{g(x)}\, d\lambda(x),$$

with equality if and only if $f_1(x) = f_2(x) = \cdots = f_n(x)$ $[\lambda]$. [Note that this implies that for discrimination against $g(x)$ the "mixture" of $f_1(x), \cdots, f_n(x)$ given by $f(x)$ provides less information than the mean of the information of the components of the mixture. See example 2.1 of chapter 3.]

8.26. Let $f(x)$ be the probability density of a random variable limited to a certain volume V in its space and such that $f(x) \log f(x)$ is summable ($t \log t$ is defined to be zero if $t = 0$). Show that $\int_V f(x) \log f(x)\, dx \geq \log (1/V)$, with equality if and only if $f(x)$ is equal almost everywhere to the constant $1/V$ in the volume. [Cf. Shannon (1948, p. 629).]

8.27. Let $f(x)$ be the probability density of a nonnegative random variable with mean μ, and such that $f(x) \log f(x)$ is summable ($t \log t$ is defined to be zero if $t = 0$). Show that $\int_0^{\infty} f(x) \log f(x)\, dx \geq -\log \mu e$, with equality if and only if $f(x)$ is equal almost everywhere to $(1/\mu)e^{-x/\mu}$, $x \geq 0$. [Cf. Shannon (1948, pp. 630–631).]

8.28. Consider the discrete random variable x that takes the values $x_1, x_2, \cdots, x_n$, and has the mean value μ, that is, $p_j = \text{Prob}\,(x = x_j)$, $\sum_{j=1}^{n} x_j p_j = \mu$. Show that $\sum_{j=1}^{n} p_j \log p_j \geq \beta\mu - \log M(\beta)$, with equality if and only if $p_j = \frac{e^{\beta x_j}}{M(\beta)}$, where $M(\beta) = \sum_{j=1}^{n} e^{\beta x_j}$, $\mu = \sum_{j=1}^{n} x_j p_j = \sum_{j=1}^{n} \frac{x_j e^{\beta x_j}}{M(\beta)} = \frac{d}{d\beta} \log M(\beta)$. (See problem 8.36.) [Cf. Brillouin (1956, pp. 41–43), Jaynes (1957, pp. 621–623).]

8.29. Show that $I(1{:}2;\, Y|X) = 0$, if and only if $h_1(y|x) = h_2(y|x)$ for almost all x (see theorem 2.2 and corollary 3.3).

8.30. Consider the discrete random variables x, y, where $p_{ij} = \text{Prob}\,(x = x_i, y = y_j)$, $i = 1, 2, \cdots, m$, $j = 1, 2, \cdots, n$, $p_{i\cdot} = \sum_{j=1}^{n} p_{ij}$, $p_{\cdot j} = \sum_{i=1}^{m} p_{ij}$, $p_{ij} > 0$, $\sum_{i=1}^{m} \sum_{j=1}^{n} p_{ij} = \sum_{i=1}^{m} p_{i\cdot} = \sum_{j=1}^{n} p_{\cdot j} = 1$, and the entropies defined by $\mathscr{H}(x, y) = -\sum_i \sum_j p_{ij} \log p_{ij}$, $\mathscr{H}(x) = -\sum_i p_{i\cdot} \log p_{i\cdot}$, $\mathscr{H}(y) = -\sum_j p_{\cdot j} \log p_{\cdot j}$, $\mathscr{H}(y|x_i) = -\sum_j \frac{p_{ij}}{p_{i\cdot}} \log \frac{p_{ij}}{p_{i\cdot}}$, $\mathscr{H}(y|x) = \sum_i p_{i\cdot} \mathscr{H}(y|x_i) = -\sum_i \sum_j p_{ij} \log \frac{p_{ij}}{p_{i\cdot}}$. Show that

(*a*) $\mathscr{H}(x, y) = \mathscr{H}(x) + \mathscr{H}(y|x)$.
(*b*) $\mathscr{H}(x, y) \leq \mathscr{H}(x) + \mathscr{H}(y)$.
(*c*) $\mathscr{H}(y) \geq \mathscr{H}(y|x)$.

[Cf. Shannon (1948, pp. 392–396).]

8.31. A real-valued function $f(x)$ defined for all values of x in an interval $a \leq x \leq b$ is said to be *convex* if for every pair $a \leq (x_1, x_2) \leq b$ and all $\lambda_1 + \lambda_2 = 1$, $\lambda_i \geq 0$, $i = 1, 2$, $\lambda_1 f(x_1) + \lambda_2 f(x_2) \geq f(\lambda_1 x_1 + \lambda_2 x_2)$.
The function is said to be *concave* if $\lambda_1 f(x_1) + \lambda_2 f(x_2) \leq f(\lambda_1 x_1 + \lambda_2 x_2)$.
The function is *strictly convex* or *strictly concave* if the equalities hold only when $x_1 = x_2$. Show that

(*a*) If $\frac{d^2 f(x)}{dx^2}$ exists at every point in $a \leq x \leq b$, a necessary and sufficient condition for $f(x)$ to be convex is that $\frac{d^2 f(x)}{dx^2} \geq 0$.

(*b*) If $f(x)$ is a convex function and $a \leq (x_1, x_2, \cdots, x_n) \leq b$, then $\lambda_1 f(x_1) + \cdots + \lambda_n f(x_n) \geq f(\lambda_1 x_1 + \cdots + \lambda_n x_n)$, $\lambda_1 + \lambda_2 + \cdots + \lambda_n = 1$, $\lambda_i \geq 0$, $i = 1, 2, \cdots, n$.

(*c*) If $f(x)$ is a convex function, $p(x) \geq 0$, $\int_a^b p(x)\, dx = 1$, then

$$\int_a^b f(x) p(x)\, dx \geq f\left(\int_a^b x p(x)\, dx\right).$$

8.32. Suppose $p_{i1} + p_{i2} + \cdots + p_{ic} = 1, p_{ij} > 0, i = 1, 2; j = 1, 2, \cdots, c$, and $q_{ij} = a_{j1}p_{i1} + a_{j2}p_{i2} + \cdots + a_{jc}p_{ic}$, $i = 1, 2$; $j = 1, 2, \cdots, c$, with $a_{j1} + a_{j2} + \cdots + a_{jc} = 1$, $j = 1, 2, \cdots, c$, and $a_{1k} + a_{2k} + \cdots + a_{ck} = 1$, $k = 1, 2, \cdots, c$, $a_{jk} \geq 0$.

Show that $I(1:2; p) = \sum_{j=1}^{c} p_{1j} \log \frac{p_{1j}}{p_{2j}} \geq \sum_{j=1}^{c} q_{1j} \log \frac{q_{1j}}{q_{2j}} = I(1:2; q)$, with equality if and only if $p_{1j}/p_{2j} = p_{1k}/p_{2k}$, $j, k = 1, 2, \cdots, c$.

8.33. Suppose that $x_1, x_2, \cdots, x_n$ is a random sample of a discrete random variable, $Y = Y(x_1, x_2, \cdots, x_n)$ is a statistic, and $\text{Prob}(x_1, x_2, \cdots, x_n|H_i) \neq 0$, $i = 1, 2$. Show that

$$E\left(\log \frac{\text{Prob}(x_1, x_2, \cdots, x_n|H_1)}{\text{Prob}(x_1, x_2, \cdots, x_n|H_2)} \middle| H_1, Y = y\right) \geq \log \frac{\text{Prob}(Y = y|H_1)}{\text{Prob}(Y = y|H_2)}.$$

When does the equality hold? [Cf. Savage (1954, p. 235).]

8.34. Consider the Poisson populations with parameters $m_1 = 1$, $m_2 = 2$, $m_3 = 3$. Show that [see problem 5.2(*c*) of chapter 1 and the last paragraph of section 3 of chapter 1]

(*a*) $J(1, 3) > J(1, 2) + J(2, 3)$.
(*b*) $\sqrt{J(1, 3)} > \sqrt{J(1, 2)} + \sqrt{J(2, 3)}$.

8.35. Show that $F(p_1, p_2) = p_1 \log \frac{p_1}{p_2} + q_1 \log \frac{q_1}{q_2}$, $0 \leq p_i \leq 1$, $p_i + q_i = 1$, $i = 1, 2$ is a convex function of $p_1(p_2)$ for fixed $p_2(p_1)$.

8.36. Suppose that in problem 8.28 the x_j's are positive integers, and $\beta_1 > 0$ such that $\sum_{j=1}^{n} e^{-\beta_1 x_j} = 1$. Show that $\mu \geq \mathscr{H}(p)/\beta_1$, where $\mathscr{H}(p) = -\sum_{j=1}^{n} p_j \log p_j$. In particular, if $x_j = j$, $n = \infty$, find β_1 and the values of p_j and μ for equality. [Note that this is related to the noiseless coding theorem. See, for example, Feinstein (1958, pp. 17–20), Shannon (1948, pp. 401–403).]

8.37. Let $0 \leq \phi(a_i|x) \leq 1$, $\sum_i \phi(a_i|x) = 1$ for all $x \in \mathscr{X}\,[\lambda]$, $p_j(a_i) = \int \phi(a_i|x) f_j(x)\, d\lambda(x)$, that is, $\phi(a_i|x)$ is the probability for "action" a_i given x, and $p_j(a_i)$ is the probability for "action" a_i under H_j, $j = 1, 2$. Show that

$$\int f_1(x) \log \frac{f_1(x)}{f_2(x)}\, d\lambda(x) \geq \sum_i p_1(a_i) \log \frac{p_1(a_i)}{p_2(a_i)},$$

and give the necessary and sufficient condition for equality. Derive corollary 3.2 as a particular case of this problem.

CHAPTER 3

Inequalities of Information Theory

1. INTRODUCTION

The Cramér–Rao inequality, which provides, under certain regularity conditions, a lower bound for the variance of an estimator, is well known to statisticians from the theory of estimation. Savage (1954, p. 238) has recommended the name "information inequality" since results on the inequality were given by Fréchet (1943) and Darmois (1945), as well as by Rao (1945) and Cramér (1946a, 1946b). Various extensions have been made by Barankin (1949, 1951), Bhattacharyya (1946b, 1947, 1948), Chapman and Robbins (1951), Chernoff (1956, with an acknowledgment to Charles Stein and Herman Rubin), Fraser and Guttman (1952), Kiefer (1952), Seth (1949), Wolfowitz (1947).

We shall derive in theorem 2.1 an inequality for the discrimination information that may be considered a generalization of the Cramér–Rao or information inequality (using the name recommended by Savage). [Cf. Kullback (1954).] Theorem 2.1 will play an important part in subsequent applications to testing statistical hypotheses. We relate theorem 2.1 (and its consequences) and the classical information inequality of the theory of estimation in sections 5 and 6.

2. MINIMUM DISCRIMINATION INFORMATION

Suppose that $f_1(x)$ and $f_2(x)$ are generalized densities of a dominated set of probability measures on the measurable space $(\mathscr{X}, \mathscr{S})$, so that (see sections 2, 4, and 7 of chapter 2)

$$\mu_i(E) = \int_E f_i(x)\, d\lambda(x), \qquad E \in \mathscr{S},\ i = 1, 2.$$

For a given $f_2(x)$ we seek the member of the dominated set of probability measures that is "nearest" to or most closely resembles the probability measure μ_2 in the sense of smallest directed divergence (see the last part of section 3 of chapter 1)

$$I(1:2) = \int f_1(x) \log \frac{f_1(x)}{f_2(x)}\, d\lambda(x).$$

Since $I(1:2) \geqq 0$, with equality if and only if $f_1(x) = f_2(x)$ $[\lambda]$ (see theorem 3.1 of chapter 2), it is clear that we must impose some additional restriction on $f_1(x)$ if the desired "nearest" probability measure is to be some other than the probability measure μ_2 itself. We shall require $f_1(x)$ to be such that $I(1:2)$ is a minimum subject to $\int T(x)f_1(x)\,d\lambda(x) = \theta$, where θ is a constant and $Y = T(x)$ a measurable statistic (see section 4 of chapter 2). In most cases, θ is a multidimensional parameter of the populations. It may also represent some other desired characteristic of the populations. In chapter 5 we examine in detail the relation of θ to observed sample values and the implications for the testing of statistical hypotheses. The underlying principle is that $f_2(x)$ will be associated with the set of populations of the null hypothesis and $f_1(x)$ will range over the set of populations of the alternative hypothesis. The sample values will be used to determine the "resemblance" between the sample, as a possible member of the set of populations of the alternative hypothesis, and the closest population of the set of populations of the null hypothesis by an estimate of the smallest directed divergence or minimum discrimination information. The null hypothesis will be rejected if the estimated minimum discrimination information is significantly large. We remark that the approach here is very similar to Shannon's rate for a source relative to a fidelity evaluation [Kolmogorov (1956, p. 104), Shannon (1948, pp. 649–650)]. [Compare with the concept of "least favorable" distribution (Fraser (1957, p. 79)), and "maximum-entropy" estimates (Jaynes (1957)).]

The requirement is then equivalent to minimizing

$$\int \left(f_1(x) \log \frac{f_1(x)}{f_2(x)} + kT(x)f_1(x) + lf_1(x) \right) d\lambda(x), \tag{2.1}$$

with k and l arbitrary constant multipliers. Following a procedure similar to that in section 3 of chapter 2, set $g(x) = f_1(x)/f_2(x)$ so that (2.1) may be written as

$$\int (g(x) \log g(x) + kT(x)g(x) + lg(x))\,d\mu_2(x). \tag{2.2}$$

If we write $\phi(t) = t \log t + kTt + lt$, $t_0 = e^{-kT-l-1}$, then $\phi(t) = \phi(t_0) + (t - t_0)\phi'(t_0) + \frac{1}{2}(t - t_0)^2\phi''(t_1)$, where t_1 lies between t and t_0. But, as may be verified, $\phi(t_0) = -t_0$, $\phi'(t_0) = 0$, $\phi''(t_1) = 1/t_1 > 0$, so that

$$\int \phi(g(x))\,d\mu_2(x) = -\int e^{-kT(x)-l-1}\,d\mu_2(x) + \tfrac{1}{2}\int (g(x) - e^{-kT(x)-l-1})^2 \frac{d\mu_2(x)}{h(x)}, \tag{2.3}$$

where $h(x)$ lies between $g(x)$ and $e^{-kT(x)-l-1}$. We see then from (2.3) that

$$\int \phi(g(x))\, d\mu_2(x) \geqq -\int e^{-kT(x)-l-1}\, d\mu_2(x), \tag{2.4}$$

with equality if and only if

$$g(x) = e^{-kT(x)-l-1}\ [\lambda]. \tag{2.5}$$

The minimum value of (2.1) thus occurs for

$$f_1(x) = f^*(x) = f_2(x)e^{-kT(x)-l-1}\ [\lambda], \tag{2.6}$$

in which case (2.1) and (2.4) yield

$$I(*:2) + k\theta + l = -\int f_2(x)e^{-kT(x)-l-1}\, d\lambda(x). \tag{2.7}$$

If we replace $-k$ by τ, for notational convenience, and set $M_2(\tau) = \int f_2(x)e^{+\tau T(x)}\, d\lambda(x)$, $M_2(\tau) < \infty$, we see from (2.6) that $1 = e^{-l-1}M_2(\tau)$, and from (2.7) that the minimum discrimination information is

$$I(*:2) = \theta\tau - \log M_2(\tau), \tag{2.8}$$

where

$$\theta = \int T(x)f^*(x)\, d\lambda(x) = \int \frac{T(x)f_2(x)e^{\tau T(x)}\, d\lambda(x)}{M_2(\tau)} = \frac{(d/d\tau)M_2(\tau)}{M_2(\tau)}, \tag{2.9}$$

for all τ in the interior of the interval in which $M_2(\tau)$ is finite. Hereafter we shall denote τ by $\tau(\theta)$ when it is important to indicate τ as a function of θ.

We can now state [cf. Kullback (1954), Sanov (1957, pp. 23–24)]:

THEOREM 2.1.† *If $f_1(x)$ and a given $f_2(x)$ are generalized densities of a dominated set of probability measures, $Y = T(x)$ is a measurable statistic such that $\theta = \int T(x)f_1(x)\, d\lambda(x)$ exists, and $M_2(\tau) = \int f_2(x)e^{\tau T(x)}\, d\lambda(x)$ exists for τ in some interval; then*

$$I(1:2) \geqq \theta\tau - \log M_2(\tau) = I(*:2), \qquad \theta = \frac{d}{d\tau}\log M_2(\tau), \tag{2.10}$$

with equality in (2.10) *if and only if*

$$f_1(x) = f^*(x) = e^{\tau T(x)}f_2(x)/M_2(\tau)\ [\lambda]. \tag{2.11}$$

We remark that $f^*(x) = f_2(x)e^{\tau T(x)}/M_2(\tau)$ is said to generate an exponential family of distributions, the family of exponential type determined by $f_2(x)$, as τ ranges over its values. The exponential family

† see Appendix page 389

is a slight extension of that introduced by Koopman (1936) and Pitman (1936) in the investigation of sufficient statistics. Many of the common distributions of statistical interest such as the normal, χ^2, Poisson, binomial, multinomial, negative binomial, etc., are of exponential type. [Cf. Aitken and Silverstone (1941), Blackwell and Girshick (1954), Brunk (1958), Girshick and Savage (1951).]

For $f^*(x)$ defined in (2.11), it is readily calculated that

$$(2.12)\quad J(*, 2) = \int (f^*(x) - f_2(x)) \log \frac{f^*(x)}{f_2(x)}\, d\lambda(x) = (\theta - E_2(T(x)))\tau,$$

where $E_2(T(x)) = \int T(x) f_2(x)\, d\lambda(x)$.

In subsequent applications of theorem 2.1, we shall have occasion to limit the populations in the dominated set over which $f^*(x)$ may range. We shall call such $f^*(x)$, and the corresponding values of τ, admissible. If there is no admissible value of τ satisfying the equation $\theta = (d/d\tau) \log M_2(\tau)$, the minimum discrimination information value is zero.

Before we look at some examples illustrating theorem 2.1, we want to examine the following results that are also related to theorem 2.1. [Cf. Chernoff (1952, 1956, pp. 17–18), Kullback (1954).] Suppose $f_1(x)$, $f_2(x)$, $f(x)$ are generalized densities of a *homogeneous* set of probability measures. Using theorem 3.1 of chapter 2, we have (see problem 8.4 in chapter 2)

$$(2.13)\quad \int f(x) \log \frac{f(x)}{f_2(x)}\, d\lambda(x) + \int f(x) \log \frac{f_2(x)}{f_1(x)}\, d\lambda(x) = \int f(x) \log \frac{f(x)}{f_1(x)}\, d\lambda(x) \geqq 0,$$

or

$$(2.14)\quad \int f(x) \log \frac{f(x)}{f_2(x)}\, d\lambda(x) \geqq \int f(x) \log \frac{f_1(x)}{f_2(x)}\, d\lambda(x),$$

with equality if and only if $f(x) = f_1(x)$ $[\lambda]$. If in theorem 2.1 we take $T(x) = \log [f_1(x)/f_2(x)]$, the minimum value of $I(f:f_2) = \int f(x) \log \frac{f(x)}{f_2(x)}\, d\lambda(x)$, subject to $\theta = \int T(x) f(x)\, d\lambda(x) = \int f(x) \log \frac{f_1(x)}{f_2(x)}\, d\lambda(x)$, is

$$(2.15)\quad \min I(f:f_2) = \theta\tau - \log M_2(\tau),$$

$$(2.16)\quad M_2(\tau) = \int f_2(x) \exp\left(\tau \log \frac{f_1(x)}{f_2(x)}\right) d\lambda(x) = \int (f_1(x))^\tau (f_2(x))^{1-\tau}\, d\lambda(x),$$

$$(2.17)\qquad \theta = \frac{d}{d\tau}\log M_2(\tau) = \frac{\displaystyle\int (f_1(x))^\tau (f_2(x))^{1-\tau}\log\frac{f_1(x)}{f_2(x)}\,d\lambda(x)}{\displaystyle\int (f_1(x))^\tau (f_2(x))^{1-\tau}\,d\lambda(x)},$$

$$(2.18)\qquad f(x) = \frac{\exp\left(\tau\log\dfrac{f_1(x)}{f_2(x)}\right) f_2(x)}{M_2(\tau)} = \frac{(f_1(x))^\tau (f_2(x))^{1-\tau}}{M_2(\tau)}.$$

We remark that if $f_1(x)$ and $f_2(x)$ are members of a family of exponential type determined by the same generalized density, then $f(x)$ is a member of the same family.

We note the following values, from (2.15)–(2.18):

τ	$M_2(\tau)$	$f(x)$	θ	$\frac{d}{d\tau}M_2(\tau)$	$\theta\tau - \log M_2(\tau)$
0	1	$f_2(x)$	$-I(2:1)$	$-I(2:1)$	0
1	1	$f_1(x)$	$I(1:2)$	$I(1:2)$	$I(1:2)$

Anticipating the discussion in section 4 [cf. Chernoff (1952), Sanov (1957, p. 18)], we now state that:

(*a*) as θ varies from $-I(2:1)$ to $I(1:2)$, τ varies continuously and strictly monotonically from 0 to 1;

(*b*) $M_2(\tau)$, $\log M_2(\tau)$ are strictly convex functions of τ;

(*c*) for θ and τ satisfying (2.17), $\theta\tau - \log M_2(\tau)$ varies continuously and strictly monotonically from 0 to $I(1:2)$ as τ varies from 0 to 1;

(*d*) $0 \leqq M_2(\tau) \leqq 1$, for $0 \leqq \tau \leqq 1$.

When $\theta = 0$ there is therefore a value τ_0, $0 < \tau_0 < 1$, such that

$$(2.19)\qquad f_0(x) = \frac{(f_1(x))^{\tau_0}(f_2(x))^{1-\tau_0}}{M_2(\tau_0)},$$

$$(2.20)\qquad I(f_0:f_2) = -\log M_2(\tau_0) = -\log m_2, \qquad m_2 = \inf_{0<\tau<1} M_2(\tau),$$

$$0 = \int f_0(x)\log\frac{f_1(x)}{f_2(x)}\,d\lambda(x) = \int f_0(x)\log\frac{f_0(x)}{f_2(x)}\,d\lambda(x) + \int f_0(x)\log\frac{f_1(x)}{f_0(x)}\,d\lambda(x),$$

or

$$(2.21)\quad I(f_0:f_2) = \int f_0(x)\log\frac{f_0(x)}{f_2(x)}\,d\lambda(x) = \int f_0(x)\log\frac{f_0(x)}{f_1(x)}\,d\lambda(x) = I(f_0:f_1).$$

Bhattacharyya (1943, 1946a) considered $M_2(\tau)$ in (2.16) for $\tau = \frac{1}{2}$ as a measure of divergence between the populations. Chernoff (1952, 1956) proposes $-\log\,(\inf_{0<\tau<1} E(e^{\tau x}))$ as a measure of the information in an experiment. Chernoff points out that this information measure is such that the information derived from n independent observations on a chance variable is n times the information from one observation, whereas the information derived from observations on several independent chance variables is less than or equal to the sum of the corresponding information measures. It is interesting to note that Schützenberger (1954, p. 65) defines the logarithm of the moment generating function (the cumulant generating function) as a *pseudo information* because it does not have all the properties of an information measure.

Example 2.1. We illustrate theorem 2.1 with a simple numerical example. Let $f_2(x) = \binom{N}{x} p^x q^{N-x}$, the binomial distribution for $N = 2$, $p = 0.4$, and $T(x) = x$. Since $M_2(\tau) = \sum_{x=0}^{2} e^{\tau x} f_2(x) = (pe^\tau + q)^2$, $f^*(x) = e^{\tau x} f_2(x)/M_2(\tau) = \binom{2}{x}(p^*)^x(q^*)^{2-x}$, where $p^* = pe^\tau/(pe^\tau + q)$, $q^* = q/(pe^\tau + q)$. Note that $f^*(x)$ is also a binomial distribution. If we want $E_1(x) = \theta = 1$, then $1 = 2pe^\tau/(pe^\tau + q) = 2p^*$ and $p^* = \frac{1}{2}$. As possible distributions with $E_1(x) = 1$, we shall take the hypergeometric distribution $f_1(x) = \binom{np}{x}\binom{nq}{N-x}\Big/\binom{n}{N}$, $n = 4$, $p = \frac{1}{2} = q$, $N = 2$; the discrete uniform distribution $f_3(x) = \frac{1}{3}$, $x = 0, 1, 2$; the discrete uniform distribution $f_4(x) = \frac{1}{2}$, $x = 0, 2$, $f_4(x) = 0$, $x = 1$; the distribution $f_5(x) = 1$, $x = 1$, $f_5(x) = 0$, $x = 0, 2$. The appropriate numerical values are given in table 2.1.

TABLE 2.1

x	f_1	f_3	f_4	f_5	f^*	f_2	$f_1 \log \frac{f_1}{f_2}$	$f_3 \log \frac{f_3}{f_2}$	$f_4 \log \frac{f_4}{f_2}$	$f_5 \log \frac{f_5}{f_2}$	$f^* \log \frac{f^*}{f_2}$
0	$\frac{1}{6}$	$\frac{1}{3}$	$\frac{1}{2}$	0	$\frac{1}{4}$	0.36	−0.12835	−0.02565	0.16425	0	−0.09116
1	$\frac{2}{3}$	$\frac{1}{3}$	0	1	$\frac{1}{2}$	0.48	0.21900	−0.12155	0	0.73397	0.02041
2	$\frac{1}{6}$	$\frac{1}{3}$	$\frac{1}{2}$	0	$\frac{1}{4}$	0.16	0.00680	0.24466	0.56972	0	0.11157
							0.09745	0.09746	0.73397	0.73397	0.04082

Note that $I(*:2)$ is the smallest value in table 2.1, and that $\tau = \log(q/p) = \log 1.5$, $\log M_2(\tau) = 2\log(pe^\tau + q) = 2\log 2q = 2\log 1.2$, $\theta = 1$, $\theta\tau - \log M_2(\tau) = \log 1.5 - 2\log 1.2 = 0.405465 - 0.364643 = 0.04082 = I(*:2)$.

This example also illustrates problem 8.25 of chapter 2 with $f^*(x) = \frac{1}{2}f_4(x) + \frac{1}{2}f_5(x)$ and $g(x) = f_2(x)$.

Example 2.2. Using the statistic $Y = \min(x_1, x_2, \cdots, x_n)$ and the populations and results in example 7.2 of chapter 2, we find that

$$M_2(\tau) = \int_{\theta_2}^{\infty} ne^{-n(y-\theta_2)+\tau y}\, dy = \frac{e^{\tau\theta_2}}{1-\tau/n}, \qquad n > \tau,$$

$$g^*(y) = (n-\tau)e^{-(n-\tau)(y-\theta_2)}, \qquad \theta_2 \le y < \infty.$$

Since $\int_{\theta}^{\infty} nye^{-n(y-\theta)}\, dy = \theta + 1/n$, $I(*:2; \mathscr{Y}) = \left(\theta_1 + \frac{1}{n}\right)\tau - \tau\theta_2 + \log\left(1 - \frac{\tau}{n}\right)$, with $\theta_1 + \frac{1}{n} = \theta_2 + \frac{1}{n(1-\tau/n)}$, or $\tau = \frac{n(\theta_1-\theta_2)}{(\theta_1-\theta_2)+1/n} < n$, and $I(*:2; \mathscr{Y}) = n(\theta_1-\theta_2) - \log(1 + n(\theta_1 - \theta_2))$. Since $\theta_1 > \theta_2$ and $x \ge \log(1+x)$ for $x > -1$, with equality if and only if $x = 0$ [Hardy, Littlewood, and Pólya (1934, theorem 142, p. 103)], we see that

$$I(1:2; \mathscr{X}) = n(\theta_1-\theta_2) \ge I(*:2; \mathscr{Y}) = n(\theta_1-\theta_2) - \log(1 + n(\theta_1-\theta_2)) \ge 0,$$

with equality for finite n if and only if $\theta_1 = \theta_2$.

Example 2.3. We take [cf. Fraser (1957, p. 145)] $T(x) = \chi_E(x)$, where $\chi_E(x) = 1$ for $x \in E$ and $\chi_E(x) = 0$ for $x \in \mathscr{X} - E = \bar{E}$; that is, $\chi_E(x)$ is the characteristic function or indicator of the set $E \in \mathscr{S}$, and

$$\int \chi_E(x) f_1(x)\, d\lambda(x) = \int_E f_1(x)\, d\lambda(x) = \mu_1(E) = \theta.$$

We now have

$$M_2(\tau) = \int e^{\tau\chi_E(x)} f_2(x)\, d\lambda(x) = \int_E e^{\tau} f_2(x)\, d\lambda(x) + \int_{\bar{E}} f_2(x)\, d\lambda(x)$$

$$= e^{\tau}\mu_2(E) + \mu_2(\bar{E}),$$

$$f^*(x) = \frac{e^{\tau\chi_E(x)} f_2(x)}{M_2(\tau)} = \frac{e^{\tau} f_2(x)}{e^{\tau}\mu_2(E) + \mu_2(\bar{E})}, \qquad x \in E,$$

$$= \frac{f_2(x)}{e^{\tau}\mu_2(E) + \mu_2(\bar{E})}, \qquad x \in \bar{E},$$

$$\theta = \mu_1(E) = \frac{e^{\tau}\mu_2(E)}{e^{\tau}\mu_2(E) + \mu_2(\bar{E})}, \qquad \tau = \log\frac{\mu_1(E)\mu_2(\bar{E})}{\mu_2(E)\mu_1(\bar{E})},$$

$$I(*:2) = \mu_1(E)\tau - \log(e^{\tau}\mu_2(E) + \mu_2(\bar{E}))$$

$$= \mu_1(E)\log\frac{\mu_1(E)\mu_2(\bar{E})}{\mu_2(E)\mu_1(\bar{E})} - \log\frac{\mu_2(\bar{E})}{\mu_1(\bar{E})}$$

$$= \mu_1(E)\log\frac{\mu_1(E)}{\mu_2(E)} + \mu_1(\bar{E})\log\frac{\mu_1(\bar{E})}{\mu_2(\bar{E})}.$$

We thus have

$$I(1:2) \ge \mu_1(E)\log\frac{\mu_1(E)}{\mu_2(E)} + \mu_1(\bar{E})\log\frac{\mu_1(\bar{E})}{\mu_2(\bar{E})},$$

with equality if and only if

$$f_1(x) = f^*(x) = \frac{\mu_1(E)}{\mu_2(E)} f_2(x)\ [\lambda], \qquad x \in E,$$

$$= \frac{\mu_1(\bar{E})}{\mu_2(\bar{E})} f_2(x)\ [\lambda], \qquad x \in \bar{E}.$$

Note that the foregoing is a special case of corollary 3.2 of chapter 2, with $E_1 = E$, $E_2 = \bar{E}$. (See problem 7.19.) We remark here that the techniques of sequential analysis [Wald (1947)] in effect determine such a partitioning of the space $\mathscr{X}$, with $\mu_1(E) = 1 - \beta$, $\mu_2(E) = \alpha$, and with no loss of information, since the partitioning is sufficient.

3. SUFFICIENT STATISTICS

We shall show that $T(x)$ is a sufficient statistic for the family of exponential type determined by $f_2(x)$. We follow the notation and concepts of section 4 of chapter 2 and shall need the following lemmas.

LEMMA 3.1. *If λ is a measure on $\mathscr{S}$, if g is a nonnegative function on $\mathscr{Y}$, integrable with respect to $\lambda T^{-1} = \gamma$, and if μ is the measure on $\mathscr{S}$ defined by $d\mu = gT\,d\lambda$, then $d\mu T^{-1} = d\nu = g\,d\lambda T^{-1} = g\,d\gamma$, or equivalently, $E_\lambda(gT|y) = g(y)\ [\gamma]$.*

Proof. From $\mu(E) = \int_E gT(x)\,d\lambda(x)$ and lemma 4.1 of chapter 2, it follows that $\nu(G) = \mu T^{-1}(G) = \mu(T^{-1}(G)) = \int_G g(y)\,d\gamma(y)$. [See Halmos (1950, p. 209), Halmos and Savage (1949), Kolmogorov (1950, p. 53), Loève (1955, p. 340).]

LEMMA 3.2. *If λ is a measure on $\mathscr{S}$, if f and h are nonnegative functions on $\mathscr{X}$ and $\mathscr{Y}$ respectively, and if f, hT, and $f\cdot hT$ are all integrable with respect to λ, then $E_\lambda(f\cdot hT|y) = E_\lambda(f|y)\cdot h(y)\ [\gamma]$.*

Proof. If $d\mu = f\,d\lambda$, then $\nu(G) = \int_G E_\lambda(f|y)\,d\gamma(y)$. From lemma 3.1 above and lemma 4.1 in chapter 2, we have

$$\int_G E_\lambda(f|y)h(y)\,d\gamma(y) = \int_G h(y)\,d\nu(y) = \int_{T^{-1}(G)} hT(x)\,d\mu(x)$$

$$= \int_{T^{-1}(G)} f(x)hT(x)\,d\lambda(x) = \int_G E_\lambda(f\cdot hT|y)\,d\gamma(y),$$

and the conclusion (uniqueness) follows from the Radon–Nikodym theorem. [See Halmos and Savage (1949), Kolmogorov (1950, p. 56), Loève (1955, p. 350).]

LEMMA 3.3. *The distribution of the statistic $Y = T(x)$ for values of x from the respective populations $\mu^*(E) = \int_E f^*(x)\, d\lambda(x)$, $\mu_2(E) = \int_E f_2(x)\, d\lambda(x)$, for $E \in \mathscr{S}$, $f^*(x) = e^{\tau T(x)} f_2(x)/M_2(\tau)$ is given respectively by*

$$(3.1) \qquad \nu^*(G) = \int_G g^*(y)\, d\gamma(y), \qquad \nu_2(G) = \int_G g_2(y)\, d\gamma(y), \qquad G \in \mathscr{T},$$

where $g^*(y) = e^{\tau y} g_2(y)/M_2(\tau)$ $[\gamma]$.

Proof. Since $d\mu^* = f^*\, d\lambda = \dfrac{e^{\tau T(x)} f_2(x)}{M_2(\tau)}\, d\lambda$, $\nu^*(G) = \int_G E_\lambda\left(\dfrac{e^{\tau T(x)} f_2(x)}{M_2(\tau)} \Big| y\right) d\gamma(y)$ $= \int_G E_\lambda(f_2(x)|y) \dfrac{e^{\tau y}}{M_2(\tau)}\, d\gamma(y) = \int_G \dfrac{g_2(y) e^{\tau y}}{M_2(\tau)}\, d\gamma(y)$, by lemma 3.2 above, and the conclusion (uniqueness) follows from the Radon–Nikodym theorem.

Note that the generalized density $g^*(y)$ of the distribution of the statistic $Y = T(x)$ generates an exponential family, the family of exponential type determined by $g_2(y)$. Hereafter, $T(x)$ will be understood to be a measurable function without further comment.

THEOREM 3.1. *The statistic $Y = T(x)$ is a sufficient statistic for the family of exponential type generated by $f_2(x)$.*

Proof. Let τ_1 and τ_2 be any values in the range of τ for which $M_2(\tau)$ is finite and let $f_i^*(x)$ and $g_i^*(y)$ be the generalized densities corresponding to τ_i, $i = 1, 2$. From lemma 3.3 we see that

$$(3.2) \qquad \frac{f_1^*(x)}{f_2^*(x)} = \frac{e^{\tau_1 T(x)}}{e^{\tau_2 T(x)}} \cdot \frac{M_2(\tau_2)}{M_2(\tau_1)} = \frac{g_1^*(T(x))}{g_2^*(T(x))} \; [\lambda],$$

or

$$(3.3) \qquad \frac{f_1^*(x)}{g_1^*(T(x))} = \frac{f_2^*(x)}{g_2^*(T(x))} \; [\lambda],$$

the necessary and sufficient condition (4.5) of chapter 2 that $Y = T(x)$ be a sufficient statistic.

Fixing μ_2 in the homogeneous set of measures in theorem 4.2 of chapter 2, and letting μ_1 range over the homogeneous set, the necessary and sufficient condition (4.5) of chapter 2 that $Y = T(x)$ be a sufficient statistic may be written as [cf. Fraser (1957, p. 20), Rao (1952, p. 135)]

$$(3.4) \qquad f_1(x) = \frac{g_1(T(x))}{g_2(T(x))} f_2(x) = h_{12}(T(x)) f_2(x) \; [\lambda],$$

with $h_{12}(T(x)) = g_1(T(x))/g_2(T(x))$ a function of $T(x)$. We see that $f^*(x)$ has the form for $f_1(x)$ in (3.4). Hence we have an alternative proof of theorem 3.1.

Note that (3.4) *by itself* is not sufficient for $T(x)$ to be a sufficient statistic. The condition that μ_1 and μ_2 are measures of a homogeneous set, or, more strictly, that μ_1 is absolutely continuous with respect to μ_2, is essential for this criterion for a sufficient statistic, known as the Neyman criterion. Since $h_{12} = f_1/f_2$, unless $f_1 = 0$ whenever $f_2 = 0$, h_{12} is not defined, and if, always for some set E, $f_1 = 0$ whenever $f_2 = 0$, then $\mu_1(E) = \int_E f_1 \, d\lambda = 0$, whenever $\mu_2(E) = \int_E f_2 \, d\lambda = 0$, or μ_1 is absolutely continuous with respect to μ_2. An illustration with rectangular distributions may be helpful. (See example 7.1 of chapter 2.)

Example 3.1. Let $f_1(x) = 1/\theta_1,\ 0 \le x \le \theta_1, \quad f_2(x) = 1/\theta_2,\ 0 \le x \le \theta_2,$
$= 0,\ x < 0,\ x > \theta_1, \quad = 0,\ x < 0,\ x > \theta_2.$

Suppose that $\theta_1 < \theta_2$, and set $T(x) = 1$, $0 \le x \le \theta_1$, $T(x) = 0$, $x < 0$, $x > \theta_1$. Now $f_1(x) = 0$ whenever $f_2(x) = 0$ so that μ_1 is absolutely continuous with respect to μ_2. It is clear that $f_1(x) = h_{12}(T(x))f_2(x)$, where $h_{12}(T(x)) = (\theta_2/\theta_1) \cdot 1$, $0 \le x \le \theta_1$, $h_{12}(T(x)) = (\theta_2/\theta_1) \cdot 0$, $x < 0$, $x > \theta_1$. Hence $T(x)$ is a sufficient statistic (cf. example 2.3). However, when $\theta_1 > \theta_2$, $f_1(x)$ is not zero whenever $f_2(x) = 0$, μ_1 is not absolutely continuous with respect to μ_2, and we cannot write $f_1(x) = h_{12}(T(x))f_2(x)$ for $\theta_2 < x \le \theta_1$.

COROLLARY 3.1. *If* $I(\tau_1 : \tau_2; \mathscr{X}) = \int f_1^*(x) \log \frac{f_1^*(x)}{f_2^*(x)} \, d\lambda(x)$, *and* $I(\tau_1 : \tau_2; \mathscr{Y}) = \int g_1^*(y) \log \frac{g_1^*(y)}{g_2^*(y)} \, d\gamma(y)$, *then* $I(\tau_1 : \tau_2; \mathscr{X}) = I(\tau_1 : \tau_2; \mathscr{Y})$.

Proof. A consequence of theorem 3.1 above and theorem 4.2 of chapter 2.

COROLLARY 3.2. *If* $\theta(\tau_i) = \int T(x) f_i^*(x) \, d\lambda(x) = E(T(x)|\tau_i)$, $i = 1, 2$, *then*

$$I(\tau_1 : \tau_2) = \theta(\tau_1)(\tau_1 - \tau_2) - \log \frac{M_2(\tau_1)}{M_2(\tau_2)} \text{ and } J(\tau_1, \tau_2) = (\theta(\tau_1) - \theta(\tau_2))(\tau_1 - \tau_2).$$

Proof. Verified by straightforward computation. (Cf. problem 5.6 of chapter 1.)

4. EXPONENTIAL FAMILY

We now want to investigate the behavior of $I(*:2) = \theta\tau - \log M_2(\tau)$ as τ and θ vary. [See Blackwell and Girshick (1954), Blanc-Lapierre and Tortrat (1956), Brunk (1958), Chernoff (1952), Girshick and Savage (1951), Khinchin (1949, pp. 76–81), Kullback (1954), Le Cam (1956).] Proofs of the following lemmas are left to the reader.

LEMMA 4.1. *For all τ in the interval of finite existence of $M_2(\tau)$, $M_2(\tau)$ is nonnegative, analytic, and*

$$\text{(4.1)}\qquad \frac{dM_2(\tau)}{d\tau} = \int T(x)e^{\tau T(x)}f_2(x)\,d\lambda(x) = \int ye^{\tau y}g_2(y)\,d\gamma(y),$$

$$\text{(4.2)}\qquad \frac{d^2M_2(\tau)}{d\tau^2} = \int (T(x))^2e^{\tau T(x)}f_2(x)\,d\lambda(x) = \int y^2e^{\tau y}g_2(y)\,d\gamma(y) \geqq 0,$$

with equality if and only if $\mu_2(x:T(x) = 0) = 1$.

LEMMA 4.2. $\theta(\tau) = \int T(x)f^*(x)\,d\lambda(x) = \int yg^*(y)\,d\gamma(y) = \frac{d}{d\tau}\log M_2(\tau)$ $= \frac{M_2'(\tau)}{M_2(\tau)}$, *where $f^*(x)$ and $g^*(y)$ are defined in lemma 3.3.*

We shall also indicate by $\tau(\theta)$ the value of τ for which $\theta = \frac{d}{d\tau}\log M_2(\tau)$ $= \frac{M_2'(\tau(\theta))}{M_2(\tau(\theta))}$.

LEMMA 4.3. $E((T(x) - \theta)^2|\tau) = E((y - \theta)^2|\tau) = \text{var}\,(y|\tau) =$ $\int f^*(x)\left(\frac{1}{f^*(x)}\frac{\partial f^*(x)}{\partial\tau}\right)^2 d\lambda(x) = \int g^*(y)\left(\frac{1}{g^*(y)}\frac{\partial g^*(y)}{\partial\tau}\right)^2 d\gamma(y) = \theta'(\tau) = \frac{d\theta(\tau)}{d\tau} =$ $\frac{d^2}{d\tau^2}\log M_2(\tau) = \frac{M_2''(\tau)}{M_2(\tau)} - \left(\frac{M_2'(\tau)}{M_2(\tau)}\right)^2.$

LEMMA 4.4. $\int (T(x) - \theta)^2 f^*(x)\,d\lambda(x)\cdot\int f^*(x)\left(\frac{1}{f^*(x)}\frac{\partial f^*(x)}{\partial\theta}\right)^2 d\lambda(x) = 1.$

LEMMA 4.5. *If $\mu_2(x:T(x) = \theta) \neq 1$, then $\theta(\tau)$ is a strictly increasing function of τ and* $\log M_2(\tau)$ *is strictly convex. For a fixed value of θ, $\theta\tau - \log M_2(\tau)$ is a concave function of τ, with maximum value $\theta\tau(\theta) - \log M_2(\tau(\theta))$, which is a convex function of θ.*

LEMMA 4.6. *If $\theta(0) = \int T(x)f_2(x)\,d\lambda(x) = \int yg_2(y)\,d\gamma(y)$, then $\theta(0) = M_2'(0)$, $M_2(0) = 1$, $\theta'(0) = E((y - \theta(0))^2|\tau = 0) = \text{var}\,(y|\tau = 0)$.*

LEMMA 4.7. *If $\theta = \frac{M_2'(\tau(\theta))}{M_2(\tau(\theta))}$ and $\mu_2(x:T(x) = \theta) \neq 1$, then*

$$\tau'(\theta) = \frac{d\tau(\theta)}{d\theta} = \frac{1}{\frac{M_2''(\tau(\theta))}{M_2(\tau(\theta))} - \left(\frac{M_2'(\tau(\theta))}{M_2(\tau(\theta))}\right)^2} = \frac{1}{\frac{d^2}{d\tau^2}\log M_2(\tau(\theta))} > 0,$$

and $\tau(\theta)$ is a strictly increasing function of θ.

LEMMA 4.8. *$I(*:2) = \theta\tau(\theta) - \log M_2(\tau(\theta)) \geqq 0$, with equality if and only if $\tau(\theta) = 0$, that is, $\theta = \theta(0) = \int y g_2(y)\,d\gamma(y)$.*

LEMMA 4.9. *$I(*:2) = \theta\tau(\theta) - \log M_2(\tau(\theta))$ is monotonically increasing for $\theta \geqq \theta(0)$ and monotonically decreasing for $\theta \leqq \theta(0)$.*

THEOREM 4.1. *$I(*:2) = (\theta(\tau) - \theta(0))^2/2\,\mathrm{var}\,(y|\tau(\xi))$, ξ between $\theta(\tau)$ and $\theta(0)$.*

Proof. Let $I(*:2) = m(\theta) = \theta\tau(\theta) - \log M_2(\tau(\theta))$; then $m'(\theta) = \frac{d}{d\theta}m(\theta)$ $= \tau(\theta)$, $m''(\theta) = \frac{d^2}{d\theta^2}m(\theta) = \tau'(\theta)$, $m(\theta(0)) = 0$, $m'(\theta(0)) = 0$, and $m(\theta)$ $= m(\theta(0)) + (\theta(\tau) - \theta(0))m'(\theta(0)) + \frac{1}{2}(\theta(\tau) - \theta(0))^2 m''(\xi)$, from which the desired conclusion follows.

In view of theorem 2.1, and theorem 4.1 of chapter 2, we may now state:

COROLLARY 4.1. *$I(1:2;\mathscr{X}) \geqq I(1:2;\mathscr{Y}) \geqq (E_1(y) - E_2(y))^2/2\,\mathrm{var}\,(y|\tau(\xi))$, where $\mathrm{var}\,(y|\tau(\xi))$ is the variance of y in the distribution defined by $e^{y\tau(\xi)}g_2(y)/M_2(\tau(\xi))$, and ξ lies between $E_1(y)$ and $E_2(y)$, with equality between the first pair if and only if $Y = T(x)$ is sufficient, and with equality between the second pair if and only if $g_1(y) = e^{\tau y}g_2(y)/M_2(\tau)$ $[\lambda]$.*

In particular, if $y = \alpha_1 y_1 + \alpha_2 y_2 + \cdots + \alpha_k y_k$, where the y_i, $i = 1, 2, \cdots, k$, are linearly independent, λ-measurable functions of $x \in \mathscr{X}$, and $\delta_i = E_1(y_i) - E_2(y_i)$, $i = 1, 2, \cdots, k$, and $\mathrm{cov}\,(y_i, y_j|\tau(\xi))$ is the covariance of y_i and y_j, $i, j = 1, 2, \cdots, k$, in the distribution defined by $\tau = \tau(\xi)$, then in terms of the matrices (and usual matrix notation) $\boldsymbol{\Sigma}(\tau(\xi)) = (\mathrm{cov}\,(y_i, y_j|\tau(\xi)))$, $\boldsymbol{\alpha}' = (\alpha_1, \alpha_2, \cdots, \alpha_k)$, $\boldsymbol{\delta}' = (\delta_1, \delta_2, \cdots, \delta_k)$, $(E_1(y) - E_2(y))^2 = \boldsymbol{\alpha}'\boldsymbol{\delta}\boldsymbol{\delta}'\boldsymbol{\alpha}$, $\mathrm{var}\,(y|\tau(\xi)) = \boldsymbol{\alpha}'\boldsymbol{\Sigma}(\tau(\xi))\boldsymbol{\alpha}$. It can be shown (see section 5 of chapter 9) that $\max\,(\boldsymbol{\alpha}'\boldsymbol{\delta}\boldsymbol{\delta}'\boldsymbol{\alpha}/\boldsymbol{\alpha}'\boldsymbol{\Sigma}(\tau(\xi))\boldsymbol{\alpha})$ for possible values of the α_i, $i = 1, 2, \cdots, k$, is $\boldsymbol{\delta}'\boldsymbol{\Sigma}^{-1}(\tau(\xi))\boldsymbol{\delta}$. We can therefore state:

COROLLARY 4.2. *$I(1:2;\mathscr{X}) \geqq I(1:2;\mathscr{Y}) \geqq \frac{1}{2}\boldsymbol{\delta}'\boldsymbol{\Sigma}^{-1}(\tau(\xi))\boldsymbol{\delta}$.*

We remark that the right-hand member in corollary 4.2 is the discrimination information measure for two multivariate normal populations with respective means $E_1(y_i)$, $E_2(y_i)$, $i = 1, 2, \cdots, k$, and common covariance matrix $\boldsymbol{\Sigma}(\tau(\xi))$ (see section 1 of chapter 9).

COROLLARY 4.3. *$J(*, 2) = (\theta(\tau) - \theta(0))^2/\mathrm{var}\,(y|\tau(\xi))$.*

Proof. Apply the procedure in the proof of theorem 4.1 to $J(*, 2)$ $= (\theta(\tau) - \theta(0))\tau(\theta)$.

COROLLARY 4.4. $I(\tau_1:\tau_2) = (E(y|\tau_1) - E(y|\tau_2))^2/2 \operatorname{var}(y|\tau(\xi))$, *where* ξ *lies between* $\theta(\tau_1)$ *and* $\theta(\tau_2)$.

Proof. Apply the procedure in the proof of theorem 4.1 to corollary 3.2.

COROLLARY 4.5. $J(\tau_1, \tau_2) = (E(y|\tau_1) - E(y|\tau_2))^2/\operatorname{var}(y|\tau(\xi))$, *where* ξ *lies between* $\theta(\tau_1)$ *and* $\theta(\tau_2)$.

Proof. Apply the procedure in the proof of theorem 4.1 to corollary 3.2.

It should be remarked that the preceding results are not only multidimensional in the variates but also in the parameters, that is, $\theta = (\theta_1, \theta_2, \cdots, \theta_k)$, $\tau = (\tau_1, \tau_2, \cdots, \tau_k)$, $Y = (Y_1, Y_2, \cdots, Y_k) = (T_1(x), T_2(x), \cdots, T_k(x)) = T(x)$, and $\theta\tau$, $\tau T(x)$, and τY are to be understood as $\theta_1\tau_1 + \theta_2\tau_2 + \cdots + \theta_k\tau_k$, $\tau_1 T_1(x) + \tau_2 T_2(x) + \cdots + \tau_k T_k(x)$, and $\tau_1 Y_1 + \tau_2 Y_2 + \cdots + \tau_k Y_k$ respectively. It will be useful to rewrite some of the preceding in an appropriate matrix notation. Let us write

$$g_{ij}^*(\pi) = \int f^*(x)\left(\frac{1}{f^*(x)}\frac{\partial f^*(x)}{\partial \pi_i}\right)\left(\frac{1}{f^*(x)}\frac{\partial f^*(x)}{\partial \pi_j}\right) d\lambda(x),$$

$$h_{ij}^*(\pi) = \int g^*(y)\left(\frac{1}{g^*(y)}\frac{\partial g^*(y)}{\partial \pi_i}\right)\left(\frac{1}{g^*(y)}\frac{\partial g^*(y)}{\partial \pi_j}\right) d\gamma(y),$$

and define the nonsingular matrices

$$\mathbf{G}^*(\pi) = (g_{ij}^*(\pi)), \qquad \mathbf{H}^*(\pi) = (h_{ij}^*(\pi)),$$

where π represents any appropriate set of parameters and i, j range over the number of components of π, for example, $i, j = 1, 2, \cdots, k$, when π is τ or θ.

Since

$$\frac{\partial f^*(x)}{\partial \tau_j} = \frac{\partial f^*(x)}{\partial \theta_1}\frac{\partial \theta_1}{\partial \tau_j} + \frac{\partial f^*(x)}{\partial \theta_2}\frac{\partial \theta_2}{\partial \tau_j} + \cdots + \frac{\partial f^*(x)}{\partial \theta_k}\frac{\partial \theta_k}{\partial \tau_j}, \qquad j = 1, 2, \cdots, k,$$

setting $a_{ij} = \partial\theta_i/\partial\tau_j$, the nonsingular matrix $\mathbf{A} = (a_{ij})$, $i, j = 1, 2, \cdots, k$,

$$\left(\frac{1}{\mathbf{f}^*}\frac{\partial \mathbf{f}^*}{\partial \boldsymbol{\tau}}\right)' = \left(\frac{1}{f^*}\frac{\partial f^*}{\partial \tau_1}, \frac{1}{f^*}\frac{\partial f^*}{\partial \tau_2}, \cdots, \frac{1}{f^*}\frac{\partial f^*}{\partial \tau_k}\right),$$

similarly, the matrix $\left(\frac{1}{\mathbf{f}^*}\frac{\partial \mathbf{f}^*}{\partial \boldsymbol{\theta}}\right)'$, we have

$$\left(\frac{1}{\mathbf{f}^*}\frac{\partial \mathbf{f}^*}{\partial \boldsymbol{\tau}}\right) = \mathbf{A}'\left(\frac{1}{\mathbf{f}^*}\frac{\partial \mathbf{f}^*}{\partial \boldsymbol{\theta}}\right), \quad \left(\frac{1}{\mathbf{f}^*}\frac{\partial \mathbf{f}^*}{\partial \boldsymbol{\tau}}\right)\left(\frac{1}{\mathbf{f}^*}\frac{\partial \mathbf{f}^*}{\partial \boldsymbol{\tau}}\right)' = \mathbf{A}'\left(\frac{1}{\mathbf{f}^*}\frac{\partial \mathbf{f}^*}{\partial \boldsymbol{\theta}}\right)\left(\frac{1}{\mathbf{f}^*}\frac{\partial \mathbf{f}^*}{\partial \boldsymbol{\theta}}\right)'\mathbf{A},$$

and taking expected values $\mathbf{G}^*(\tau) = \mathbf{A}'\mathbf{G}^*(\theta)\mathbf{A}$. In a similar fashion we also have $\mathbf{H}^*(\tau) = \mathbf{A}'\mathbf{H}^*(\theta)\mathbf{A}$. Lemma 4.3 may now be written as

LEMMA 4.10. $\mathbf{\Sigma}(\tau(\theta)) = \mathbf{G}^*(\tau) = \mathbf{H}^*(\tau) = \mathbf{A}$,

and lemma 4.4 as

LEMMA 4.11. $\mathbf{\Sigma}(\tau(\theta)) \cdot \mathbf{G}^*(\theta) = \mathbf{I} = \mathbf{G}^*(\tau)\mathbf{G}^*(\theta)$.

Since

$$d\theta_i = \frac{\partial\theta_i}{\partial\tau_1} d\tau_1 + \frac{\partial\theta_i}{\partial\tau_2} d\tau_2 + \cdots + \frac{\partial\theta_i}{\partial\tau_k} d\tau_k, \qquad i = 1, 2, \cdots, k,$$

setting the matrix $(\mathbf{d\theta})' = (d\theta_1, d\theta_2, \cdots, d\theta_k)$, similarly, the matrix $(\mathbf{d\tau})'$, we have $(\mathbf{d\theta}) = \mathbf{A}(\mathbf{d\tau})$ or $(\mathbf{d\tau}) = \mathbf{A}^{-1}(\mathbf{d\theta})$. Since

$$d\tau_i = \frac{\partial\tau_i}{\partial\theta_1} d\theta_1 + \frac{\partial\tau_i}{\partial\theta_2} d\theta_2 + \cdots + \frac{\partial\tau_i}{\partial\theta_k} d\theta_k, \qquad i = 1, 2, \cdots, k,$$

we may write $a^{ij} = \partial\tau_i/\partial\theta_j$, $i, j = 1, 2, \cdots, k$, and $(a^{ij}) = \mathbf{A}^{-1}$. Thus lemma 4.7 may now be written as

LEMMA 4.12. $(a^{ij}) = \mathbf{A}^{-1} = \mathbf{\Sigma}^{-1}(\tau(\theta))$.

As was noted in section 6 of chapter 2, the matrices $\mathbf{G}^*(\pi)$, $\mathbf{H}^*(\pi)$ are Fisher information matrices. [Cf. Fisher (1956, p. 155).]

We illustrate the foregoing with a number of examples.

Example 4.1. $\mathcal{X}$ is the space of n independent observations O_n on the two-valued variate success or failure, $Y = T(x)$ is the number of successes in the n observations, and p_i, $q_i = 1 - p_i$, $i = 1, 2$, are the respective probabilities of success corresponding to H_i, $i = 1, 2$. It is found that [cf. problem 5.2(*b*) of chapter 1 and problem 8.12 of chapter 2]

$$I(1:2; O_n) = nI(1:2; O_1) = n\left(p_1 \log\frac{p_1}{p_2} + q_1 \log\frac{q_1}{q_2}\right), \tag{4.3}$$

$$I(1:2; \mathcal{Y}) = \sum_{y=0}^{n} \frac{n!}{y!(n-y)!} p_1^y q_1^{n-y} \log\frac{p_1^y q_1^{n-y}}{p_2^y q_2^{n-y}} \tag{4.4}$$

$$= n\left(p_1 \log\frac{p_1}{p_2} + q_1 \log\frac{q_1}{q_2}\right),$$

$$g^*(y) = \frac{e^{\tau y} g_2(y)}{M_2(\tau)} = \frac{n!}{y!(n-y)!}(p^*)^y(q^*)^{n-y}, \qquad M_2(\tau) = (p_2e^\tau + q_2)^n, \tag{4.5}$$

$$p^* = \frac{p_2e^\tau}{p_2e^\tau + q_2}, \qquad q^* = \frac{q_2}{p_2e^\tau + q_2}, \qquad \tau(p_2) = 0, \qquad \tau(p_1) = \log\frac{p_1q_2}{q_1p_2},$$

$$I(*:2) = np_1\tau(p_1) - n\log(p_2e^{\tau(p_1)} + q_2) = n\left(p_1 \log\frac{p_1}{p_2} + q_1 \log\frac{q_1}{q_2}\right) \tag{4.6}$$

$$= (np_1 - np_2)^2/2npq = n(p_1 - p_2)^2/2pq,$$

where $p = \dfrac{p_2e^\tau}{p_2e^\tau + q_2}$, $q = \dfrac{q_2}{p_2e^\tau + q_2}$, for some value of τ between $\tau(p_2) = 0$

and $\tau(p_1) = \log \frac{p_1 q_2}{q_1 p_2}$; that is, p lies between p_1 and p_2. Note that in this example $I(1:2; \mathscr{X}) = I(1:2; \mathscr{Y}) = I(*:2)$.

Example 4.2. $\mathscr{X}$ is the space of n independent observations O_n from the normal populations $N(\theta_i, \sigma_i^2)$, $i = 1, 2$, $Y = T(x) = \bar{x}$, the average of the n observations. It is found that [cf. problem 5.2(*f*) of chapter 1 and problem 8.13 of chapter 2]

$$(4.7) \qquad I(1:2; O_n) = nI(1:2; O_1) = \frac{n}{2}\left(\log \frac{\sigma_2^2}{\sigma_1^2} - 1 + \frac{\sigma_1^2}{\sigma_2^2} + \frac{(\theta_1 - \theta_2)^2}{\sigma_2^2}\right),$$

$$(4.8) \qquad I(1:2; \bar{x}) = \frac{1}{2}\log \frac{\sigma_2^2}{\sigma_1^2} - \frac{1}{2} + \frac{\sigma_1^2}{2\sigma_2^2} + \frac{n(\theta_1 - \theta_2)^2}{2\sigma_2^2},$$

$$(4.9) \qquad g^*(\bar{x}) = \frac{e^{\tau\bar{x}} g_2(\bar{x})}{M_2(\tau)} = \frac{\exp\left[-\frac{n}{2\sigma_2^2}\left(\bar{x} - \theta_2 - \frac{\tau\sigma_2^2}{n}\right)^2\right]}{\sigma_2\sqrt{2\pi}/\sqrt{n}},$$

$$M_2(\tau) = \exp\left(\tau\theta_2 + \frac{\tau^2\sigma_2^2}{2n}\right), \qquad \theta^* = \theta_2 + \frac{\tau\sigma_2^2}{n},$$

where $\theta^* = (d/d\tau)\log M_2(\tau)$ is the mean of the distribution with density $g^*(\bar{x})$, the values of τ for $\theta^* = \theta_2$ and $\theta^* = \theta_1$ are respectively

$$\tau(\theta_2) = 0, \qquad \tau(\theta_1) = n(\theta_1 - \theta_2)/\sigma_2^2,$$

$$(4.10) \qquad I(*:2) = \theta_1\tau(\theta_1) - \theta_2\tau(\theta_1) - \frac{\tau^2(\theta_1)\sigma_2^2}{2n} = \frac{n(\theta_1 - \theta_2)^2}{2\sigma_2^2}.$$

Note that in this example $I(1:2; \mathscr{X}) > I(1:2; \mathscr{Y}) > I(*:2)$. (See problem 7.21.)

Example 4.3. $\mathscr{X}$ is the same as in example 4.2, $Y = T(x) = (\bar{x}, s^2)$, where $\bar{x}$ is the average and $s^2 = \frac{1}{n-1}\sum_{i=1}^{n}(x_i - \bar{x})^2$ is the unbiased sample variance of the n observations. It is found that

(4.11) $I(1:2; \mathscr{X})$ is the same as in (4.7),

(4.12) $I(1:2; \bar{x})$ is the same as in (4.8),

$$(4.13) \qquad I(1:2; s^2) = \frac{n-1}{2}\left(\log \frac{\sigma_2^2}{\sigma_1^2} - 1 + \frac{\sigma_1^2}{\sigma_2^2}\right),$$

(4.14) $I(1:2; \mathscr{Y}) = I(1:2; \bar{x}) + I(1:2; s^2)$ (cf. theorem 2.1 in chapter 2),

$$(4.15) \qquad g^*(y) = \frac{\exp(\tau_1\bar{x} + \tau_2 s^2) g_2(\bar{x}, s^2)}{M_2(\tau_1, \tau_2)}$$

$$= \frac{\exp[-n(\bar{x} - \theta^*)^2/2\sigma_2^2]}{\sigma_2\sqrt{2\pi}/\sqrt{n}} \cdot \frac{(n-1)}{2\sigma_*^2\Gamma\left(\frac{n-1}{2}\right)}\left(\frac{(n-1)s^2}{2\sigma_*^2}\right)^{\frac{n-3}{2}}$$

$$\times \exp\left(-\frac{(n-1)s^2}{2\sigma_*^2}\right),$$

$$M_2(\tau_1, \tau_2) = \left[\exp\left(\theta_2\tau_1 + \frac{\tau_1^2\sigma_2^2}{2n}\right)\right]\left(1 - \frac{2\tau_2\sigma_2^2}{n-1}\right)^{-\frac{n-1}{2}},$$

$$\theta^* = \frac{\partial}{\partial\tau_1}\log M_2(\tau_1, \tau_2) = \theta_2 + \tau_1\sigma_2^2/n,$$

$$\sigma_*^2 = \frac{\partial}{\partial\tau_2}\log M_2(\tau_1, \tau_2) = \sigma_2^2/[1 - 2\tau_2\sigma_2^2/(n-1)],$$

the values of τ_1 and τ_2 for $\theta^* = \theta_2$, $\theta^* = \theta_1$, $\sigma_*^2 = \sigma_2^2$, and $\sigma_*^2 = \sigma_1^2$ are respectively

$$\tau_1(\theta_2) = 0, \quad \tau_1(\theta_1) = n(\theta_1 - \theta_2)/\sigma_2^2, \quad \tau_2(\sigma_2^2) = 0, \tau_2(\sigma_1^2) = \frac{n-1}{2}\left(\frac{1}{\sigma_2^2} - \frac{1}{\sigma_1^2}\right),$$

(4.16) $$I(*:2) = \theta_1\tau_1(\theta_1) - \theta_2\tau_1(\theta_1) - \sigma_2^2\tau_1^2(\theta_1)/2n + \sigma_1^2\tau_2(\sigma_1^2) + \frac{n-1}{2}\log\left(1 - \frac{2\sigma_2^2\tau_2(\sigma_1^2)}{n-1}\right)$$

$$= \frac{n(\theta_1 - \theta_2)^2}{2\sigma_2^2} + \frac{n-1}{2}\left(\log\frac{\sigma_2^2}{\sigma_1^2} - 1 + \frac{\sigma_1^2}{\sigma_2^2}\right)$$

$$= \frac{n(\theta_1 - \theta_2)^2}{2\sigma_2^2} + \frac{(\sigma_1^2 - \sigma_2^2)^2}{2}\frac{n-1}{2\sigma^4} = \tfrac{1}{2}\boldsymbol{\delta}'\boldsymbol{\Sigma}^{-1}(\tau(\xi))\boldsymbol{\delta},$$

where σ^2 lies between σ_1^2 and σ_2^2, and

$$\boldsymbol{\Sigma}(\tau(\xi)) = \begin{pmatrix}\sigma_2^2/n & 0\\ 0 & 2\sigma^4/(n-1)\end{pmatrix}, \qquad \boldsymbol{\delta}' = ((\theta_1 - \theta_2), (\sigma_1^2 - \sigma_2^2)).$$

Note that in this example $I(1:2; \mathscr{X}) = I(1:2; \mathscr{Y}) > I(*:2)$, and that the statistic $Y = T(x) = (\bar{x}, s^2)$ is sufficient.

Example 4.4. $\mathscr{X}$ is the space of n independent observations O_n from the normal populations $N(0, \sigma_i^2)$, $i = 1, 2$, $Y = T(x) = s^2$, where $(n-1)s^2 = \sum_{i=1}^{n}(x_i - \bar{x})^2$. It is found that

(4.17) $$I(1:2; O_n) = nI(1:2; O_1) = \frac{n}{2}\left(\log\frac{\sigma_2^2}{\sigma_1^2} - 1 + \frac{\sigma_1^2}{\sigma_2^2}\right),$$

(4.18) $I(1:2; s^2)$ is the same as in (4.13),

(4.19) $$g^*(s^2) = e^{\tau s^2}g_2(s^2)/M_2(\tau)$$

$$= \frac{(n-1)}{2\sigma_*^2\,\Gamma\left(\frac{n-1}{2}\right)}\left(\frac{(n-1)s^2}{2\sigma_*^2}\right)^{\frac{n-3}{2}}\exp\left(-\frac{(n-1)s^2}{2\sigma_*^2}\right),$$

$$M_2(\tau) = \left(1 - \frac{2\tau\sigma_2^2}{n-1}\right)^{-\frac{n-1}{2}},$$

$$\sigma_*^2 = \frac{\partial}{\partial\tau}\log M_2(\tau) = \sigma_2^2/(1 - 2\tau\sigma_2^2/(n-1)),$$

$$\tau(\sigma_2^2) = 0, \quad \tau(\sigma_1^2) = \frac{n-1}{2}\left(\frac{1}{\sigma_2^2} - \frac{1}{\sigma_1^2}\right),$$

$$(4.20)\qquad I(*:2) = \sigma_1^2\tau(\sigma_1^2) + \frac{n-1}{2}\log\left(1 - \frac{2\sigma_2^2\tau(\sigma_1^2)}{n-1}\right)$$

$$= \frac{n-1}{2}\left(\log\frac{\sigma_2^2}{\sigma_1^2} - 1 + \frac{\sigma_1^2}{\sigma_2^2}\right)$$

$$= \frac{(\sigma_1^2 - \sigma_2^2)^2}{2}\cdot\frac{n-1}{2\sigma^4},$$

where σ^2 lies between σ_1^2 and σ_2^2. Note that in this example $I(1:2; \mathscr{X}) > I(1:2; \mathscr{Y}) = I(*:2)$, and that s^2 is *not* a sufficient statistic.

Example 4.5. $\mathscr{X}$ is the same as in example 4.4, $Y = T(x) = \frac{1}{n}\sum_{i=1}^{n} x_i^2$. It is found that

(4.21) $I(1:2; \mathscr{X})$ is the same as in (4.17),

$$(4.22)\qquad I(1:2; \mathscr{Y}) = \frac{n}{2}\left(\log\frac{\sigma_2^2}{\sigma_1^2} - 1 + \frac{\sigma_1^2}{\sigma_2^2}\right),$$

$$(4.23)\qquad g^*(y) = e^{\tau y}g_2(y)/M_2(\tau)$$

$$= \frac{n}{2\sigma_*^2\Gamma\left(\frac{n}{2}\right)}\left(\frac{ny}{2\sigma_*^2}\right)^{\frac{n-2}{2}}\exp\left(-\frac{ny}{2\sigma_*^2}\right),$$

$$M_2(\tau) = (1 - 2\tau\sigma_2^2/n)^{-n/2},$$

$$\sigma_*^2 = \frac{\partial}{\partial\tau}\log M_2(\tau) = \sigma_2^2/(1 - 2\tau\sigma_2^2/n),$$

$$\tau(\sigma_2^2) = 0, \qquad \tau(\sigma_1^2) = \frac{n}{2}\left(\frac{1}{\sigma_2^2} - \frac{1}{\sigma_1^2}\right),$$

$$(4.24)\qquad I(*:2) = \sigma_1^2\tau(\sigma_1^2) + \frac{n}{2}\log\left(1 - \frac{2\sigma_2^2\tau(\sigma_1^2)}{n}\right)$$

$$= \frac{n}{2}\left(\log\frac{\sigma_2^2}{\sigma_1^2} - 1 + \frac{\sigma_1^2}{\sigma_2^2}\right)$$

$$= \frac{(\sigma_1^2 - \sigma_2^2)^2}{2}\cdot\frac{n}{2\sigma^4},$$

where σ^2 lies between σ_1^2 and σ_2^2. Note that in this example $I(1:2; \mathscr{X}) = I(1:2; \mathscr{Y}) = I(*:2)$, and that $\frac{1}{n}\sum_{i=1}^{n} x_i^2$ is a sufficient statistic.

Example 4.6. $\mathscr{X}$ is the space of n independent observations O_n from bivariate normal populations. We shall consider bivariate normal populations with zero means, unit variances, and correlation coefficients ρ_1 and ρ_2 respectively. It is found that (see example 2.1 in chapter 2)

$$(4.25)\qquad I(1:2; O_n) = nI(1:2; O_1) = n\left(\frac{1}{2}\log\frac{1-\rho_2^2}{1-\rho_1^2} + \frac{\rho_2^2 - \rho_1\rho_2}{1-\rho_2^2}\right).$$

The nonsingular transformation

$$(4.26)\qquad u = x_1 - x_2, \qquad v = x_1 + x_2$$

transforms the bivariate normal density

$$(4.27)\qquad \frac{1}{2\pi(1-\rho^2)^{1/2}}\exp\left(-\frac{1}{2(1-\rho^2)}(x_1^2 - 2\rho x_1x_2 + x_2^2)\right)$$

into a product of independent normal densities with zero means and variances $2(1-\rho)$ and $2(1+\rho)$,

$$(4.28)\qquad \frac{1}{\sqrt{2\pi}(2(1-\rho))^{1/2}}\exp\left(-\frac{u^2}{4(1-\rho)}\right) \times \frac{1}{\sqrt{2\pi}(2(1+\rho))^{1/2}}\exp\left(-\frac{v^2}{4(1+\rho)}\right).$$

It is found [the derivation of (4.17) and the fact that $I(1:2; u, v) = I(1:2; u) + I(1:2; v)$ are applicable] that

$$(4.29)\qquad I(1:2; u, v) = \frac{1}{2}\left(\log\frac{1-\rho_2}{1-\rho_1} - 1 + \frac{1-\rho_1}{1-\rho_2}\right) + \frac{1}{2}\left(\log\frac{1+\rho_2}{1+\rho_1} - 1 + \frac{1+\rho_1}{1+\rho_2}\right)$$
$$= \frac{1}{2}\log\frac{1-\rho_2^2}{1-\rho_1^2} + \frac{\rho_2^2 - \rho_1\rho_2}{1-\rho_2^2},$$

illustrating the additivity for independent random variables (see section 2 of chapter 2) and the invariance under nonsingular transformations (see corollary 4.1 of chapter 2). We now take $Y = T(x) = (y_1, y_2)$, where

$$(4.30)\qquad y_1 = \frac{1}{n}\sum_{i=1}^{n} u_i^2 = \frac{1}{n}\sum_{i=1}^{n}(x_{1i} - x_{2i})^2, \qquad y_2 = \frac{1}{n}\sum_{i=1}^{n} v_i^2 = \frac{1}{n}\sum_{i=1}^{n}(x_{1i} + x_{2i})^2,$$

and find that (cf. example 4.5)

$$(4.31)\qquad I(1:2; \mathscr{Y}) = nI(1:2; u, v) = I(1:2; \mathscr{X}),$$

$$(4.32)\qquad g^*(y) = e^{\tau_1 y_1 + \tau_2 y_2} g_2(y_1, y_2)/M_2(\tau_1, \tau_2)$$
$$= \frac{n}{4(1-\rho^*)\,\Gamma\left(\frac{n}{2}\right)}\left(\frac{ny_1}{4(1-\rho^*)}\right)^{\frac{n-2}{2}}\exp\left(-\frac{ny_1}{4(1-\rho^*)}\right)$$
$$\times \frac{n}{4(1+\rho^*)\,\Gamma\left(\frac{n}{2}\right)}\left(\frac{ny_2}{4(1+\rho^*)}\right)^{\frac{n-2}{2}}\exp\left(-\frac{ny_2}{4(1+\rho^*)}\right),$$

$$M_2(\tau_1, \tau_2) = (1 - 4(1 - \rho_2)\tau_1/n)^{-n/2}(1 - 4(1 + \rho_2)\tau_2/n)^{-n/2},$$

$$\theta_1^* = 2(1 - \rho^*), \qquad \theta_2^* = 2(1 + \rho^*), \qquad \theta_1^* = \frac{\partial}{\partial \tau_1} \log M_2(\tau_1, \tau_2),$$

$$\theta_2^* = \frac{\partial}{\partial \tau_2} \log M_2(\tau_1, \tau_2),$$

$$2(1 - \rho^*) = \frac{2(1 - \rho_2)}{1 - 4(1 - \rho_2)\tau_1/n}, \qquad 2(1 + \rho^*) = \frac{2(1 + \rho_2)}{1 - 4(1 + \rho_2)\tau_2/n},$$

$$\tau_1(\rho_2) = 0, \qquad \tau_1(\rho_1) = \frac{n}{4}\left(\frac{1}{1 - \rho_2} - \frac{1}{1 - \rho_1}\right),$$

$$\tau_2(\rho_2) = 0, \qquad \tau_2(\rho_1) = \frac{n}{4}\left(\frac{1}{1 + \rho_2} - \frac{1}{1 + \rho_1}\right),$$

$$(4.33) \quad I(*:2) = 2(1 - \rho_1)\tau_1(\rho_1) + \frac{n}{2}\log\left(1 - \frac{4(1 - \rho_2)\tau_1(\rho_1)}{n}\right)$$

$$+ 2(1 + \rho_1)\tau_2(\rho_1) + \frac{n}{2}\log\left(1 - \frac{4(1 + \rho_2)\tau_2(\rho_1)}{n}\right)$$

$$= n\left(\frac{1}{2}\log\frac{1 - \rho_2^2}{1 - \rho_1^2} + \frac{\rho_2^2 - \rho_1\rho_2}{1 - \rho_2^2}\right)$$

$$= \frac{n}{2}(\rho_1 - \rho_2)^2 \cdot \frac{1 + \rho^2}{(1 - \rho^2)^2},$$

where ρ lies between ρ_1 and ρ_2. Note that in this example $I(1:2; \mathscr{X}) = I(1:2; \mathscr{Y}) = I(*:2)$, and that $Y = T(x) = (y_1, y_2)$ is sufficient.

Example 4.7. We shall use results derived in example 7.2 of chapter 2 and example 2.2. In order to use an unbiased estimate, let us consider the statistic $Y = T(x) = \min(x_1, x_2, \cdots, x_n) - 1/n$. We find that

$$g_2(y) = n \exp(-n(y + 1/n - \theta_2)), \qquad \theta_2 - 1/n \le y < \infty,$$

$$I(1:2; \mathscr{Y}) = n(\theta_1 - \theta_2), \qquad \theta_1 \ge \theta_2 \quad \text{(as in example 7.2 of chapter 2)},$$

$$g^*(y) = e^{\tau y} g_2(y)/M_2(\tau), \qquad \theta_2 - 1/n \le y < \infty,$$

$$= (n - \tau)\exp(-(n - \tau)(y + 1/n - \theta_2)), \text{ if } \quad n > \tau,$$

$$M_2(\tau) = (\exp(\tau\theta_2 - \tau/n))/(1 - \tau/n), \; n > \tau,$$

$$\frac{\partial}{\partial \tau}\log M_2(\tau) = \theta^* = \theta_2 - 1/n + 1/(n - \tau), \qquad \tau(\theta_2) = 0,$$

and $$\tau(\theta_1) = \frac{n(\theta_1 - \theta_2)}{(\theta_1 - \theta_2) + 1/n} < n, \text{ as required,}$$

$$I(*:2) = \theta_1\tau(\theta_1) - \theta_2\tau(\theta_1) + \tau(\theta_1)/n + \log(1 - \tau(\theta_1)/n)$$

$$= n(\theta_1 - \theta_2) - \log(1 + n(\theta_1 - \theta_2)) \quad \text{(as in example 2.2)},$$

$$= \frac{(\theta_1 - \theta_2)^2}{2}\,\frac{n^2}{(1 + n(\theta - \theta_2))^2} = (\theta_1 - \theta_2)^2/2 \operatorname{var}(y|\tau(\theta)),$$

where θ lies between θ_1 and θ_2, $\theta_1 \geq \theta \geq \theta_2$. Note that in this example $I(1:2; \mathscr{X}) = I(1:2; \mathscr{Y}) > I(*:2)$.

5. NEIGHBORING PARAMETERS

In section 6 of chapter 2 we examined the relation between Fisher's information measure and those we have been studying. We now continue that examination to study the relation between the inequality of theorem 2.1 and its consequences and the classical information inequality of the theory of estimation. Let us suppose that y_i, $i = 1, 2, \cdots, k$, in corollary 4.2 are unbiased estimators of the parameters. We saw in section 6 of chapter 2 that under suitable regularity conditions, to within terms of higher order,

$$(5.1) \qquad 2I(\theta + \Delta\theta : \theta; \mathscr{X}) = (\mathbf{\Delta\theta})'\mathbf{G}(\theta)(\mathbf{\Delta\theta}) = J(\theta + \Delta\theta, \theta; \mathscr{X}),$$

where $(\mathbf{\Delta\theta})' = (\Delta\theta_1, \Delta\theta_2, \cdots, \Delta\theta_k)$, and $\mathbf{G}(\theta)$ is the positive definite matrix $(g_{ij}(\theta))$,

$$(5.2) \qquad g_{ij}(\theta) = \int f(x)\left(\frac{\partial}{\partial\theta_i}\log f(x)\right)\left(\frac{\partial}{\partial\theta_j}\log f(x)\right) d\lambda(x), \quad i, j = 1, 2, \cdots, k.$$

Similarly, we also have

$$(5.3) \qquad 2I(\theta + \Delta\theta : \theta; \mathscr{Y}) = (\mathbf{\Delta\theta})'\mathbf{H}(\theta)(\mathbf{\Delta\theta}) = J(\theta + \Delta\theta, \theta; \mathscr{Y}),$$

where $(\Delta\mathbf{\theta})'$ is defined above, and $\mathbf{H}(\theta)$ is the positive definite matrix $(h_{ij}(\theta))$,

$$(5.4) \qquad h_{ij}(\theta) = \int g(y)\left(\frac{\partial}{\partial\theta_i}\log g(y)\right)\left(\frac{\partial}{\partial\theta_j}\log g(y)\right) d\gamma(y), \quad i, j = 1, 2, \cdots, k.$$

We can now state [cf. Barankin (1951), Cramér (1946b), Darmois (1945)]:

THEOREM 5.1. *Under suitable regularity conditions*

$$(5.5) \qquad (\mathbf{\Delta\theta})'\mathbf{G}(\theta)(\mathbf{\Delta\theta}) \geqq (\mathbf{\Delta\theta})'\mathbf{H}(\theta)(\mathbf{\Delta\theta}) \geqq (\mathbf{\Delta\theta})'\mathbf{\Sigma}^{-1}(\mathbf{\Delta\theta}),$$

where $(\mathbf{\Delta\theta})$, $\mathbf{G}(\theta)$, $\mathbf{H}(\theta)$ *are defined in* (5.1)–(5.4) *and* $\mathbf{\Sigma}$ *is the covariance matrix of the unbiased estimators. The first two members are equal if and only if the unbiased estimators are sufficient and the last two members are equal if and only if* $g(y)$ *in* (5.4) *is of the form* $e^{\tau(\theta)y}h(y)/M(\tau(\theta))$, *where* $h(y)$ *does not contain* θ *and* $M(\tau(\theta)) = \int e^{\tau(\theta)y}h(y)\, d\gamma(y)$.

Proof. Use corollaries 4.1, 4.2, 4.4.

Certain useful results about quadratic forms will be needed and are given in the following lemmas. [Cf. Barankin and Gurland (1951, pp. 109–110), Fraser (1957, pp. 55–56), Kullback (1954, p. 749), Roy and Bose (1953, p. 531).]

LEMMA 5.1. *If both* $\mathbf{x}'\mathbf{A}\mathbf{x}$ *and* $\mathbf{x}'\mathbf{C}\mathbf{x}$ *are positive definite quadratic forms (matrix notation) such that* $\mathbf{x}'\mathbf{A}\mathbf{x} \geqq \mathbf{x}'\mathbf{C}\mathbf{x}$, *then*

(*a*) *the roots of* $|\mathbf{A} - \lambda\mathbf{C}| = 0$ *are real and* $\geqq 1$;

(*b*) $|\mathbf{A}| \geqq |\mathbf{C}|$;

(*c*) *any principal minor of* **A** *is not less than the corresponding principal minor of* **C** *(determinant or quadratic form)*;

(*d*) $\mathbf{y}'\mathbf{C}^{-1}\mathbf{y} \geqq \mathbf{y}'\mathbf{A}^{-1}\mathbf{y}$;

(*e*) *any principal minor of* $\mathbf{C}^{-1}$ *is not less than the corresponding principal minor of* $\mathbf{A}^{-1}$ *(determinant or quadratic form).*

Proof. Statements (*a*), (*b*), and (*c*) are immediate corollaries of known theorems on positive definite quadratic forms, for example, theorems 44 and 48 in Ferrar (1941). Since $\mathbf{A}^{-1} = \mathbf{C}^{-1}\mathbf{C}\mathbf{A}^{-1}$ and $\mathbf{C}^{-1} = \mathbf{C}^{-1}\mathbf{A}\mathbf{A}^{-1}$, there exists a nonsingular matrix **B** such that [Bôcher (1924, p. 301)] $\mathbf{C}^{-1} = \mathbf{B}'\mathbf{A}\mathbf{B}$ and $\mathbf{A}^{-1} = \mathbf{B}'\mathbf{C}\mathbf{B}$. Thus, applying the transformation $\mathbf{x} = \mathbf{B}\mathbf{y}$ gives $\mathbf{x}'\mathbf{A}\mathbf{x} = \mathbf{y}'\mathbf{B}'\mathbf{A}\mathbf{B}\mathbf{y} = \mathbf{y}'\mathbf{C}^{-1}\mathbf{y}$, $\mathbf{x}'\mathbf{C}\mathbf{x} = \mathbf{y}'\mathbf{B}'\mathbf{C}\mathbf{B}\mathbf{y} = \mathbf{y}'\mathbf{A}^{-1}\mathbf{y}$, and (*d*) and (*e*) then follow.

We remark that $\mathbf{A} \geqq \mathbf{C}$ may be defined as meaning that $\mathbf{x}'\mathbf{A}\mathbf{x} \geqq \mathbf{x}'\mathbf{C}\mathbf{x}$ for all real vectors (matrices) $\mathbf{x} \neq \mathbf{0}$.

LEMMA 5.2. *If* $\mathbf{A} = (a_{ij})$, $i, j = 1, 2, \cdots, k$, *is a positive definite matrix, then* $a^{11} \geqq a^{11.2} \geqq a^{11.23} \geqq \cdots \geqq a^{11.23\cdots(k-1)} \geqq 1/a_{11}$, *where* $a^{11.23\cdots j}$ *is the element in the first row and first column of the inverse of the matrix obtained by deleting rows and columns* $2, 3, \cdots, j$, *in* **A**.

Proof. Consider two multivariate normal populations with common covariance matrix **A** and difference of means $\boldsymbol{\alpha}' = (\alpha_1, \alpha_2, \cdots, \alpha_k)$. As already noted in connection with corollary 4.2, and shown in chapter 9, the discrimination information measure for the two multivariate populations is $I(1:2;\mathscr{X}) = \frac{1}{2}\boldsymbol{\alpha}'\mathbf{A}^{-1}\boldsymbol{\alpha}$. The variates $y_1 = x_1$, $y_2 = x_3$, $y_3 = x_4$, $\cdots, y_{k-1} = x_k$ are also multivariate normal with covariance matrix **B**, where **B** is the matrix **A** with the second row and second column deleted [Wilks (1943, p. 68)]. For the distributions of the y's we then have $I(1:2;\mathscr{Y}) = \frac{1}{2}\boldsymbol{\beta}'\mathbf{B}^{-1}\boldsymbol{\beta}$, where $\boldsymbol{\beta}' = (\beta_1, \beta_2, \cdots, \beta_{k-1})$, $\beta_1 = \alpha_1$, $\beta_2 = \alpha_3$, $\cdots, \beta_{k-1} = \alpha_k$. But according to section 4 of chapter 2, $I(1:2;\mathscr{X}) \geqq I(1:2;\mathscr{Y})$, or $\boldsymbol{\alpha}'\mathbf{A}^{-1}\boldsymbol{\alpha} \geqq \boldsymbol{\beta}'\mathbf{B}^{-1}\boldsymbol{\beta}$ for all $\alpha_1, \alpha_2, \cdots, \alpha_k$, and therefore in particular for $\alpha_2 = 0$, $\boldsymbol{\beta}'\mathbf{C}\boldsymbol{\beta} \geqq \boldsymbol{\beta}'\mathbf{B}^{-1}\boldsymbol{\beta}$, where **C** is the matrix $\mathbf{A}^{-1}$ with the second row and second column deleted. From lemma 5.1 we can then conclude that $a^{11} \geqq a^{11.2}$. Successive application of the procedure then leads to the desired conclusion.

LEMMA 5.3. *If* $\mathbf{A}$ *is a* $k \times k$ *positive definite matrix, and* $\mathbf{U}$ *an* $r \times k$ *matrix,* $r \leqq k$, *of rank* r, *then* $\boldsymbol{\alpha}'\mathbf{A}^{-1}\boldsymbol{\alpha} \geqq \boldsymbol{\alpha}'\mathbf{U}'(\mathbf{UAU}')^{-1}\mathbf{U}\boldsymbol{\alpha}$, *where* $\boldsymbol{\alpha}' = (\alpha_1, \alpha_2, \cdots, \alpha_k)$.

Proof. Consider the two multivariate normal populations in lemma 5.2 for which $I(1:2;\mathcal{X}) = \frac{1}{2}\boldsymbol{\alpha}'\mathbf{A}^{-1}\boldsymbol{\alpha}$. The variates $y_1, y_2, \cdots, y_r$, defined by $\mathbf{y} = \mathbf{Ux}$, with $\mathbf{y}' = (y_1, y_2, \cdots, y_r)$, $\mathbf{x}' = (x_1, x_2, \cdots, x_k)$, and $\mathbf{U}$ the $r \times k$ matrix of the lemma, are also multivariate normal with a common covariance matrix $\mathbf{UAU}'$ and difference of means $\mathbf{U}\boldsymbol{\alpha}$ [Wilks (1943, p. 71)]. For the distributions of the y's we then have $I(1:2;\mathcal{Y}) = \frac{1}{2}\boldsymbol{\alpha}'\mathbf{U}'(\mathbf{UAU}')^{-1}\mathbf{U}\boldsymbol{\alpha}$. But according to section 4 of chapter 2, $I(1:2;\mathcal{X}) \geqq I(1:2;\mathcal{Y})$ and the desired conclusion follows.

LEMMA 5.4. *If* $\mathbf{B}$ *is a* $k \times k$ *positive definite matrix,* $\mathbf{U}$ *an* $r \times k$ *matrix,* $r \leqq k$, *of rank* r, *and* $\mathbf{C}$ *a* $k \times m$ *matrix of rank* $m \leqq k$, *then* $\boldsymbol{\beta}'\mathbf{C}'\mathbf{BC}\boldsymbol{\beta} \geqq \boldsymbol{\beta}'\mathbf{C}'\mathbf{U}'(\mathbf{UB}^{-1}\mathbf{U}')^{-1}\,\mathbf{UC}\boldsymbol{\beta}$, *where* $\boldsymbol{\beta}' = (\beta_1, \beta_2, \cdots, \beta_m)$.

Proof. In lemma 5.3 set $\mathbf{B} = \mathbf{A}^{-1}$ and $\boldsymbol{\alpha} = \mathbf{C}\boldsymbol{\beta}$.

COROLLARY 5.1. *For arbitrary* $\boldsymbol{\alpha}' = (\alpha_1, \alpha_2, \cdots, \alpha_k)$, α_i, $i = 1, 2, \cdots, k$, *real,* $\boldsymbol{\alpha}'\mathbf{G}(\theta)\boldsymbol{\alpha} \geqq \boldsymbol{\alpha}'\mathbf{H}(\theta)\boldsymbol{\alpha} \geqq \boldsymbol{\alpha}'\boldsymbol{\Sigma}^{-1}\boldsymbol{\alpha}$, $\boldsymbol{\alpha}'\boldsymbol{\Sigma}\boldsymbol{\alpha} \geqq \boldsymbol{\alpha}'\mathbf{H}^{-1}(\theta)\boldsymbol{\alpha} \geqq \boldsymbol{\alpha}'\mathbf{G}^{-1}(\theta)\boldsymbol{\alpha}$, *where* $\mathbf{G}(\theta)$, $\mathbf{H}(\theta)$, $\boldsymbol{\Sigma}$, *and the conditions for equality are given in theorem* 5.1.

Proof. $\mathbf{G}(\theta)$, $\mathbf{H}(\theta)$, $\boldsymbol{\Sigma}$ are positive definite since they are covariance matrices of linearly independent variables. The first set of inequalities is simply a repetition of theorem 5.1 and the second set of inequalities follows by applying lemma 5.1.

COROLLARY 5.2. *If* y_i *is an unbiased estimator of* θ_i, *then* ${\sigma_{y_i}}^2 \geqq h^{ii}(\theta) \geqq g^{ii}(\theta)$, $i = 1, 2, \cdots, k$, *where* $h^{ii}(\theta)$ and $g^{ii}(\theta)$ *are respectively the elements in the ith row and ith column of* $\mathbf{H}^{-1}(\theta)$ *and* $\mathbf{G}^{-1}(\theta)$.

Proof. Use corollary 5.1 and lemma 5.1.

COROLLARY 5.3. *If* y_1 *is an unbiased estimator of* θ_1, *then* ${\sigma_{y_1}}^2 \geqq$

$$h^{11}(\theta) \geqq g^{11}(\theta) \geqq g^{11.2}(\theta) \geqq g^{11.23}(\theta) \geqq \cdots \geqq g^{11.23\cdots(k-1)} \geqq \frac{1}{g_{11}(\theta)},$$

where $g^{11.23\cdots j}$ *is the element in the first row and first column of the inverse of the matrix obtained by deleting rows and columns* $2, 3, \cdots, j$, *in* $\mathbf{G}(\theta)$. *A similar result holds for unbiased estimators of the other parameters.*

Proof. Use corollary 5.1 and lemma 5.2. Note that $g^{ii}(\theta) = 1/g_{ii}(\theta)$ when $\mathbf{G}(\theta)$ is a diagonal matrix.

Example 5.1. In example 4.1 set $p_1 = p + \Delta p$, $p_2 = p$. The lower bound for the variance of an unbiased estimator of p, pq/n, is attained for the estimator $\hat{p} = y/n$.

Example 5.2. In example 4.2 set $\theta_1 = \theta + \Delta\theta$, $\theta_2 = \theta$, $\sigma_1^2 = \sigma^2 + \Delta\sigma^2$, $\sigma_2^2 = \sigma^2$. We find that $\mathbf{G} = \begin{pmatrix} \frac{n}{\sigma^2} & 0 \\ 0 & \frac{n}{2\sigma^4} \end{pmatrix}$, $\mathbf{H} = \begin{pmatrix} \frac{n}{\sigma^2} & 0 \\ 0 & \frac{1}{2\sigma^4} \end{pmatrix}$, and the lower bound for the variance of an unbiased estimator of θ, σ^2/n, is attained for the estimator $\hat{\theta} = \bar{x}$.

Example 5.3. In example 4.3 set $\theta_1 = \theta + \Delta\theta$, $\theta_2 = \theta$, $\sigma_1^2 = \sigma^2 + \Delta\sigma^2$, $\sigma_2^2 = \sigma^2$. We find that $\mathbf{G} = \begin{pmatrix} \frac{n}{\sigma^2} & 0 \\ 0 & \frac{n}{2\sigma^4} \end{pmatrix}$, $\mathbf{H} = \begin{pmatrix} \frac{n}{\sigma^2} & 0 \\ 0 & \frac{n}{2\sigma^4} \end{pmatrix}$, and $\mathbf{\Sigma} = \begin{pmatrix} \frac{\sigma^2}{n} & 0 \\ 0 & \frac{2\sigma^4}{n-1} \end{pmatrix}$. The lower bound for the variance of an unbiased estimator of σ^2, $2\sigma^4/n$, that is, g^{22}, is not attained by the estimator s^2 with a variance $2\sigma^4/(n-1)$. From examples 4.4 and 4.5 we see that when the population mean is known (we used the mean zero) the lower bound for the variance of an unbiased estimator of σ^2 is attained for the estimator $\frac{1}{n}\sum_{i=1}^{n} x_i^2$.

Example 5.4. In example 4.6 set $\rho_1 = \rho + \Delta\rho$, $\rho_2 = \rho$. We find that $\mathbf{G}(\rho) = \left(\frac{n(1+\rho^2)}{(1-\rho^2)^2}\right) = \mathbf{H}(\rho)$ and the lower bound for the variance of an unbiased estimator of ρ is $(1-\rho^2)^2/n(1+\rho^2)$. [Cf. Kendall (1946, pp. 33–34).]

We shall now change the assumption that the y_i, $i = 1, 2, \cdots, k$, are unbiased estimators of the parameters. Instead suppose that $E(y_i) = \theta_i(\phi_1, \phi_2, \cdots, \phi_r)$, $i = 1, 2, \cdots, k$, $k \geqq r$, that is, the parameters are $\phi_1, \phi_2, \cdots, \phi_r$, and the y's are no longer unbiased estimates of these parameters, which may be fewer in number than the y's. We now define

$$(5.6) \qquad \mu_{ij} = \frac{\partial\theta_i}{\partial\phi_j}, \qquad \mathbf{U} = (\mu_{ij}), \qquad i = 1, 2, \cdots, k, j = 1, 2, \cdots, r,$$

where the matrix $\mathbf{U}$ is assumed to be of rank r. The differences of the expected values of the y_i for neighboring values of the parameters are now given by $\Delta\theta_i = \theta_i(\phi + \Delta\phi) - \theta_i(\phi) = \mu_{i1}\,\Delta\phi_1 + \cdots + \mu_{ir}\,\Delta\phi_r + o(\Delta\phi)$, or in matrix notation, neglecting terms of higher order,

$$(5.7) \qquad (\mathbf{\Delta\theta}) = \mathbf{U}(\mathbf{\Delta\phi}).$$

We also have

$$(5.8) \qquad \frac{\partial}{\partial\phi_j}\log f(x) = \mu_{1j}\frac{\partial}{\partial\theta_1}\log f(x) + \cdots + \mu_{kj}\frac{\partial}{\partial\theta_k}\log f(x), \qquad j = 1, 2, \cdots, r,$$

or in matrix notation

$$(5.9) \qquad \left(\frac{\partial}{\partial\boldsymbol{\phi}}\log \mathbf{f}(\mathbf{x})\right) = \mathbf{U}'\left(\frac{\partial}{\partial\boldsymbol{\theta}}\log \mathbf{f}(\mathbf{x})\right).$$

Similarly, we have

$$(5.10)\qquad \left(\frac{\partial}{\partial \boldsymbol{\phi}} \log \mathbf{g}(\mathbf{y})\right) = \mathbf{U}'\left(\frac{\partial}{\partial \boldsymbol{\theta}} \log \mathbf{g}(\mathbf{y})\right).$$

We thus have

$$(5.11)\qquad \left(\frac{\partial}{\partial \boldsymbol{\phi}} \log \mathbf{f}(\mathbf{x})\right)\left(\frac{\partial}{\partial \boldsymbol{\phi}} \log \mathbf{f}(\mathbf{x})\right)' = \mathbf{U}'\left(\frac{\partial}{\partial \boldsymbol{\theta}} \log \mathbf{f}(\mathbf{x})\right)\left(\frac{\partial}{\partial \boldsymbol{\theta}} \log \mathbf{f}(\mathbf{x})\right)'\mathbf{U},$$

and taking expected values [cf. Fisher (1956, p. 155), also section 4]

$$(5.12)\qquad \mathbf{G}(\phi) = \mathbf{U}'\mathbf{G}(\theta)\mathbf{U},$$

where $\mathbf{G}(\theta)$ is the matrix defined in (5.2) and $\mathbf{G}(\phi) = (g_{ij}(\phi))$ is the matrix with

$$(5.13)\qquad g_{ij}(\phi) = \int f(x)\left(\frac{\partial}{\partial \phi_i} \log f(x)\right)\left(\frac{\partial}{\partial \phi_j} \log f(x)\right) d\lambda(x), \qquad i, j = 1, 2, \cdots, r.$$

Similarly, we have

$$(5.14)\qquad \mathbf{H}(\phi) = \mathbf{U}'\mathbf{H}(\theta)\mathbf{U},$$

where $\mathbf{H}(\theta)$ is the matrix defined in (5.4) and $\mathbf{H}(\phi) = (h_{ij}(\phi))$ is the matrix with

$$(5.15)\qquad h_{ij}(\phi) = \int g(y)\left(\frac{\partial}{\partial \phi_i} \log g(y)\right)\left(\frac{\partial}{\partial \phi_j} \log g(y)\right) d\gamma(y), \qquad i, j = 1, 2, \cdots, r.$$

We now state:

THEOREM 5.2. *Under suitable regularity conditions*

$$(5.16)\qquad (\boldsymbol{\Delta\phi})'\mathbf{G}(\phi)(\boldsymbol{\Delta\phi}) \geqq (\boldsymbol{\Delta\phi})'\mathbf{H}(\phi)(\boldsymbol{\Delta\phi}) \geqq (\boldsymbol{\Delta\phi})'\mathbf{U}'\boldsymbol{\Sigma}^{-1}\mathbf{U}(\boldsymbol{\Delta\phi}),$$

where $\mathbf{U}$, $(\boldsymbol{\Delta\phi})$, $\mathbf{G}(\phi)$, $\mathbf{H}(\phi)$ *are defined in* (5.6), (5.7), (5.12), (5.14) *and* $\boldsymbol{\Sigma} = (\sigma_{ij})$, $i, j = 1, 2, \cdots, k$, *is the covariance matrix with* $\sigma_{ij} = E(y_i - \theta_i(\phi_1, \phi_2, \cdots, \phi_r))(y_j - \theta_j(\phi_1, \phi_2, \cdots, \phi_r))$. *The first two members are equal if and only if the statistics* $y_1, y_2, \cdots, y_k$ *are sufficient. The last two members are equal if* (5.25) *below is satisfied.*

Proof. Use (5.6), (5.7), (5.12), (5.14) in (5.5) to obtain (5.16) and the condition for equality of the first two members.

We now consider conditions for equality of the last two members in (5.16).

Suppose there exist functions $z_i(x)$, $i = 1, 2, \cdots, r$, such that

$$(5.17)\qquad \mathbf{z} = \mathbf{C}\mathbf{y},$$

where

(5.18)

$$\mathbf{z} = \begin{pmatrix} z_1 - \phi_1 \\ \cdot \\ \cdot \\ \cdot \\ z_r - \phi_r \end{pmatrix}, \quad \mathbf{y} = \begin{pmatrix} y_1 - \theta_1 \\ \cdot \\ \cdot \\ \cdot \\ y_k - \theta_k \end{pmatrix}, \quad \mathbf{C} = (c_{ij}), \quad \begin{matrix} i = 1, 2, \cdots, r, \\ j = 1, 2, \cdots, k, \end{matrix}$$

and $\mathbf{C}$ is of rank r. The expected value of $\mathbf{zz}' = \mathbf{Cyy}'\mathbf{C}'$ yields

$$\boldsymbol{\Sigma}_1 = \mathbf{C}\boldsymbol{\Sigma}\mathbf{C}', \tag{5.19}$$

where $\boldsymbol{\Sigma}_1$ is the covariance matrix of the z's, which are unbiased estimators of the ϕ's. Letting $\mathbf{R} = (\mathbf{U}'\boldsymbol{\Sigma}^{-1}\mathbf{U})^{-1}$, lemmas 5.4 and 5.1 yield

$$\boldsymbol{\alpha}'\mathbf{C}\boldsymbol{\Sigma}\mathbf{C}'\boldsymbol{\alpha} \geqq \boldsymbol{\alpha}'\mathbf{CURU}'\mathbf{C}'\boldsymbol{\alpha}, \tag{5.20}$$

$$|\mathbf{C}\boldsymbol{\Sigma}\mathbf{C}'| \geqq |\mathbf{CURU}'\mathbf{C}'| = |\mathbf{CU}|^2|\mathbf{R}|, \tag{5.21}$$

$$|\mathbf{CU}|^2 \leqq |\mathbf{C}\boldsymbol{\Sigma}\mathbf{C}'| \cdot |\mathbf{U}'\boldsymbol{\Sigma}^{-1}\mathbf{U}|. \tag{5.22}$$

If $\mathbf{CU} = \mathbf{I}$, then from (5.20) and corollary 5.4,

$$\boldsymbol{\alpha}'\mathbf{C}\boldsymbol{\Sigma}\mathbf{C}'\boldsymbol{\alpha} \geqq \boldsymbol{\alpha}'\mathbf{R}\boldsymbol{\alpha} \geqq \boldsymbol{\alpha}'\mathbf{H}^{-1}(\phi)\boldsymbol{\alpha} \geqq \boldsymbol{\alpha}'\mathbf{G}^{-1}(\phi)\boldsymbol{\alpha}. \tag{5.23}$$

Note that when the matrix $\mathbf{C}$ in (5.17) consists of constants independent of the parameters,

$$(\Delta\boldsymbol{\phi}) = \mathbf{C}(\Delta\boldsymbol{\theta}) = \mathbf{CU}(\Delta\boldsymbol{\phi}), \tag{5.24}$$

using (5.7), or $\mathbf{CU} = \mathbf{I}$.

When the generalized density of the y's is $g(y) = e^{\tau(\phi)y}h(y)/M_2(\tau(\phi))$, with $\tau(\phi)y = \sum_{i=1}^{k} y_i\tau_i(\theta_1(\phi_1, \phi_2, \cdots, \phi_r), \cdots, \theta_k(\phi_1, \phi_2, \cdots, \phi_r))$, $h(y)$ independent of the parameters (cf. theorem 5.1), and the matrix $\mathbf{B} = (b_{ij})$, $b_{ij} = \partial\tau_j/\partial\phi_i$, $i = 1, 2, \cdots, r$, $j = 1, 2, \cdots, k$, of rank r, that is, if

$$\left(\frac{\partial}{\partial\boldsymbol{\phi}} \log \mathrm{g}(\mathbf{y})\right) = \mathbf{By}, \tag{5.25}$$

where $\mathbf{y}$ is defined in (5.18), then

$$\mathbf{H}(\phi) = \mathbf{B}\boldsymbol{\Sigma}\mathbf{B}'. \tag{5.26}$$

Since $\left(\frac{\partial}{\partial\boldsymbol{\phi}} \log \mathrm{g}(\mathbf{y})\right)\left(\frac{\partial}{\partial\boldsymbol{\phi}} \log \mathrm{g}(\mathbf{y})\right)' = \mathbf{Byy}'\mathbf{B}'$, (5.26) follows by taking expected values. Since $a_{ij} = \partial\theta_i/\partial\tau_j$, $i, j = 1, 2, \cdots, k$, and $\mathbf{A} = (a_{ij})$, we have $\mathbf{AB}' = \mathbf{U}$ and, by lemma 4.10, this is the same as

$$\boldsymbol{\Sigma}\mathbf{B}' = \mathbf{U}. \tag{5.27}$$

From (5.26) and (5.27) we then have

$$\mathbf{H}^{-1}(\phi)\mathbf{B}\mathbf{U} = \mathbf{I}. \tag{5.28}$$

With $\mathbf{H}^{-1}(\phi)\mathbf{B}$ as the matrix $\mathbf{C}$ in (5.23), we have

$$\boldsymbol{\alpha}'\mathbf{H}^{-1}(\phi)\mathbf{B}\boldsymbol{\Sigma}\mathbf{B}'\mathbf{H}^{-1}(\phi)\boldsymbol{\alpha} \geqq \boldsymbol{\alpha}'\mathbf{R}\boldsymbol{\alpha} \geqq \boldsymbol{\alpha}'\mathbf{H}^{-1}(\phi)\boldsymbol{\alpha}. \tag{5.29}$$

Using (5.26) in (5.29) yields

$$\boldsymbol{\alpha}'\mathbf{H}^{-1}(\phi)\boldsymbol{\alpha} \geqq \boldsymbol{\alpha}'\mathbf{R}\boldsymbol{\alpha} \geqq \boldsymbol{\alpha}'\mathbf{H}^{-1}(\phi)\boldsymbol{\alpha}, \tag{5.30}$$

or $\mathbf{H}(\phi) = \mathbf{U}'\boldsymbol{\Sigma}^{-1}\mathbf{U}$, and we have equality in the last two members of (5.16).

COROLLARY 5.4. *For arbitrary* $\boldsymbol{\alpha}' = (\alpha_1, \alpha_2, \cdots, \alpha_r)$, α_i, $i = 1, 2, \cdots, r$, *real*, $\boldsymbol{\alpha}'\mathbf{G}(\phi)\boldsymbol{\alpha} \geqq \boldsymbol{\alpha}'\mathbf{H}(\phi)\boldsymbol{\alpha} \geqq \boldsymbol{\alpha}'\mathbf{U}'\boldsymbol{\Sigma}^{-1}\mathbf{U}\boldsymbol{\alpha}$, $\boldsymbol{\alpha}'(\mathbf{U}'\boldsymbol{\Sigma}^{-1}\mathbf{U})^{-1}\boldsymbol{\alpha} \geqq \boldsymbol{\alpha}'\mathbf{H}^{-1}(\phi)\boldsymbol{\alpha} \geqq \boldsymbol{\alpha}'\mathbf{G}^{-1}(\phi)\boldsymbol{\alpha}$, *where the matrices* $\mathbf{U}$, $\mathbf{G}(\phi)$, $\mathbf{H}(\phi)$, $\boldsymbol{\Sigma}$ *and the conditions for equality are given in theorem* 5.2.

Proof. Proceed as in corollary 5.1.

Example 5.5. This example is a continuation of example 5.4. Take the generalized density $g^*(y)$ in (4.32) to be $g(y)$ by letting $\rho^* = \rho_2 = \rho$. Since $E(y_1) = 2(1 - \rho)$, $E(y_2) = 2(1 + \rho)$, we have $\theta_1 = 2(1 - \rho)$, $\theta_2 = 2(1 + \rho)$, $\phi_1 = \rho$, $k = 2$, $r = 1$, $\mathbf{U}' = (-2, 2)$. With $\rho_1 = \rho + \Delta\rho$, $\rho_2 = \rho$, $1 - \rho_2 = \theta_1/2$, $1 + \rho_2 = \theta_2/2$, $1 - \rho_1 = (\theta_1 + \Delta\theta_1)/2$, $1 + \rho_1 = (\theta_2 + \Delta\theta_2)/2$, we see from the first version in (4.29) that

$$I(\theta + \Delta\theta{:}\theta; \mathscr{Y}) = \frac{n}{2}\left(\log\frac{\theta_1}{\theta_1 + \Delta\theta_1} - 1 + \frac{\theta_1 + \Delta\theta_1}{\theta_1}\right) + \frac{n}{2}\left(\log\frac{\theta_2}{\theta_2 + \Delta\theta_2} - 1 + \frac{\theta_2 + \Delta\theta_2}{\theta_2}\right),$$

and to within terms of higher order

$$I(\theta + \Delta\theta{:}\theta; \mathscr{Y}) = \frac{n}{2}\left(\frac{(\Delta\theta_1)^2}{2\theta_1^2} + \frac{(\Delta\theta_2)^2}{2\theta_2^2}\right).$$

We thus have, since y is a sufficient statistic,

$$\mathbf{G}(\theta) = \begin{pmatrix} \dfrac{n}{2\theta_1^2} & 0 \\ 0 & \dfrac{n}{2\theta_2^2} \end{pmatrix} = \mathbf{H}(\theta),$$

$$\mathbf{G}(\phi) = (-2, 2)\begin{pmatrix} \dfrac{n}{2\theta_1^2} & 0 \\ 0 & \dfrac{n}{2\theta_2^2} \end{pmatrix}\begin{pmatrix} -2 \\ 2 \end{pmatrix} = \mathbf{H}(\phi) = \frac{2n}{\theta_1^2} + \frac{2n}{\theta_2^2}$$

$$= \frac{2n}{4(1 - \rho)^2} + \frac{2n}{4(1 + \rho)^2} = \frac{n(1 + \rho^2)}{(1 - \rho^2)^2},$$

the value derived in example 5.4. Since $\operatorname{var}(y_1) = 2\theta_1^2/n$, $\operatorname{var}(y_2) = 2\theta_2^2/n$, and $\operatorname{cov}(y_1, y_2) = 0$, we find here that $\mathbf{G}(\phi) = \mathbf{H}(\phi) = \mathbf{U}'\mathbf{\Sigma}^{-1}\mathbf{U}$. Corresponding to (5.17) we have $z_1 - \rho = -\frac{1}{4}(y_1 - \theta_1) + \frac{1}{4}(y_2 - \theta_2)$, that is,

$$\mathbf{C} = (-\tfrac{1}{4}, +\tfrac{1}{4})$$

and

$$\mathbf{\Sigma}_1 = (-\tfrac{1}{4}, +\tfrac{1}{4})\begin{pmatrix} \frac{2\theta_1^2}{n} & 0 \\ 0 & \frac{2\theta_2^2}{n} \end{pmatrix}\begin{pmatrix} -\frac{1}{4} \\ +\frac{1}{4} \end{pmatrix} = \frac{\theta_1^2 + \theta_2^2}{8n} = \frac{1 + \rho^2}{n}.$$

Note that $\mathbf{CU} = 1$. We see that the variance of the unbiased estimator of ρ, $z_1 = (y_2 - y_1)/4 = \frac{1}{n}\sum_{i=1}^{n} x_{1i}x_{2i}$, is $(1 + \rho^2)/n > (1 - \rho^2)^2/n(1 + \rho^2)$, the lower bound for an unbiased estimator of ρ [cf. Stuart (1955b, p. 528)]. The estimate z_1, the product moment form with the population means and variances, may take values that exceed 1 in absolute value. From (4.32) we see that the matrix $\mathbf{B}$ of (5.25) is

$$\mathbf{B} = \left(-\frac{n}{4(1 - \rho)^2}, \frac{n}{4(1 + \rho)^2}\right),$$

since $\tau_1 = -n/4(1 - \rho)$, $\tau_2 = -n/4(1 + \rho)$, and therefore that

$$\mathbf{\Sigma B}' = \begin{pmatrix} \frac{8(1 - \rho)^2}{n} & 0 \\ 0 & \frac{8(1 + \rho)^2}{n} \end{pmatrix}\begin{pmatrix} -\frac{n}{4(1 - \rho)^2} \\ \frac{n}{4(1 + \rho)^2} \end{pmatrix} = \begin{pmatrix} -2 \\ 2 \end{pmatrix} = \mathbf{U},$$

verifying (5.27). We find that

$$\mathbf{H}^{-1}(\phi)\mathbf{B} = \frac{(1 - \rho^2)^2}{n(1 + \rho^2)}\left(-\frac{n}{4(1 - \rho)^2}, \frac{n}{4(1 + \rho)^2}\right) = \left(-\frac{(1 + \rho)^2}{4(1 + \rho^2)}, \frac{(1 - \rho)^2}{4(1 + \rho^2)}\right),$$

and with $\mathbf{H}^{-1}(\phi)\mathbf{B}$ as the matrix $\mathbf{C}$ in (5.17) we have

$$\begin{aligned} z_1 - \rho &= -\frac{(1 + \rho)^2}{4(1 + \rho^2)}(y_1 - \theta_1) + \frac{(1 - \rho)^2}{4(1 + \rho^2)}(y_2 - \theta_2) \\ &= \left(-\frac{1}{4} - \frac{\rho}{2(1 + \rho^2)}\right)(y_1 - 2(1 - \rho)) + \left(\frac{1}{4} - \frac{\rho}{2(1 + \rho^2)}\right)(y_2 - 2(1 + \rho)) \\ &= \frac{y_2 - y_1}{4} - \rho - \frac{\rho}{2(1 + \rho^2)}(y_1 + y_2 - 4). \end{aligned}$$

Since $E(y_1) = 2(1 - \rho)$ and $E(y_2) = 2(1 + \rho)$, so that $E(y_1 + y_2) = 4$, let us consider the estimator $r(y_1, y_2) = (y_2 - y_1)/(y_2 + y_1)$. Since $|r(y_1, y_2)| \leq 1$, and $r(y_1, y_2)$ is continuous and has continuous derivatives of the first and second order with respect to y_1 and y_2 in a neighborhood of the point $(E(y_1), E(y_2))$, we may apply the result on p. 354 of Cramér (1946a), that is, $E(r(y_1, y_2)) = r(E(y_1), E(y_2)) + O(1/n)$, $\operatorname{var}(r(y_1, y_2)) = a^2 \operatorname{var}(y_1) + 2ab \operatorname{cov}(y_1, y_2) + b^2 \operatorname{var}(y_2) + O(1/n^{3/2})$, where a and b are respectively $\frac{\partial}{\partial y_1} r(y_1, y_2)$, $\frac{\partial}{\partial y_2} r(y_1, y_2)$, evaluated at the point $(E(y_1), E(y_2))$. Since $a = -(1 + \rho)/4$,

$b = (1-\rho)/4$, we find that $E(r(y_1, y_2)) = \rho + O(1/n)$ and var $(r(y_1, y_2)) = (1-\rho^2)^2/n + O(1/n^{3/2})$. The estimate $r(y_1, y_2)$ is consistent [Wilks (1943, Theorem (A), p. 134)] and its variance, which is less than var $((y_2 - y_1)/4)$, does not attain the lower bound. Taking the bivariate normal population with the five parameters $(\theta_1, \theta_2, \sigma_1^2, \sigma_2^2, \rho)$, we find [cf. Kendall (1946, p. 38) who considers the parameters as $(\theta_1, \theta_2, \sigma_1, \sigma_2, \rho)$]

$$\mathbf{G} = n\begin{pmatrix} \frac{1}{\sigma_1^2(1-\rho^2)} & -\frac{\rho}{\sigma_1\sigma_2(1-\rho^2)} & 0 & 0 & 0 \\ -\frac{\rho}{\sigma_1\sigma_2(1-\rho^2)} & \frac{1}{\sigma_2^2(1-\rho^2)} & 0 & 0 & 0 \\ 0 & 0 & \frac{2-\rho^2}{4\sigma_1^4(1-\rho^2)} & -\frac{\rho^2}{4\sigma_1^2\sigma_2^2(1-\rho^2)} & -\frac{\rho}{2\sigma_1^2(1-\rho^2)} \\ 0 & 0 & -\frac{\rho^2}{4\sigma_1^2\sigma_2^2(1-\rho^2)} & \frac{2-\rho^2}{4\sigma_2^4(1-\rho^2)} & -\frac{\rho}{2\sigma_2^2(1-\rho^2)} \\ 0 & 0 & -\frac{\rho}{2\sigma_1^2(1-\rho^2)} & -\frac{\rho}{2\sigma_2^2(1-\rho^2)} & \frac{1+\rho^2}{(1-\rho^2)^2} \end{pmatrix}$$

We find that $g^{55} = (1-\rho^2)^2/n = g^{55.1} = g^{55.12}$, $g^{55.123} = (2-\rho^2)(1-\rho^2)^2/2n$, $g^{55.1234} = 1/g_{55} = (1-\rho^2)^2/n(1+\rho^2)$, verifying corollary 5.3. Note that var $(r(y_1, y_2))$ approaches the lower bound g^{55} (the greatest lower bound for the variance of an unbiased estimator of ρ) as $n \to \infty$. We also see that $g^{11} = \sigma_1^2/n$ and $g^{11.2} = g^{11.23} = g^{11.234} = g^{11.2345} = 1/g_{11} = \sigma_1^2(1-\rho^2)/n$, verifying corollary 5.3. We also see that $g^{33} = \dfrac{2\sigma_1^4}{n} = g^{33.1} = g^{33.12}$, $g^{33.124} = 2\sigma_1^4(1-\rho^4)/n$, $g^{33.1245} = 1/g_{33} = 2\sigma_1^4(1-\rho^2)/n(1-\rho^2/2)$, verifying corollary 5.3.

6. EFFICIENCY

We define the *discrimination efficiency* of the statistic $Y = T(x)$ by the ratio $I(1:2;\mathcal{Y})/I(1:2;\mathcal{X})$. From the properties discussed in chapter 2, this ratio is nonnegative and ≤ 1 with equality if and only if $Y = T(x)$ is a sufficient statistic. When the generalized densities of the populations are of the same functional form but differ according to the values of the k-dimensional parameter $\theta = (\theta_1, \theta_2, \cdots, \theta_k)$, we define the discrimination efficiency of the statistic $Y = T(x)$ at the point θ in the k-dimensional parameter space by $\lim_{\Delta\theta\to 0} (I(\theta + \Delta\theta:\theta;\mathcal{Y})/I(\theta + \Delta\theta:\theta;\mathcal{X}))$.

The *discrimination efficiency* of the unbiased estimators y_i, $i = 1, 2, \cdots, k$, of theorem 5.1, at a point $\theta = (\theta_1, \theta_2, \cdots, \theta_k)$ in the k-dimensional parameter space may therefore be defined by

$$(6.1) \qquad \lambda = (\mathbf{d\theta})'\mathbf{H}(\theta)(\mathbf{d\theta})/(\mathbf{d\theta})'\mathbf{G}(\theta)(\mathbf{d\theta}).$$

We take $(\mathbf{d\theta})'\mathbf{G}(\theta)(\mathbf{d\theta})$ as the basis of the metric of the parameter space [cf. Rao (1945)]. The $g_{ij}(\theta)$ in (5.2) are the components of a covariant tensor of the second order, the *fundamental tensor* of the metric [Eisenhart (1926, p. 35)]. Since $(\mathbf{d\theta})'\mathbf{H}(\theta)(\mathbf{d\theta}) \leqq (\mathbf{d\theta})'\mathbf{G}(\theta)(\mathbf{d\theta})$, and both quadratic forms are positive definite, the roots of $|\mathbf{H}(\theta) - \lambda\mathbf{G}(\theta)| = 0$ are real, positive, and all $\leqq 1$. (See lemma 5.1.) Accordingly, there exists a real transformation of the θ's such that at a point θ in the parameter space the quadratic forms in (6.1) may be written as

$$(6.2) \qquad \lambda = (\lambda_1\, d\psi_1^2 + \cdots + \lambda_k\, d\psi_k^2)/(d\psi_1^2 + \cdots + d\psi_k^2),$$

and $\lambda_1, \lambda_2, \cdots, \lambda_k$ are the roots of $|\mathbf{H}(\theta) - \lambda\mathbf{G}(\theta)| = 0$ [Eisenhart (1926, p. 108)]. Writing

$$(6.3) \qquad \cos^2\alpha_i = d\psi_i^2/(d\psi_1^2 + \cdots + d\psi_k^2), \qquad i = 1, 2, \cdots, k,$$

(6.2) may be written as

$$(6.4) \qquad \lambda = \lambda_1 \cos^2\alpha_1 + \lambda_2 \cos^2\alpha_2 + \cdots + \lambda_k \cos^2\alpha_k.$$

The directions at the point θ determined by $\cos\alpha_1 = 1$, $\cos\alpha_2 = 1, \cdots$, are known as the *principal directions* determined by the tensor $h_{ij}(\theta)$ [Eisenhart (1926, p. 110)]. Furthermore, at the point θ, the finite maxima and minima of λ defined by (6.1) are given for the principal directions at the point and are indeed the roots of $|\mathbf{H}(\theta) - \lambda\mathbf{G}(\theta)| = 0$. Since $(\mathbf{d\theta})'\mathbf{G}(\theta)(\mathbf{d\theta})$ is positive definite, λ is finite for all directions [Eisenhart (1926, par. 33)].

The *estimation efficiency* [cf. Fisher (1956, pp. 145–152)] of the unbiased estimators $y_1, y_2, \cdots, y_k$ is defined as the product of the discrimination efficiencies for the principal directions at the point θ, that is (see lemma 5.1),

$$(6.5) \qquad \text{Eff.} = \lambda_1\lambda_2 \cdots \lambda_k = |\mathbf{H}(\theta)|/|\mathbf{G}(\theta)| \leqq 1.$$

This is invariant for all nonsingular transformations of the parameters, with equality holding if and only if the estimators are sufficient.

Suppose we have n independent observations from an l-variate population with k parameters. The *asymptotic discrimination efficiency* of the unbiased estimators $\hat{y}_i$, $i = 1, 2, \cdots, k$, of theorem 5.1 at a point θ in the parameter space is defined by

$$(6.6) \qquad \lambda = (\mathbf{d\theta})'\mathbf{\Sigma}^{-1}(\mathbf{d\theta})/n(\mathbf{d\theta})'\mathbf{G}(\theta)(\mathbf{d\theta}), \qquad n \text{ large},$$

where the elements of $\mathbf{G}(\theta)$ are computed for a single observation from the l-variate population. Since $(\mathbf{d\theta})'\mathbf{\Sigma}^{-1}(\mathbf{d\theta}) \leqq n(\mathbf{d\theta})'\mathbf{G}(\theta)(\mathbf{d\theta})$, and both forms are positive definite, the roots of

$$(6.7) \qquad |\mathbf{\Sigma}^{-1} - \lambda n\mathbf{G}(\theta)| = 0$$

are real, positive, and $\leqq 1$. (See lemma 5.1.) The roots of (6.7) are the finite maxima and minima of (6.6) and are given for the principal directions determined by the tensor σ^{ij} at the point θ, where $\mathbf{\Sigma}^{-1} = (\sigma^{ij})$.

The *asymptotic estimation efficiency* of the unbiased estimators $y_1, y_2, \cdots, y_k$ [cf. Cramér (1946a, pp. 489, 494)] is defined as the product of the asymptotic discrimination efficiencies for the principal directions at the point θ, that is,

$$\text{(6.8)} \quad \text{Asymp. eff.} = \lambda_1\lambda_2 \cdots \lambda_k = |\mathbf{\Sigma}^{-1}|/|n\mathbf{G}(\theta)| \leqq 1, \qquad n \text{ large,}$$

the equality holding for all n if the conditions for equality in theorem 5.1 are satisfied. If $|\mathbf{\Sigma}||\mathbf{G}(\theta)| \to n^{-k}$, the asymptotic estimation efficiency approaches unity and the roots of (6.7) approach 1.

The *discrimination efficiency* of the biased estimators y_i, $i = 1, 2, \cdots, k$, of theorem 5.2, at a point $\phi = (\phi_1, \phi_2, \cdots, \phi_r)$ in the r-dimensional parameter space may be defined by

$$\text{(6.9)} \qquad \lambda = (\mathbf{d}\boldsymbol{\phi})'\mathbf{H}(\phi)(\mathbf{d}\boldsymbol{\phi})/(\mathbf{d}\boldsymbol{\phi})'\mathbf{G}(\phi)(\mathbf{d}\boldsymbol{\phi}),$$

where the matrices $\mathbf{G}(\phi)$, $\mathbf{H}(\phi)$ are defined in (5.12) and (5.14) respectively. A discussion similar to that covering (6.1)–(6.4) permits us to state that λ defined by (6.9) is finite for all directions, the finite maxima and minima of λ are the roots of $|\mathbf{H}(\phi) - \lambda\mathbf{G}(\phi)| = |\mathbf{U}'\mathbf{H}(\theta)\mathbf{U} - \lambda\mathbf{U}'\mathbf{G}(\theta)\mathbf{U}| = 0$, and are given for the principal directions at the point ϕ determined by the tensor $h_{ij}(\phi)$, with $(\mathbf{d}\boldsymbol{\phi})'\mathbf{G}(\phi)(\mathbf{d}\boldsymbol{\phi})$ as the basis of the metric of the parameter space. Note from theorem 5.2 that if the statistic $Y = T(x) = (y_1, y_2, \cdots, y_k)$ is sufficient, the discrimination efficiency is 1.

The *estimation efficiency* of the biased estimators y_i, $i = 1, 2, \cdots, k$, of theorem 5.2, at a point $\phi = (\phi_1, \phi_2, \cdots, \phi_r)$ in the r-dimensional parameter space, may be defined as the product of the discrimination efficiencies for the principal directions at the point, that is,

$$\text{(6.10)} \quad \text{Eff.} = \lambda_1\lambda_2 \cdots \lambda_r = |\mathbf{H}(\phi)|/|\mathbf{G}(\phi)| = |\mathbf{U}'\mathbf{H}(\theta)\mathbf{U}|/|\mathbf{U}'\mathbf{G}(\theta)\mathbf{U}| \leqq 1,$$

with equality if and only if the statistics are sufficient.

The *asymptotic discrimination efficiency* at a point $\phi = (\phi_1, \phi_2, \cdots, \phi_r)$ in the r-dimensional parameter space is defined by (see theorem 5.2)

$$\text{(6.11)} \qquad \lambda = (\mathbf{d}\boldsymbol{\phi})'\mathbf{U}'\mathbf{\Sigma}^{-1}\mathbf{U}(\mathbf{d}\boldsymbol{\phi})/n(\mathbf{d}\boldsymbol{\phi})'\mathbf{G}(\phi)(\mathbf{d}\boldsymbol{\phi}), \qquad n \text{ large,}$$

where the elements of $\mathbf{G}(\phi)$ are computed for a single observation from the population. The value of λ in (6.11) is finite for all directions and the finite maxima and minima of λ are given by the roots of

$$\text{(6.12)} \qquad |\mathbf{U}'\mathbf{\Sigma}^{-1}\mathbf{U} - \lambda n\mathbf{G}(\phi)| = |\mathbf{U}'\mathbf{\Sigma}^{-1}\mathbf{U} - \lambda n\mathbf{U}'\mathbf{G}(\theta)\mathbf{U}| = 0,$$

that is, for the principal directions determined by the tensor with components those of the matrix $\mathbf{U}'\mathbf{\Sigma}^{-1}\mathbf{U}$.

The *asymptotic estimation efficiency* of the biased estimators $y_1, y_2, \cdots, y_k$ of theorem 5.2 is defined as the product of the asymptotic discrimination efficiencies for the principal directions at the point $\phi = (\phi_1, \phi_2, \cdots, \phi_r)$, that is,

$$(6.13)\quad \text{Asymp. eff.} = \lambda_1 \cdots \lambda_r = |\mathbf{U}'\mathbf{\Sigma}^{-1}\mathbf{U}|/|n\mathbf{G}(\phi)| = |\mathbf{U}'\mathbf{\Sigma}^{-1}\mathbf{U}|/|n\mathbf{U}'\mathbf{G}(\theta)\mathbf{U}| \leqq 1, \quad n \text{ large.}$$

For unbiased estimators of $(\phi_1, \phi_2, \cdots, \phi_r)$ with covariance matrix $\mathbf{C\Sigma C}'$ such that $\mathbf{CU} = \mathbf{I}$, we see from (5.22) that $|(\mathbf{C\Sigma C}')^{-1}| \leqq |\mathbf{U}'\mathbf{\Sigma}^{-1}\mathbf{U}|$ and therefore such unbiased estimators are not more efficient asymptotically than the biased estimators we have been considering. Furthermore, if (5.25) is satisfied, and $Y = T(x) = (y_1, y_2, \cdots, y_k)$ is a sufficient statistic, the asymptotic efficiency in (6.13) is 1 for all n.

Example 6.1. From example 5.3 we see that the discrimination efficiency of $(\bar{x}, s^2)$ is unity, as is also the estimation efficiency. However, since the roots of $|\mathbf{\Sigma}^{-1} - \lambda n\mathbf{G}| = \begin{vmatrix} \dfrac{n}{\sigma^2} - \lambda\dfrac{n}{\sigma^2} & 0 \\ 0 & \dfrac{n-1}{2\sigma^4} - \lambda\dfrac{n}{2\sigma^4} \end{vmatrix} = 0$ are $\lambda_1 = 1$, $\lambda_2 = \dfrac{n-1}{n}$, the asymptotic discrimination efficiency for σ^2 fixed, that is, in the direction of the mean, is unity, whereas the asymptotic discrimination efficiency for θ fixed, that is, in the direction of the variance, is $(n-1)/n$, and the asymptotic estimation efficiency is $(n-1)/n$.

Example 6.2. From example 5.5 we see that the discrimination efficiency of (y_1, y_2) is unity, as is also the estimation efficiency, with similar values for the asymptotic discrimination efficiency and the asymptotic estimation efficiency. The asymptotic discrimination efficiency and the asymptotic estimation efficiency of the unbiased estimator $z_1 = (y_2 - y_1)/4$ are both $(n/(1+\rho^2))/(n(1+\rho^2)/(1-\rho^2)^2) = (1-\rho^2)^2/(1+\rho^2)^2$, which is less than 1 unless $\rho^2 = 0$. The consistent estimator $r(y_1, y_2)$ has an asymptotic discrimination efficiency, as well as an asymptotic estimation efficiency, $(n/(1-\rho^2)^2)/(n(1+\rho^2)/(1-\rho^2)^2) = 1/(1+\rho^2)$, which is less than 1 unless $\rho^2 = 0$. $r(y_1, y_2)$ is more efficient than $(y_2 - y_1)/4$. The results in the last part of example 5.5 and corollary 5.3 indicate that there cannot exist an unbiased estimator of ρ with asymptotic estimation efficiency greater than that of $r(y_1, y_2)$. Note, however, that for $z = (y_2 - y_1)/4 - \rho(y_1 + y_2 - 4)/2(1+\rho^2)$, $E(z) = \rho$, $\text{var}(z) = (1-\rho^2)^2/n(1+\rho^2)$.

7. PROBLEMS

7.1. Prove the statement (attributed to Chernoff) about the behavior of $-\log(\inf_{0<\tau<1} E(e^{\tau x}))$ (as an information measure) in the remarks following (2.21).

7.2. Prove corollary 3.2.

7.3. Prove the lemmas in section 4.

7.4. Show that $I(1:2; \mathscr{X}) = I(1:2; \mathscr{Y}) = I(*:2)$, for Poisson distributions, when $Y = T(x) = x_1 + x_2 + \cdots + x_n$.

7.5. Prove corollary 4.3.

7.6. Prove theorem 5.1.

7.7. Prove corollary 5.4.

7.8. In the related examples 4.6, 5.4, and 5.5 we discuss a sufficient statistic. Is there a sufficient estimate for the parameter ρ?

7.9. Prove the invariance of the efficiency defined in (6.5).

7.10. Express (6.6) as the limit of a ratio involving $I(*:2)$ and $I(1:2; \mathscr{X})$.

7.11. Can we determine the discrimination efficiency and the estimation efficiency for the statistic and populations of example 4.7?

7.12. Compare the results in example 4.7 with those obtained using the sample average as the statistic.

7.13. Compute $J(*, 2)$ for example 2.3.

7.14. Compute $J(1, 2; O_n)$, $J(*, 2)$: (*a*) for example 4.1; (*b*) for example 4.2; (*c*) for example 4.3; (*d*) for example 4.4; (*e*) for example 4.7.

7.15. Consider the minimum value of $I(f:f_1) = \int f(x) \log \frac{f(x)}{f_1(x)}\, d\lambda(x)$, subject to $\theta = \int T(x) f(x)\, d\lambda(x) = \int f(x) \log \frac{f_1(x)}{f_2(x)}\, d\lambda(x)$. Show that for $\theta = 0$, $\min I(f:f_1)$ satisfies (2.21). [Cf. example 3.1 of chapter 5; Chernoff (1952, p. 504).]

7.16. Show that $\int_E (f_1(x))^\tau (f_2(x))^{1-\tau}\, d\lambda(x) \leq (\mu_1(E))^\tau (\mu_2(E))^{1-\tau}$, for $E \in \mathscr{S}$ and $0 < \tau < 1$. [Cf. Adhikari and Joshi (1956), Joshi (1957).]

7.17. Show that $2(p_1 - p_2)^2 + \frac{4}{3}(p_1 - p_2)^4 \leq p_1 \log \frac{p_1}{p_2} + q_1 \log \frac{q_1}{q_2} \leq \frac{(p_1 - p_2)^2}{2pq}$, with pq the smaller of $p_i q_i$, $q_i = 1 - p_i$, $i = 1, 2$. [Cf. Schützenberger (1954, pp. 58–59).]

7.18. Show that $J(f_0, f_2) = \tau_0 I(2:1)$, with τ_0 and $f_0(x)$ defined in (2.19).

7.19. Extend the procedure of example 2.3 to derive corollary 3.2 of chapter 2.

7.20. Consider the discrete random variable x that takes the values $x_1, x_2, \cdots, x_n$, with $\text{Prob}\,(x = x_j | H_1) = p_j$, $\text{Prob}\,(x = x_j | H_2) = 1/n$. With $T(x) = x$, show how problem 8.28 of chapter 2 follows from theorem 2.1.

7.21. Re-examine example 4.2 when $\sigma_1^2 = \sigma_2^2$.

7.22. If $\mu^*(E) = \int_E f^*(x)\, d\lambda(x)$, $\mu_2(E) = \int_E f_2(x)\, d\lambda(x)$, $E \in \mathcal{S}$ with $f^*(x)$ defined in (2.11), show that

$$\tau \max_{x\epsilon E} T(x) - \log M_2(\tau) \geq \log \frac{\mu^*(E)}{\mu_2(E)} \geq \tau \min_{x\epsilon E} T(x) - \log M_2(\tau), \qquad \tau > 0,$$

$$\tau \min_{x\epsilon E} T(x) - \log M_2(\tau) \geq \log \frac{\mu^*(E)}{\mu_2(E)} \geq \tau \max_{x\epsilon E} T(x) - \log M_2(\tau), \qquad \tau < 0.$$

[Cf. Chernoff (1952, 1956), Kolmogorov (1950, p. 42).]

7.23. In problem 7.22 let $f_2(x) = \binom{n}{x}\left(\frac{1}{2}\right)^n$, $x = 0, 1, 2, \cdots, n$, and $T(x) = x$, then:

(*a*) $M_2(\tau) = (\frac{1}{2} + \frac{1}{2}e^{\tau})^n$.

(*b*) $f^*(x) = \binom{n}{x}(p^*)^x(q^*)^{n-x}$, $x = 0, 1, \cdots, n$, $p^* = \dfrac{e^{\tau}}{1 + e^{\tau}}$, $q^* = 1 - p^*$.

(*c*) $$n \log 2q^* \geq \log \frac{\sum_{x=0}^{r}\binom{n}{x}(p^*)^x(q^*)^{n-x}}{\sum_{x=0}^{r}\binom{n}{x}\left(\frac{1}{2}\right)^n} \geq r \log 2p^* + (n - r)\log 2q^*, \quad p^* < \tfrac{1}{2}.$$

(*d*) $$r \log 2p^* + (n - r)\log 2q^* \geq \log \frac{\sum_{x=0}^{r}\binom{n}{x}(p^*)^x(q^*)^{n-x}}{\sum_{x=0}^{r}\binom{n}{x}\left(\frac{1}{2}\right)^n} \geq n \log 2q^*, \quad p^* > \tfrac{1}{2}.$$

(*e*) $$\log \frac{\sum_{x=0}^{r}\binom{n}{x}p^x q^{n-x}}{\sum_{x=0}^{r}\binom{n}{x}\left(\frac{1}{2}\right)^n} \geq n\left[\frac{r}{n}\log\frac{2r}{n} + \left(1 - \frac{r}{n}\right)\log 2\left(1 - \frac{r}{n}\right)\right], \quad p < \frac{r}{n} < \frac{1}{2}.$$

7.24. In problem 7.22 let $f_2(x) = \dfrac{1}{\sqrt{2\pi}} e^{-x^2/2}$, $-\infty < x < \infty$, and $T(x) = x$, then:

(*a*) $M_2(\tau) = e^{\tau^2/2}$.

(*b*) $f^*(x) = \dfrac{1}{\sqrt{2\pi}} e^{-(x-\tau)^2/2}$.

(*c*) $$\log \frac{\int_{a-\tau}^{\infty} e^{-y^2/2}\, dy}{\int_{a}^{\infty} e^{-y^2/2}\, dy} \geq a\tau - \frac{\tau^2}{2}, \quad \tau > 0.$$

7.25. Show that $|\mathbf{G}(\theta)| \cdot |\mathbf{\Sigma}| \geq 1$, where $\mathbf{\Sigma}$ and $\mathbf{G}(\theta)$ are defined in theorem 5.1. When does the equality hold?

7.26. Show that $|\mathbf{G}(\phi)| \cdot |\mathbf{U}'\mathbf{\Sigma}^{-1}\mathbf{U}|^{-1} \geq 1$, where $\mathbf{\Sigma}$, $\mathbf{G}(\phi)$, and $\mathbf{U}$ are defined in theorem 5.2. When does the equality hold?

7.27. Find the value of $f^*(x) = e^{\tau T(x)} f_2(x)/M_2(\tau)$, and $I(*:2)$ when $T(x) = 1 + x$, $f_2(x) = p_2 q_2^{\,x}$, $x = 0, 1, 2, \cdots$, $q_2 = 1 - p_2$, $E_1(T(x)) = \theta = 1/p_1$. (Cf. problem 5.5 in chapter 1.)

7.28. Show that for $M_2(\tau)$ defined in (2.16), m_2 defined in (2.20),

$$E_1 = \left\{x : \frac{f_1(x)}{f_2(x)} \geq \frac{p}{q}\right\},\ E_2 = \left\{x : \frac{f_1(x)}{f_2(x)} < \frac{p}{q}\right\},\ p + q = 1, p > 0:$$

(*a*) $M_2(\tau) \geq (p/q)^\tau \mu_2(E_1) + (q/p)^{1-\tau}\mu_1(E_2)$.

(*b*) $p\mu_2(E_1) + q\mu_1(E_2) \leq m_2$.

7.29. Show that $I(1:2) \geqq -2 \log \int (f_1(x) f_2(x))^{1/2} d\lambda(x)$. When does the equality hold?

7.30. Show that $\int (f_1(x) f_2(x))^{1/2} d\lambda(x) \leqq 1$. When does the equality hold?

7.31. Show that $-2 \log \int (f_1(x) f_2(x))^{1/2} d\lambda(x) \geqq 2(1 - \int (f_1(x) f_2(x))^{1/2} d\lambda(x)) = \int ((f_1(x))^{1/2} - (f_2(x))^{1/2})^2\, d\lambda(x)$. When does the equality hold?

7.32. Show that $\frac{1}{4}(\int |f_1(x) - f_2(x)| d\lambda(x))^2 \leqq \int ((f(x))^{1/2} - (f_2(x))^{1/2})^2\, d\lambda(x)$. When does the equality hold?

CHAPTER 4

Limiting Properties

1. INTRODUCTION

The fundamental properties (other than additivity) of the information measures discussed in the preceding chapters are described by inequalities. The law of large numbers and the central limit theorem make it possible to derive good approximations for large-sample results. The asymptotic behavior is often illuminating for smaller size samples also. In this chapter we shall consider some limiting properties and in the next chapter we shall study asymptotic distribution properties of estimates of the information measures. These ideas will also be applied by the reader in solving a number of problems set for him at the end of several of the succeeding chapters.

2. LIMITING PROPERTIES

The following theorem 2.1 is essentially a continuation of theorem 4.1 of chapter 2. Consider the measurable transformations $T_N(x)$ of the probability spaces $(\mathscr{X}, \mathscr{S}, \mu_i)$ onto the probability spaces $(\mathscr{Y}, \mathscr{T}, \nu_i^{(N)})$, where $T_N^{-1}(G) = \{x : T_N(x) \in G\}$, $\nu_i^{(N)}(G) = \mu_i(T_N^{-1}(G))$, for $G \in \mathscr{T}$, $i = 1, 2$; that is, $T_N(x)$ is a statistic and N may be the sample size.

THEOREM 2.1.† *If the $T_N(x)$ are such that*

$$\lim_{N\to\infty} \nu_i^{(N)}(G) = \nu_i(G), \qquad i = 1, 2, \qquad G \in \mathscr{T}, \tag{2.1}$$

where $\nu_i(G)$ is a probability measure, then

$$I(1:2; \mathscr{X}) \geqq \liminf_{N\to\infty} I(1^{(N)}:2^{(N)}; \mathscr{Y}) \geqq I(1:2; \mathscr{Y}).$$

The expression $I(1^{(N)}:2^{(N)}; \mathscr{Y})$ is the discrimination information measure corresponding to $\nu_i^{(N)}(G)$, $G \in \mathscr{T}$, $i = 1, 2$.

† see Appendix page 389

Proof. We first derive a result that is similar to a lemma used by Doob (1936). From corollary 3.2 of chapter 2, we have

$$I(1^{(N)}:2^{(N)};\mathscr{Y}) \geqq \sum_j \nu_1^{(N)}(G_j) \log \frac{\nu_1^{(N)}(G_j)}{\nu_2^{(N)}(G_j)}, \tag{2.2}$$

where the sum is taken over any set of pairwise disjoint G_j such that $\mathscr{Y} = \cup_j G_j$. Accordingly,

$$\liminf_{N\to\infty} I(1^{(N)}:2^{(N)};\mathscr{Y}) \geqq \sum_j \nu_1(G_j) \log \frac{\nu_1(G_j)}{\nu_2(G_j)}, \tag{2.3}$$

and therefore

$$\liminf_{N\to\infty} I(1^{(N)}:2^{(N)};\mathscr{Y}) \geqq I(1:2;\mathscr{Y}), \tag{2.4}$$

since the right-hand member of (2.4) is the l.u.b. (sup) of the right-hand member of (2.3) over all such partitions of $\mathscr{Y}$. Combining theorem 4.1 of chapter 2 and (2.4) completes the proof [cf. Gel'fand, Kolmogorov, and Iaglom (1956), Kullback (1954)].

As a particular case of the foregoing, take the probability measure spaces $(\mathscr{X}, \mathscr{S}, \mu_1^{(N)}, \mu_1, \mu_2)$, and assume that $\lim_{N\to\infty} \mu_1^{(N)}(E) = \mu_1(E)$ for all $E \in \mathscr{S}$. We have:

COROLLARY 2.1. $\liminf_{N\to\infty} I(1^{(N)}:2) \geqq I(1:2)$.

Proof. The proof is similar to that of theorem 2.1. We list some of the steps primarily to clarify the symbols. For any partition of $\mathscr{X}$ into pairwise disjoint E_j,

$$I(1^{(N)}:2) \geqq \sum_j \mu_1^{(N)}(E_j) \log \frac{\mu_1^{(N)}(E_j)}{\mu_2(E_j)},$$

$$\liminf_{N\to\infty} I(1^{(N)}:2) \geqq \Sigma \mu_1(E_j) \log \frac{\mu_1(E_j)}{\mu_2(E_j)},$$

$$\liminf_{N\to\infty} I(1^{(N)}:2) \geqq I(1:2).$$

Consider again the probability measure spaces of corollary 2.1 with the generalized densities

$$\mu_1^{(N)}(E) = \int_E f_1^{(N)}(x)\, d\lambda(x), \qquad \mu_i(E) = \int_E f_i(x)\, d\lambda(x), \qquad i = 1, 2, \quad E \in \mathscr{S}.$$

We have:

LEMMA 2.1. $\lim_{N\to\infty} I(1^{(N)}:1) = 0$, *if* $\lim_{N\to\infty} (f_1^{(N)}(x)/f_1(x)) = 1\ [\lambda]$, *uniformly.*

Proof. Letting $g^{(N)}(x) = f_1^{(N)}(x)/f_1(x)$, then, as in theorem 3.1 of chapter 2,

$$I(1^{(N)}:1) = \int f_1^{(N)}(x) \log \frac{f_1^{(N)}(x)}{f_1(x)}\, d\lambda(x) \tag{2.5}$$

$$= \frac{1}{2}\int (g^{(N)}(x) - 1)^2 \frac{1}{h^{(N)}(x)}\, d\mu_1(x),$$

where $h^{(N)}(x)$ lies between $g^{(N)}(x)$ and 1. For sufficiently large N, for all $x\ [\lambda]$, $|g^{(N)}(x) - 1| < \epsilon$, $\dfrac{1}{h^{(N)}(x)} < \dfrac{1}{1-\epsilon}$, $\epsilon > 0$, so that $0 \leqq I(1^{(N)}:1) < \dfrac{1}{2}\dfrac{\epsilon^2}{1-\epsilon}$ and therefore $\lim_{N\to\infty} I(1^{(N)}:1) = 0$.

LEMMA 2.2. *If* $\lim_{N\to\infty} I(1^{(N)}:1) = 0$, *then* $f_1^{(N)}(x) \to f_1(x)$ *in the mean with respect to the measure* λ, *or* $\mu_1^{(N)}(E) \to \mu_1(E)$ *uniformly in* $E \in \mathcal{S}$, *or* $f_1^{(N)}(x) \to f_1(x)$ *in probability.*†

THEOREM 2.2. *If* $\lim_{N\to\infty} \dfrac{f_1^{(N)}(x)}{f_1(x)} = 1\ [\lambda]$, *uniformly, then* $\lim_{N\to\infty} I(1^{(N)}:2) = I(1:2)$ *if* $I(1:2)$ *is finite.*

Proof.

$$I(1^{(N)}:2) = \int f_1^{(N)}(x) \log \frac{f_1^{(N)}(x)}{f_2(x)}\, d\lambda(x)$$

$$= \int f_1^{(N)}(x) \log \frac{f_1^{(N)}(x)}{f_1(x)}\, d\lambda(x) + \int f_1^{(N)}(x) \log \frac{f_1(x)}{f_2(x)}\, d\lambda(x),$$

$$I(1^{(N)}:2) - I(1:2) = \int f_1^{(N)}(x) \log \frac{f_1^{(N)}(x)}{f_1(x)}\, d\lambda(x)$$

$$+ \int (f_1^{(N)}(x) - f_1(x)) \log \frac{f_1(x)}{f_2(x)}\, d\lambda(x).$$

For sufficiently large N,

$$|I(1^{(N)}:2) - I(1:2)| \leqq I(1^{(N)}:1) + \epsilon \int f_1(x) \left| \log \frac{f_1(x)}{f_2(x)} \right| d\lambda(x),$$

and therefore $\lim_{N\to\infty} I(1^{(N)}:2) = I(1:2)$. (See problem 4.17.)

Example 2.1. As an illustration of theorem 2.1, consider N independent observations from binomial distributions with parameters p_i, $q_i = 1 - p_i$, $i = 1, 2$. As $N \to \infty$, $p_i \to 0$, $Np_i \to m_i < \infty$, the binomial distributions

† see Appendix page 390

approach as limits the Poisson distributions with parameters $m_i = Np_i$, $i = 1, 2$. We find that

(2.6)
$$I(1^{(N)}:2^{(N)}) = \sum_{y=0}^{N} \frac{N!}{y!(N-y)!} p_1{}^y q_1^{N-y} \log \frac{p_1{}^y q_1^{N-y}}{p_2{}^y q_2^{N-y}}$$
$$= N\left(p_1 \log \frac{p_1}{p_2} + q_1 \log \frac{q_1}{q_2}\right),$$

(2.7)
$$I(1:2) = \sum_{y=0}^{\infty} \frac{m_1{}^y e^{-m_1}}{y!} \log \frac{m_1{}^y e^{-m_1}}{m_2{}^y e^{-m_2}}$$
$$= (m_2 - m_1) + m_1 \log \frac{m_1}{m_2}.$$

From the inequality $x_1 \log (x_1/x_2) \geq x_1 - x_2$ (the right-hand member of (2.7) is nonnegative) and $m_i = Np_i$, $i = 1, 2$, it follows that

(2.8)
$$Np_1 \log \frac{p_1}{p_2} + Nq_1 \log \frac{q_1}{q_2} = m_1 \log \frac{m_1}{m_2} + N\left(1 - \frac{m_1}{N}\right) \log \frac{1 - m_1/N}{1 - m_2/N}$$
$$\geq m_1 \log \frac{m_1}{m_2} + N\left(\frac{m_2}{N} - \frac{m_1}{N}\right)$$
$$= m_1 \log \frac{m_1}{m_2} + (m_2 - m_1),$$

or $\liminf_{N\to\infty} I(1^{(N)}:2^{(N)}) \geq I(1:2)$. As a matter of fact, as may be seen from the first two members of (2.8), it is true here that $\lim_{N\to\infty} I(1^{(N)}:2^{(N)}) = I(1:2)$.

Example 2.2. As an illustration of corollary 2.1, take for μ_1 and μ_2 the Poisson distributions with respective parameters $m_1 = 1$ and $m_2 = 1.5$ and for $\mu_1^{(N)}$, the negative binomial distribution $(\Gamma(N + x)/x!\Gamma(N))p^x q^{-N-x}$, $q = 1 + p, p > 0, N > 0, x = 0, 1, 2, \cdots$. As $N \to \infty, p \to 0, Np \to m < \infty$, the negative binomial distribution approaches as a limit the Poisson distribution with parameter m [cf. Wilks (1943, pp. 54–55)]. In table 2.1 are listed the values of the negative binomial for $N = 2$, $p = 0.5$, $q = 1.5$, those for the Poisson distributions, and the computations for $I(1^{(N)}:2)$ and $I(1:2)$. The numerical values for the negative binomial are taken from Cochran (1954, Table 1, p. 419).

TABLE 2.1

x	$p_1^{(N)}(x)$	$p_1(x)$	$p_2(x)$	$p_1^{(N)} \log (p_1^{(N)}/p_2)$	$p_1 \log (p_1/p_2)$
0	0.4444	0.3679	0.2231	0.30624	0.18402
1	0.2963	0.3679	0.3347	−0.03611	0.03479
2	0.1482	0.1839	0.2510	−0.07813	−0.05720
3	0.0658	0.0613	0.1255	−0.04249	−0.04392
4+	0.0453	0.0190	0.0657	−0.01678	−0.02357
	1.0000	1.0000	1.0000	0.13273	0.09412

All the values $x \geq 4$ were grouped in computing table 2.1. Note that $I(1^{(N)}:2) = 0.13273 > 0.09412 = I(1:2)$, and that 0.09412 is smaller than the value obtained from

$$(m_2 - m_1) + m_1 \log (m_1/m_2) = 1.5 - 1 + 1 \log (1/1.5) = 0.09453,$$

illustrating the statement in sections 3 and 4 of chapter 2 that grouping loses information. (See problem 4.3.)

3. TYPE I AND TYPE II ERRORS

Suppose that the space $\mathscr{X}$ is partitioned into the disjoint sets E_1 and E_2, that is, $E_1 \cap E_2 = 0$, $\mathscr{X} = E_1 \cup E_2$, with $\mathscr{X}$ the sample space of n independent observations. Assume a test procedure such that if the sample point $x \in E_1$ we accept the hypothesis H_1 (reject H_2), and if the sample point $x \in E_2$ we accept the hypothesis H_2 (reject H_1). We treat H_2 as the null hypothesis. E_1 is called the critical region. The probability of incorrectly accepting H_1, the type I error, is $\alpha = \text{Prob}\,(x \in E_1 | H_2) = \mu_2(E_1)$, and the probability of incorrectly accepting H_2, the type II error, is $\beta = \text{Prob}\,(x \in E_2 | H_1) = \mu_1(E_2)$. [Cf. Hoel (1954, pp. 30–35).]

We now state:

THEOREM 3.1.

$$(a)\quad I(1:2; O_n) = nI(1:2; O_1) \geq \beta \log \frac{\beta}{1-\alpha} + (1-\beta) \log \frac{1-\beta}{\alpha},$$

$$(b)\quad I(2:1; O_n) = nI(2:1; O_1) \geq \alpha \log \frac{\alpha}{1-\beta} + (1-\alpha) \log \frac{1-\alpha}{\beta},$$

where O_n indicates a sample of n independent observations and O_1 a single observation.

Proof. A consequence of the additivity property (theorem 2.1 of chapter 2), corollary 3.2 of chapter 2, and $1 - \alpha = \mu_2(E_2)$, $1 - \beta = \mu_1(E_1)$. (Cf. example 2.3 of chapter 3.)

Note that the right-hand sides of the inequalities in theorem 3.1 are the values of $I(1:2)$ and $I(2:1)$ for binomial distributions with $p_1 = \beta$, $q_1 = 1 - \beta$, $p_2 = 1 - \alpha$, $q_2 = \alpha$ [see (2.6), for example, with $N = 1$]. These values also appear in Wald's theorem on the efficiency of sequential tests [Wald (1947, pp. 196–199)]. We remark that (see problem 8.35 in chapter 2) $F(p_1, p_2) = p_1 \log (p_1/p_2) + q_1 \log (q_1/q_2)$ is a convex function of p_2 for fixed p_1, $F(p_1, p_2) = 0$ for $p_2 = p_1$, and $F(p_1, p_2)$ is monotonically decreasing for $0 \leqq p_2 \leqq p_1$ and monotonically increasing for $p_1 \leqq p_2 \leqq 1$.

Table 3.1 lists illustrative values of $F(p_1, p_2)$ for $p_1 = 0.05$. (For a more extensive table see Table II on pages 378–379.)

TABLE 3.1. $F(p_1, p_2)$, $p_1 = 0.05$

p_2		p_2		p_2		p_2	
0.01	0.04129	0.20	0.09394	0.55	0.58996	0.90	1.99422
0.02	0.01628	0.25	0.14410	0.60	0.69751	0.95	2.65000
0.03	0.00575	0.30	0.20052	0.65	0.82036	0.96	2.86147
0.04	0.00121	0.35	0.26322	0.70	0.96309	0.97	3.13424
0.05	0.00000	0.40	0.33259	0.75	1.13285	0.98	3.51892
0.10	0.01671	0.45	0.40936	0.80	1.34161	0.99	4.17690
0.15	0.05074	0.50	0.49464	0.85	1.61188		

For a fixed value of α, say α_0, $0 < \alpha_0 < 1$, a lower bound to the minimum possible β, say β_n^*, is obtained from

$$(3.1) \qquad I(2:1; O_1) \geqq \frac{1}{n}\left(\alpha_0 \log \frac{\alpha_0}{1 - \beta_n^*} + (1 - \alpha_0) \log \frac{1 - \alpha_0}{\beta_n^*}\right),$$

by using theorem 3.1(*b*). Similarly, for a fixed value of β, say β_0, $0 < \beta_0 < 1$, a lower bound to the minimum possible α, say α_n^*, is obtained from

$$(3.2) \qquad I(1:2; O_1) \geqq \frac{1}{n}\left(\beta_0 \log \frac{\beta_0}{1 - \alpha_n^*} + (1 - \beta_0) \log \frac{1 - \beta_0}{\alpha_n^*}\right).$$

Thus, for example, if $nI(1:2; O_1) = 4.17690$ and $\beta_0 = 0.05$, we see from table 3.1 that $\alpha_n^* \geqq 0.01$.

To examine the behavior of (3.1) and (3.2) for $n \to \infty$ we shall make use of the weak law of large numbers or Khintchine's theorem [see, for example, Cramér (1946a, p. 253), Feller (1950, p. 191)]. If $I(1:2; O_1)$ is finite, and we have a sample of n independent observations from the population under H_1, then

$$\frac{1}{n}\left(\log \frac{f_1(x_1)}{f_2(x_1)} + \cdot \cdot \cdot + \log \frac{f_1(x_n)}{f_2(x_n)}\right),$$

converges in probability to $I(1:2; O_1)$, that is, for any $\epsilon > 0$, $\delta > 0$, and $\beta > 0$, for sufficiently large n

$$(3.3) \qquad \text{Prob}\left\{\frac{1}{n} \log \frac{f_1(x_1) \cdot \cdot \cdot f_1(x_n)}{f_2(x_1) \cdot \cdot \cdot f_2(x_n)} < I(1:2; O_1) - \epsilon \middle| H_1\right\} < \beta,$$

$$\text{Prob}\left\{\frac{1}{n} \log \frac{f_1(x_1) \cdot \cdot \cdot f_1(x_n)}{f_2(x_1) \cdot \cdot \cdot f_2(x_n)} > I(1:2; O_1) + \epsilon \middle| H_1\right\} < \delta,$$

or

$$\text{Prob}\left\{e^{n(I(1:2;\,O_1)-\epsilon)} \leqq \frac{f_1(x_1)\cdots f_1(x_n)}{f_2(x_1)\cdots f_2(x_n)}\,\Big|\,H_1\right\} \geqq 1-\beta. \tag{3.4}$$

We may therefore classify the samples under H_1 into two disjoint groups, E_1 and E_2, such that the samples of E_1 satisfy the inequality

$$f_1(x_1)\cdots f_1(x_n) \geqq e^{n(I(1:2;\,O_1)-\epsilon)} f_2(x_1)\cdots f_2(x_n), \tag{3.5}$$

and the samples of E_2 occur with a probability (under H_1) less than β for sufficiently large n. Integrating (3.5) over E_1, we get

$$1 \geqq \text{Prob}\,(E_1|H_1) \geqq e^{n(I(1:2;\,O_1)-\epsilon)}\text{Prob}\,(E_1|H_2), \tag{3.6}$$

or, for any value of β, say β_0, $0 < \beta_0 < 1$,

$$\lim_{n\to\infty} \frac{1}{n}\log\frac{1}{\alpha_n{}^*} \geqq I(1:2;\,O_1). \tag{3.7}$$

Combining (3.7) with the value that may be derived from (3.2), we have

$$\lim_{n\to\infty} \frac{1}{n}\log\frac{1}{\alpha_n{}^*} \geqq I(1:2;\,O_1) \geqq \lim_{n\to\infty}\frac{1}{n}\left((1-\beta_0)\log\frac{1-\beta_0}{\alpha_n{}^*} + \beta_0\log\frac{\beta_0}{1-\alpha_n{}^*}\right). \tag{3.8}$$

We now state:

THEOREM 3.2. *For any value of β, say β_0, $0 < \beta_0 < 1$,*

$$\lim_{n\to\infty} (\alpha_n{}^*)^{1/n} = e^{-I(1:2;\,O_1)}, \quad or \quad \lim_{n\to\infty}\left(\frac{1}{n}\log\frac{1}{\alpha_n{}^*}\right) = I(1:2;\,O_1).$$

Proof. Let E_3 denote the samples satisfying

$$e^{n(I(1:2;\,O_1)-\epsilon)} \leqq \frac{f_1(x_1)\cdots f_1(x_n)}{f_2(x_1)\cdots f_2(x_n)} \leqq e^{n(I(1:2;\,O_1)+\epsilon)}. \tag{3.9}$$

We see from (3.3) that Prob $(E_3|H_1) \geqq 1-\beta-\delta$. Integrating the right-hand inequality in (3.9) over E_3, we find that

$$\text{Prob}\,(E_3|H_1) \leqq e^{n(I(1:2;\,O_1)+\epsilon)}\,\text{Prob}\,(E_3|H_2). \tag{3.10}$$

Since $E_3 \subset E_1$, where E_1 is defined by (3.5), Prob $(E_3|H_2) \leqq$ Prob $(E_1|H_2)$, and (3.10) yields

$$1-\beta-\delta \leqq e^{n(I(1:2;\,O_1)+\epsilon)}\alpha_n{}^*. \tag{3.11}$$

Combining (3.6) and (3.11), we now have [cf. Joshi (1957)],

$$(1-\beta-\delta)e^{-n(I(1:2;\,O_1)+\epsilon)} \leqq \alpha_n{}^* \leqq e^{-n(I(1:2;\,O_1)-\epsilon)}. \tag{3.12}$$

The desired result follows from (3.8) and (3.12).

Similarly, we may derive:

THEOREM 3.3. *For any value of* α, *say* α_0, $0 < \alpha_0 < 1$,

$$\lim_{n\to\infty} (\beta_n^*)^{1/n} = e^{-I(2:1;\,O_1)}, \quad or \quad \lim_{n\to\infty}\left(\frac{1}{n}\log\frac{1}{\beta_n^*}\right) = I(2:1;\,O_1).$$

Chernoff (1956) derived theorems 3.2 and 3.3 by using an extension of the central limit theorem given by Cramér (1938). Chernoff attributes the results to unpublished work of C. Stein. [Cf. Sanov (1957, p. 40).]

Note that from theorems 3.2 and 3.3, at least for large samples, the ratios

$$I(1:2;\,X)/I(1:2;\,Y) \qquad \text{and} \qquad I(2:1;\,X)/I(2:1;\,Y)$$

may be used as measures of the relative efficiencies of competitive variables X and Y in the sense that

$$\frac{I(1:2;\,X)}{I(1:2;\,Y)} = \frac{n_y}{n_x}, \qquad \frac{I(2:1;\,X)}{I(2:1;\,Y)} = \frac{N_y}{N_x},$$

where n_y, n_x, and N_y, N_x are respectively the sample sizes needed to attain for given β_0 the same α_n^*, and for given α_0 the same β_n^* [cf. Chernoff (1956)].

Discussions that express the type I and type II errors asymptotically in terms of $J(1, 2)$, were given by Mourier (1946, 1951), and Sakaguchi (1955). Mourier and Sakaguchi show that if the region E_1^* is defined by

$$\frac{1}{n}\left(\log\frac{f_1(x_1)}{f_2(x_1)} + \cdots + \log\frac{f_1(x_n)}{f_2(x_n)}\right) > \frac{\sigma_2 I(1:2;\,O_1) - \sigma_1 I(2:1;\,O_1)}{\sigma_1 + \sigma_2},$$

$$\text{where } \sigma_1^2 = \int\left(\log\frac{f_1(x)}{f_2(x)}\right)^2 d\mu_1(x) - (I(1:2;\,O_1))^2,$$

$$\sigma_2^2 = \int\left(\log\frac{f_1(x)}{f_2(x)}\right)^2 d\mu_2(x) - (I(2:1;\,O_1))^2,$$

so that $\alpha_n^* = \text{Prob}\,(O_n \in E_1^*|H_2)$, $1 - \beta_n^* = \text{Prob}\,(O_n \in E_1^*|H_1)$, then

$$\lim_{n\to\infty}\frac{\min\max\,(\alpha_n, \beta_n)}{\max\,(\alpha_n^*, \beta_n^*)} = 1,$$

$$\lim_{n\to\infty}\frac{\max\,(\alpha_n^*, \beta_n^*)}{\phi(\sqrt{n}J(1, 2;\,O_1)/(\sigma_1 + \sigma_2))} = 1,$$

and

$$\lim_{n\to\infty} \frac{\min \max (\alpha_n, \beta_n)}{\phi(\sqrt{n}J(1, 2; O_1)/(\sigma_1 + \sigma_2))} = 1,$$

with $\phi(x) = \int_x^\infty \frac{e^{-t^2/2}\, dt}{\sqrt{2\pi}}$ and α_n, β_n the errors for any other region E_1.

4. PROBLEMS

4.1. Consider the probability measure spaces $(\mathscr{X}, \mathscr{S}, \mu_2^{(N)}, \mu_1, \mu_2)$ and assume that $\lim_{N\to\infty} \mu_2^{(N)}(E) = \mu_2(E)$ for all $E \in \mathscr{S}$. Prove that $\liminf_{N\to\infty} I(1:2^{(N)}) \geq I(1:2)$.

4.2. Show that for the negative binomial distributions $(\Gamma(N + x)/x!\Gamma(N)) p_i^x q_i^{-N-x}$, $q_i = 1 + p_i, p_i > 0$, $N > 0$, $x = 0, 1, 2, \cdots$, $i = 1, 2$, $I(1:2) = Np_1 \log (p_1/p_2) - Nq_1 \log (q_1/q_2)$.

4.3. As $N \to \infty$, $p_i \to 0$, $Np_i \to m_i < \infty$, the negative binomial distributions in problem 4.2 approach the Poisson distributions with parameters m_i, $i = 1, 2$, as a limit. Show

(*a*) That theorem 2.1 is satisfied.
(*b*) That $\lim_{N\to\infty} I(1^{(N)}:2^{(N)}) = I(1:2)$.
(*c*) That corollary 2.1 is satisfied.

4.4. Show that the distributions in example 2.1 satisfy lemma 2.1.

4.5. Show that the distributions in problem 4.3 satisfy lemma 2.1.

4.6. Show that the distributions in example 2.1 satisfy theorem 2.2.

4.7. Show that the distributions in problem 4.3 satisfy theorem 2.2.

4.8. Compute the results for table 2.1 grouping

(*a*) All values $x \geq 3$.
(*b*) All values $x \geq 2$.
(*c*) All values $x \geq 1$.

4.9. (*a*) Show that for a sample of n independent observations from the normal populations $N(\mu_i, \sigma^2)$, $i = 1, 2$, $J(1, 2; O_n) = n(\mu_1 - \mu_2)^2/\sigma^2$.

(*b*) Consider the quantizing transformation (or grouping) of the normal variables in (*a*) above, $y = 1$ for $x < g$ and $y = 0$ for $x \geq g$, so that y is a binomial variable with $p_i, q_i = 1 - p_i$, $p_i = \int_{-\infty}^{g} \frac{e^{-(x-\mu_i)^2/2\sigma^2}}{\sigma\sqrt{2\pi}}\, dx$, $i = 1, 2$. Show that $J(1, 2; \mathscr{Y}) = n((p_1 - p_2) \log (p_1/p_2) + (q_1 - q_2) \log (q_1/q_2))$.

(*c*) Show that $J(1, 2; \mathscr{Y})$ is a maximum when $g = (\mu_1 + \mu_2)/2$, and that

$$\max J(1, 2; \mathscr{Y}) = 2n(\hat{p}_1 - \hat{q}_1) \log \frac{\hat{p}_1}{\hat{q}_1}, \text{ with } \hat{p}_1 = \int_{-\infty}^{\frac{\mu_2-\mu_1}{2\sigma}} \frac{e^{-t^2/2}}{\sqrt{2\pi}}\, dt.$$

(*d*) Show that $\max J(1, 2; \mathscr{Y})/J(1, 2; O_n) \to 2/\pi$ as $(\mu_2 - \mu_1)/\sigma \to 0$.

(*e*) Show that $\max J(1, 2; \mathscr{Y})/J(1, 2; O_n) \to \frac{1}{4}$ as $(\mu_2 - \mu_1)/\sigma \to \infty$. [See Questions and Answers, *Am. Statistician*, Vol. 7 (1953, pp. 14–15).]

4.10. If in theorem 2.1 $T(x)$ is a sufficient statistic, with $\nu_i(G) = \mu_i(T^{-1}(G))$, for $G \in \mathscr{T}$, $i = 1, 2$, and $T^{-1}(G) = \{x: T(x) \in G\}$, then $\liminf_{N\to\infty} I(1^{(N)}:2^{(N)}; \mathscr{Y}) = I(1:2; \mathscr{Y})$.

4.11. In the notation of theorem 3.1, show that

$$J(1, 2; O_n) = nJ(1, 2; O_1) \geq (1 - \alpha - \beta) \log \frac{(1-\alpha)(1-\beta)}{\alpha\beta}$$
$$\geq 2\left[\frac{\alpha+\beta}{2} \log \frac{(\alpha+\beta)/2}{1-(\alpha+\beta)/2} + \left(1 - \frac{\alpha+\beta}{2}\right) \log \frac{1-(\alpha+\beta)/2}{(\alpha+\beta)/2}\right].$$

4.12. If $I(2:1; O_1)$ is finite, show that for any value of α, say α_0, $0 < \alpha_0 < 1$,

$$\lim_{n\to\infty} \frac{1}{n} \log \frac{1}{\beta_n{}^*} \geq I(2:1; O_1) \geq \lim_{n\to\infty} \frac{1}{n}\left((1-\alpha_0)\log\frac{1-\alpha_0}{\beta_n{}^*} + \alpha_0 \log \frac{\alpha_0}{1-\beta_n{}^*}\right).$$

4.13. Prove theorem 3.3.

4.14. Show that $n(I(1:2; O_1) - \epsilon) \leq I(1:2; O_n, E_3) \leq n(I(1:2; O_1) + \epsilon)$, with the region E_3 defined in (3.9) and $I(1:2; O_n, E_3)$ defined in accordance with (2.4) of chapter 1. [Cf. Joshi (1957).]

4.15. In the notation of theorem 3.2, show that $\text{Prob}\,(E_3|H_2) \leq e^{-n(I(1:2;O_1)-\epsilon)}$, and thus that $\lim_{n\to\infty} \text{Prob}\,(E_3|H_2) = 0$ if $I(1:2; O_1) > \epsilon$. [Cf. Joshi (1957), Savage (1954, pp. 46–50).]

4.16. Show that $\displaystyle\lim_{\substack{n\to\infty \\ \frac{r}{n}\to p}} \frac{1}{n} \log \frac{2^n}{\sum_{x=0}^{r}\binom{n}{x}} = p \log \frac{p}{\frac{1}{2}} + q \log \frac{q}{\frac{1}{2}}, p < \frac{r}{n} < \frac{1}{2}, q = 1 - p.$ (Cf. problem 7.23 in chapter 3.)

4.17. If $\displaystyle\lim_{N\to\infty} \frac{f_2^{(N)}(x)}{f_2(x)} = 1\ [\lambda]$ uniformly, then $\lim_{N\to\infty} I(1:2^{(N)}) = I(1:2)$ if $I(1:2)$ is finite.

4.18. Let $\mathscr{X} = E_1 \cup E_2 = E_1{}^* \cup E_2{}^*$, with $\mathscr{X}$ the sample space in section 3, $E_1 \cap E_2 = 0 = E_1{}^* \cap E_2{}^*$, $\alpha = \mu_2(E_1) = \mu_2(E_1{}^*)$, $\beta = \mu_1(E_2)$, and $\beta^* = \mu_1(E_2{}^*)$. Show that for $\beta^* < \beta < 1 - \alpha$:

(*a*) $(1-\alpha)\log\dfrac{1-\alpha}{\beta^*} > (1-\alpha)\log\dfrac{1-\alpha}{\beta}$.

(*b*) $\alpha \log \dfrac{\alpha}{1-\beta^*} < \alpha \log \dfrac{\alpha}{1-\beta}$.

(*c*) $\alpha \log \dfrac{\alpha}{1-\beta^*} + (1-\alpha)\log\dfrac{1-\alpha}{\beta^*} > \alpha \log \dfrac{\alpha}{1-\beta} + (1-\alpha)\log\dfrac{1-\alpha}{\beta}$.

4.19. In the notation of problem 4.18, show that for $1 - \alpha < \beta^* < \beta$:

(*a*) $(1 - \alpha) \log \frac{1 - \alpha}{\beta^*} > (1 - \alpha) \log \frac{1 - \alpha}{\beta}$.

(*b*) $\alpha \log \frac{\alpha}{1 - \beta^*} < \alpha \log \frac{\alpha}{1 - \beta}$.

(*c*) $\alpha \log \frac{\alpha}{1 - \beta^*} + (1 - \alpha) \log \frac{1 - \alpha}{\beta^*} < \alpha \log \frac{\alpha}{1 - \beta} + (1 - \alpha) \log \frac{1 - \alpha}{\beta}$.

4.20. In the notation of problem 4.18, show that if

$$\alpha \log \frac{\alpha}{1 - \beta^*} + (1 - \alpha) \log \frac{1 - \alpha}{\beta^*} > \alpha \log \frac{\alpha}{1 - \beta} + (1 - \alpha) \log \frac{1 - \alpha}{\beta},$$

then $\beta^* < \beta < 1 - \alpha$, or $1 - \alpha < \beta < \beta^*$.

4.21. Suppose $p_{10} + p_{20} + \cdots + p_{c0} = 1$, $p_{i0} > 0$, $p_{ij} = a_{i1}p_{1,j-1} + a_{i2}p_{2,j-1} + \cdots + a_{ic}p_{c,j-1}$, $a_{i1} + a_{i2} + \cdots + a_{ic} = 1$, $a_{1k} + a_{2k} + \cdots + a_{ck} = 1$, $a_{jk} \geq 0$, $i, k = 1, 2, \cdots, c; j = 1, 2, \cdots$, show that $\lim_{N \to \infty} \sum_{i=1}^{c} p_{iN} \log \frac{p_{iN}}{1/c} = 0$. (Cf. problem 8.32 in chapter 2.)

4.22. If the sample space in problem 7.28 of chapter 3 is that of n independent observations, and we write $\alpha = \mu_2(E_1)$, and $\beta = \mu_1(E_2)$, where the regions E_1 and E_2 are defined in problem 7.28 of chapter 3, then $\lim_{n \to \infty} (p\alpha + q\beta) = 0$. [Cf. Joshi (1957).]

CHAPTER 5

Information Statistics

1. ESTIMATE OF $I(*:2)$

We have thus far studied the information measures as parameters or functionals of the populations. We shall now examine estimators of these measures, information statistics, and investigate the general asymptotic distribution theory of these estimators (statistics). We shall obtain exact distributions, or better approximations than given by the general theory, in particular applications in the chapters following.

In chapter 3 we introduced the minimum discrimination information $I(*:2)$ as the minimum value of

$$I(1:2) = \int f_1(x) \log \frac{f_1(x)}{f_2(x)}\, d\lambda(x),$$

for a given $f_2(x)$, and all $f_1(x)$ such that

$$\theta = \int T(x) f_1(x)\, d\lambda(x).$$

The minimum value $I(*:2) = \theta\tau(\theta) - \log M_2(\tau(\theta))$ [see the remark following (2.9) in chapter 3] occurs for the conjugate distribution [to use a term introduced by Khinchin (1949, p. 79)] with generalized density given by [cf. Cramér (1938)]

$$f^*(x) = \frac{e^{\tau T(x)} f_2(x)}{M_2(\tau)}, \qquad M_2(\tau) = \int e^{\tau T(x)} f_2(x)\, d\lambda(x), \qquad \theta = \frac{d}{d\tau} \log M_2(\tau).$$

When $f_2(x)$ is the generalized density of n independent observations, we shall estimate $I(*:2)$ by using the observed value of $T(x)$ in a sample O_n as an estimate of θ, $\hat{\theta}(x)$, and a related estimate of τ, $\hat{\tau}(x) = \tau(\hat{\theta}(x))$, such that

(1.1) $$T(x) = \hat{\theta}(x) = \left[\frac{d}{d\tau} \log M_2(\tau)\right]_{\tau = \hat{\tau}(x) = \tau(\hat{\theta}(x))}.$$

Note that (1.1) is $T(x) = [E(T(x))]_{\tau=\hat{\tau}}$. [Cf. Barton (1956), Kupperman (1958, p. 573).] If there are several different functions of x that are unbiased estimators of θ, we shall use as $\hat{\theta}(x)$ the one yielding the largest value of $I(*:2)$. The estimate of $I(*:2)$ is then

$$(1.2)\quad \hat{I}(*:2;\, O_n) = \hat{\theta}(x)\hat{\tau}(x) - \log M_2(\hat{\tau}(x)) = \hat{\theta}\tau(\hat{\theta}) - \log M_2(\tau(\hat{\theta})).$$

$\hat{I}(*:2;\, O_n)$ in (1.2) is the minimum discrimination information between a population with generalized density of the form $f^*(x)$ above, with the value of the parameter θ the same as the value $\hat{\theta}$ of the sample, and the population with generalized density $f_2(x)$. Since $\hat{I}(*:2;\, O_n) \geqq 0$, with equality if and only if $\hat{\tau} = 0$, that is, when $\hat{\theta}$ is equal to the value of the parameter in the population with generalized density $f_2(x)$, $\hat{I}(*:2;\, O_n)$ is a measure of the *directed divergence* (cf. section 3 of chapter 1) between the sample and $f_2(x)$. The larger the value of $\hat{I}(*:2;\, O_n)$, the worse is the "resemblance" between the sample and the population with generalized density $f_2(x)$. Samples yielding the same value of $\hat{I}(*:2;\, O_n)$ are therefore *equivalent* insofar as directed divergence is concerned. Note that equivalent samples do not necessarily imply the same value of $\hat{\theta}$. [Cf. Bulmer (1957).] Before continuing the argument we shall illustrate the foregoing by another look at some of the examples in chapter 3.

Example 1.1. In example 4.1 of chapter 3, $\theta = np_1$, so that

$$T(x) = y = \hat{\theta} = n\hat{p}_1, \qquad \hat{\tau}(x) = \tau(\hat{p}_1) = \log\frac{yq_2}{(n-y)p_2},$$

$$\hat{I}(*:2;\, O_n) = y\log\frac{y}{np_2} + (n-y)\log\frac{n-y}{nq_2} = n\left(\hat{p}_1\log\frac{\hat{p}_1}{p_2} + \hat{q}_1\log\frac{\hat{q}_1}{q_2}\right).$$

We see from the values of $F(p_1, p_2)$ in Table II, pages 378–379, that only when $p_2 = 0.5$ do equivalent samples have values $\hat{p}_1$ such that $|p_2 - \hat{p}_1| =$ constant.

Example 1.2. In example 4.2 of chapter 3, $\theta = \theta_1$, $T(x) = \bar{x} = \hat{\theta}$, $\hat{\tau}(x) = \tau(\hat{\theta}) = n(\bar{x} - \theta_2)/\sigma_2^2$, and $\hat{I}(*:2;\, O_n) = n(\bar{x} - \theta_2)^2/2\sigma_2^2$. Note that equivalent values of $\bar{x}$ are situated symmetrically about θ_2.

Example 1.3. In example 4.3 of chapter 3, $\theta = (\theta_1, \sigma_1^2)$, so that

$$T(x) = (\bar{x}, s^2) = (\hat{\theta}_1, \hat{\sigma}_1^2),$$

$$\tau_1(\hat{\theta}_1) = \frac{n(\bar{x} - \theta_2)}{\sigma_2^2}, \qquad \tau_2(\hat{\sigma}_1^2) = \frac{n-1}{2}\left(\frac{1}{\sigma_2^2} - \frac{1}{s^2}\right),$$

$$\hat{I}(*:2;\, O_n) = \frac{n(\bar{x} - \theta_2)^2}{2\sigma_2^2} + \frac{n-1}{2}\left(\log\frac{\sigma_2^2}{s^2} - 1 + \frac{s^2}{\sigma_2^2}\right).$$

Note that here equivalent samples are those for which the values of $\bar{x}$ and s^2 lie on the curve in the $(\bar{x}, s^2)$-plane for which $\hat{I}(*:2;\, O_n) =$ constant.

Example 1.4. In examples 4.4 and 4.5 of chapter 3 we saw that $y = (1/n)\sum_{i=1}^{n} x_i^2$ provided an unbiased estimator of $\theta = \sigma_1^2$ with a larger value of $I(*:2)$ than the unbiased estimator s^2, where $(n-1)s^2 = \sum_{i=1}^{n}(x_i - \bar{x})^2$, when the hypotheses specified the normal distributions $N(0, \sigma_i^2)$. From example 4.5 we see that

$$\tau(\hat{\theta}) = \frac{n}{2}\left(\frac{1}{\sigma_2^2} - \frac{1}{y}\right) \quad \text{and} \quad \hat{I}(*:2; O_n) = \frac{n}{2}\left(\log\frac{\sigma_2^2}{y} - 1 + \frac{y}{\sigma_2^2}\right).$$

Note that equivalent values of y are not situated symmetrically about σ_2^2.

Example 1.5. In example 4.6 of chapter 3, for the transformed variates u and v defined in (4.26) of that example, $\theta = (2(1-\rho_1), 2(1+\rho_1))$, so that

$$T(x) = (y_1, y_2) = (2(1-\hat{\rho}_1), 2(1+\hat{\rho}_1)),$$

$$\tau_1(2(1-\hat{\rho}_1)) = \frac{n}{4}\left(\frac{1}{1-\rho_2} - \frac{2}{y_1}\right), \qquad \tau_2(2(1+\hat{\rho}_1)) = \frac{n}{4}\left(\frac{1}{1+\rho_2} - \frac{2}{y_2}\right),$$

$$\hat{I}(*:2; O_n) = \frac{n}{2}\left(\log\frac{2(1-\rho_2)}{y_1} - 1 + \frac{y_1}{2(1-\rho_2)} + \log\frac{2(1+\rho_2)}{y_2} - 1 + \frac{y_2}{2(1+\rho_2)}\right).$$

Note that equivalent samples are those for which y_1 and y_2 lie on the curve in the (y_1, y_2)-plane for which $\hat{I}(*:2; O_n) = $ constant.

Example 1.6. In example 4.7 of chapter 3, $\theta = \theta_1 \geq \theta_2$ and $\hat{\theta}_1 = L - 1/n$, where $L = \min(x_1, x_2, \cdots, x_n)$, $\tau(\hat{\theta}_1) = n(L - 1/n - \theta_2)/(L - \theta_2)$, and

$$\hat{I}(*:2; O_n) = n(L - 1/n - \theta_2) - \log(1 + n(L - 1/n - \theta_2))$$
$$= n(L - \theta_2) - 1 - \log n(L - \theta_2).$$

Note that $\hat{I}(*:2; O_n)$ is not defined for $L < \theta_2$. For any value $n(L - \theta_2) > 1$ there is an equivalent value L' such that $n(L' - \theta_2) < 1$; also $\hat{I}(*:2; O_n) = 0$ if and only if $n(L - \theta_2) = 1$.

2. CLASSIFICATION

We shall introduce the problem of classifying or assigning a sample to one of several possible populations with a result essentially due to Kupperman (1957, 1958), relating a priori and a posteriori probabilities of hypotheses with information statistics. Suppose that a sample O_n can occur only if one of the set of r exhaustive and mutually exclusive events $H_1, H_2, \cdots, H_r$ occurs. The a priori probabilities of these latter events (which we may call hypotheses) are denoted by $P(H_1), P(H_2), \cdots, P(H_r)$ respectively, where $P(H_m) > 0$ and $\sum_{m=1}^{r} P(H_m) = 1$. The conditional probabilities for O_n to occur are denoted by $P(O_n|H_m)$, $m = 1, 2, \cdots, r$.

The a posteriori probability of H_m, given that O_n has occurred, is denoted by $P(H_m|O_n)$. From Bayes' theorem (cf. section 2 of chapter 1), we have that

$$(2.1)\quad P(H_m|O_n) = P(H_m)P(O_n|H_m)/\sum_{j=1}^{r} P(H_j)P(O_n|H_j), \qquad m = 1, 2, \cdots, r.$$

Suppose now that the conditional probabilities for O_n to occur are the probability measures of an exponential family (see section 4 of chapter 3) with respective generalized densities for a given H_i

$$(2.2)\quad f_i(x) = e^{\tau_i T(x)}f(x)/M(\tau_i), \qquad M(\tau_i) = \int e^{\tau_i T(x)}f(x)\, d\lambda(x), \qquad i = 1, 2, \cdots, m.$$

For any pair of the generalized densities (2.2), say $f_1(x)$ and $f_2(x)$, we have by corollary 3.2 of chapter 3,

$$(2.3)\quad I(1:2; O_n) = I(\tau_1:\tau_2; O_n) = \theta_1\tau_1 - \theta_1\tau_2 - \log M(\tau_1) + \log M(\tau_2),$$

where $\theta_1 = E_1(T(x)) = \int T(x)f_1(x)\, d\lambda(x)$. The estimate defined in (1.2) therefore is

$$(2.4)\quad \hat{I}(*:2; O_n) = \hat{\tau}T(x) - \log M(\hat{\tau}) - \tau_2 T(x) + \log M(\tau_2),$$

where $T(x) = (d/d\tau)\log M(\tau)|_{\tau=\hat{\tau}}$. Similarly, the directed divergence between the sample and the population defined by $f_m(x)$, $m = 1, 2, \cdots, r$, is

$$(2.5)\quad \hat{I}(*:m; O_n) = \hat{\tau}T(x) - \log M(\hat{\tau}) - \tau_m T(x) + \log M(\tau_m), \qquad m = 1, 2, \cdots, r,$$

where $T(x) = (d/d\tau)\log M(\tau)|_{\tau=\hat{\tau}}$. The difference between any pair of the estimates in (2.5) accordingly is, using (2.2),

$$\begin{aligned}(2.6)\quad \hat{I}(*:i; O_n) - \hat{I}(*:j; O_n) &= \tau_j T(x) - \tau_i T(x) - \log M(\tau_j) + \log M(\tau_i)\\ &= \log (f_j(x)/f_i(x)), \qquad i \neq j, \quad i, j = 1, 2, \cdots, m.\end{aligned}$$

But from (2.1) [cf. (2.3) in chapter 1],

$$(2.7)\quad \log \frac{f_j(x)}{f_i(x)} = \log \frac{P(H_j|O_n)}{P(H_i|O_n)} - \log \frac{P(H_j)}{P(H_i)},$$

or, using (2.6),

$$(2.8)\quad \hat{I}(*:i; O_n) - \hat{I}(*:j; O_n) = \log \frac{P(H_j|O_n)}{P(H_i|O_n)} - \log \frac{P(H_j)}{P(H_i)}.$$

If we assign the sample to the population which it best resembles, that is, for which $\hat{I}(*:j;\, O_n)$ is smallest, then we see from (2.8) that

$$(2.9)\qquad \hat{I}(*:i;\, O_n) - \hat{I}(*:j;\, O_n) = \log \frac{P(H_j|O_n)}{P(H_i|O_n)} - \log \frac{P(H_j)}{P(H_i)} \geqq 0,$$

or

$$(2.10)\qquad \log \frac{P(H_j|O_n)}{P(H_j)} \geqq \log \frac{P(H_i|O_n)}{P(H_i)}, \qquad i \neq i, \qquad i = 1, 2, \cdots, r.$$

The procedure thus selects the exponential population for which the ratio of the a posteriori probability of H_j to the a priori probability of H_j is greatest. (See problem 7.11.) We remark that the conclusion is true for multivariate exponential populations with parameters in an h-dimensional Euclidean parameter space. This is the same as a maximum-likelihood procedure. [Cf. Good (1950, pp. 62–64, 68–73, 82–83), Savage (1954, pp. 46–50, 134–135, 234–235).] (See section 4.)

Note that the left-hand side of (2.10) is the information in O_n in favor of H_j (see example 4.1 in chapter 1).

In many problems of interest to the statistician, the generalized density $f_2(x)$ implicit in the definition of $\hat{I}(*:2;\, O_n)$ in (1.2), ranges over a family of populations we denote by the symbol H. Let $\hat{I}(*:H)$ represent the minimum of $\hat{I}(*:2;\, O_n)$ as $f_2(x)$ ranges over the populations of H, that is, $\hat{I}(*:H) = \min_{f_2 \epsilon H} \hat{I}(*:2;\, O_n)$. The value of $\hat{I}(*:H)$ is thus a measure of the *directed divergence* between the sample and that member of the family of populations H that the sample most closely resembles. If the value of $\hat{\theta}$ in the sample is the same as the value of the parameter θ for one of the members of the family of populations H, then of course $\hat{I}(*:H) = 0$, that is, the sample yields no information for discrimination against H.

When there are two or more groups of populations, for convenience denoted by H_1, H_2, H_3, $\cdots$, we shall assign the sample to the group with the smallest value among $\hat{I}(*:H_1)$, $\hat{I}(*:H_2)$, $\hat{I}(*:H_3)$, $\cdots$. This means that we shall assign the sample to that group of populations among which there is one that the sample best resembles, or against which the sample provides least information for discrimination. (See the remarks at the end of section 3 of chapter 1.)

3. TESTING HYPOTHESES

We shall call $\hat{I}(*:H)$ the *minimum discrimination information statistic*, and test a null hypothesis H_2 against an alternative hypothesis H_1 by rejecting H_2 if Prob $\{\hat{I}(*:H_2) - \hat{I}(*:H_1) \geqq c|H_2\} \leqq \alpha$. By appropriate choice of the constant c by which we require $\hat{I}(*:H_2)$ to exceed $\hat{I}(*:H_1)$

before we reject the hypothesis H_2, we can control the magnitude of the type I error (the probability of rejecting the null hypothesis H_2 when the sample is from a population of H_2). We shall see that this procedure also provides a test with desirable properties so far as the magnitude of the type II error is concerned (the probability of accepting the null hypothesis when the sample is from a population of H_1). [For the theory of hypothesis testing see, for example, Fraser (1957, pp. 69–108), Hoel (1954, pp. 30–38, 182–196).]

Before we examine the properties of the minimum discrimination information statistic it may be helpful to illustrate the procedure. In the following examples we shall ignore the probabilities involved and consider only the expression $\hat{I}(*:H_2) - \hat{I}(*:H_1) \geqq c$, that is, the critical region or the sample values on the basis of which we reject the null hypothesis.

Example 3.1. Suppose we have an observation x, which may indeed be a sample of n independent observations, and we want to test a simple null hypothesis H_2, the observation is from a population with generalized density $f_2(x)$, against the simple alternative hypothesis H_1, the observation is from a population with generalized density $f_1(x)$. With the statistic $T(x) = \log (f_1(x)/f_2(x))$, we have in accordance with the estimation procedure mentioned in section 1, $\hat{\theta} = \log (f_1(x)/f_2(x))$. From (2.16) and (2.17) of chapter 3 and (1.1) and (1.2) of this chapter, defining $N_2(\hat{\tau}_2)$ and $N_1(\hat{\tau}_1)$ below by context, we have

$$\hat{I}(*:H_2) = \hat{\tau}_2 \log \frac{f_1(x)}{f_2(x)} - \log M_2(\hat{\tau}_2), \qquad M_2(\tau) = \int (f_1(x))^{\tau}(f_2(x))^{1-\tau}\, d\lambda(x),$$

$$\log \frac{f_1(x)}{f_2(x)} = \frac{\displaystyle\int (f_1(x))^{\hat{\tau}_2}(f_2(x))^{1-\hat{\tau}_2} \log \frac{f_1(x)}{f_2(x)}\, d\lambda(x)}{\displaystyle\int (f_1(x))^{\hat{\tau}_2}(f_2(x))^{1-\hat{\tau}_2}\, d\lambda(x)} = \frac{N_2(\hat{\tau}_2)}{M_2(\hat{\tau}_2)}.$$

Similarly, we have

$$\hat{I}(*:H_1) = \hat{\tau}_1 \log \frac{f_1(x)}{f_2(x)} - \log M_1(\hat{\tau}_1), \qquad M_1(\tau) = \int (f_1(x))^{1+\tau}(f_2(x))^{-\tau}\, d\lambda(x),$$

$$\log \frac{f_1(x)}{f_2(x)} = \frac{\displaystyle\int (f_1(x))^{1+\hat{\tau}_1}(f_2(x))^{-\hat{\tau}_1} \log \frac{f_1(x)}{f_2(x)}\, d\lambda(x)}{\displaystyle\int (f_1(x))^{1+\hat{\tau}_1}(f_2(x))^{-\hat{\tau}_1}\, d\lambda(x)} = \frac{N_1(\hat{\tau}_1)}{M_1(\hat{\tau}_1)}.$$

Since $N_2(\hat{\tau}_2)/M_2(\hat{\tau}_2) = N_1(\hat{\tau}_1)/M_1(\hat{\tau}_1) = \log (f_1(x)/f_2(x))$, we have, as shown by Chernoff (1952, p. 504), $\hat{\tau}_2 = \hat{\tau}_1 + 1$, $M_2(\hat{\tau}_2) = M_1(\hat{\tau}_1)$. Accordingly,

$$\hat{I}(*:H_2) - \hat{I}(*:H_1) = \hat{\tau}_2 \log \frac{f_1(x)}{f_2(x)} - \log M_2(\hat{\tau}_2) - (\hat{\tau}_2 - 1) \log \frac{f_1(x)}{f_2(x)} + \log M_1(\hat{\tau}_1) = \log \frac{f_1(x)}{f_2(x)},$$

and the critical region is therefore of the form $\log \frac{f_1(x)}{f_2(x)} \geq c$. This is the most powerful critical region, as yielded by the fundamental lemma of Neyman and Pearson (1933). [Cf. Fraser (1957, p. 73).]

Example 3.2. We shall need some of the results in example 1.1 of this chapter and example 4.1 of chapter 3. Suppose the hypothesis H_1 specifies the binomial distribution with $p = p_1$, $q_1 = 1 - p_1$, and the hypothesis H_2 specifies the binomial distribution with $p = p_2$, $q_2 = 1 - p_2$. We estimate $\theta = np^*$ by $n\hat{p}$, where $n\hat{p} = y$, $\hat{q} = 1 - \hat{p}$, and $y = T(x)$ is the number of observed successes in a sample of n independent observations O_n. From the results in example 1.1 we see that

$$\hat{I}(*:H_1) = n\left(\hat{p} \log \frac{\hat{p}}{p_1} + \hat{q} \log \frac{\hat{q}}{q_1}\right), \qquad \hat{I}(*:H_2) = n\left(\hat{p} \log \frac{\hat{p}}{p_2} + \hat{q} \log \frac{\hat{q}}{q_2}\right).$$

We therefore reject H_2 if (merging constants as they occur so that c is not necessarily the same constant throughout)

$$n\left(\hat{p} \log \frac{\hat{p}}{p_2} + \hat{q} \log \frac{\hat{q}}{q_2}\right) - n\left(\hat{p} \log \frac{\hat{p}}{p_1} + \hat{q} \log \frac{\hat{q}}{q_1}\right) \geq c,$$

or

$$\hat{p} \log \frac{p_1}{p_2} + \hat{q} \log \frac{q_1}{q_2} \geq c,$$

or

$$\hat{p} \log \frac{p_1 q_2}{p_2 q_1} \geq c.$$

When $p_1 > p_2$, $\log (p_1q_2/p_2q_1) > 0$ and we reject H_2 if $\hat{p} \geq c$. On the other hand, when $p_1 < p_2$, $\log (p_1q_2/p_2q_1) < 0$ and we reject H_2 if $\hat{p} \leq c$. (This example is a special case of example 3.1.) See figure 3.1.

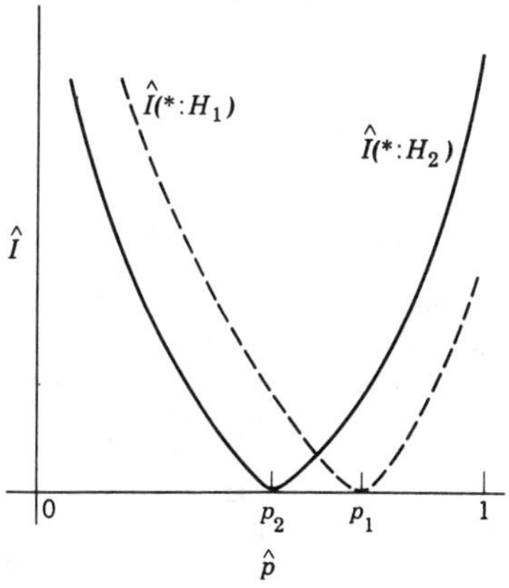

Figure 3.1

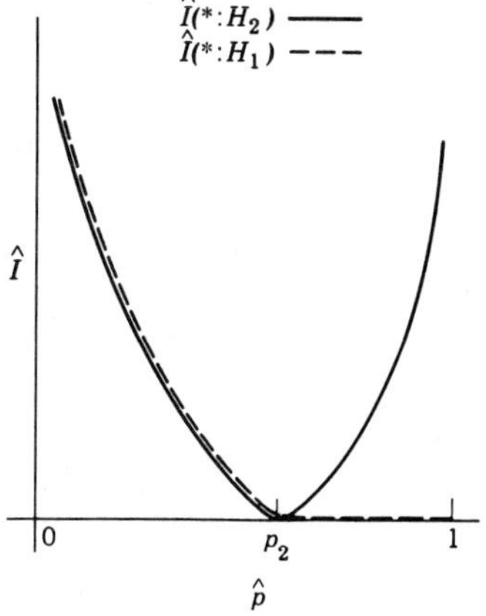

Figure 3.2

Example 3.3. We continue example 3.2, but now the hypothesis H_2 specifies the binomial distribution with $p = p_2$, $q_2 = 1 - p_2$, and the hypothesis H_1 specifies the binomial distribution with $p > p_2$, $q = 1 - p$. As before, we estimate $\theta = np^*$ by $n\hat{p} = y$, and $\hat{I}(*:H_2) = n(\hat{p}\log(\hat{p}/p_2) + \hat{q}\log(\hat{q}/q_2))$. In section 3 of chapter 4 we noted that $F(\hat{p}, p) = \hat{p}\log(\hat{p}/p) + \hat{q}\log(\hat{q}/q)$ is a convex function of p for given $\hat{p}$, $F(\hat{p}, p) = 0$ for $p = \hat{p}$, and $F(\hat{p}, p)$ is monotonically decreasing for $0 \leq p \leq \hat{p}$ and monotonically increasing for $\hat{p} \leq p \leq 1$; therefore $\hat{I}(*:H_1) = 0$, if $\hat{p} > p_2$, and $\hat{I}(*:H_1) = n(\hat{p}\log(\hat{p}/p_2) + \hat{q}\log(\hat{q}/q_2))$, if $\hat{p} < p_2$. We therefore reject H_2 if $\hat{p} > p_2$ and $\hat{p}\log(\hat{p}/p_2) + \hat{q}\log(\hat{q}/q_2) \geq c$, that is, if $\hat{p} \geq c > p_2$. See figure 3.2. Here we have a uniformly most powerful critical region. [Cf. Neyman (1950, p. 325, pp. 326–327).]

Example 3.4. Continuing examples 3.2 and 3.3, H_2 now specifies the family of binomial distributions with $p \geq p_2$, $q = 1 - p$, and H_1 specifies the family of binomial distributions with $p \leq p_1 < p_2$. As before, we estimate $\theta = np^*$ by $n\hat{p} = y$. From the behavior of $F(\hat{p}, p)$ described in example 3.3, $\hat{I}(*:H_1)$ and $\hat{I}(*:H_2)$ are as follows:

	$\hat{I}(*:H_1)$	$\hat{I}(*:H_2)$
$\hat{p} > p_2$	$n\left(\hat{p}\log\frac{\hat{p}}{p_1} + \hat{q}\log\frac{\hat{q}}{q_1}\right)$	0
$p_1 \leq \hat{p} \leq p_2$	$n\left(\hat{p}\log\frac{\hat{p}}{p_1} + \hat{q}\log\frac{\hat{q}}{q_1}\right)$	$n\left(\hat{p}\log\frac{\hat{p}}{p_2} + \hat{q}\log\frac{\hat{q}}{q_2}\right)$
$\hat{p} < p_1$	0	$n\left(\hat{p}\log\frac{\hat{p}}{p_2} + \hat{q}\log\frac{\hat{q}}{q_2}\right)$

We therefore assign the sample to the family of populations H_2 if $\hat{p} > p$, where p $(q = 1 - p)$ satisfies $p\log\frac{p}{p_1} + q\log\frac{q}{q_1} = p\log\frac{p}{p_2} + q\log\frac{q}{q_2}$, that is, $p = \left(\log\frac{q_2}{q_1}\right)\Big/\log\frac{p_1 q_2}{p_2 q_1}$. [Cf. Chernoff (1952, p. 502).]

If $\hat{p} = p$, $\hat{I}(*:H_2) = \hat{I}(*:H_1)$. See figure 3.3.

Example 3.5. Suppose we have a random sample O_n of n independent observations and we take the set E in example 2.3 of chapter 3 as the interval $0 \leq x < \infty$ and its complement $\bar{E}$ as the interval $-\infty < x < 0$. Consider the null hypothesis H_2 that $f_2(x)$ is the generalized density of an absolutely continuous distribution such that $\mu_2(E) = \mu_2(\bar{E}) = \frac{1}{2}$. We shall use $\hat{p} = y/n$ as an estimate of $\mu_1(E)$ (here $\mu_1(E)$ is θ), where $y = \sum_{i=1}^{n} T(x_i) = \sum_{i=1}^{n} \chi_E(x_i)$, that is, y is the number of nonnegative observations in the sample, and $\hat{q} = 1 - \hat{p}$. If the alternative hypothesis H_1 is that $f_2(x)$ is the generalized density of any absolutely continuous distribution such that $\mu_2(E) = p \neq \frac{1}{2}$, $\mu_2(\bar{E}) = q = 1 - p$, then $\hat{I}(*:H_2) = n(\hat{p}\log 2\hat{p} + \hat{q}\log 2\hat{q})$ and $\hat{I}(*:H_1) = 0$. We therefore reject H_2 if $\hat{p}\log 2\hat{p} + \hat{q}\log 2\hat{q} \geq c$, or $\hat{p}\log\hat{p} + \hat{q}\log\hat{q} \geq c$, that is, if $|\hat{p} - \frac{1}{2}| \geq c$. See figure 3.4. [Cf. Fraser (1957, pp. 167–169).]

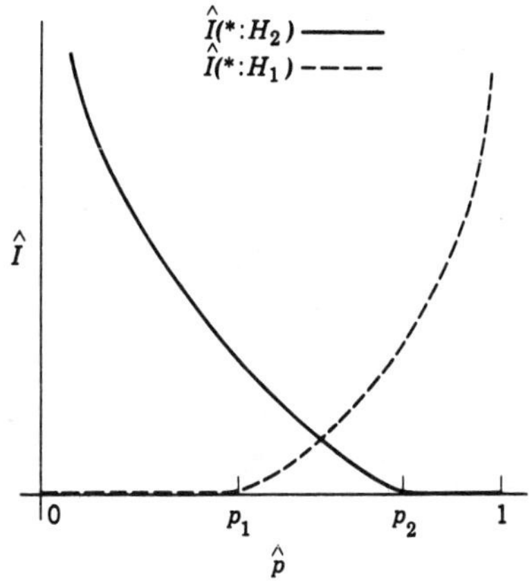

Figure 3.3

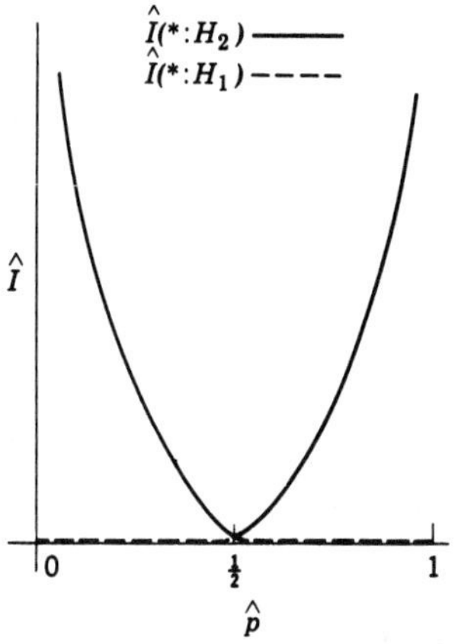

Figure 3.4

Example 3.6. We shall need some of the results in example 1.4 above and examples 4.4 and 4.5 of chapter 3. Suppose the hypothesis H_i specifies the normal distribution $N(0, \sigma_i^2)$, $i = 1, 2$. We shall estimate $\theta = \sigma^{*2}$ by the statistic $y = T(x) = (1/n)\Sigma x_i^2$ of example 4.5 (rather than by the statistic s^2 of example 4.4). From example 1.4, we see that

$$\hat{I}(*:H_1) = \frac{n}{2}\left(\log\frac{\sigma_1^2}{y} - 1 + \frac{y}{\sigma_1^2}\right), \qquad \hat{I}(*:H_2) = \frac{n}{2}\left(\log\frac{\sigma_2^2}{y} - 1 + \frac{y}{\sigma_2^2}\right).$$

We therefore reject H_2 if

$$\frac{n}{2}\left(\log\frac{\sigma_2^2}{y} - 1 + \frac{y}{\sigma_2^2}\right) - \frac{n}{2}\left(\log\frac{\sigma_1^2}{y} - 1 + \frac{y}{\sigma_1^2}\right) \geq c,$$

or

$$\frac{y}{\sigma_2^2} - \frac{y}{\sigma_1^2} \geq c,$$

or

$$(\sigma_1^2 - \sigma_2^2)\Sigma x_i^2 \geq c,$$

or

$$\Sigma x_i^2 \geq c \text{ if } \sigma_1^2 > \sigma_2^2, \qquad \Sigma x_i^2 \leq c \text{ if } \sigma_1^2 < \sigma_2^2.$$

See figure 3.5.

This is a special case of example 3.1.

Example 3.7. We continue example 3.6, but now H_2 specifies the family of normal distributions $N(0, \sigma^2)$, $\sigma^2 \geq \sigma_2^2$, and H_1 specifies the family of normal distributions $N(0, \sigma^2)$, $\sigma^2 \leq \sigma_1^2 < \sigma_2^2$. Note that $F(y, \sigma^2) = \log(\sigma^2/y) - 1 + y/\sigma^2$ is a convex function of $1/\sigma^2$ for given y, $F(y, \sigma^2) = 0$ for $\sigma^2 = y$, $F(y, \sigma^2)$ is monotonically decreasing for $0 < \sigma^2 \leq y$, and monotonically increasing for $y \leq \sigma^2 < \infty$. $\hat{I}(*:H_1)$ and $\hat{I}(*:H_2)$ are therefore as follows:

	$\hat{I}(*:H_1)$	$\hat{I}(*:H_2)$
$y > \sigma_2^2$	$\frac{n}{2}\left(\log\frac{\sigma_1^2}{y} - 1 + \frac{y}{\sigma_1^2}\right)$	0
$\sigma_1^2 \leq y \leq \sigma_2^2$	$\frac{n}{2}\left(\log\frac{\sigma_1^2}{y} - 1 + \frac{y}{\sigma_1^2}\right)$	$\frac{n}{2}\left(\log\frac{\sigma_2^2}{y} - 1 + \frac{y}{\sigma_2^2}\right)$
$y < \sigma_1^2$	0	$\frac{n}{2}\left(\log\frac{\sigma_2^2}{y} - 1 + \frac{y}{\sigma_2^2}\right)$

We therefore assign the sample to the family of populations H_2 if $y > \sigma^2$, where σ^2 satisfies $\log\frac{\sigma_2^2}{\sigma^2} - 1 + \frac{\sigma^2}{\sigma_2^2} = \log\frac{\sigma_1^2}{\sigma^2} - 1 + \frac{\sigma^2}{\sigma_1^2}$, that is,

$$\sigma^2 = (\log(\sigma_2^2/\sigma_1^2))/(1/\sigma_1^2 - 1/\sigma_2^2).$$

[Cf. Chernoff (1952, p. 502).]

If $y = \sigma^2$, $\hat{I}(*:H_2) = \hat{I}(*:H_1)$. See figure 3.6.

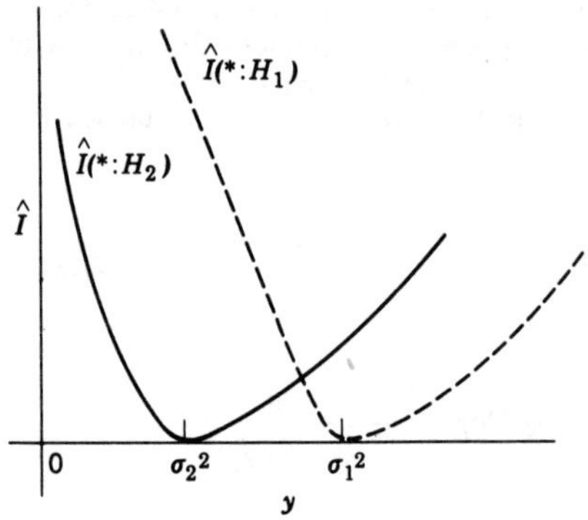

Figure 3.5

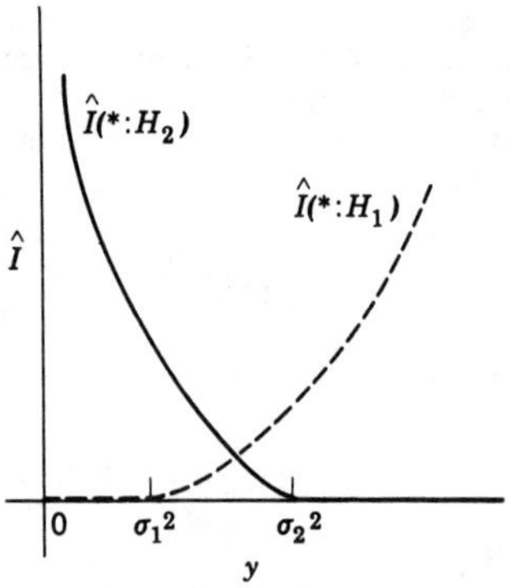

Figure 3.6

Example 3.8. We continue examples 3.6 and 3.7, but now the null hypothesis H_2 specifies the normal distributions $N(0, \sigma^2)$, $\sigma^2 \geq \sigma_2^2$, and the alternative hypothesis H_1 specifies the normal distributions $N(0, \sigma^2)$, $\sigma^2 < \sigma_2^2$. $\hat{I}(*:H_1)$ and $\hat{I}(*:H_2)$ are therefore as follows:

	$\hat{I}(*:H_1)$	$\hat{I}(*:H_2)$
$y \geq \sigma_2^2$	$\frac{n}{2}\left(\log\frac{\sigma_2^2}{y} - 1 + \frac{y}{\sigma_2^2}\right)$	0
$y < \sigma_2^2$	0	$\frac{n}{2}\left(\log\frac{\sigma_2^2}{y} - 1 + \frac{y}{\sigma_2^2}\right)$

We therefore reject H_2 if $y < \sigma_2^2$ and $\log(\sigma_2^2/y) - 1 + y/\sigma_2^2 \geq c$, that is, if $y \leq c < \sigma_2^2$. Here we have a uniformly most powerful critical region. [Cf. Fraser (1957, p. 84).] Symmetrically, if we treat H_1 as the null hypothesis and H_2 as the alternative hypothesis, we reject H_1 if $y \geq \sigma_2^2$ and $\log(\sigma_2^2/y) - 1 + y/\sigma_2^2 \geq c$, that is, if $y \geq c > \sigma_2^2$. If σ_2^2 is not specified, this suggests a confidence interval for the parameter σ^2 determined by $\log(\sigma^2/y) - 1 + y/\sigma^2 \leq c$, with confidence coefficient Prob $[\log(\sigma^2/y) - 1 + y/\sigma^2 \leq c|\sigma^2] = 1 - \alpha$. We may also say that the sample provides less than a desired amount of information for discriminating against a hypothetical value of σ^2 falling within the confidence interval. See figure 3.7.

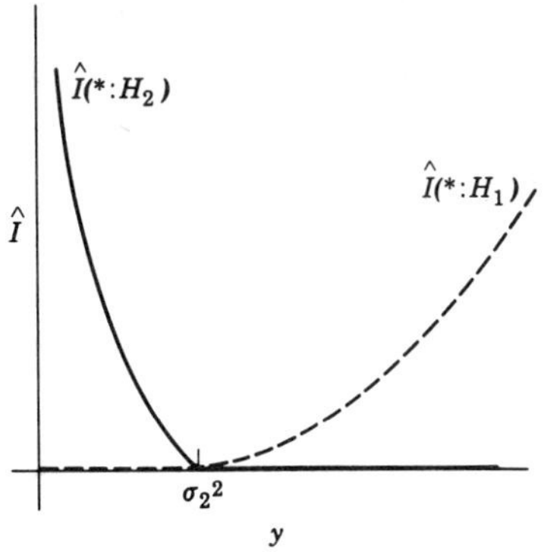

Figure 3.7

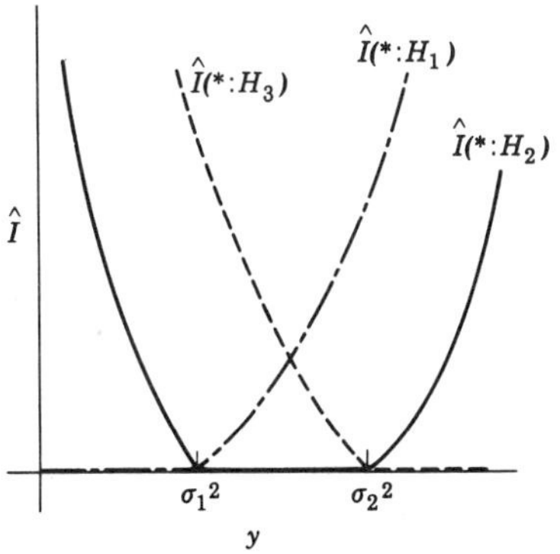

Figure 3.8

Example 3.9. We continue examples 3.6 through 3.8, but now we must assign the sample either to: H_1, the family of normal distributions $N(0, \sigma^2)$, $\sigma^2 < \sigma_1^2$; H_2, the family of normal distributions $N(0, \sigma^2)$, $\sigma_1^2 \leq \sigma^2 \leq \sigma_2^2$; or H_3, the family of normal distributions $N(0, \sigma^2)$, $\sigma^2 > \sigma_2^2$. $\hat{I}(*:H_1)$, $\hat{I}(*:H_2)$, and $\hat{I}(*:H_3)$ are as follows:

	$\hat{I}(*:H_1)$	$\hat{I}(*:H_2)$	$\hat{I}(*:H_3)$
$y > \sigma_2^2$	$\frac{n}{2}\left(\log\frac{\sigma_1^2}{y} - 1 + \frac{y}{\sigma_1^2}\right)$	$\frac{n}{2}\left(\log\frac{\sigma_2^2}{y} - 1 + \frac{y}{\sigma_2^2}\right)$	0
$\sigma_1^2 \leq y \leq \sigma_2^2$	$\frac{n}{2}\left(\log\frac{\sigma_1^2}{y} - 1 + \frac{y}{\sigma_1^2}\right)$	0	$\frac{n}{2}\left(\log\frac{\sigma_2^2}{y} - 1 + \frac{y}{\sigma_2^2}\right)$
$y < \sigma_1^2$	0	$\frac{n}{2}\left(\log\frac{\sigma_1^2}{y} - 1 + \frac{y}{\sigma_1^2}\right)$	$\frac{n}{2}\left(\log\frac{\sigma_2^2}{y} - 1 + \frac{y}{\sigma_2^2}\right)$

We therefore assign the sample to the family H_i, $i = 1, 2, 3$, for which $\hat{I}(*:H_i) = 0$. See figure 3.8.

Example 3.10. Let us reconsider example 3.9 but with a null hypothesis H_2 specifying the family H_2, and an alternative hypothesis H_4 specifying the family H_1 or H_3, that is, $H_4 = H_1 \cup H_3$. We see that $\hat{I}(*:H_4) = 0$ for $y > \sigma_2^2$ or $y < \sigma_1^2$, and $\hat{I}(*:H_4) = \min(\hat{I}(*:H_1), \hat{I}(*:H_3))$ for $\sigma_1^2 \leq y \leq \sigma_2^2$ (see example 3.9).

We therefore reject H_2 if $y > \sigma_2^2$ and $\log(\sigma_2^2/y) - 1 + y/\sigma_2^2 \geq c$, that is, if $y \geq c > \sigma_2^2$; or if $y < \sigma_1^2$ and $\log(\sigma_1^2/y) - 1 + y/\sigma_1^2 \geq c$, that is, if $y \leq c < \sigma_1^2$. The constants are to be determined by the significance level desired. See figure 3.9.

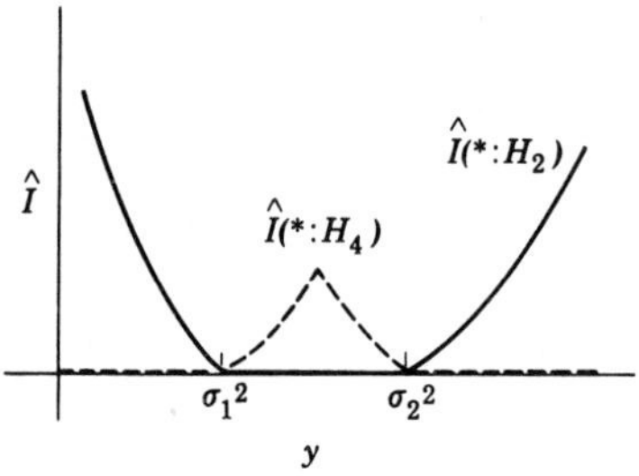

Figure 3.9

Example 3.11. We shall need the results of example 1.2. Suppose the alternative hypothesis H_1 specifies the normal distribution $N(\mu, 1)$, $\mu = \mu_1 > \mu_2$, and the null hypothesis H_2 specifies the normal distribution $N(\mu, 1)$, $\mu \leq \mu_2$. We estimate θ by $\hat{\theta} = \bar{x}$ and $\hat{I}(*:\mu) = n(\bar{x} - \mu)^2/2$. $\hat{I}(*:H_1)$ and $\hat{I}(*:H_2)$ are as follows:

	$\hat{I}(*:H_1)$	$\hat{I}(*:H_2)$
$\bar{x} \leq \mu_2$	$\dfrac{n(\bar{x} - \mu_1)^2}{2}$	0
$\mu_2 < \bar{x} < \mu_1$	$\dfrac{n(\bar{x} - \mu_1)^2}{2}$	$\dfrac{n(\bar{x} - \mu_2)^2}{2}$
$\mu_1 \leq \bar{x}$	0	$\dfrac{n(\bar{x} - \mu_2)^2}{2}$

We therefore reject H_2 if $\mu_2 < \bar{x} < \mu_1$ and $n(\bar{x} - \mu_2)^2/2 - n(\bar{x} - \mu_1)^2/2 \geq c$, or if $\mu_1 \leq \bar{x}$ and $n(\bar{x} - \mu_2)^2/2 \geq c$, that is, if $\bar{x} \geq c > \mu_2$. See figure 3.10. Lehmann (1949, p. 2–17) shows that this critical region is uniformly most powerful.

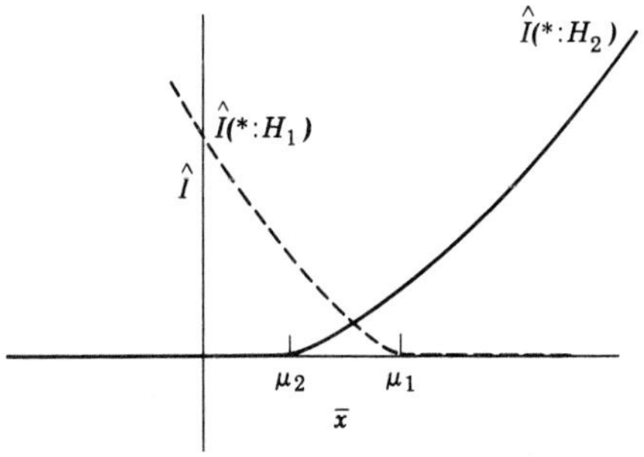

Figure 3.10

Example 3.12. We continue example 3.11 but now the alternative hypothesis H_1 specifies the normal distribution $N(\mu, 1)$, $\mu = 0$, and the null hypothesis H_2 specifies the normal distribution $N(\mu, 1)$, $\mu \leq -\mu_2$, $\mu \geq \mu_2$. $\hat{I}(*:H_1)$ and $\hat{I}(*:H_2)$ are as follows:

	$\hat{I}(*:H_1)$	$\hat{I}(*:H_2)$
$\bar{x} \leq -\mu_2$	$n\bar{x}^2/2$	0
$-\mu_2 < \bar{x} < \mu_2$	$n\bar{x}^2/2$	$n(\bar{x} - \mu_2)^2/2$
$\bar{x} \geq \mu_2$	$n\bar{x}^2/2$	0

We therefore reject H_2 if $-\mu_2 < \bar{x} < \mu_2$ and $n(\bar{x} - \mu_2)^2/2 - n\bar{x}^2/2 \geq c$, that is, if $|\bar{x}| \leq c$. See figure 3.11. Lehmann (1949, p. 2–18) shows this to be a most powerful critical region, or test procedure.

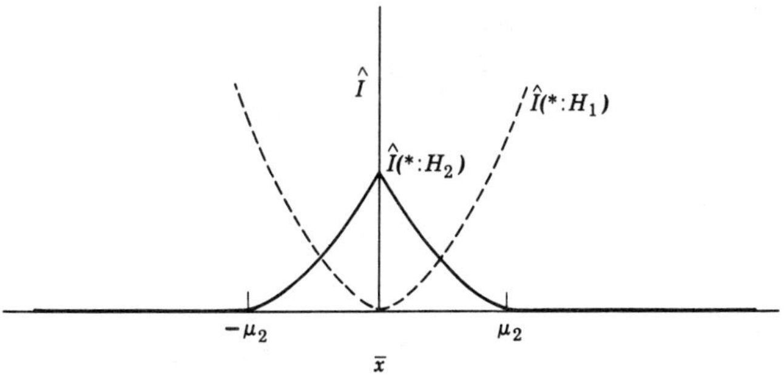

Figure 3.11

4. DISCUSSION

The reader may have noted that $\hat{\tau}(x)$ is the maximum-likelihood estimate of τ as a parameter of the generalized density $f^*(x)$ [cf. Barton (1956)]. In fact, since

$$\frac{d}{d\tau}\log f^*(x) = T(x) - \frac{d}{d\tau}\log M_2(\tau),$$

with $(d/d\tau)\log M_2(\tau)$ a strictly increasing function of τ (see lemmas 4.2 and 4.5 of chapter 3), the value of τ for which $(d/d\tau)\log f^*(x) = 0$ is unique and given in (1.1). [Cf. Khinchin (1949, pp. 79–81).]

Furthermore, as might be expected from the general argument, the minimum discrimination information statistic is related to the likelihood-ratio test of Neyman and Pearson (1928). As a matter of fact, we may write [cf. Barnard (1949), Fisher (1956, pp. 71–73)]

$$\hat{I}(*:2;O_n) = \hat{\theta}\tau(\hat{\theta}) - \log M_2(\tau(\hat{\theta})) = \log\frac{\max_{\tau} f^*(x)}{f_2(x)}, \tag{4.1}$$

where we recall that $f_2(x) = f^*(x)$ for $\tau = 0$, and

$$\hat{I}(*:H) = \min_{f_2\in H}\log\frac{\max_{\tau} f^*(x)}{f_2(x)}. \tag{4.2}$$

If the populations of H are members of the exponential family over which $f^*(x)$ ranges, and we denote the range of values of τ by Ω, and the range of values of τ corresponding to H by ω, then $\max_{f_2\in H} f_2(x) = \max_{\tau\in\omega} f^*(x)$ and

$$\hat{I}(*:H) = \log\frac{\max_{\tau\in\Omega} f^*(x)}{\max_{\tau\in\omega} f^*(x)} = -\log\lambda, \tag{4.3}$$

where λ is the Neyman-Pearson likelihood ratio [see, for example, Hoel (1954, pp. 189–192), Wilks (1943, p. 150)],

$$\lambda = \frac{P^*(\max\omega)}{P^*(\max\Omega)} = \frac{\max_{\tau\in\omega} f^*(x)}{\max_{\tau\in\Omega} f^*(x)}. \tag{4.4}$$

If H_2 implies that $\tau \in \omega_2$ and H_1 that $\tau \in \omega_1$, then

$$\hat{I}(*:H_2) = \log\frac{\max_{\tau\in\Omega} f^*(x)}{\max_{\tau\in\omega_2} f^*(x)}, \tag{4.5}$$

$$\hat{I}(*:H_1) = \log\frac{\max_{\tau\in\Omega} f^*(x)}{\max_{\tau\in\omega_1} f^*(x)}, \tag{4.6}$$

and

$$(4.7)\qquad \hat{I}(*:H_2) - \hat{I}(*:H_1) = \log \frac{\max_{\tau\in\omega_1} f^*(x)}{\max_{\tau\in\omega_2} f^*(x)}$$

$$= \log \frac{\max_{\tau\in\omega_1} f^*(x)}{f_2(x)} - \log \frac{\max_{\tau\in\omega_2} f^*(x)}{f_2(x)}.$$

We remark that here $\hat{I}(*:H_2) - \hat{I}(*:H_1) = -\log \lambda^*$, where likelihood ratios of the form

$$\lambda^* = \max_{\tau\in\omega_2} f^*(x)/\max_{\tau\in\omega_1} f^*(x)$$

have been studied by Chernoff (1954) for certain hypotheses.

If H_2 implies that $\tau \in \omega$ and H_1 that $\tau \in \Omega - \omega$, then $\hat{I}(*:H_1) = 0$ if $\hat{I}(*:H_2) > 0$, since $\hat{I}(*:2;O_n)$ is convex and nonnegative. The test of the null hypothesis H_2 now depends only on the value of $\hat{I}(*:H_2)$, because when $\hat{I}(*:H_2) = 0$ we accept the null hypothesis H_2 with no further test.

Some simple examples follow. We shall apply these notions to a wider variety of important statistical problems in subsequent chapters.

Example 4.1. Suppose we want to test a null hypothesis of homogeneity that n independent observations in a sample O_n are from the same normal population, with specified variance σ^2, against an alternative hypothesis that the observations are from normal populations with different means but the same specified variance σ^2. We denote the null hypothesis by $H_2(\mu|\sigma^2)$ or $H_2(\cdot|\sigma^2)$ according as the common mean is, or is not, specified, and the alternative hypothesis by $H_1(\mu_i|\sigma^2)$ or $H_1(\cdot|\sigma^2)$ according as the different means are, or are not, specified.

With $T(x) = (x_1, x_2, \cdots, x_n)$ and $f_2(x) = \prod_{i=1}^{n} \frac{\exp[-(x_i - \mu)^2/2\sigma^2]}{\sigma\sqrt{2\pi}}$, we have

$$(4.8)\qquad \hat{I}(*:2;O_n) = \sum_{i=1}^{n} (x_i\hat{\tau}_i - \mu\hat{\tau}_i - \frac{\sigma^2}{2}\hat{\tau}_i^2),$$

where $\hat{\tau}_i$ satisfies $x_i = \mu + \sigma^2\hat{\tau}_i$. We thus have

$$(4.9)\qquad \hat{I}(*:H_2(\mu|\sigma^2)) = \sum_{i=1}^{n} (x_i - \mu)^2/2\sigma^2.$$

If μ is not specified, $\hat{I}(*:H_2(\cdot|\sigma^2)) = \min_{\mu} \hat{I}(*:H_2(\mu|\sigma^2))$ is

$$(4.10)\qquad \hat{I}(*:H_2(\cdot|\sigma^2)) = \sum_{i=1}^{n} (x_i - \bar{x})^2/2\sigma^2, \qquad \bar{x} = (x_1 + x_2 + \cdots + x_n)/n.$$

On the other hand, with $T(x) = (x_1, x_2, \cdots, x_n)$ but

$$f_2(x) = \prod_{i=1}^{n} \frac{\exp[-(x_i - \mu_i)^2/2\sigma^2]}{\sigma\sqrt{2\pi}},$$

we have

$$(4.11)\qquad \hat{I}(*:2; O_n) = \sum_{i=1}^{n} \left(x_i\hat{\tau}_i - \mu_i\hat{\tau}_i - \frac{\sigma^2}{2}\hat{\tau}_i^2\right),$$

where $\hat{\tau}_i$ satisfies $x_i = \mu_i + \sigma^2\hat{\tau}_i$. We thus have

$$(4.12)\qquad \hat{I}(*:H_1(\mu_i|\sigma^2)) = \sum_{i=1}^{n} (x_i - \mu_i)^2/2\sigma^2.$$

If the μ_i are not specified, $\hat{I}(*:H_1(\cdot|\sigma^2)) = \min_{\mu_i} \hat{I}(*:H_1(\mu_i|\sigma^2))$ is

$$(4.13)\qquad \hat{I}(*:H_1(\cdot|\sigma^2)) = 0.$$

If we require that the conjugate distribution in (4.8), that is,

$$f^*(x) = \frac{f_2(x)\exp(\tau_1x_1 + \cdots + \tau_nx_n)}{M_2(\tau_1, \tau_2, \cdots, \tau_n)} = \prod_{i=1}^{n} \frac{\exp[-(x_i - \mu - \sigma^2\tau_i)^2/2\sigma^2]}{\sigma\sqrt{2\pi}},$$

range over normal populations with a common mean, then $\mu_1^* = \mu_2^* = \cdots = \mu_n^*$ implies that $\mu + \sigma^2\tau_1 = \mu + \sigma^2\tau_2 = \cdots = \mu + \sigma^2\tau_n$, or only values $\tau_1 = \tau_2 = \cdots = \tau_n = \tau$ are admissible. With this restriction, (4.8) yields

$$(4.14)\qquad \hat{I}(H_2(\cdot|\sigma^2):2; O_n) = n\bar{x}\hat{\tau} - n\mu\hat{\tau} - n\frac{\sigma^2}{2}\hat{\tau}^2,$$

where $\hat{\tau}$ satisfies $\bar{x} = \mu + \sigma^2\hat{\tau}$, and (4.14) becomes

$$(4.15)\qquad \hat{I}(H_2(\cdot|\sigma^2):2; O_n) = n(\bar{x} - \mu)^2/2\sigma^2.$$

Note that if $\omega_1 = \Omega$ is the n-dimensional space of $\tau_1, \tau_2, \cdots, \tau_n$, then (4.9) is $\log(\max_{\tau\in\omega_1} f^*(x)/f_2(x))$, and that if ω_2 is the subspace of Ω with $\tau_1 = \tau_2 = \cdots = \tau_n$, then (4.15) is $\log(\max_{\tau\in\omega_2} f^*(x)/f_2(x))$. From (4.10), (4.13), and the foregoing we see that (4.7) becomes

$$(4.16)\qquad \sum_{i=1}^{n} (x_i - \bar{x})^2/2\sigma^2 = \sum_{i=1}^{n} (x_i - \mu)^2/2\sigma^2 - n(\bar{x} - \mu)^2/2\sigma^2.$$

The hypothesis $H_2(\mu|\sigma^2)$ is the intersection of two hypotheses, (i) that the sample is homogeneous, and (ii) that the mean of the homogeneous sample is μ. Rewriting (4.16) as

$$(4.17)\qquad \sum_{i=1}^{n} (x_i - \mu)^2/2\sigma^2 = \sum_{i=1}^{n} (x_i - \bar{x})^2/2\sigma^2 + n(\bar{x} - \mu)^2/2\sigma^2$$

or

$$\hat{I}(*:H_2(\mu|\sigma^2)) = \hat{I}(*:H_2(\cdot|\sigma^2)) + \hat{I}(H_2(\cdot|\sigma^2):2; O_n)$$

reflects the fact that the first term on the right is the minimum discrimination information statistic to test the homogeneity and the second term on the right is the minimum discrimination information statistic to test the value of the mean for a homogeneous sample.

Example 4.2. Suppose we have a homogeneous random sample O_n, namely, one from the same normal population, and we want to test a hypothesis about the mean with no specification of the variance. Let the hypothesis $H_2(\mu, \sigma^2)$ imply that the sample is from a specified normal population $N(\mu, \sigma^2)$, and the hypothesis $H_2(\mu)$ imply that the sample is from a normal population with specified mean μ and unspecified variance. Suppose the alternative hypothesis H_1 implies that the sample is from an unspecified normal population.

With $T(x) = (\bar{x}, s^2)$, where s^2 is the unbiased sample variance, and

$$f_2(x) = \prod_{i=1}^{n} (1/\sigma\sqrt{2\pi}) \exp\left[-(x_i - \mu)^2/2\sigma^2\right],$$

we see from example 1.3 in this chapter and example 4.3 in chapter 3 that

$$\hat{I}(*:H_2(\mu, \sigma^2)) = \hat{\tau}_1\bar{x} - \hat{\tau}_1\mu - \frac{\sigma^2}{2n}\hat{\tau}_1^{\,2} + \frac{n-1}{2}\log\left(1 - \frac{2\sigma^2\hat{\tau}_2}{n-1}\right) + s^2\hat{\tau}_2,$$

with $\bar{x} = \mu + \hat{\tau}_1(\sigma^2/n)$, $s^2 = \sigma^2/(1 - 2\hat{\tau}_2\sigma^2/(n-1))$, or

$$\hat{I}(*:H_2(\mu, \sigma^2)) = \frac{n(\bar{x} - \mu)^2}{2\sigma^2} + \frac{n-1}{2}\left(\log\frac{\sigma^2}{s^2} - 1 + \frac{s^2}{\sigma^2}\right). \tag{4.18}$$

We note from examples 4.2 and 4.3 in chapter 3 that if the normal populations have the same variances under H_1 and H_2, that is, $\sigma_1^{\,2} = \sigma_2^{\,2} = \sigma_*^{\,2}$, then $\tau_2 = 0$ is the only admissible value. We reach the same conclusion by requiring that in the generalized density $g^*(y)$ in (4.15) of chapter 3 the variance parameters in the distribution of $\bar{x}$ and s^2 be the same. Accordingly, for $\hat{I}(*:H_2(\mu))$ we have the same expression as above for $\hat{I}(*:H_2(\mu, \sigma^2))$ except that $\bar{x} = \mu + \hat{\tau}_1(\sigma^2/n)$ and $\hat{\tau}_2 = 0$, or $s^2 = \sigma^2$, so that

$$\hat{I}(*:H_2(\mu)) = n(\bar{x} - \mu)^2/2s^2. \tag{4.19}$$

We see that $\hat{I}(*:H_1) = 0$, and the test of the hypothesis $H_2(\mu)$ depends only on the value of $\hat{I}(*:H_2(\mu))$. This is the familiar Student t-test. (See problem 7.8.)

Example 4.3. Suppose we want to test a null hypothesis about the variance of a normal population from which a random sample O_n has been drawn. Let the hypothesis $H_2(\sigma^2)$ imply that the sample is from a normal population with specified variance σ^2. We see from (4.18) that

$$\hat{I}(*:H_2(\sigma^2)) = \min_{\mu} \hat{I}(*:H_2(\mu, \sigma^2)),$$

or

$$\hat{I}(*:H_2(\sigma^2)) = \frac{n-1}{2}\left(\log\frac{\sigma^2}{s^2} - 1 + \frac{s^2}{\sigma^2}\right). \tag{4.20}$$

The hypothesis $H_2(\mu, \sigma^2)$ in example 4.2 is the intersection of two hypotheses, (i) that the mean of the homogeneous sample is μ, given σ^2, and (ii) $H_2(\sigma^2)$. Rewriting (4.18) as

$$\hat{I}(*:H_2(\mu, \sigma^2)) = \hat{I}(H_2(\cdot|\sigma^2):2; O_n) + \hat{I}(*:H_2(\sigma^2)) \tag{4.21}$$

reflects this because of (4.17).

5. ASYMPTOTIC PROPERTIES

The asymptotic distribution of the likelihood ratio λ is known for certain cases. Wilks (1938a) showed that, under suitable regularity conditions, $-2\log\lambda$ is asymptotically distributed as χ^2 with $(k - r)$ degrees of freedom, under the null hypothesis that a (vector) parameter lies on an r-dimensional hyperplane of k-dimensional space. Wald (1943) generalized Wilks' theorem to more general subsets of the parameter space than linear subspaces and showed that the likelihood-ratio test has asymptotically best average power and asymptotically best constant power over certain families of surfaces in the parameter space and that it is an asymptotically most stringent test. [For the concept of stringency see, for example, Fraser (1957, pp. 103–107).] Wald (1943) also showed that under the alternative hypothesis the distribution of $-2 \log \lambda$ asymptotically approaches that of noncentral χ^2. Chernoff (1954) derived, under suitable regularity conditions, the asymptotic distribution of $-2 \log \lambda^*$ [see the remark following (4.7)]. In many cases $-2 \log \lambda^*$ behaves like a random variable that is sometimes zero and sometimes χ^2. [See, for example, Bartlett (1955, pp. 225–226), Fraser (1957, pp. 196–200), Hoel (1954, pp. 189–196), Wilks (1943, pp. 150–152), for the likelihood-ratio test and its asymptotic χ^2 properties.]

Kupperman (1957) showed that for a random sample of n observations, under regularity conditions given below,

$$(5.1) \qquad 2n\hat{I} = 2n\left[\int f(x, \boldsymbol{\theta}) \log \frac{f(x, \boldsymbol{\theta})}{f(x, \boldsymbol{\theta}_2)}\, d\lambda(x)\right]_{\boldsymbol{\theta}=\hat{\boldsymbol{\theta}}}$$

is asymptotically distributed as χ^2 with k degrees of freedom [k is the number of components of the (vector) parameter] under the null hypothesis, where $f(x, \boldsymbol{\theta})$ is the generalized density of a multivariate, multiparameter population, the random vector $\hat{\boldsymbol{\theta}}$ is any consistent, asymptotically multivariate normal, efficient estimator of $\boldsymbol{\theta}$, and the vector $\boldsymbol{\theta}_2$ is specified by the null hypothesis. The regularity conditions are (cf. section 6 of chapter 2):

1. $\boldsymbol{\theta} = (\theta_1, \theta_2, \cdots, \theta_k)$ is a point of the parameter space Θ, which is assumed to be an open convex set in a k-dimensional Euclidean space.
2. The family of populations defined by $f(x, \boldsymbol{\theta})$, $\boldsymbol{\theta} \in \Theta$, is homogeneous.
3. $f(x, \boldsymbol{\theta})$ has continuous first- and second-order partial derivatives with respect to the θ's in Θ, for $x \in \mathscr{X}$ [λ].
4. For all $\boldsymbol{\theta} \in \Theta$,

$$\int \frac{\partial f(x, \boldsymbol{\theta})}{\partial \theta_i}\, d\lambda(x) = 0, \qquad \int \frac{\partial^2 f(x, \boldsymbol{\theta})}{\partial \theta_i\, \partial \theta_j}\, d\lambda(x) = 0, \qquad i, j = 1, 2, \cdots, k.$$

5. The integrals

$$c_{ij}(\boldsymbol{\theta}) = \int \frac{\partial \log f(x, \boldsymbol{\theta})}{\partial \theta_i} \cdot \frac{\partial \log f(x, \boldsymbol{\theta})}{\partial \theta_j} f(x, \boldsymbol{\theta})\, d\lambda(x), \qquad i, j = 1, 2, \cdots, k,$$

are finite for all $\boldsymbol{\theta} \in \Theta$.

6. For all $\boldsymbol{\theta} \in \Theta$, the matrix $\mathbf{C}(\boldsymbol{\theta}) = (c_{ij}(\boldsymbol{\theta}))$ is positive-definite.

If instead of a single sample, as above, we have r independent samples of size n_i, $i = 1, 2, \cdots, r$, and each with a consistent, asymptotically multivariate normal, efficient estimator $\hat{\boldsymbol{\theta}}_i = (\hat{\theta}_{i1}, \hat{\theta}_{i2}, \cdots, \hat{\theta}_{ik})$, $i = 1, 2, \cdots, r$, then under the regularity conditions above, Kupperman (1957) showed that

$$2\sum_{i=1}^{r} n_i \hat{I}_i(\boldsymbol{\theta}) = 2\sum_{i=1}^{r} n_i \left[\int f(x, \boldsymbol{\theta}_i) \log \frac{f(x, \boldsymbol{\theta}_i)}{f(x, \boldsymbol{\theta})} \, d\lambda(x)\right]_{\boldsymbol{\theta}_i = \hat{\boldsymbol{\theta}}_i}$$

is asymptotically distributed as χ^2 with rk degrees of freedom under the null hypothesis that the r samples are all from the same population specified by $f(x, \boldsymbol{\theta})$. Kupperman (1957) showed that under the null hypothesis that the r samples are from the same population whose functional form is known, but with unspecified parameters,

$$(5.2) \qquad 2\sum_{i=1}^{r} n_i \hat{I}_i = 2\sum_{i=1}^{r} n_i \left[\int f(x, \boldsymbol{\theta}_i) \log \frac{f(x, \boldsymbol{\theta}_i)}{f(x, \boldsymbol{\theta})} \, d\lambda(x)\right]_{\substack{\boldsymbol{\theta}_i = \hat{\boldsymbol{\theta}}_i \\ \boldsymbol{\theta} = \hat{\boldsymbol{\theta}}}}$$

is asymptotically distributed as χ^2 with $(r - 1)k$ degrees of freedom, where n_i is the number of independent observations in the ith sample, $\hat{\boldsymbol{\theta}}_i$ is a consistent, asymptotically multivariate normal, efficient estimator of the k parameters for the ith sample, and $n\hat{\boldsymbol{\theta}} = n_1\hat{\boldsymbol{\theta}}_1 + n_2\hat{\boldsymbol{\theta}}_2 + \cdots + n_r\hat{\boldsymbol{\theta}}_r$, $n = n_1 + n_2 + \cdots + n_r$. When the null hypothesis is not true, Kupperman (1957) showed that $2n\hat{I}$, $2\sum_{i=1}^{r} n_i \hat{I}_i(\boldsymbol{\theta})$, and $2\sum_{i=1}^{r} n_i \hat{I}_i$ converge in probability to an indefinitely large number and that the large-sample distribution may be approximated by a distribution related to the noncentral χ^2-distribution with a large noncentrality parameter and the same number of degrees of freedom as the χ^2-distribution under the null hypothesis. Kupperman (1957) also showed that, under the same regularity conditions as above, similar results hold for the estimates of the divergence. Thus, with the same notation as above,

$$n\hat{J} = n\left[\int (f(x, \boldsymbol{\theta}) - f(x, \boldsymbol{\theta}_2)) \log \frac{f(x, \boldsymbol{\theta})}{f(x, \boldsymbol{\theta}_2)} \, d\lambda(x)\right]_{\boldsymbol{\theta} = \hat{\boldsymbol{\theta}}}$$

is asymptotically distributed as χ^2 with k degrees of freedom when the sample is from the population specified by $f(x, \boldsymbol{\theta}_2)$;

$$\sum_{i=1}^{r} n_i \hat{J}_i(\boldsymbol{\theta}) = \sum_{i=1}^{r} n_i \left[\int (f(x, \boldsymbol{\theta}_i) - f(x, \boldsymbol{\theta})) \log \frac{f(x, \boldsymbol{\theta}_i)}{f(x, \boldsymbol{\theta})} \, d\lambda(x)\right]_{\boldsymbol{\theta}_i = \hat{\boldsymbol{\theta}}_i}$$

is asymptotically distributed as χ^2 with rk degrees of freedom if the r samples are from the population specified by $f(x, \boldsymbol{\theta})$;

$$\sum_{i=1}^{r} n_i \hat{J}_i = \sum_{i=1}^{r} n_i \left[\int (f(x, \boldsymbol{\theta}_i) - f(x, \boldsymbol{\theta})) \log \frac{f(x, \boldsymbol{\theta}_i)}{f(x, \boldsymbol{\theta})} \, d\lambda(x)\right]_{\substack{\boldsymbol{\theta}_i = \hat{\boldsymbol{\theta}}_i \\ \boldsymbol{\theta} = \hat{\boldsymbol{\theta}}}}$$

is asymptotically distributed as χ^2 with $(r-1)k$ degrees of freedom if the r samples are from the same population.

For two samples, Kupperman (1957) showed that

$$\frac{n_1 n_2}{n_1+n_2}\left[\int (f(x,\boldsymbol{\theta}_1)-f(x,\boldsymbol{\theta}_2))\log\frac{f(x,\boldsymbol{\theta}_1)}{f(x,\boldsymbol{\theta}_2)}\,d\lambda(x)\right]_{\substack{\boldsymbol{\theta}_1=\hat{\boldsymbol{\theta}}_1\\ \boldsymbol{\theta}_2=\hat{\boldsymbol{\theta}}_2}}$$

is asymptotically distributed as χ^2 with k degrees of freedom when the two independent samples are from the same population with unspecified vector parameter $\boldsymbol{\theta}$.

The behavior of the estimates of the divergence when the null hypothesis is not true is similar to that of the estimates of the discrimination information.

These tests are consistent, the power tends to 1 for large samples. [See, for example, Fraser (1957, p. 108).]

Example 5.1. We may infer that $2\hat{I}(*:H_2(\mu|\sigma^2)) = \sum_{i=1}^{n}(x_i-\mu)^2/\sigma^2$, in (4.9), asymptotically has a χ^2 distribution with n degrees of freedom. (It can of course be shown that this is true for all n.) We may reach this conclusion by Wilks' theorem, since there are n parameters $\tau_1, \tau_2, \cdots, \tau_n$, and the null hypothesis specifies the point $\tau_1 = \tau_2 = \cdots = \tau_n = 0$.

Example 5.2. We may infer that $2\hat{I}(*:H_2(\cdot|\sigma^2)) = \sum_{i=1}^{n}(x_i-\bar{x})^2/\sigma^2$, in (4.10), asymptotically has a χ^2 distribution with $(n-1)$ degrees of freedom. (It can of course be shown that this is true for all n.) We may reach this conclusion by Kupperman's result in (5.2), since $2I(1:2) = (\mu_1-\mu_2)^2/\sigma^2$ for normal distributions with different means and the same variance, and each observation is a sample of size 1, so that $\hat{\mu}_i = x_i$, $\mu_2 = \bar{x}$, $k = 1$, and $r = n$.

Example 5.3. We may infer that

$$2\hat{I}(*:H_2(\mu,\sigma^2)) = n(\bar{x}-\mu)^2/\sigma^2 + (n-1)(\log(\sigma^2/s^2) - 1 + s^2/\sigma^2),$$

in (4.18), asymptotically has a χ^2 distribution with 2 degrees of freedom. We may reach this conclusion by using Wilks' theorem, since there are two parameters τ_1, τ_2 and the null hypothesis implies $\tau_1 = \tau_2 = 0$.

Example 5.4. Suppose we have a sample of n independent observations from a normal population with zero mean and unknown variance. From example 3.8, and the asymptotic properties, we may determine a confidence interval for the parameter σ^2 with asymptotic confidence coefficient $(1-\alpha)$ from

$$n(\log(\sigma^2/y) - 1 + y/\sigma^2) \leq \chi_\alpha^2, \tag{5.3}$$

where $y = (1/n)\sum_{i=1}^{n} x_i^2$ and χ_α^2 is the tabulated value of χ^2 for 1 degree of freedom at the $100\alpha\,\%$ significance level. Since the left-hand side of (5.3) is a convex function of $1/\sigma^2$ for given y, the equality in (5.3) is satisfied for two values of σ^2. (See examples 3.8 and 5.6.)

We shall supplement the preceding statements by a more detailed examination of the asymptotic behavior of $2\hat{I}(*:H)$. First, let us examine more explicitly the relation between $T(x) = \hat{\theta}$ and the estimate of τ, $\hat{\tau}(x) = \tau(\hat{\theta})$, in (1.1). Since

$$(5.4)\quad \left[\frac{d}{d\tau}\log M_2(\tau)\right]_{\tau=\tau(\hat{\theta})} = \left[\frac{d}{d\tau}\log M_2(\tau)\right]_{\tau=\tau(\theta)} + (\tau(\hat{\theta}) - \tau(\theta))\left[\frac{d^2}{d\tau^2}\log M_2(\tau)\right]_{\tau=\tau(\bar{\theta})},$$

where $\tau(\bar{\theta})$ lies between $\tau(\hat{\theta})$ and $\tau(\theta)$, with $\theta = [(d/d\tau)\log M_2(\tau)]_{\tau=\tau(\theta)}$, we get from lemma 4.3 of chapter 3, (1.1), and (5.4) the relation

$$(5.5)\qquad \hat{\theta} - \theta = (\tau(\hat{\theta}) - \tau(\theta))\ \mathrm{var}\,(\hat{\theta}|\tau(\bar{\theta})).$$

We recall to the reader's attention the inherent multidimensionality of the variables and the parameters, as already mentioned for lemmas 4.10 through 4.12 of chapter 3. In terms of the matrices (vectors) $\boldsymbol{\theta}' = (\theta_1, \theta_2, \cdots, \theta_k)$, $\hat{\boldsymbol{\theta}}' = (\hat{\theta}_1, \hat{\theta}_2, \cdots, \hat{\theta}_k)$, $\boldsymbol{\tau}' = (\tau_1, \tau_2, \cdots, \tau_k)$, $\hat{\boldsymbol{\tau}}' = (\hat{\tau}_1, \hat{\tau}_2, \cdots, \hat{\tau}_k)$, we may write instead of (5.5):

$$(5.6)\qquad \hat{\boldsymbol{\theta}} - \boldsymbol{\theta} = \boldsymbol{\Sigma}(\tau(\bar{\theta}))(\hat{\boldsymbol{\tau}} - \boldsymbol{\tau}),$$

or

$$(5.7)\qquad \hat{\boldsymbol{\tau}} - \boldsymbol{\tau} = \boldsymbol{\Sigma}^{-1}(\tau(\bar{\theta}))(\hat{\boldsymbol{\theta}} - \boldsymbol{\theta}),$$

where $\boldsymbol{\Sigma}(\tau(\bar{\theta}))$ is the covariance matrix of the $\hat{\theta}$'s in the conjugate distribution with parameter $\tau(\bar{\theta})$. We may also derive (5.7) directly from $\tau(\hat{\theta}) = \tau(\theta) + (\hat{\theta} - \theta)[d\tau(\theta)/d\theta]_{\theta=\bar{\theta}}$ and lemmas 4.7 and 4.12 of chapter 3.

If we write $\hat{I}(*:2; O_n) = m(\hat{\theta}) = \hat{\theta}\tau(\hat{\theta}) - \log M_2(\tau(\hat{\theta}))$ and follow the procedure in the proof of theorem 4.1 of chapter 3, we see that

$$(5.8)\quad \hat{I}(*:2; O_n) = I(*:2; O_n) + (\hat{\theta} - \theta)\tau(\theta) + (\hat{\theta} - \theta)^2/2\ \mathrm{var}\,(\hat{\theta}|\tau(\bar{\theta})),$$

where $\bar{\theta}$ lies between $\hat{\theta}$ and θ. In terms of the matrices defined above for (5.6), we have

$$(5.9)\quad \hat{I}(*:2; O_n) = I(*:2; O_n) + (\hat{\boldsymbol{\theta}} - \boldsymbol{\theta})'\boldsymbol{\tau} + \tfrac{1}{2}(\hat{\boldsymbol{\theta}} - \boldsymbol{\theta})'\boldsymbol{\Sigma}^{-1}(\tau(\bar{\theta}))(\hat{\boldsymbol{\theta}} - \boldsymbol{\theta}).$$

If $\hat{\boldsymbol{\theta}}$ is of the form $(1/n)$ times the sum of n independent, identically distributed random vectors with finite covariance matrix $\boldsymbol{\Sigma}_1(\tau(\theta))$, then by the *central limit theorem* [Cramér (1937, pp. 112–113; 1955, pp. 114–116)], the distribution of $\sqrt{n}(\hat{\boldsymbol{\theta}} - \boldsymbol{\theta})$ tends to the multivariate normal distribution with zero means and covariance matrix $\boldsymbol{\Sigma}_1(\tau(\theta)) = n\boldsymbol{\Sigma}(\tau(\theta))$, and in particular $\hat{\boldsymbol{\theta}}$ converges to $\boldsymbol{\theta}$ in probability. [See, for example, Fraser (1957, pp. 208–215).]

We see from lemma 4.7 in chapter 3 that $\tau(\theta)$ is a continuous function of θ for all τ in the interval of finite existence of $M_2(\tau)$. We may therefore apply a theorem of Mann and Wald (1943) on stochastic limits, to conclude that the convergence in probability of $\hat{\boldsymbol{\theta}}$ to $\boldsymbol{\theta}$ implies the convergence in probability of $\boldsymbol{\tau}(\hat{\theta})$ to $\boldsymbol{\tau}(\theta)$. [Cf. Cramér (1946a, pp. 252–255).] Since $\boldsymbol{\tau}(\bar{\theta})$ lies between $\boldsymbol{\tau}(\hat{\theta})$ and $\boldsymbol{\tau}(\theta)$ [that is, each component of $\boldsymbol{\tau}(\bar{\theta})$ lies between the corresponding component of $\boldsymbol{\tau}(\hat{\theta})$ and $\boldsymbol{\tau}(\theta)$], $\boldsymbol{\tau}(\bar{\theta})$ converges in probability to $\boldsymbol{\tau}(\theta)$, and from lemmas 4.3 and 4.10 in chapter 3 and the Mann and Wald (1943) theorem, $\boldsymbol{\Sigma}(\tau(\bar{\theta}))$ converges in probability to $\boldsymbol{\Sigma}(\tau(\theta))$. From (5.7) we see that the distribution of $\hat{\boldsymbol{\tau}} - \boldsymbol{\tau}$ tends to the multivariate normal distribution with zero means and covariance matrix $\mathbf{G}^*(\theta) = \mathbf{G}^{*-1}(\tau) = \boldsymbol{\Sigma}^{-1}(\tau(\theta))$, where the matrices are defined in lemmas 4.10 and 4.11 of chapter 3. This is a well-known classical property of maximum-likelihood estimates.

At this point it is appropriate to remind the reader that the results in (5.6), (5.7), (5.9), and the previous paragraph are in terms of the parameters of the distribution of $\hat{\boldsymbol{\theta}}$, and not explicitly in terms of the parameters for a single observation. We must therefore remember that

$$\text{(5.10)} \qquad \boldsymbol{\Sigma}(\tau(\theta)) = O(1/n), \qquad \mathbf{G}^*(\theta) = O(n).$$

If the sample O_n is from the population with generalized density $f_2(x)$, then $\boldsymbol{\theta} = \boldsymbol{\theta}(0)$, $\boldsymbol{\tau} = \mathbf{0}$, $I(*:2;\, O_n) = 0$, and $2\hat{I}(*:2;\, O_n)$, as may be seen from (5.9), is asymptotically the quadratic form of the exponent of a multivariate normal distribution and therefore is distributed as χ^2 with k degrees of freedom [cf. Rao (1952, p. 55), problem 10.21 in chapter 9]. Note the similarity between (5.9) with $\boldsymbol{\tau} = \mathbf{0}$ and (6.4) of chapter 2 with $\hat{\boldsymbol{\theta}} - \boldsymbol{\theta}$ as $(\boldsymbol{\Delta\theta})$.

We may now determine a confidence region with asymptotic confidence coefficient $1 - \alpha$ for the parameters of $f_2(x)$ from the inequality

$$\text{(5.11)} \qquad 2\hat{I}(*:2;\, O_n) \leqq \chi^2(\alpha, k),$$

where $\chi^2(\alpha, k)$ is the value for which the χ^2-distribution with k degrees of freedom yields Prob $(\chi^2 \geqq \chi^2(\alpha, k)) = \alpha$. Since $2\hat{I}(*:2;\, O_n)$ is a convex function, the inequality (5.11) yields two limiting values for a single parameter, values within a closed curve for two parameters, values within a closed surface for three parameters, etc. We shall give some examples before we take up the distribution under the alternative hypothesis.

Example 5.5. We saw in example 1.1 that for the binomial distribution, $2\hat{I}(*:2;\, O_n) = 2n(\hat{p}\log(\hat{p}/p_2) + \hat{q}\log(\hat{q}/q_2))$, where $y = n\hat{p}$ is the observed number of successes. We thus have a 95% confidence interval for p_2 determined by the inequality

$$\text{(5.12)} \qquad 2n\left(\hat{p}\log\frac{\hat{p}}{p_2} + \hat{q}\log\frac{\hat{q}}{q_2}\right) \leq 3.84.$$

In table 5.1 are some 95% confidence intervals for the binomial computed by Howard R. Roberts. [See Roberts (1957) for a chart of confidence belts.]

TABLE 5.1

n \ $\hat{p}$	0	0.1	0.2	0.3	0.4	0.5	0.6	0.7	0.8	0.9	1.0
10	0	0.006	0.036	0.085	0.146	0.217	0.300	0.393	0.501	0.628	0.826
	0.174	0.372	0.499	0.607	0.700	0.783	0.854	0.915	0.964	0.994	1.000
20	0	0.017	0.067	0.132	0.207	0.291	0.383	0.484	0.595	0.722	0.909
	0.091	0.278	0.405	0.516	0.617	0.709	0.793	0.868	0.933	0.983	1.000
30	0	0.025	0.085	0.157	0.238	0.327	0.422	0.524	0.636	0.760	0.938
	0.062	0.240	0.364	0.476	0.578	0.673	0.762	0.843	0.915	0.975	1.000
50	0	0.037	0.106	0.185	0.272	0.364	0.462	0.565	0.676	0.797	0.962
	0.038	0.203	0.324	0.435	0.538	0.636	0.728	0.815	0.894	0.963	1.000
100	0	0.051	0.130	0.216	0.307	0.403	0.502	0.606	0.715	0.831	0.981
	0.019	0.169	0.285	0.394	0.498	0.597	0.693	0.784	0.870	0.949	1.000
250	0	0.067	0.154	0.246	0.341	0.438	0.538	0.641	0.747	0.859	0.992
	0.008	0.141	0.253	0.359	0.462	0.562	0.659	0.756	0.846	0.933	1.000
1000	0	0.082	0.176	0.272	0.370	0.469	0.570	0.671	0.774	0.880	0.998
	0.002	0.120	0.226	0.329	0.430	0.531	0.630	0.728	0.824	0.918	1.000

Example 5.6. We saw in example 1.4 that for a sample from a normal distribution with zero mean, $2\hat{I}(*:2; O_n) = n(\log(\sigma_2^2/y) - 1 + y/\sigma_2^2)$, where $y = (1/n)\sum_{i=1}^{n} x_i^2$. We thus have a 95% confidence interval for σ_2^2 determined by the inequality (cf. example 3.8)

$$n\left(\log\frac{\sigma_2^2}{y} - 1 + \frac{y}{\sigma_2^2}\right) \leq 3.84. \tag{5.13}$$

For $n = 10$ we get $y/2.15 \leq \sigma_2^2 \leq y/0.359$, and for $n = 100$ we get $y/1.303 \leq \sigma_2^2 \leq y/0.748$.

Example 5.7. We get from example 1.5 that for samples from a bivariate normal distribution with zero means and unit variances

$$2\hat{I}(*:2; O_n) = n\left(\log\frac{4(1-\rho_2^2)}{y_1y_2} - 2 + \frac{y_1 + y_2 - \rho_2(y_2 - y_1)}{2(1-\rho_2^2)}\right),$$

where $y_1 = \frac{1}{n}\sum_{i=1}^{n}(x_{1i} - x_{2i})^2$, $y_2 = \frac{1}{n}\sum_{i=1}^{n}(x_{1i} + x_{2i})^2$. We thus have a 95% confidence interval for ρ_2 determined by the inequality

$$n\left(\log\frac{4(1-\rho_2^2)}{y_1y_2} - 2 + \frac{y_1 + y_2 - \rho_2(y_2 - y_1)}{2(1-\rho_2^2)}\right) \leq 3.84. \tag{5.14}$$

We remark here that according to section 3.4 of chapter 12, for a sample O_n from a bivariate normal distribution with no specification of the means and variances, a 95% confidence interval for ρ is determined by the inequality

$$(5.15) \qquad (n-1)\left(\log\frac{1-\rho^2}{1-r^2} - 2 + \frac{2(1-r\rho)}{1-\rho^2}\right) \leq 3.84,$$

where r is the usual sample product-moment correlation coefficient.

Example 5.8. We saw in example 5.3 that $2\hat{I}(*:H_2(\mu, \sigma^2)) = n(\bar{x}-\mu)^2/\sigma^2 + (n-1)(\log(\sigma^2/s^2) - 1 + s^2/\sigma^2)$, with s^2 the unbiased sample variance, is asymptotically distributed as χ^2 with 2 degrees of freedom if the normal population parameters are μ and σ^2. Accordingly, for a sample O_n from a normal distribution, a 95% confidence region for (μ, σ^2) is determined by the inequality

$$(5.16) \qquad \frac{n(\bar{x}-\mu)^2}{\sigma^2} + (n-1)\left(\log\frac{\sigma^2}{s^2} - 1 + \frac{s^2}{\sigma^2}\right) \leq 5.99.$$

Example 5.9. We saw in example 1.6 that

$$2\hat{I}(*:2; O_n) = 2(n(L-\theta_2) - 1 - \log n(L-\theta_2)),$$

with $L = \min(x_1, x_2, \cdots, x_n)$, for a sample from the population defined by $f_2(x) = \exp[-(x-\theta_2)]$, $\theta_2 \leq x < \infty$. Accordingly, for a sample O_n from the population defined by the density $f_2(x)$, a 95% confidence interval for θ_2 is determined by the inequality

$$(5.17) \qquad n(L-\theta_2) - 1 - \log n(L-\theta_2) \leq 1.92.$$

We find that $0.057 \leq n(L-\theta_2) \leq 4.40$, that is, $L - 4.40/n \leq \theta_2 \leq L - 0.057/n$.

On the other hand, if the sample O_n is not from the population with generalized density $f_2(x)$, then, as may be seen from (5.9), asymptotically,

$$(5.18) \qquad E(2\hat{I}(*:2; O_n)) = 2I(*:2; O_n) + k = O(n) + k,$$

and

$$(5.19) \qquad 2\hat{I}(*:2; O_n) - 2I(*:2; O_n) - 2(\hat{\boldsymbol{\theta}} - \boldsymbol{\theta})'\boldsymbol{\tau}$$

is distributed as χ^2 with k degrees of freedom.

We shall now show that (5.19) is twice the logarithm of a likelihood ratio. Since

$$\begin{aligned} I(*:2; O_n) + (\hat{\theta} - \theta)\tau(\theta) &= \theta\tau(\theta) - \log M_2(\tau(\theta)) + (\hat{\theta}-\theta)\tau(\theta) \\ &= \hat{\theta}\tau(\theta) - \log M_2(\tau(\theta)) = \log(f^*(x)/f_2(x)), \end{aligned}$$

we may write [see (5.9)]

$$\begin{aligned} (5.20) \quad & 2\hat{I}(*:2; O_n) - 2I(*:2; O_n) - 2(\hat{\boldsymbol{\theta}} - \boldsymbol{\theta})'\boldsymbol{\tau} \\ &= 2\hat{I}(*:2; O_n) - 2(\hat{\boldsymbol{\theta}}'\boldsymbol{\tau} - \log M_2(\boldsymbol{\tau}')) \\ &= 2\log\frac{\max_{\tau} f^*(x)}{f_2(x)} - 2\log\frac{f^*(x)}{f_2(x)} = 2\log\frac{\max_{\tau} f^*(x)}{f^*(x)} \\ &= (\hat{\boldsymbol{\theta}} - \boldsymbol{\theta})'\boldsymbol{\Sigma}^{-1}(\tau(\bar{\theta}))(\hat{\boldsymbol{\theta}} - \boldsymbol{\theta}) = (\hat{\boldsymbol{\tau}} - \boldsymbol{\tau})'\boldsymbol{\Sigma}(\tau(\bar{\theta}))(\hat{\boldsymbol{\tau}} - \boldsymbol{\tau}). \end{aligned}$$

The test that rejects the null hypothesis [the sample is from the population with generalized density $f_2(x)$] if the value of $2\hat{I}(*:2; O_n)$ is large is consistent (has a power that tends to 1 as the sample size increases indefinitely). We see this by noting that if the sample is from the population with generalized density $f_2(x)$, then for large samples Prob $[2\hat{I}(*:2; O_n) \geqq \chi^2(\alpha, k)] = \alpha$, where $\chi^2(\alpha, k)$ depends only on α and the degrees of freedom k. On the other hand, if the sample is not from the population with generalized density $f_2(x)$, then from the weak law of large numbers, or Khintchine's theorem [cf. section 3 of chapter 4; Cramér (1946a, p. 253), Feller (1950, p. 191)], for any $\epsilon > 0$, $\beta > 0$, for sufficiently large n [see (5.18)]: Prob $[2\hat{I}(*:2; O_n) \geqq 2I(*:2; O_n) + k - \epsilon] \geqq 1 - \beta$. Note that for large enough n, $2I(*:2; O_n) + k - \epsilon \geqq \chi^2(\alpha, k)$, even for alternatives very close to the null hypothesis, close in the sense of small $I(*:2; O_1)$, since $I(*:2; O_n) = nI(*:2; O_1)$.

In order to derive a more useful statement about the asymptotic distribution under the alternative hypothesis than that about the expression in (5.19), we proceed as follows. Since

$$(5.21)\quad \begin{aligned}&(\hat{\boldsymbol{\theta}} - \boldsymbol{\theta} + \boldsymbol{\Sigma}(\tau(\bar{\theta}))\boldsymbol{\tau})'\boldsymbol{\Sigma}^{-1}(\tau(\bar{\theta}))(\hat{\boldsymbol{\theta}} - \boldsymbol{\theta} + \boldsymbol{\Sigma}(\tau(\bar{\theta}))\boldsymbol{\tau})\\ &\quad = (\hat{\boldsymbol{\theta}} - \boldsymbol{\theta})'\boldsymbol{\Sigma}^{-1}(\tau(\bar{\theta}))(\hat{\boldsymbol{\theta}} - \boldsymbol{\theta}) + 2(\hat{\boldsymbol{\theta}} - \boldsymbol{\theta})'\boldsymbol{\tau} + \boldsymbol{\tau}'\boldsymbol{\Sigma}(\tau(\bar{\theta}))\boldsymbol{\tau},\end{aligned}$$

we have from (5.6), (5.9), and (5.21):

$$(5.22)\quad \begin{aligned}&2\hat{I}(*:2; O_n) - 2I(*:2; O_n) + \boldsymbol{\tau}'\boldsymbol{\Sigma}(\tau(\bar{\theta}))\boldsymbol{\tau}\\ &\quad = (\hat{\boldsymbol{\theta}} - \boldsymbol{\theta} + \boldsymbol{\Sigma}(\tau(\bar{\theta}))\boldsymbol{\tau})'\boldsymbol{\Sigma}^{-1}(\tau(\bar{\theta}))(\hat{\boldsymbol{\theta}} - \boldsymbol{\theta} + \boldsymbol{\Sigma}(\tau(\bar{\theta}))\boldsymbol{\tau})\\ &\quad = \hat{\boldsymbol{\tau}}'\boldsymbol{\Sigma}(\tau(\bar{\theta}))\hat{\boldsymbol{\tau}}.\end{aligned}$$

We saw by the *central limit theorem* that the distribution of $\sqrt{n}(\hat{\boldsymbol{\theta}} - \boldsymbol{\theta})$ tends to a multivariate normal distribution with zero means and covariance matrix $\boldsymbol{\Sigma}_1(\tau(\theta)) = n\boldsymbol{\Sigma}(\tau(\theta))$. Consequently, asymptotically (cf. section 3 in chapter 12),

$$(5.23)\quad \hat{I}(*:2; O_n) = \hat{\boldsymbol{\theta}}'\hat{\boldsymbol{\tau}} - \boldsymbol{\theta}(0)'\hat{\boldsymbol{\tau}} - \tfrac{1}{2}\hat{\boldsymbol{\tau}}'\boldsymbol{\Sigma}(0)\hat{\boldsymbol{\tau}},$$

where $\hat{\boldsymbol{\theta}} = \boldsymbol{\theta}(0) + \boldsymbol{\Sigma}(0)\hat{\boldsymbol{\tau}}$ [cf. (5.6) with $\boldsymbol{\tau} = \mathbf{0}$], so that

$$(5.24)\quad \begin{aligned}2\hat{I}(*:2; O_n) &= (\hat{\boldsymbol{\theta}} - \boldsymbol{\theta}(0))'\boldsymbol{\Sigma}^{-1}(0)(\hat{\boldsymbol{\theta}} - \boldsymbol{\theta}(0))\\ &= n(\hat{\boldsymbol{\theta}} - \boldsymbol{\theta}(0))'\boldsymbol{\Sigma}_1^{-1}(0)(\hat{\boldsymbol{\theta}} - \boldsymbol{\theta}(0)) = \hat{\boldsymbol{\tau}}'\boldsymbol{\Sigma}(0)\hat{\boldsymbol{\tau}},\end{aligned}$$

and similarly,

$$(5.25)\quad \begin{aligned}2I(*:2; O_n) &= (\boldsymbol{\theta}(\tau) - \boldsymbol{\theta}(0))'\boldsymbol{\Sigma}^{-1}(0)(\boldsymbol{\theta}(\tau) - \boldsymbol{\theta}(0))\\ &= n(\boldsymbol{\theta}(\tau) - \boldsymbol{\theta}(0))'\boldsymbol{\Sigma}_1^{-1}(0)(\boldsymbol{\theta}(\tau) - \boldsymbol{\theta}(0)) = \boldsymbol{\tau}'\boldsymbol{\Sigma}(0)\boldsymbol{\tau}.\end{aligned}$$

We conclude from (5.22), (5.24), and (5.25) that $\boldsymbol{\Sigma}(\tau(\bar{\theta})) = \boldsymbol{\Sigma}(0)$ and therefore that $2\hat{I}(*:2; O_n)$ asymptotically is distributed as noncentral χ^2 with k degrees of freedom and noncentrality parameter $2I(*:2; O_n)$.

Note that this is consistent with (5.18) since the expected value of noncentral χ^2 is the sum of the noncentrality parameter and the degrees of freedom. (See problem 10.22 in chapter 9 and section 6.1 in chapter 12.)

Accordingly, whenever $f_2(x)$ is itself a member of an exponential family, as will be the case in most of the applications in the subsequent chapters, we see that

$$(5.26)\quad 2\hat{I}(*:H_2) = 2\log\frac{\max_{\tau\in\Omega} f^*(x)}{\max_{\tau\in\omega_2} f^*(x)} = \min_{\tau\in\omega_2}(\hat{\boldsymbol{\tau}} - \boldsymbol{\tau})'\boldsymbol{\Sigma}(0)(\hat{\boldsymbol{\tau}} - \boldsymbol{\tau}),$$

where Ω is the k-dimensional space of the τ's and ω_2 is the subspace of Ω for which $f^*(x)$ ranges over the populations of H_2. If ω_2 is an r-dimensional subspace of Ω, we may then infer from Wilks (1938a) and Wald (1943) that $2\hat{I}(*:H_2)$ is distributed asymptotically as χ^2 with $k - r$ degrees of freedom if the sample is from a population belonging to those specified by H_2, and that $2\hat{I}(*:H_2)$ is asymptotically distributed as noncentral χ^2 with $k - r$ degrees of freedom and noncentrality parameter $2I(*:H_2)$ in the contrary case. [Cf. Bartlett (1955, pp. 225–226), Bateman (1949), Cramér (1946a, pp. 424–434, 506), Fisher (1922a, 1924), Neyman (1949), Rao (1952, pp. 55–62), Weibull (1953).] We compare the exact probabilities that may be computed with the approximations from the asymptotic theory for particular illustrations in section 4 of chapter 6 and section 4 of chapter 7.

We remark that for many of the subsequent applications exact distributions are available, or better approximations may be found than those provided by the general theory. In each instance the asymptotic behavior agrees with the conclusions from the general theory.

6. ESTIMATE OF $J(*, 2)$

For the conjugate distribution $f^*(x) = e^{\tau T(x)}f_2(x)/M_2(\tau)$ defined in section 1, we find that

$$(6.1)\qquad J(*, 2) = \int (f^*(x) - f_2(x)) \log\frac{f^*(x)}{f_2(x)}\,d\lambda(x)$$
$$= (\theta - \theta(0))\tau(\theta).$$

Note that this is corollary 3.2 of chapter 3 with $\tau_1 = \tau$, $\tau_2 = 0$. We estimate $J(*, 2)$ by

$$(6.2)\qquad \hat{J}(*, 2) = (\hat{\theta} - \theta(0))\tau(\hat{\theta}),$$

where $T(x) = \hat{\theta} = \left[\dfrac{d}{d\tau}\log M_2(\tau)\right]_{\tau=\tau(\hat{\theta})}$. (See section 1.)

The implicit multidimensionality may be exhibited by writing

$$\hat{J}(*, 2) = (\hat{\boldsymbol{\theta}} - \boldsymbol{\theta}(0))'\hat{\boldsymbol{\tau}}, \tag{6.3}$$

where the matrices are defined in (5.6).

By proceeding as in section 5, we see that if the sample is from the population $f_2(x)$ specified by the null hypothesis, asymptotically

$$\hat{J}(*, 2) = (\hat{\boldsymbol{\theta}} - \boldsymbol{\theta}(0))'\boldsymbol{\Sigma}^{-1}(0)(\hat{\boldsymbol{\theta}} - \boldsymbol{\theta}(0)) \tag{6.4}$$

is distributed as χ^2 with k degrees of freedom.

On the other hand, from (5.23)

$$\hat{J}(*, 2) = \hat{\boldsymbol{\tau}}'\boldsymbol{\Sigma}(0)\hat{\boldsymbol{\tau}}, \tag{6.5}$$

that is, asymptotically $\hat{J}(*, 2)$ is equal to $2\hat{I}(*:2)$ and therefore the conclusions about the asymptotic behavior of $\hat{J}(*, 2)$ are the same as for $2\hat{I}(*:2)$. Note the similarity with the relation between $J(\theta, \theta + \Delta\theta)$ and $2I(\theta:\theta + \Delta\theta)$ in section 6 of chapter 2.

We shall denote the minimum value of $\hat{J}(*, 2)$ as f_2 ranges over the populations of H_2 by $\hat{J}(*, H_2)$. The asymptotic behavior of $\hat{J}(*, H_2)$ is the same as that of $2\hat{I}(*:H_2)$.

7. PROBLEMS

7.1. Consider the normal distributions $N(\mu_i, \sigma^2)$, $i = 1, 2$, $\mu_1 < \mu_2$. Show that for all regions A for which $\int_A f_1(x)\,dx = 1 - \alpha$, the maximum of $\int_A f_1(x) \log \frac{f_1(x)}{f_2(x)}\,dx$ occurs for the region $A = \{x: -\infty < x < g\}$.

7.2. Show that the critical region in example 3.3 is uniformly most powerful.

7.3. If in example 3.4 $p_1 = 0.20$, $p_2 = 0.80$, what is the critical value p? If $n = 25$, what are the errors of classification?

7.4. Show that the critical region in example 3.8 is uniformly most powerful.

7.5. Show that the critical region in example 3.11 is uniformly most powerful.

7.6. Show that the critical region in example 3.12 is most powerful.

7.7. Sketch the confidence region of (5.16) for $n = 100$, $\mu = 0$, $\sigma^2 = 1$.

7.8. Show that the unrestricted minimum of (4.18) with respect to σ^2 is $\frac{n-1}{2}\log\left(1 + \frac{n(\bar{x} - \mu)^2}{(n-1)s^2}\right)$ which for large n is approximately $\frac{n(\bar{x} - \mu)^2}{2s^2}$.

7.9. Prove the statement at the end of example 1.1.

7.10. Suppose the hypothesis H_i specifies the normal distribution $N(\mu_i, \sigma_i^2)$, $i = 1, 2$. Develop the test for the null hypothesis H_2 paralleling the procedures in the examples in section 3. [Cf. Kupperman (1957, pp. 94–96).]

7.11. Show that the classification procedure described in the first half of section 2, when $r = 2$, is such that the probability of misclassification tends to zero as the sample size tends to infinity. (Cf. problem 7.28 in chapter 3 and problem 4.22 in chapter 4.)

CHAPTER 6

Multinomial Populations

1. INTRODUCTION

We shall now undertake the application of the principles and results developed and derived in the preceding chapters to the analysis of samples for tests of statistical hypotheses.

In this chapter we take up the analysis of one or more samples from multinomial populations and in the next chapter the analysis for Poisson populations. The analyses in this chapter provide the basic structure for the analyses of contingency tables in chapter 8. We shall see that the analyses in chapters 6, 7, and 8 are in many respects similar to those of the analysis of variance. Indeed, we shall see in chapters 10 and 11 that the same basic technique applied to the analysis of samples from normal populations for the general linear hypothesis leads to the analysis of variance and its multivariate generalization.

We shall use the minimum discrimination information statistic obtained by replacing population parameters in the expression for the minimum discrimination information by best unbiased estimates under the various hypotheses.

For the special type of multinomial distribution that arises when sampling words or species of animals, an approximately unbiased estimate of entropy is given by Good (1953, p. 247). Miller and Madow (1954) give the maximum-likelihood estimate, and its asymptotic distribution, of the Shannon–Wiener measure of information for a multinomial.

All the formulas in chapters 6, 7, and 8 may be expressed in terms of the form $n \log n$ or $m \log n$ (all logarithms herein are to the Naperian base e). Table I on pages 367–377 gives values of $\log n$ and $n \log n$ for $n = 1$ through 1000. I am indebted to Sheldon G. Levin for the computation of the table of $n \log n$. Tables of $n \log n$ to base 2 and base 10 for $n = 1$ through 1000 may be found in a technical report by Miller and Ross (1954). Fisher (1956, pp. 137–138) lists $n \log n$ to base 10 for $n = 1$ through 150. Bartlett (1952) lists values, all to the Naperian base e, of $-\log p$, $-p \log p$, for $p = 0.00, 0.01, \cdots, 0.99, 1.00$, and $-(p \log p$

$+ q \log q), p + q = 1, p = 0.00, 0.01, \cdots, 0.50$. Klemmer, in an article on pages 71–77 of Quastler (1955), gives, all to the base 2, a table of $\log n$, $n = 1$ through 999, and a table of $-p \log p$ for $p = 0.001$ through 0.999. He also refers to AFCRC–TR 54–50 which contains, all to the base 2, a table of $\log n$ to 5 decimal places, $n = 1$ through 1000, a table of $n \log n$ to 5 decimal places, $n = 1$ through 500, and a table of $-p \log p$, $p \leqq 0.2500$ to 4 decimal places and $p \geqq 0.251$ to 3 decimal places. Dolanský and Dolanský (1952) have tabulated, all to the base 2, $-\log p$, $-p \log p$, and $-(p \log p + q \log q), p + q = 1$.

2. BACKGROUND

Suppose two simple statistical hypotheses, say H_1 and H_2, specify the probabilities of two hypothetical c-valued populations (c categories or classes),

$$(2.1) \qquad H_i: p_{i1}, p_{i2}, \cdots, p_{ic}, \qquad p_{i1} + p_{i2} + \cdots + p_{ic} = 1, \qquad i = 1, 2.$$

The mean information per observation from the population hypothesized by H_1, for discriminating for H_1 against H_2, is (see section 2 of chapter 1 for the general populations, of which this is a special case)

$$(2.2) \qquad I(1:2) = p_{11} \log \frac{p_{11}}{p_{21}} + p_{12} \log \frac{p_{12}}{p_{22}} + \cdots + p_{1c} \log \frac{p_{1c}}{p_{2c}}.$$

The mean information per observation from the population hypothesized by H_2, for discriminating for H_2 against H_1, is (see section 3 of chapter 1)

$$(2.3) \qquad I(2:1) = p_{21} \log \frac{p_{21}}{p_{11}} + p_{22} \log \frac{p_{22}}{p_{12}} + \cdots + p_{2c} \log \frac{p_{2c}}{p_{1c}}.$$

The divergence between H_1 and H_2, a measure of the difficulty of discriminating between them, is (see section 3 of chapter 1)

$$(2.4) \qquad J(1, 2) = I(1:2) + I(2:1) = (p_{11} - p_{21}) \log \frac{p_{11}}{p_{21}} + (p_{12} - p_{22}) \log \frac{p_{12}}{p_{22}} + \cdots + (p_{1c} - p_{2c}) \log \frac{p_{1c}}{p_{2c}}.$$

According to the general conclusions in chapter 2,

$$(2.5) \qquad I(1:2) \geqq 0, \qquad I(2:1) \geqq 0, \qquad J(1, 2) \geqq 0,$$

where the equality in (2.5) is satisfied in each case, if and only if $p_{1i} = p_{2i}$, $i = 1, 2, \cdots, c$, that is, the hypotheses imply the same population.

The mean discrimination information and divergence for a random sample of N independent observations, O_N, are,

$$(2.6)\qquad I(1:2; O_N) = NI(1:2) = N\sum_{i=1}^{c} p_{1i}\log(p_{1i}/p_{2i}),$$

$$(2.7)\qquad I(2:1; O_N) = NI(2:1) = N\sum_{i=1}^{c} p_{2i}\log(p_{2i}/p_{1i}),$$

$$(2.8)\qquad J(1,2; O_N) = NJ(1,2) = N\sum_{i=1}^{c}(p_{1i}-p_{2i})\log(p_{1i}/p_{2i}).$$

3. CONJUGATE DISTRIBUTIONS

Consider the N-total multinomial distribution on a c-valued population (c categories or classes),

$$(3.1)\qquad p(x) = p(x_1, x_2, \cdots, x_c) = \frac{N!}{x_1!\cdots x_c!}p_1^{x_1}p_2^{x_2}\cdots p_c^{x_c},$$

where $p_i > 0$, $i = 1, 2, \cdots, c$, $p_1 + p_2 + \cdots + p_c = 1$, $x_1 + x_2 + \cdots + x_c = N$. Suppose that $p^*(x)$ is any distribution on the c-valued population such that every possible observation from $p^*(x)$ is also a possible observation from $p(x)$. This is to avoid the contingency that $p^*(x) \neq 0$ and $p(x) = 0$. (See section 7 of chapter 2.)

Theorem 2.1 of chapter 3 permits us to assert:

LEMMA 3.1. *The least informative distribution on the c-valued population, with given expected values, for discrimination against the multinomial distribution* $p(x)$ *in* (3.1), *namely the distribution* $p^*(x)$ *such that* $E^*(x_i) = \theta_i$ *and* $\sum_{x_1+\cdots+x_c=N} p^*(x)\log\frac{p^*(x)}{p(x)}$ *is a minimum, is the distribution*

$$(3.2)\qquad p^*(x) = e^{\tau_1x_1+\cdots+\tau_cx_c}p(x)/(p_1e^{\tau_1}+\cdots+p_ce^{\tau_c})^N$$

$$= \frac{N!}{x_1!\cdots x_c!}(p_1^*)^{x_1}\cdots(p_c^*)^{x_c},$$

where $p_i^* = p_ie^{\tau_i}/(p_1e^{\tau_1}+\cdots+p_ce^{\tau_c})$, $i = 1, 2, \cdots, c$, *the* τ*'s are real parameters, and* $\theta_i = (\partial/\partial\tau_i)\log(p_1e^{\tau_1}+\cdots+p_ce^{\tau_c})^N$.

Note that the least informative distribution $p^*(x)$ here is a multinomial distribution. A simple numerical illustration of lemma 3.1 is in example 2.1 of chapter 3.

The multinomial distribution $p^*(x)$ in (3.2) is the conjugate distribution (see section 1 of chapter 5) of the multinomial distribution $p(x)$. The following are derived by the procedures exemplified in chapter 3:

$$(3.3)\quad \theta_i = Np_i^* = Np_ie^{\tau_i}/(p_1e^{\tau_1} + \cdots + p_ce^{\tau_c}), \qquad i = 1, 2, \cdots, c,$$

$$(3.4)\quad \theta_i/\theta_j = p_ie^{\tau_i}/p_je^{\tau_j}, \qquad i, j = 1, 2, \cdots, c,$$

$$(3.5)\quad \tau_i = \log(\theta_i/Np_i) + \log k, \qquad i = 1, 2, \cdots, c,$$
$$k = p_1e^{\tau_1} + \cdots + p_ce^{\tau_c} > 0,$$

$$(3.6)\quad I(*:2; O_N) = \sum_{x_1+\cdots+x_c=N} p^*(x)\log\frac{p^*(x)}{p(x)}$$
$$= \tau_1\theta_1 + \cdots + \tau_c\theta_c - N\log(p_1e^{\tau_1} + \cdots + p_ce^{\tau_c})$$
$$= \theta_1\log\frac{\theta_1}{Np_1} + \cdots + \theta_c\log\frac{\theta_c}{Np_c},$$

$$(3.7)\quad J(*, 2; O_N) = \sum_{x_1+\cdots+x_c=N} (p^*(x) - p(x))\log\frac{p^*(x)}{p(x)}$$
$$= \tau_1(\theta_1 - Np_1) + \cdots + \tau_c(\theta_c - Np_c)$$
$$= (\theta_1 - Np_1)\log\frac{\theta_1}{Np_1} + \cdots + (\theta_c - Np_c)\log\frac{\theta_c}{Np_c}.$$

Since the value of k in (3.5) is arbitrary, we shall take $k = 1$ for convenience so that in a homogeneous notation

$$(3.8)\quad \tau_i = \log(\theta_i/Np_i), \qquad i = 1, 2, \cdots, c.$$

On the other hand, since $x_c = N - x_1 - x_2 - \cdots - x_{c-1}$, we may also set $\tau_c = 0$, or $\log k = -\log(\theta_c/Np_c)$, in which case

$$(3.9)\quad \tau_i = \log\frac{\theta_ip_c}{p_i\theta_c}, \qquad i = 1, 2, \cdots, c - 1,$$
$$\tau_c = 0.$$

For applications to problems of tests of hypotheses about multinomial populations, the basic distribution in (3.1) will be that of the null hypothesis H_2, whereas the conjugate distribution will range over the populations of the alternative hypothesis H_1.

4. SINGLE SAMPLE

4.1. Basic Problem

Suppose we have a random sample of N independent observations, $x_1, x_2, \cdots, x_c$, $x_1 + x_2 + \cdots + x_c = N$, with a multinomial distribution on a c-valued population (c categories or classes), and we want to test the null hypothesis H_2 that the sample is from the population specified by

$$H_2:(p) = (p_1, p_2, \cdots, p_c), \qquad p_1 + p_2 + \cdots + p_c = 1, \tag{4.1}$$

against the alternative hypothesis H_1 that the sample is from any possible c-valued multinomial population.

We take for the conjugate distribution (3.2) the one with parameters the same as the observed best unbiased sample estimates, that is, $\hat{\theta}_i = N\hat{p}_i{}^* = x_i$, $i = 1, 2, \cdots, c$. From (3.8)

$$\hat{\tau}_i = \log \frac{x_i}{Np_i}, \qquad i = 1, 2, \cdots, c, \tag{4.2}$$

and the minimum discrimination information statistic is

$$\hat{I}(*:2; O_N) = x_1 \log \frac{x_1}{Np_1} + \cdots + x_c \log \frac{x_c}{Np_c}, \tag{4.3}$$

and the corresponding estimate of the divergence is

$$\hat{J}(*, 2; O_N) = N\left[\left(\frac{x_1}{N} - p_1\right) \log \frac{x_1}{Np_1} + \cdots + \left(\frac{x_c}{N} - p_c\right) \log \frac{x_c}{Np_c}\right]. \tag{4.4}$$

Note that (4.3) is (2.6) with the substitution of x_i/N for p_{1i} and p_i for p_{2i}, and that (4.4) is (2.8) with the same substitutions. (See problem 7.15.)

Under the null hypothesis H_2 of (4.1), it follows from sections 5 and 6 of chapter 5 that $2\hat{I}(*:2; O_N)$ and $\hat{J}(*, 2; O_N)$ are asymptotically distributed as χ^2 with $(c - 1)$ degrees of freedom. Under an alternative, $2\hat{I}(*:2; O_N)$ and $\hat{J}(*, 2; O_N)$ are asymptotically distributed as noncentral χ^2 with $(c - 1)$ degrees of freedom and noncentrality parameters $2I(*:2; O_N)$ and $J(*, 2; O_N)$ respectively, where $I(*:2; O_N)$ and $J(*, 2; O_N)$ are (4.3) and (4.4) with x_i/N, $i = 1, 2, \cdots, c$, replaced by the alternative probability. [See the last member of (3.6).]

Note that we may also write (4.3) as

$$\hat{I}(*:2; O_N) = \sum_{i=1}^{c} x_i \log x_i - \sum_{i=1}^{c} x_i \log p_i - N \log N,$$

for computational convenience with the table of $n \log n$.

Since $\log x \leqq x - 1$, $x > 0$, and the equality holds if and only if $x = 1$ [see Hardy, Littlewood, and Pólya (1934, p. 106, th. 150), or the statement following (2.7) in chapter 4], it follows that $(a - b)/a \leqq \log (a/b) \leqq (a - b)/b$, $a/b > 0$, and the equalities hold if and only if $a = b$. We may therefore use as a first approximation to $\log (a/b)$ the mean of its upper and lower bounds, that is, $\log (a/b) \approx \frac{1}{2}[(a - b)/a + (a - b)/b] = (a^2 - b^2)/2ab$, the approximation being better the closer a/b is to 1. This approximation in (4.3) and (4.4) yields

$$(4.5) \qquad 2\hat{I}(*:2; O_N) \approx \sum_{i=1}^{c} \frac{(x_i - Np_i)^2}{Np_i} = \chi^2,$$

$$(4.6) \qquad \hat{J}(*, 2; O_N) \approx \frac{1}{2}\sum_{i=1}^{c} \frac{(x_i - Np_i)^2}{Np_i} + \frac{1}{2}\sum_{i=1}^{c} \frac{(x_i - Np_i)^2}{x_i} = \tfrac{1}{2}(\chi^2 + \chi'^2),$$

where the first sum in (4.6) is K. Pearson's χ^2, and the second sum in (4.6) is Neyman's χ'^2 [Haldane (1955), Jeffreys (1948, pp. 170–173), Neyman (1929)].

We remark that $2\hat{I}(*:2; O_N)$ is $-2 \log \lambda$, with λ the likelihood-ratio test [see, for example, Fisher (1922b, pp. 357–358), Good (1957, p. 863), Wilks (1935a, p. 191)]. It is interesting to recall that Wilks (1935a) remarked that there was no theoretical reason why χ^2 should be preferred to $-2 \log \lambda$ and that $-2 \log \lambda$ can be computed with fewer operations than χ^2. Good (1957, p. 863) remarks that (I use the notation of this section) (i) $2\hat{I}(*:2; O_N)$ more closely puts the possible samples in order of their likelihoods under the null hypothesis, as compared with χ^2, for given $N, c, p_1, p_2, \cdots, p_c$, (ii) the calculation of $2\hat{I}(*:2; O_N)$ can be done by additions, subtractions, and table-lookups only, when tables of $2n \log n$ (to base e) are available, but the calculation is less "well-conditioned" than for χ^2, in the sense that more significant figures must be held, (iii) χ^2 is a simpler mathematical function of the observations and it should be easier to approximate closely to its distribution, given the null hypothesis.

4.2. Analysis of $\hat{I}(*:2; O_N)$

Significant values of $\hat{I}(*:2; O_N)$ may imply groupings of the categories as suggested by the nature of the data. $\hat{I}(*:2; O_N)$ in (4.3) can be *additively* analyzed to check such hypothetical groupings.

We consider first an analysis into $(c - 1)$ dichotomous comparisons of each category with the pool of all its successor categories. [Cf. Cochran (1954), Lancaster (1949).] Let us define

$$N_i = N - x_1 - x_2 - \cdots - x_i, \qquad i = 1, 2, \cdots, c - 1,$$
$$q_i = 1 - p_1 - p_2 - \cdots - p_i, \qquad i = 1, 2, \cdots, c - 1.$$

The analysis in table 4.1 is derived in a straightforward fashion from these definitions and the properties of the logarithm. The convexity property

$$a_1 \log \frac{a_1}{b_1} + \cdots + a_n \log \frac{a_n}{b_n} \geqq (a_1 + \cdots + a_n) \log \frac{a_1 + \cdots + a_n}{b_1 + \cdots + b_n},$$

where $a_i > 0$, $b_i > 0$, $i = 1, \cdots, n$, and the equality holds if and only if a_i/b_i = constant, $i = 1, 2, \cdots, n$ [see Hardy et al. (1934, p. 97, th. 117); also example 3.2 of chapter 2], ensures that the dichotomous comparisons are made with the minimum discrimination information statistic, that is, each "between component" is the minimum value of the "within component" below it in table 4.1 for the given grouping.

TABLE 4.1

Component due to	Information	D.F.
Within categories $c-1$ to $c\|x_1, \cdots, x_{c-2}$	$2\left(x_{c-1}\log\frac{x_{c-1}q_{c-2}}{N_{c-2}p_{c-1}} + x_c\log\frac{x_cq_{c-2}}{N_{c-2}p_c}\right)$	1
Between category $c-2$ and categories $(c-1)+c\|x_1, \cdots, x_{c-3}$	$2\left(x_{c-2}\log\frac{x_{c-2}q_{c-3}}{N_{c-3}p_{c-2}} + (N_{c-3}-x_{c-2})\log\frac{(N_{c-3}-x_{c-2})q_{c-3}}{N_{c-3}q_{c-2}}\right)$	1
$\cdot$	$\cdot$	$\cdot$
$\cdot$	$\cdot$	$\cdot$
$\cdot$	$\cdot$	$\cdot$
Within categories 3 to $c\|x_1, x_2$	$2\left(x_3\log\frac{x_3q_2}{N_2p_3} + x_4\log\frac{x_4q_2}{N_2p_4} + \cdots + x_c\log\frac{x_cq_2}{N_2p_c}\right)$	$c-3$
Between category 2 and categories $3+\cdots+c\|x_1$	$2\left(x_2\log\frac{x_2q_1}{N_1p_2} + (N_1-x_2)\log\frac{(N_1-x_2)q_1}{N_1q_2}\right)$	1
Within categories 2 to $c\|x_1$	$2\left(x_2\log\frac{x_2q_1}{N_1p_2} + x_3\log\frac{x_3q_1}{N_1p_3} + \cdots + x_c\log\frac{x_cq_1}{N_1p_c}\right)$	$c-2$
Between category 1 and categories $2+\cdots+c$	$2\left(x_1\log\frac{x_1}{Np_1} + (N-x_1)\log\frac{N-x_1}{N(1-p_1)}\right)$	1
Total, $2\hat{I}(*:2; O_N)$	$2\left(x_1\log\frac{x_1}{Np_1} + \cdots + x_c\log\frac{x_c}{Np_c}\right)$	$c-1$

We remark that the analysis in table 4.1 is a reflection of two facts:

1. A multinomial distribution may be expressed as the product of a

marginal binomial distribution and a conditional multinomial distribution of the other categories (cf. section 2 of chapter 2), for example,

$$\frac{N!}{x_1!x_2!\cdots x_c!}p_1^{x_1}\cdots p_c^{x_c} = \frac{N!}{x_1!(N-x_1)!}p_1^{x_1}(1-p_1)^{N-x_1} \times \frac{(N-x_1)!}{x_2!\cdots x_c!}\left(\frac{p_2}{q_1}\right)^{x_2}\cdots\left(\frac{p_c}{q_1}\right)^{x_c},$$

$$\frac{(N-x_1)!}{x_2!\cdots x_c!}\left(\frac{p_2}{q_1}\right)^{x_2}\cdots\left(\frac{p_c}{q_1}\right)^{x_c} = \frac{(N-x_1)!}{x_2!(N-x_1-x_2)!}\left(\frac{p_2}{q_1}\right)^{x_2}\left(1-\frac{p_2}{q_1}\right)^{N-x_1-x_2} \times \frac{(N-x_1-x_2)!}{x_3!\cdots x_c!}\left(\frac{p_3}{q_2}\right)^{x_3}\cdots\left(\frac{p_c}{q_2}\right)^{x_c},$$

$$\vdots \qquad\qquad \vdots$$

where $N_1 = N - x_1$, $N_2 = N - x_1 - x_2, \cdots, q_1 = 1 - p_1$, $q_2 = 1 - p_1 - p_2 = q_1 - p_2, \cdots$, $p_2/q_1 + \cdots + p_c/q_1 = 1$, $p_3/q_2 + \cdots + p_c/q_2 = 1, \cdots$.

2. The hypothesis H_2 is equivalent to the intersection of $c - 1$ hypotheses $H_{21}, \cdots, H_{2(c-1)}$, $H_2 = H_{21} \cap H_{22} \cap \cdots \cap H_{2(c-1)}$, where H_{21} is the hypothesis that the probability of occurrence of the first category is p_1, H_{22} is the hypothesis that the probability of occurrence of the second category is p_2 given that the probability of the first category is p_1, H_{23} is the hypothesis that the probability of occurrence of the third category is p_3 given that those of the first two categories are p_1 and p_2 respectively, etc.

The degrees of freedom in table 4.1 are those of the asymptotic χ^2-distributions under the null hypothesis H_2 of (4.1). We leave to the reader the estimation of the corresponding divergences. Note that the divergence in (4.4) does not permit a corresponding additive analysis.

We next consider a grouping or partitioning of the categories into two sets, say categories 1 to i, and $i + 1$ to c. Let us define

$$y_1 = x_1 + x_2 + \cdots + x_i, \qquad y_2 = x_{i+1} + x_{i+2} + \cdots + x_c,$$
$$p_{11} = p_1 + p_2 + \cdots + p_i, \qquad p_{22} = p_{i+1} + p_{i+2} + \cdots + p_c.$$

The analysis in table 4.2 is derived in a straightforward fashion from these definitions and the properties of the logarithm. The degrees of freedom in table 4.2 are those of the asymptotic χ^2-distributions under the null hypothesis H_2 of (4.1). We leave to the reader the estimation of the corresponding divergences. Note that the convexity property ensures that the "between component" is the minimum value of $2\hat{I}(*:2; O_N)$ for the given partitioning.

Without repeating all the details as for table 4.1, we note, for example, that in "within categories 1 to i," y_1 is the total (corresponding to N of the multinomial), and the conditional probabilities are $p_1/p_{11}, \cdots, p_i/p_{11}$.

TABLE 4.2

Component due to	Information	D.F.
Between categories $1 + \cdots + i$ and categories $(i+1) + \cdots + c$	$2\left(y_1 \log \frac{y_1}{Np_{11}} + y_2 \log \frac{y_2}{Np_{22}}\right)$	1
Within categories $(i+1)$ to c	$2\left(x_{i+1} \log \frac{x_{i+1}p_{22}}{y_2 p_{i+1}} + \cdots + x_c \log \frac{x_c p_{22}}{y_2 p_c}\right)$	$c - i - 1$
Within categories 1 to i	$2\left(x_1 \log \frac{x_1 p_{11}}{y_1 p_1} + \cdots + x_i \log \frac{x_i p_{11}}{y_1 p_i}\right)$	$i - 1$
Total, $2\hat{I}(*:2; O_N)$	$2\left(x_1 \log \frac{x_1}{Np_1} + \cdots + x_c \log \frac{x_c}{Np_c}\right)$	$c - 1$

4.3. Parametric Case

Let us now consider an analysis of $\hat{I}(*:2; O_N)$ assuming that $p_1, \cdots, p_c$ are known functions of independent parameters $\phi_1, \phi_2, \cdots, \phi_k$, $k < c$, and "fitting" the multinomial distribution by estimating the ϕ's. Suppose we have estimates $\tilde{\phi}_j(x_1, x_2, \cdots, x_c)$, $j = 1, 2, \cdots, k$ (by some procedure to be determined), and we write $\tilde{p}_i = p_i(\tilde{\phi}_1, \tilde{\phi}_2, \cdots, \tilde{\phi}_k)$, $i = 1, 2, \cdots, c$, $\tilde{p}_1 + \tilde{p}_2 + \cdots + \tilde{p}_c = 1$. We may write (4.3) as

$$(4.7) \quad \hat{I}(*:2; O_N) = \sum_{i=1}^{c} x_i \log \frac{x_i}{N\tilde{p}_i} + N \sum_{i=1}^{c} \tilde{p}_i \log \frac{\tilde{p}_i}{p_i} + N \sum_{i=1}^{c} \left(\frac{x_i}{N} - \tilde{p}_i\right) \log \frac{\tilde{p}_i}{p_i}.$$

For the decomposition of $\hat{I}(*:2; O_N)$ in (4.7) to be additive information-wise, that is, all terms to be of the form (2.2), the last term in (4.7) should be zero. We therefore require that the $\tilde{\phi}$'s be such that identically in the ϕ's

$$(4.8) \quad \sum_{i=1}^{c} x_i \log \frac{\tilde{p}_i}{p_i} = N \sum_{i=1}^{c} \tilde{p}_i \log \frac{\tilde{p}_i}{p_i}.$$

Note that the left-hand side of (4.8) is the *observed* value, $\log (\tilde{p}(x)/p(x))$, and the right-hand side of (4.8) is the *expected* value of the information in a sample of N observations from a population $(\tilde{p})$ for discriminating for $(\tilde{p})$ against (p). [Cf. (1.1) in chapter 5.] From (4.8), which is an identity in the ϕ's, we get

$$\sum_{i=1}^{c} \frac{x_i}{p_i} \frac{\partial p_i}{\partial \phi_j} = N \sum_{i=1}^{c} \frac{\tilde{p}_i}{p_i} \frac{\partial p_i}{\partial \phi_j}, \qquad i = 1, 2, \cdots, k, \tag{4.9}$$

and in particular when $(\phi) = (\tilde{\phi})$,

$$\sum_{i=1}^{c} \frac{x_i}{\tilde{p}_i} \left(\frac{\partial p_i}{\partial \phi_j}\right)_{\phi_j = \tilde{\phi}_j} = N \sum_{i=1}^{c} \frac{\tilde{p}_i}{\tilde{p}_i} \left(\frac{\partial p_i}{\partial \phi_j}\right)_{\phi_j = \tilde{\phi}_j} = 0, \qquad j = 1, 2, \cdots, k, \tag{4.10}$$

since $\sum_{i=1}^{c} \left(\frac{\partial p_i}{\partial \phi_j}\right)_{\phi_j = \tilde{\phi}_j} = 0$, or the $\tilde{\phi}$'s are the solutions of

$$\sum_{i=1}^{c} \frac{x_i}{p_i} \frac{\partial p_i}{\partial \phi_j} = 0, \qquad j = 1, 2, \cdots, k. \tag{4.11}$$

The equations (4.11) are the maximum-likelihood equations for estimating the ϕ's, and are also those which solve the problem of finding the ϕ's for which $\hat{I}(*:2: O_N)$ in (4.3) is a minimum. (See section 4 of chapter 5.) The properties of the estimates may be found, for example, in Cramér (1946a, pp. 426–434). (See problem 7.14.)

With estimates of the ϕ's satisfying (4.11), we have the analysis of $2\hat{I}(*:2; O_N)$ into additive components summarized in table 4.3. The degrees of freedom are those of the asymptotic χ^2-distributions under the null hypothesis H_2 in (4.1) with $p_i = p_i(\phi_1, \phi_2, \cdots, \phi_k)$, $i = 1, 2, \cdots, c$. [Cf. (4.17) in chapter 5.]

The divergences do not provide a similar additive analysis (with these estimates), but the estimate of the divergence corresponding to the error component is

$$\hat{J}(*, \tilde{p}) = N \sum_{i=1}^{c} \left(\frac{x_i}{N} - \tilde{p}_i\right) \log \frac{x_i}{N\tilde{p}_i}. \tag{4.12}$$

TABLE 4.3

Component due to	Information	D.F.
$\tilde{\phi}$'s or $(\tilde{p})$ against (p), $2\hat{I}(\tilde{p}:p)$	$2N \sum_{i=1}^{c} \tilde{p}_i \log \frac{\tilde{p}_i}{p_i}$	k
Error, (x/N) against $(\tilde{p})$, $2\hat{I}(*:\tilde{p})$	$2 \sum_{i=1}^{c} x_i \log \frac{x_i}{N\tilde{p}_i}$	$c - k - 1$
Total, $2\hat{I}(*:p)$	$2 \sum_{i=1}^{c} x_i \log \frac{x_i}{Np_i}$	$c - 1$

Under the null hypothesis H_2 of (4.1), $2\hat{I}(*:\tilde{p})$ and $\hat{J}(*, \tilde{p})$ are asymptotically distributed as χ^2 with $c - k - 1$ degrees of freedom. [For notational convenience we write $2\hat{I}(*:p) = 2\hat{I}(*:2; O_n)$.]

An example of this procedure is given by Fisher (1950), who considers a series of observations in which the number i occurs x_i times with a null hypothesis that the probabilities are given by the Poisson values $p_i = e^{-m}m^i/i!$. (Here m plays the role of the parameter ϕ.) The equation corresponding to (4.11) is $\sum_i x_i(-1 + i/m) = 0$, or $\tilde{m} = \sum_i ix_i/\sum_i x_i = \bar{\imath}$. The particular values [Fisher (1950, p. 18)] are:

i	x_i	$N\tilde{p}_i$	
0	124	119.6415	$\tilde{m} = 11/70$,
1	12	18.8008	
2	2	1.4772	$2\hat{I}(*:\tilde{p}) = 12.318$,
3	2	0.0774	$c - k - 1 = 2$.
4+	0	0.0031	
	140	140.0000	

Fisher compares this test, the usual χ^2 procedure, and the test for discrepancy of the variance, with the exact probabilities calculated from the conditional distribution for samples of the same size and average as the one in question. He concludes that [Fisher (1950, p. 24)] (in the notation of this section) $2\hat{I}(*:\tilde{p})$ "which is essentially the logarithmic difference in likelihood between the most likely Poisson series and the most likely theoretical series" is a measure that "seems well fitted to take the place of the conventional χ^2, when class expectations are small." [Cf. Cramér (1946a, pp. 434–437).]

4.4. "One-Sided" Binomial Hypothesis

We shall now examine a problem which is in some respects a special case of section 4.1, and is in some important respects different. Specifically, we want to test a "one-sided" hypothesis about a sample from a binomial population. Suppose we have a random sample of x "successes" and $N - x$ "failures" from a binomial population. We are interested in testing the two hypotheses:

(4.13)
H_1: the binomial population probability of success is $p_1 > p$,
H_2: the binomial population probability of success is equal to p.

See example 3.3 of chapter 5.

The results in section 3 apply to the binomial if we set $c = 2$, $p_1 = p$, $p_2 = q = 1 - p$, $x_1 = x$, $x_2 = N - x$, $\tau_1 = \tau$, $\tau_2 = 0$. The conjugate distribution [cf. (3.2)] ranges over the binomial distributions of H_1 in (4.13), if $p^* = (pe^\tau/(pe^\tau + q)) > p$. Only values of $\tau > 0$ are therefore admissible [see the paragraph following (2.12) in chapter 3]. With the value of the observed best unbiased sample estimate as the parameter of the conjugate distribution, that is, $\hat{\theta} = N\hat{p}^* = x$, we have

$$\hat{I}(p^*:p) = \hat{\tau}x - N\log(pe^{\hat{\tau}} + q), \tag{4.14}$$

$$\hat{\tau} = \log(xq/p(N-x)). \tag{4.15}$$

If $x > Np$, $\hat{\tau} = \log(xq/p(N-x)) > 0$ is admissible. If $x < Np$, $\hat{\tau} < 0$ is not admissible. We thus have the minimum discrimination information statistic (see example 3.3 of chapter 5, also the discussion following theorem 2.1 in chapter 3),

$$(4.16)\quad \hat{I}(H_1:H_2;O_N) = x\log\frac{x}{Np} + (N-x)\log\frac{N-x}{Nq}, \qquad x > Np,$$
$$= 0, \qquad x \leq Np.$$

Asymptotically, $2\hat{I}(H_1:H_2;O_N)$ has a χ^2 distribution with 1 degree of freedom under the null hypothesis H_2 of (4.13), but the α significance level must be taken from the usual χ^2 tables at the 2α level, since we do not consider values of $x < Np$ for which $\hat{I}(H_1:H_2;O_N)$ is the same as for some value of $x > Np$.

Instead of the simple null hypothesis H_2 of (4.13), let us consider the composite null hypothesis H_2':

(4.17)
H_1: the binomial population probability of success is $p_1 > p_0$,
H_2': the binomial population probability of success is $p \leq p_0$.

It may be verified from the behavior of $F(\hat{p}, p)$ in section 3 of chapter 4 and example 3.3 of chapter 5 that (see problem 7.17)

$$(4.18)\quad \inf_{p \leq p_0}\left(x\log\frac{x}{Np} + (N-x)\log\frac{N-x}{Nq}\right)$$
$$= x\log\frac{x}{Np_0} + (N-x)\log\frac{N-x}{Nq_0}, \qquad x > Np_0.$$

The minimum discrimination information statistic for the least informative distribution against the distributions of the composite null hypothesis is therefore

$$(4.19)\quad \hat{I}(H_1:H_2';O_N) = x\log\frac{x}{Np_0} + (N-x)\log\frac{N-x}{Nq_0}, \qquad x > Np_0,$$
$$= 0, \qquad x \leq Np_0.$$

Under the null hypothesis H_2' of (4.17), asymptotically,

$$\text{Prob}\{2\hat{I}(H_1:H_2'; O_N) \geqq \chi_{2\alpha}^2\} \leqq \alpha,$$

where $\chi_{2\alpha}^2$ is the usual χ^2 value at the 2α level for 1 degree of freedom.

Similarly, for the hypotheses

(4.20) H_3: the binomial population probability of success is $p_1 < p_0$,
H_2'': the binomial population probability of success is $p \geqq p_0$,

we have

$$\text{(4.21)}\quad \hat{I}(H_3:H_2''; O_N) = x \log \frac{x}{Np_0} + (N - x) \log \frac{N - x}{Nq_0}, \qquad x < Np_0,$$
$$= 0, \qquad x \geqq Np_0.$$

Under the null hypothesis H_2'' of (4.20), asymptotically,

$$\text{Prob}\{2\hat{I}(H_3:H_2''; O_N) \geqq \chi_{2\alpha}^2\} \leqq \alpha,$$

where $\chi_{2\alpha}^2$ is as above.

The two-sided hypothesis

(4.22) H_4: the binomial population probability of success is $p_1 \neq p_0$,
H_2: the binomial population probability of success is $p = p_0$,

is a special case of section 4.1, and

$$\text{(4.23)}\qquad 2\hat{I}(H_4:H_2; O_N) = 2\left(x \log \frac{x}{Np_0} + (N - x) \log \frac{N - x}{Nq_0}\right)$$

is asymptotically distributed as χ^2 with 1 degree of freedom under the null hypothesis H_2 of (4.22).

Note that H_2, H_2', and H_2'', respectively of (4.22), (4.17), and (4.20), satisfy $H_2 \rightleftharpoons H_2' \cap H_2''$, that is, $(p = p_0)$ if and only if $(p \leqq p_0)$ and $(p \geqq p_0)$; also H_4, H_1, and H_3, respectively of (4.22), (4.17), and (4.20), satisfy $H_4 \rightleftharpoons H_1 \cup H_3$, that is, $(p \neq p_0)$ if and only if $(p_1 > p_0)$ or $(p_1 < p_0)$. The region of acceptance common to the hypotheses H_2' and H_2'',

$$\text{(4.24)}\qquad x \log (x/Np_0) + (N - x) \log ((N - x)/Nq_0) \leqq \text{constant},$$

is also the region of acceptance of H_2.

4.5. "One-Sided" Multinomial Hypotheses

We now examine "one-sided" hypotheses for some problems on a c-valued population (c mutually exclusive categories).

The first problem tests a hypothesis H_1 that the first category occurs with a probability greater than $1/c$, against the null hypothesis H_2 of uniformity, that is,

$$(4.25)\qquad \begin{aligned} &H_1\colon p_1 > 1/c, \qquad p_1 + p_2 + \cdots + p_c = 1,\\ &H_2\colon p_1 = p_2 = \cdots = p_c = 1/c. \end{aligned}$$

Suppose we have a random sample of N independent observations as in section 4.1. From section 3, we see that the conjugate distribution ranges over the populations of H_1 in (4.25) if $p_1^* = e^{\tau_1}/(e^{\tau_1} + e^{\tau_2} + \cdots + e^{\tau_c}) > 1/c$. Only values of the τ_i, $i = 1, 2, \cdots, c$, such that $(c-1)e^{\tau_1} > e^{\tau_2} + \cdots + e^{\tau_c}$ are therefore admissible. With the values of the observed best unbiased sample estimates as the parameters of the conjugate distribution, that is, $\hat{\theta}_i = N\hat{p}_i^* = x_i$, we have

$$(4.26)\qquad \hat{I}(p^*:p) = \hat{\tau}_1 x_1 + \hat{\tau}_2 x_2 + \cdots + \hat{\tau}_c x_c - N \log ((e^{\hat{\tau}_1} + e^{\hat{\tau}_2} + \cdots + e^{\hat{\tau}_c})/c),$$

$$(4.27)\qquad x_i = \frac{Ne^{\hat{\tau}_i}}{e^{\hat{\tau}_1} + e^{\hat{\tau}_2} + \cdots + e^{\hat{\tau}_c}}, \qquad i = 1, 2, \cdots, c.$$

Since $e^{\hat{\tau}_i} = x_i/N$, $i = 1, 2, \cdots, c$ [we take $ck = 1$ in (3.5)], the $\hat{\tau}_i$ are in the admissible region if

$$(4.28)\qquad (c-1)\frac{x_1}{N} > \frac{x_2 + x_3 + \cdots + x_c}{N} = \frac{N - x_1}{N},$$

that is, if $x_1 > N/c$. If $x_1 \leqq N/c$, we must find the value of $\hat{I}(p^*:p)$ along the boundary of the admissible region, $(c-1)e^{\hat{\tau}_1} = e^{\hat{\tau}_2} + \cdots + e^{\hat{\tau}_c}$, the only other possible region for which $\hat{I}(p^*:p)$ may differ from zero, in which case [cf. Brunk (1958, p. 438)]

$$(4.29)\qquad \begin{aligned} \hat{I}(p^*:p) &= x_1 \log \frac{e^{\hat{\tau}_2} + \cdots + e^{\hat{\tau}_c}}{c-1} + x_2\hat{\tau}_2 + \cdots + x_c\hat{\tau}_c - N \log e^{\hat{\tau}_1}\\ &= x_2\hat{\tau}_2 + \cdots + x_c\hat{\tau}_c - (N - x_1) \log \frac{e^{\hat{\tau}_2} + \cdots + e^{\hat{\tau}_c}}{c-1}. \end{aligned}$$

The last expression is that for an $(N - x_1)$-total multinomial distribution over a $(c-1)$-valued population [by analogy with (4.26)]. We have, therefore,

$$(4.30)\qquad \hat{I}(H_1:H_2;O_N) = \sum_{i=1}^{c} x_i \log \frac{cx_i}{N}, \qquad x_1 > \frac{N}{c},$$

$$(4.31)\qquad \hat{I}(H_1:H_2;O_N) = \sum_{i=2}^{c} x_i \log \frac{(c-1)x_i}{N - x_1}, \qquad x_1 \leqq \frac{N}{c},$$

that is, when $x_1 \leqq N/c$, the rejection of the null hypothesis depends on the conditional values of $x_2, \cdots, x_c$.

If we set $p_i = 1/c$, $i = 1, 2, \cdots, c$, in table 4.1, the last three rows yield table 4.4, where the degrees of freedom are those of the asymptotic χ^2-distributions under the null hypothesis H_2 of (4.25).

TABLE 4.4

Component due to	Information	D.F.
Within categories 2 to $c \mid x_1$	$2\left(x_2 \log \frac{(c-1)x_2}{N-x_1} + \cdots + x_c \log \frac{(c-1)x_c}{N-x_1}\right)$	$c-2$
Between category 1 and categories $(2 + \cdots + c)$	$2\left(x_1 \log \frac{cx_1}{N} + (N-x_1) \log \frac{c(N-x_1)}{N(c-1)}\right)$	1
Total, $2\hat{I}(*:2; O_N)$	$2\left(x_1 \log \frac{cx_1}{N} + \cdots + x_c \log \frac{cx_c}{N}\right)$	$c-1$

Note that twice (4.30) is the total in table 4.4, and twice (4.31) is the component due to within categories 2 to c, given x_1, in table 4.4. The α significance level must be taken from the usual χ^2 tables at the 2α level.

The second problem restricts the hypothesis H_1 of (4.25) to equal probabilities of occurrence for all categories but the first, that is,

$$(4.32)\qquad \begin{aligned} &H_1'\colon\ p_1 = p > \frac{1}{c}, \qquad p_2 = p_3 = \cdots = p_c = \frac{1-p}{c-1}, \\ &H_2\colon\ p_1 = p_2 = \cdots = p_c = 1/c. \end{aligned}$$

The conjugate distribution ranges over the populations of H_1' in (4.32) if

$$p_1^* = \frac{e^{\tau_1}}{e^{\tau_1} + \cdots + e^{\tau_c}} > \frac{1}{c}, \qquad p_2^* = \frac{e^{\tau_2}}{e^{\tau_1} + \cdots + e^{\tau_c}}$$
$$= \cdots = p_c^* = \frac{e^{\tau_c}}{e^{\tau_1} + \cdots + e^{\tau_c}},$$

or $\tau_2 = \tau_3 = \cdots = \tau_c = \tau$, $\tau_1 > \tau$, are the only admissible values, and (4.26) is now

$$(4.33)\qquad \hat{I}(p^*:p) = \hat{\tau}_1 x_1 + (N - x_1)\hat{\tau} - N \log \frac{e^{\hat{\tau}_1} + (c-1)e^{\hat{\tau}}}{c},$$

$$(4.34)\qquad x_1 = \frac{Ne^{\hat{\tau}_1}}{e^{\hat{\tau}_1} + (c-1)e^{\hat{\tau}}}, \qquad N - x_1 = \frac{N(c-1)e^{\hat{\tau}}}{e^{\hat{\tau}_1} + (c-1)e^{\hat{\tau}}}.$$

the lower class limit as the corrected value. [See Cochran (1952).] The central probabilities in table 4.11 were obtained from the upper tail only of χ_0 as a normal variate $N(0, 1)$. The noncentral probabilities in table 4.11

TABLE 4.7

$$H_1':p_1 = p > \frac{1}{c}, \qquad p_2 = p_3 = \cdots = p_c = \frac{1-p}{c-1}$$

$$H_2:p_1 = p_2 = \cdots = p_c = 1/c$$

				Prob $(x_1 \leq N/c)$ = Prob $(2\hat{I}(H_1':H_2) = 0)$					
c	N	N/c	H_2	$H_1':p = 0.15$	$p = 0.20$	$p = 0.25$	$p = 0.30$	$p = 0.35$	$p = 0.40$
5	5	1	0.74	—	—	0.63	0.53	0.43	0.34
5	10	2	0.68	—	—	0.53	0.38	0.26	0.17
5	15	3	0.65	—	—	0.46	0.30	0.17	0.09
5	20	4	0.63	—	—	0.41	0.24	0.12	0.05
5	25	5	0.62	—	—	0.38	0.19	0.08	0.03
5	30	6	0.61	—	—	0.35	0.16	0.06	0.02
5	35	7	0.60	—	—	0.32	0.13	0.04	0.01
5	40	8	0.59	—	—	0.30	0.11	0.03	0.01
5	45	9	0.59	—	—	0.28	0.09	0.02	0.004
10	10	1	0.74	0.54	0.38	0.24	0.15	0.09	0.05
10	20	2	0.68	0.40	0.21	0.09	0.04	0.01	0.004
10	30	3	0.65	0.32	0.12	0.04	0.01	0.002	0.0003
10	40	4	0.63	0.26	0.08	0.02	0.003	0.0003	0.00003

TABLE 4.8

Prob $(x_1 \geq x_1')$

c	N	x_1'	H_2	$H_1':p = 0.15$	$p = 0.20$	$p = 0.25$	$p = 0.30$	$p = 0.35$	$p = 0.40$
5	5	4	0.0067	—	—	0.0156	0.0308	0.0540	0.0870
5	15	7	0.0181	—	—	0.0566	0.1311	0.2452	0.3902
5	25	10	0.0173	—	—	0.0713	0.1894	0.3697	0.5754
5	35	13	0.0142	—	—	0.0756	0.2271	0.4577	0.6943
5	45	16	0.0110	—	—	0.0753	0.2538	0.5248	0.7751
10	10	4	0.0128	0.0500	0.1209	0.2241	0.3504	0.4862	0.6177
10	20	6	0.0113	0.0673	0.1958	0.3828	0.5836	0.7546	0.8744
10	30	8	0.0078	0.0698	0.2392	0.4857	0.7186	0.8762	0.9565
10	40	9	0.0155	0.1354	0.4069	0.7002	0.8890	0.9697	0.9939

were obtained from the noncentral χ^2-distribution with 1 degree of freedom [see (6.9) in chapter 12]:

$$\text{Prob}\,(\chi^2 \geqq \chi_0^2) = \frac{1}{\sqrt{2\pi}}\left[\int_{\chi_0}^{\infty} e^{-\frac{1}{2}(t-\sqrt{\lambda})^2}\,dt + \int_{\chi_0}^{\infty} e^{-\frac{1}{2}(t+\sqrt{\lambda})^2}\,dt\right].$$

Note that Prob $(x_1 \geqq x_1')$ is supposed to be the same as Prob $(\chi^2 \geqq \chi_0^2)$. Indeed, χ_0^2 was selected, with a correction for continuity, to correspond

TABLE 4.9

$$2N\left(p \log cp + q \log \frac{cq}{c-1}\right) = 2I(H_1':H_2;\,O_N)$$

c	N	$p = 0.15$	0.20	0.25	0.30	0.35	0.40
5	5			0.0738	0.2817	0.6090	1.0465
5	15			0.2215	0.8450	1.8270	3.1395
5	25			0.3691	1.4084	3.0450	5.2325
5	35			0.5167	1.9717	4.2630	7.3255
5	45			0.6644	2.5351	5.4810	9.4185
10	10	0.2447	0.8881	1.8466	3.0733	4.5388	6.2248
10	20	0.4894	1.7761	3.6933	6.1465	9.0777	12.4495
10	30	0.7341	2.6642	5.5399	9.2198	13.6165	18.6743
10	40	0.9788	3.5522	7.3865	12.2931	18.1554	24.8991

TABLE 4.10

$$2\hat{I}(H_1':H_2;\,O_N) = 2\left(x_1 \log \frac{cx_1}{N} + (N - x_1)\log \frac{(N-x_1)c}{N(c-1)}\right), \qquad x_1 > \frac{N}{c}$$

c	N	x_1	$2\hat{I}$	$2\hat{I}$ (corrected) $= \chi_0^2$	c	N	x_1	$2\hat{I}$	$2\hat{I}$ (corrected) $= \chi_0^2$
5	5	3	3.819		10	10	3	3.073	
5	5	4	8.318	6.068	10	10	4	6.225	4.649
5	15	6	3.139		10	20	5	3.693	
5	15	7	5.375	4.257	10	20	6	6.147	4.920
5	25	9	3.440		10	30	7	4.486	
5	25	10	5.232	4.336	10	30	8	6.682	5.584
5	35	12	3.887		10	40	8	3.552	
5	35	13	5.484	4.686	10	40	9	5.326	4.439
5	45	15	4.385						
5	45	16	5.871	5.128					

to x_1'. Table 4.7 gives the probability of incorrectly accepting H_2 when $x_1 \leqq N/c$ under various members of H_1'. Even for the small values of N the approximation is good.

TABLE 4.11

				H_2				H_1'	
				Central Binomial				Noncentral Binomial	
c	N	x_1'	χ_0^2	Prob $(\chi^2 \geqq \chi_0^2)$	Prob $(x_1 \geqq x_1')$	p	λ	Prob $(\chi^2 \geqq \chi_0^2)$	Prob $(x_1 \geqq x_1')$
5	5	4	6.068	0.0069	0.0067	0.25	0.0738	0.0175	0.0156
5	15	7	4.257	0.0196	0.0181	0.30	0.8450	0.1285	0.1311
5	25	10	4.336	0.0187	0.0173	0.35	3.0450	0.3670	0.3697
5	35	13	4.686	0.0152	0.0142	0.40	7.3255	0.7088	0.6943
5	45	16	5.128	0.0118	0.0110	0.25	0.6644	0.0759	0.0753
10	10	4	4.649	0.0155	0.0128	0.15	0.2447	0.0526	0.0500
10	10	4	4.649			0.35	4.5388	0.4920	0.4862
10	20	6	4.920	0.0133	0.0113	0.20	1.7761	0.1869	0.1958
10	30	8	5.584	0.0091	0.0078	0.25	5.5399	0.4960	0.4857
10	40	9	4.439	0.0176	0.0155	0.15	0.9788	0.1324	0.1354
10	40	9	4.439			0.40	24.8991	0.9980	0.9939

5. TWO SAMPLES

5.1. Basic Problem

Suppose we have two independent random samples of N_1 and N_2 independent observations with multinomial distributions on a c-valued population. We denote the samples by

$$(x) = (x_1, x_2, \cdots, x_c), \qquad \sum_{i=1}^{c} x_i = N_1,$$

and

$$(y) = (y_1, y_2, \cdots, y_c), \qquad \sum_{i=1}^{c} y_i = N_2.$$

We want to test a null hypothesis of homogeneity H_2, the samples are from the same population, against the hypothesis H_1, the samples are from different populations, that is,

$$\text{(5.1)} \quad \begin{aligned} &H_1\text{: the samples are from different populations } (p_1) = (p_{11}, p_{12}, \cdots, p_{1c}), (p_2) = (p_{21}, p_{22}, \cdots, p_{2c}), \\ &H_2\text{: the samples are from the same population } (p) = (p_1, p_2, \cdots, p_c), p_{1i} = p_{2i} = p_i, i = 1, 2, \cdots, c. \end{aligned}$$

Since the samples are independent, we have in the notation of (3.1) (we omit indication of sample size, etc., except where confusion might otherwise result):

$$(5.2)\qquad I(1:2) = \sum_{(x),(y)} p_1(x)p_2(y) \log \frac{p_1(x)p_2(y)}{p(x)p(y)}$$
$$= N_1 \sum_{i=1}^{c} p_{1i} \log \frac{p_{1i}}{p_i} + N_2 \sum_{i=1}^{c} p_{2i} \log \frac{p_{2i}}{p_i},$$

$$(5.3)\qquad J(1, 2) = \sum_{(x),(y)} (p_1(x)p_2(y) - p(x)p(y)) \log \frac{p_1(x)p_2(y)}{p(x)p(y)}$$
$$= N_1 \sum_{i=1}^{c} (p_{1i} - p_i) \log \frac{p_{1i}}{p_i} + N_2 \sum_{i=1}^{c} (p_{2i} - p_i) \log \frac{p_{2i}}{p_i}.$$

The conjugate distributions are (see section 3 and section 1 in chapter 5),

$$(5.4)\qquad p_1^*(x) = \frac{p(x)e^{\tau_{11}x_1 + \cdots + \tau_{1c}x_c}}{(p_1e^{\tau_{11}} + \cdots + p_ce^{\tau_{1c}})^{N_1}},$$

$$(5.5)\qquad p_2^*(y) = \frac{p(y)e^{\tau_{21}y_1 + \tau_{22}y_2 + \cdots + \tau_{2c}y_c}}{(p_1e^{\tau_{21}} + \cdots + p_ce^{\tau_{2c}})^{N_2}}.$$

We find

$$(5.6)\qquad I(p^*:p) = \Sigma p_1^*(x)p_2^*(y) \log \frac{p_1^*(x)p_2^*(y)}{p(x)p(y)}$$
$$= \sum_{i=1}^{c} (\tau_{1i}E_1^*(x_i) + \tau_{2i}E_2^*(y_i)) - N_1 \log (p_1e^{\tau_{11}} + \cdots + p_ce^{\tau_{1c}}) - N_2 \log (p_1e^{\tau_{21}} + \cdots + p_ce^{\tau_{2c}}),$$

where $E_i^*(\ )$ denotes expected values from the population $p_i^*(\ )$.

Set $E_1^*(x_i) = N_1p_{1i}^*$, $E_2^*(y_i) = N_2p_{2i}^*$, where $p_{ji}^* = p_{ji}e^{\tau_{ji}}/(p_{j1}e^{\tau_{j1}} + \cdots + p_{jc}e^{\tau_{jc}})$, $j = 1, 2$; $i = 1, 2, \cdots, c$, and (5.6) is

$$(5.7)\qquad I(p^*:p) = N_1 \sum_{i=1}^{c} p_{1i}^* \log \frac{p_{1i}^*}{p_i} + N_2 \sum_{i=1}^{c} p_{2i}^* \log \frac{p_{2i}^*}{p_i}.$$

We take the conjugate distributions as those with parameters the same as the respective observed sample best unbiased estimates, that is, $\hat{p}_{1i}^* = x_i/N_1$, and $\hat{p}_{2i}^* = y_i/N_2$, $i = 1, 2, \cdots, c$, and

$$(5.8)\qquad \hat{I}(p^*:p) = \sum_{i=1}^{c} \left(x_i \log \frac{x_i}{N_1p_i} + y_i \log \frac{y_i}{N_2p_i}\right).$$

The null hypothesis H_2 of (5.1) usually does not specify the p_i, $i = 1, 2, \cdots, c$. We can analyze $\hat{I}(p^*:p)$ in (5.8) into two additive components, one due to the deviations between the p_i and their best unbiased estimates from the pooled samples, and the other due to what may be termed error within the samples. The analysis is summarized in table 5.1. The degrees of freedom are those of the asymptotic χ^2-distributions under the null hypothesis H_2 of (5.1). Note that the within component in table 5.1 is the minimum value of the total for variations of the p_i, $\sum_{i=1}^{c} p_i = 1$, that is, over the populations H_2.

We remark that the analysis in table 5.1 is a reflection of the fact that the hypothesis H_2 in (5.1) is the intersection of the hypotheses $H_2(\cdot)$, the samples are homogeneous, and $H_2(\cdot|(p))$, the homogeneous samples are from the population $(p) = (p_1, p_2, \cdots, p_c)$, that is, $H_2 = H_2(\cdot) \cap H_2(\cdot|(p))$. The between component in table 5.1, $2\hat{I}(\hat{p}:p)$, is a test for the hypothesis $H_2(\cdot|(p))$, and the within component in table 5.1, $2\hat{I}(p^*:\hat{p})$ or $2\hat{I}(H_1:H_2)$, is a conditional test for the hypothesis $H_2(\cdot)$, subject to the observed values of $\hat{p}_i = (x_i + y_i)/(N_1 + N_2)$, $i = 1, 2, \cdots, c$.

TABLE 5.1

Component due to	Information	D.F.
$\hat{p}_i = (x_i + y_i)/(N_1 + N_2)$ (Between), $2\hat{I}(\hat{p}:p)$	$2\sum_{i=1}^{c}(x_i + y_i)\log\frac{(x_i + y_i)}{(N_1 + N_2)p_i}$	$c - 1$
Error, $2\hat{I}(p^*:\hat{p})$ (Within)	$2\sum_{i=1}^{c}\left(x_i\log\frac{(N_1 + N_2)x_i}{N_1(x_i + y_i)} + y_i\log\frac{(N_1 + N_2)y_i}{N_2(x_i + y_i)}\right)$	$c - 1$
Total, $2\hat{I}(p^*:p)$	$2\sum_{i=1}^{c}\left(x_i\log\frac{x_i}{N_1p_i} + y_i\log\frac{y_i}{N_2p_i}\right)$	$2(c - 1)$

The error component in table 5.1 may also be expressed as

$$\text{(5.9)}\quad \hat{I}(p^*:\hat{p}) = \hat{I}(H_1:H_2) = \Sigma x_i\log x_i + \Sigma y_i\log y_i - \Sigma(x_i + y_i)\log(x_i + y_i) + (N_1 + N_2)\log(N_1 + N_2) - N_1\log N_1 - N_2\log N_2,$$

for computational convenience with the table of $n \log n$.

The divergences do not provide a similar additive analysis (with these estimates), but the estimate of the divergence corresponding to the error component is

$$\text{(5.10)}\quad \hat{J}(p^*, \hat{p}) = \hat{J}(H_1, H_2) = N_1\Sigma\left(\frac{x_i}{N_1} - \frac{x_i + y_i}{N_1 + N_2}\right)\log\frac{(N_1 + N_2)x_i}{N_1(x_i + y_i)} + N_2\Sigma\left(\frac{y_i}{N_2} - \frac{x_i + y_i}{N_1 + N_2}\right)\log\frac{(N_1 + N_2)y_i}{N_2(x_i + y_i)}$$

$$= \frac{N_1N_2}{N_1 + N_2}\Sigma\left(\frac{x_i}{N_1} - \frac{y_i}{N_2}\right)\log\frac{N_2x_i}{N_1y_i}.$$

Note that $\hat{I}(H_1:H_2) = \hat{I}(p^*:\hat{p})$ in table 5.1 is (5.2) with the substitution of x_i/N_1 for p_{1i}, y_i/N_2 for p_{2i}, and $(x_i + y_i)/(N_1 + N_2)$ for p_i, and that (5.10) is (5.3) with the same substitutions.

$2\hat{I}(H_1:H_2)$ and $\hat{J}(H_1, H_2)$ are asymptotically distributed as χ^2 with $(c - 1)$ degrees of freedom under the hypothesis H_2 of (5.1) (the samples are from the same population).

With the approximations used in (4.5) and (4.6), we find [cf. Pearson (1911)]

$$(5.11)\quad 2\hat{I}(H_1:H_2) \approx \frac{1}{N_1N_2}\sum\frac{(N_2x_i - N_1y_i)^2}{x_i + y_i} = \chi^2$$

$$\hat{J}(H_1, H_2) \approx \frac{1}{2N_1N_2}\sum\frac{(N_2x_i - N_1y_i)^2}{x_i + y_i} + \frac{1}{2(N_1 + N_2)^2}\sum\frac{(N_2x_i - N_1y_i)^2(x_i + y_i)}{x_iy_i}.$$

5.2. "One-Sided" Hypothesis for the Binomial

We now consider a "one-sided" hypothesis about two binomial distributions. Suppose we have two independent random binomial samples of N_1 and N_2 independent observations, of which, respectively, x and y are "successes." We want to test the two hypotheses:

(5.12)
H_1': the samples are from different binomial populations with respective probabilities of success $p_1, p_2, p_1 > p_2$,
H_2: the samples are from the same binomial population, $p_1 = p_2 = p$.

From the analogues of (5.4), (5.5), and (5.6) for binomial distributions (cf. section 4.4), we see that the conjugate distributions range over the binomial populations of H_1' in (5.12) if

$$p_1^* = \frac{pe^{\tau_1}}{pe^{\tau_1} + (1 - p)} > p_2^* = \frac{pe^{\tau_2}}{pe^{\tau_2} + (1 - p)}.$$

Only values $\tau_1 > \tau_2$ are therefore admissible. We take the conjugate distributions as those with parameters the same as the respective observed sample best unbiased estimates, that is, $\hat{p}_1^* = x/N_1$, $\hat{p}_2^* = y/N_2$, and [cf. (5.6)]

$$(5.13)\quad \hat{I}(p^*:p) = \hat{\tau}_1x - N_1\log(pe^{\hat{\tau}_1} + (1 - p)) + \hat{\tau}_2y - N_2\log(pe^{\hat{\tau}_2} + (1 - p)),$$

$$(5.14)\quad x = \frac{N_1pe^{\hat{\tau}_1}}{pe^{\hat{\tau}_1} + (1 - p)}, \qquad y = \frac{N_2pe^{\hat{\tau}_2}}{pe^{\hat{\tau}_2} + (1 - p)},$$

or

$$(5.15)\qquad \hat{\tau}_1 = \log \frac{(1-p)x}{p(N_1 - x)}, \qquad \hat{\tau}_2 = \log \frac{(1-p)y}{p(N_2 - y)}.$$

If $x/N_1 > y/N_2$, then $\hat{\tau}_1 > \hat{\tau}_2$, and the $\hat{\tau}$'s are admissible. However, if $x/N_1 \leqq y/N_2$, then $\hat{\tau}_1 \leqq \hat{\tau}_2$, the $\hat{\tau}$'s are not admissible, and we must find the value of $\hat{I}(p^*:p)$ along the boundary $\hat{\tau}_1 = \hat{\tau}_2$ of the admissible area. With $\hat{\tau} = \hat{\tau}_2 = \hat{\tau}_1$ in (5.13) we have

$$(5.16)\quad x + y = \frac{(N_1 + N_2)pe^{\hat{\tau}}}{pe^{\hat{\tau}} + (1-p)}, \quad \text{or} \quad \hat{\tau} = \log \frac{(1-p)(x+y)}{p(N_1 + N_2 - x - y)}.$$

We thus have

$$(5.17)\quad \hat{I}(p^*:p) = x \log \frac{x}{N_1 p} + (N_1 - x) \log \frac{N_1 - x}{N_1(1-p)} + y \log \frac{y}{N_2 p} + (N_2 - y) \log \frac{N_2 - y}{N_2(1-p)}, \qquad \frac{x}{N_1} > \frac{y}{N_2},$$

$$(5.18)\quad \hat{I}(p^*:p) = (x+y) \log \frac{x+y}{(N_1 + N_2)p} + (N_1 + N_2 - x - y) \log \frac{N_1 + N_2 - x - y}{(N_1 + N_2)(1-p)}, \qquad \frac{x}{N_1} \leqq \frac{y}{N_2}.$$

Table 5.2 gives the analysis of table 5.1 for binomial distributions, for the two-sided hypotheses:

(5.19)
H_1: the samples are from binomial populations with respective probabilities of success p_1, p_2, $p_1 \neq p_2$,
H_2: the samples are from the same binomial population, $p_1 = p_2 = p$.

We see therefore that $2\hat{I}(p^*:p)$ in (5.17) is the total of table 5.2 when $x/N_1 > y/N_2$, and $2\hat{I}(p^*:p)$ in (5.18) is the between component of table 5.2 when $x/N_1 \leqq y/N_2$.

The hypothesis H_2 of (5.12) usually does not specify the value of p, and the minimum values of the total and between component respectively (with respect to variations of p, that is, over the populations H_2) are then

$$(5.20)\quad 2\hat{I}(H_1':H_2) = \text{error component of table 5.2}, \qquad x/N_1 > y/N_2,$$
$$= 0, \qquad x/N_1 \leqq y/N_2.$$

Asymptotically, $2\hat{I}(H_1':H_2)$ in (5.20) is distributed as χ^2 with 1 degree of freedom under the null hypothesis H_2 of (5.12), but the α significance level must be taken from the usual χ^2 tables at the 2α level.

Similarly, for testing the two hypotheses

(5.21)
H_1'': the samples are from different binomial populations with respective probabilities of success $p_1, p_2, p_1 < p_2$,

H_2: the samples are from the same binomial population, $p_1 = p_2 = p$,

we have $\hat{I}(p^*:p)$ in (5.17) when $x/N_1 < y/N_2$ and $\hat{I}(p^*:p)$ in (5.18) when $x/N_1 \geqq y/N_2$. The hypothesis H_2 of (5.21) usually does not specify the value of p, and then

(5.22) $$2\hat{I}(H_1'':H_2) = \text{error component of table 5.2}, \quad x/N_1 < y/N_2, \qquad = 0, \quad x/N_1 \geqq y/N_2.$$

Asymptotically, $2\hat{I}(H_1'':H_2)$ in (5.22) is distributed as χ^2 with 1 degree of freedom under the null hypothesis H_2 of (5.21), but the α significance level must be taken from the usual χ^2 tables at the 2α level.

Note that H_1, H_1', and H_1'' of (5.19), (5.12), and (5.21) respectively satisfy $H_1 \rightleftharpoons H_1' \cup H_1''$, that is, $(p_1 \neq p_2)$ if and only if $(p_1 > p_2)$ or $(p_1 < p_2)$.

We remark that in table 5.2 when $x/N_1 = y/N_2$ the total becomes equal to the between component and the within component vanishes. In any

TABLE 5.2

Component due to	Information	D.F.
$\hat{p} = \dfrac{x+y}{N_1+N_2}$ $2\hat{I}(\hat{p}:p)$, (Between)	$2\left((x+y)\log\dfrac{x+y}{(N_1+N_2)p} + (N_1+N_2-x-y)\log\dfrac{N_1+N_2-x-y}{(N_1+N_2)(1-p)}\right)$	1
Error, $2\hat{I}(p^*:\hat{p})$ (Within)	$2\left(x\log\dfrac{(N_1+N_2)x}{N_1(x+y)} + (N_1-x)\log\dfrac{(N_1+N_2)(N_1-x)}{N_1(N_1+N_2-x-y)} + y\log\dfrac{(N_1+N_2)y}{N_2(x+y)} + (N_2-y)\log\dfrac{(N_1+N_2)(N_2-y)}{N_2(N_1+N_2-x-y)}\right)$	1
Total, $2\hat{I}(p^*:p)$	$2\left(x\log\dfrac{x}{N_1p} + (N_1-x)\log\dfrac{(N_1-x)}{N_1(1-p)} + y\log\dfrac{y}{N_2p} + (N_2-y)\log\dfrac{N_2-y}{N_2(1-p)}\right)$	2

case, the alternative hypotheses H_1', H_1'', or H_1 will be accepted if the minimum discrimination information statistic exceeds some constant.

We summarize the preceding in table 5.3, the entries describing for the different hypotheses when the test statistic is the total, within, or between component in table 5.2. For example, in the test of $H_1':(p_1 > p_2)$ against $H_2:(p_1 = p_2 = p)$, when $p = p_0$ is specified, use the total $2\hat{I}(p^*:p)$ when $x/N_1 > y/N_2$, the between $2\hat{I}(\hat{p}:p)$ when $x/N_1 \leqq y/N_2$. However, when p is not specified, use the within $2\hat{I}(p^*:\hat{p})$ when $x/N_1 > y/N_2$, accept the null hypothesis H_2 when $x/N_1 \leqq y/N_2$.

TABLE 5.3

	$H_1':H_2$ (5.12)		$H_1'':H_2$ (5.21)		$H_1:H_2$ (5.19)
Between, $2\hat{I}(\hat{p}:p)$	$\dfrac{x}{N_1} \leqq \dfrac{y}{N_2}$,	$p = p_0$	$\dfrac{x}{N_1} \geqq \dfrac{y}{N_2}$,	$p = p_0$	$p = p_0$
Within, $2\hat{I}(p^*:\hat{p})$	$\dfrac{x}{N_1} > \dfrac{y}{N_2}$,	$p = \hat{p} = \dfrac{x+y}{N_1+N_2}$	$\dfrac{x}{N_1} < \dfrac{y}{N_2}$,	$p = \hat{p} = \dfrac{x+y}{N_1+N_2}$	$p = \hat{p} = \dfrac{x+y}{N_1+N_2}$
	0,	$\dfrac{x}{N_1} \leqq \dfrac{y}{N_2}$	0,	$\dfrac{x}{N_1} \geqq \dfrac{y}{N_2}$	
Total, $2\hat{I}(p^*:p)$	$\dfrac{x}{N_1} > \dfrac{y}{N_2}$,	$p = p_0$	$\dfrac{x}{N_1} < \dfrac{y}{N_2}$,	$p = p_0$	$p = p_0$

6. *r* SAMPLES

6.1. Basic Problem

Suppose we have r independent random samples with multinomial distributions on a c-valued population, and we are interested in a test of the null hypothesis that the samples are homogeneous. We denote the samples by

$$(x_i) = (x_{i1}, x_{i2}, \cdots, x_{ic}), \quad x_{i1} + x_{i2} + \cdots + x_{ic} = N_i, \quad i = 1, 2, \cdots, r,$$

and consider the two hypotheses,

(6.1)
H_1: the samples are from different populations $(p_{i1}, p_{i2}, \cdots, p_{ic})$, $i = 1, 2, \cdots, r$,

H_2: the samples are from the same population $(p_1, p_2, \cdots, p_c)$, namely, $p_{ij} = p_j > 0$, $i = 1, 2, \cdots, r$, $j = 1, 2, \cdots, c$.

Without repeating the detailed argument, which is similar to that already used, we find here that,

$$I(1:2) = \sum_{i=1}^{r} N_i \sum_{j=1}^{c} p_{ij} \log \frac{p_{ij}}{p_j}, \tag{6.2}$$

$$J(1, 2) = \sum_{i=1}^{r} N_i \sum_{j=1}^{c} (p_{ij} - p_j) \log \frac{p_{ij}}{p_j}. \tag{6.3}$$

For the conjugate distributions with parameters the same as the respective observed sample best unbiased estimates, we have

$$\hat{I}(p^*:p) = \sum_{i=1}^{r} \sum_{j=1}^{c} x_{ij} \log \frac{x_{ij}}{N_i p_j}. \tag{6.4}$$

The hypothesis H_2 of (6.1) usually does not specify the $p_j, j = 1, 2, \cdots, c$. We can analyze $\hat{I}(p^*:p)$ in (6.4) into two additive components, one due to the deviations between the p_j and their best unbiased estimates from the pooled samples, and the other due to what may be termed error within the samples. Letting $x_j = \sum_{i=1}^{r} x_{ij}$, $N = N_1 + N_2 + \cdots + N_r$, the analysis is summarized in table 6.1. The degrees of freedom are those of the asymptotic χ^2-distributions under the null hypothesis H_2 of (6.1). Note that $\hat{I}(p^*:\hat{p})$ in table 6.1 is the minimum value of (6.4) for variations of the p_j, $\sum_{j=1}^{c} p_j = 1$, that is, over the populations of H_2, and by the convexity property (see section 4.2, and section 3 of chapter 2), $\sum_{i=1}^{r} \sum_{j=1}^{c} x_{ij} \log \frac{x_{ij}}{N_i p_j} \geq \sum_{j=1}^{c} x_j \log \frac{x_j}{N p_j}$. We shall write $\hat{I}(H_1:H_2) = \hat{I}(p^*:\hat{p})$.

We remark that the analysis in table 6.1 is a reflection of the fact that the hypothesis H_2 in (6.1) is the intersection of the hypotheses $H_2(\cdot)$, the

TABLE 6.1

Component due to	Information	D.F.
$\hat{p}_j = x_j/N$, $2\hat{I}(\hat{p}:p)$ (Between)	$2 \sum_{j=1}^{c} x_j \log \frac{x_j}{N p_j}$	$c - 1$
Error, $2\hat{I}(p^*:\hat{p})$ (Within)	$2 \sum_{i=1}^{r} \sum_{j=1}^{c} x_{ij} \log \frac{N x_{ij}}{N_i x_j}$	$(r - 1)(c - 1)$
Total, $2\hat{I}(p^*:p)$	$2 \sum_{i=1}^{r} \sum_{j=1}^{c} x_{ij} \log \frac{x_{ij}}{N_i p_j}$	$r(c - 1)$

samples are homogeneous, and $H_2(\cdot|(p))$, the homogeneous samples are from the population $(p) = (p_1, p_2, \cdots, p_c)$, that is, $H_2 = H_2(\cdot) \cap H_2(\cdot|(p))$. The between component in table 6.1, $2\hat{I}(\hat{p}:p)$, tests the hypothesis $H_2(\cdot|(p))$, and the within component in table 6.1, $2\hat{I}(p^*:\hat{p})$ or $2\hat{I}(H_1:H_2)$, conditionally tests the hypothesis $H_2(\cdot)$, subject to the observed values of $\hat{p}_j = x_j/N$, $j = 1, 2, \cdots, c$.

The error component in table 6.1 may also be expressed as

$$(6.5)\quad \hat{I}(p^*:\hat{p}) = \hat{I}(H_1:H_2) = \sum_{i=1}^{r}\sum_{j=1}^{c} x_{ij}\log x_{ij} - \sum_{j=1}^{c} x_j \log x_j + N\log N - \sum_{i=1}^{r} N_i \log N_i,$$

for computational convenience with the table of $n \log n$.

The divergences do not provide a similar additive analysis (with these estimates) but the estimate of the divergence corresponding to the error component is

$$(6.6)\qquad \hat{J}(p^*, \hat{p}) = \hat{J}(H_1, H_2) = \sum_{i=1}^{r} N_i \sum_{j=1}^{c}\left(\frac{x_{ij}}{N_i} - \frac{x_j}{N}\right)\log\frac{N x_{ij}}{N_i x_j}.$$

Note that $\hat{I}(p^*:\hat{p})$ of table 6.1 is (6.2) with the substitution of x_{ij}/N_i for p_{ij} and x_j/N for p_j and that (6.6) is (6.3) with the same substitutions.

$2\hat{I}(H_1:H_2)$ and $\hat{J}(H_1, H_2)$ are asymptotically distributed as χ^2 with $(r-1)(c-1)$ degrees of freedom under the null hypothesis H_2 of (6.1) (the samples are from the same population).

With the approximations used in (4.5) and (4.6), we find that [cf. Hsu (1949, pp. 397–398)] (see problem 7.18)

$$2\hat{I}(H_1:H_2) \approx \sum_{i=1}^{r}\sum_{j=1}^{c} N\left(x_{ij} - \frac{N_i x_j}{N}\right)^2 \Big/ N_i x_j.$$

6.2. Partition

The error component in table 6.1 can be analyzed into $(r-1)$ comparisons, each of $(c-1)$ degrees of freedom, between each sample and the pooled sample of all its predecessors. This permits an assessment of each sample as it is added, to test for an abrupt change. [Cf. Cochran (1954, pp. 422–423), Lancaster (1949).] For partitioning within the categories see section 4.2.

To indicate the successive pooling of the samples, we define,

$$(6.7)\quad y_{ij} = x_{1j} + x_{2j} + \cdots + x_{ij}, \qquad i = 2, \cdots, r-1, \qquad j = 1, 2, \cdots, c,$$
$$y_{i1} + y_{i2} + \cdots + y_{ic} = N_1 + N_2 + \cdots + N_i = M_i.$$

The analysis in table 6.2 is derived in a straightforward fashion from the definitions in (6.7) and the properties of the logarithm. Note that the convexity property (see section 4.2, and section 3 of chapter 2) ensures that each between component is the minimum value of the within component below it in table 6.2 for the given pooling.

TABLE 6.2

Component due to	Information	D.F.
Within samples 1 and 2	$2\sum_{i=1}^{2}\sum_{j=1}^{c} x_{ij}\log\frac{M_2x_{ij}}{N_iy_{2j}}$	$c-1$
Between sample 3 and samples 1 and 2	$2\sum_{j=1}^{c}\left(x_{3j}\log\frac{M_3x_{3j}}{N_3y_{3j}} + y_{2j}\log\frac{M_3y_{2j}}{M_2y_{3j}}\right)$	$c-1$
. . .	. . .	. . .
Within samples 1 to $r-2$	$2\sum_{i=1}^{r-2}\sum_{j=1}^{c} x_{ij}\log\frac{M_{r-2}x_{ij}}{N_iy_{r-2,j}}$	$(r-3)(c-1)$
Between sample $(r-1)$ and samples 1 to $r-2$	$2\sum_{j=1}^{c}\left(x_{r-1,j}\log\frac{M_{r-1}x_{r-1,j}}{N_{r-1}y_{r-1,j}} + y_{r-2,j}\log\frac{M_{r-1}y_{r-2,j}}{M_{r-2}y_{r-1,j}}\right)$	$c-1$
Within samples 1 to $r-1$	$2\sum_{i=1}^{r-1}\sum_{j=1}^{c} x_{ij}\log\frac{M_{r-1}x_{ij}}{N_iy_{r-1,j}}$	$(r-2)(c-1)$
Between sample r and samples 1 to $r-1$	$2\sum_{j=1}^{c}\left(x_{rj}\log\frac{Nx_{rj}}{N_rx_j} + y_{r-1,j}\log\frac{Ny_{r-1,j}}{M_{r-1}x_j}\right)$	$c-1$
$2\hat{I}(H_1:H_2)$ (Within)	$2\sum_{i=1}^{r}\sum_{j=1}^{c} x_{ij}\log\frac{Nx_{ij}}{N_ix_j}$	$(r-1)(c-1)$

We remark that (see the remark about the analysis in table 4.1) the analysis in table 6.2 is a reflection of two facts:

1. The hypothesis of homogeneity H_2 in (6.1) is equivalent to the intersection of $(r-1)$ hypotheses $H_2(1, 2)$, $H_2(1 + 2, 3), \cdots$, $H_2(1 + 2 + \cdots + r - 1, r)$, $H_2 = H_2(1, 2) \cap H_2(1 + 2, 3) \cap \cdots \cap$

$H_2(1 + 2 + \cdots + r - 1, r)$, where $H_2(1, 2)$ is the hypothesis that samples 1 and 2 are homogeneous, $H_2(1 + 2, 3)$ is the hypothesis that sample 3 is homogeneous with the pooled homogeneous samples 1 and 2, $H_2(1 + 2 + 3, 4)$ is the hypothesis that sample 4 is homogeneous with the pooled homogeneous samples 1, 2, and 3, etc.

2. The distribution of two independent samples may be expressed as the product of a marginal distribution of the pooled samples and a conditional distribution of the individual samples given the pooled sample, that is, in the notation of (6.7),

$$\frac{N_1!}{x_{11}!\cdots x_{1c}!}p_1^{x_{11}}\cdots p_c^{x_{1c}}\cdot\frac{N_2!}{x_{21}!\cdots x_{2c}!}p_1^{x_{21}}\cdots p_c^{x_{2c}}$$
$$=\frac{(N_1+N_2)!}{y_{21}!\cdots y_{2c}!}p_1^{y_{21}}\cdots p_c^{y_{2c}}\cdot\frac{N_1!N_2!}{(N_1+N_2)!}\frac{y_{21}!\cdots y_{2c}!}{x_{11}!\cdots x_{1c}!x_{21}!\cdots x_{2c}!},$$

with similar results for 3, 4, $\cdots$ samples. [Cf. Bartlett (1937).]

The degrees of freedom in table 6.2 are those of the asymptotic χ^2-distributions under the null hypothesis H_2 of (6.1).

We leave to the reader the estimate of the divergences, as well as the expression in terms of the form $n \log n$ for computational convenience.

There may be some basis for considering a partitioning of the r samples into two or more sets. We shall indicate the analysis for a partitioning into two sets to illustrate the procedure which is easily extended to more than two sets.

For convenience we take samples 1 to r_1 as set 1, samples $r_1 + 1$ to r as set 2, and define

$$(6.8)\qquad \begin{aligned} z_{1j} &= x_{1j} + x_{2j} + \cdots + x_{r_1j}, \\ & \qquad\qquad\qquad\qquad j = 1, 2, \cdots, c, \\ z_{2j} &= x_{r_1+1,j} + \cdots + x_{rj}, \\ T_1 &= \sum_{j=1}^{c} z_{1j}, \qquad T_2 = \sum_{j=1}^{c} z_{2j}, \qquad N = T_1 + T_2. \end{aligned}$$

The analysis in table 6.3 is derived in a straightforward fashion from the definitions in (6.8) and the properties of the logarithm. Note that the convexity property (see section 4.2) ensures that the between component in table 6.3 is the minimum value of $2\hat{I}(H_1:H_2)$ for the given partitioning. (Cf. table 4.2.) We leave to the reader the details of remarks about the analysis of the null hypothesis and the distributions that are similar to those for table 6.2.

The degrees of freedom in table 6.3 are those of the asymptotic χ^2-distributions under the null hypothesis H_2 of (6.1).

We leave to the reader the estimate of the divergences as well as the expression in terms of the form $n \log n$ for computational convenience.

TABLE 6.3

Component due to	Information	D.F.
Between set 1 and set 2	$2\sum_{j=1}^{c}\left(z_{1j}\log\frac{Nz_{1j}}{T_1x_j}+z_{2j}\log\frac{Nz_{2j}}{T_2x_j}\right)$	$c-1$
Within set 2	$2\sum_{i=r_1+1}^{r}\sum_{j=1}^{c}x_{ij}\log\frac{T_2x_{ij}}{N_iz_{2j}}$	$(r-r_1-1)(c-1)$
Within set 1	$2\sum_{i=1}^{r_1}\sum_{j=1}^{c}x_{ij}\log\frac{T_1x_{ij}}{N_iz_{1j}}$	$(r_1-1)(c-1)$
$2\hat{I}(H_1:H_2)$ Within r samples	$2\sum_{i=1}^{r}\sum_{j=1}^{c}x_{ij}\log\frac{Nx_{ij}}{N_ix_i}$	$(r-1)(c-1)$

6.3. Parametric Case

We now assume that $p_1, p_2, \cdots, p_c$ of table 6.1 are known functions of independent parameters $\phi_1, \phi_2, \cdots, \phi_k$, $k < c$. Suppose we "fit" the multinomial distribution using estimates (by some procedure to be determined) $\tilde{\phi}_l$, $l = 1, 2, \cdots, k$, of the ϕ's. We write $\tilde{p}_j = p_j(\tilde{\phi}_1, \tilde{\phi}_2, \cdots, \tilde{\phi}_k)$, $j = 1, 2, \cdots, c$, $\tilde{p}_1 + \tilde{p}_2 + \cdots + \tilde{p}_c = 1$.

If the $\tilde{p}_j$, or the $\tilde{\phi}_l$, are such that identically in the ϕ's

$$\sum_{j=1}^{c}\frac{x_j}{N}\log\frac{\tilde{p}_j}{p_j}=\sum_{j=1}^{c}\tilde{p}_j\log\frac{\tilde{p}_j}{p_j}, \tag{6.9}$$

we get the further analysis of table 6.1 summarized in table 6.4. The condition (6.9) is to ensure that the analysis in table 6.4 is additive informationwise, and is analogous to (4.8). Table 6.4 includes a further analysis of table 5.1 when $r = 2$. We see [cf. (4.8)–(4.11)] that (6.9) implies that the $\tilde{\phi}_l$'s are the solutions of

$$\sum_{j=1}^{c}\frac{x_j}{p_j}\frac{\partial p_j}{\partial \phi_l}=0, \qquad l = 1, 2, \cdots, k. \tag{6.10}$$

The equations (6.10) are the maximum-likelihood equations for estimating the ϕ's, and are also those which solve the problem of finding the ϕ's that minimize the between component or the total in table 6.1. We leave to the reader the estimate of the divergences, as well as the expression in terms of the form $n \log n$ for computational convenience. The degrees of freedom in table 6.4 are those of the asymptotic χ^2-distributions under the null hypothesis H_2 of (6.1), the p's being taken as functions of the ϕ's. (See problem 7.16.)

TABLE 6.4

Component due to	Information	D.F.
Between $\hat{p}_j = x_j/N$ and $\tilde{p}_j = p_j(\tilde{\phi}_1, \cdots, \tilde{\phi}_k)$, $2\hat{I}(\hat{p}:\tilde{p})$	$2\sum_{j=1}^{c} x_j \log \frac{x_j}{N\tilde{p}_j}$	$c - k - 1$
Error, $2\hat{I}(p^*:\hat{p})$ (Within)	$2\sum_{i=1}^{r}\sum_{j=1}^{c} x_{ij} \log \frac{Nx_{ij}}{N_i x_j}$	$(r-1)(c-1)$
$2\hat{I}(p^*:\tilde{p})$ or between x_{ij}/N_i and $\tilde{p}$	$2\sum_{i=1}^{r}\sum_{j=1}^{c} x_{ij} \log \frac{x_{ij}}{N_i \tilde{p}_j}$	$r(c-1) - k$
$\tilde{\phi}$'s, $2\hat{I}(\tilde{p}:p)$ $(\tilde{p})$ against (p)	$2N\sum_{j=1}^{c} \tilde{p}_j \log \frac{\tilde{p}_j}{p_j}$	k
Total, $2\hat{I}(p^*:p)$	$2\sum_{i=1}^{r}\sum_{j=1}^{c} x_{ij} \log \frac{x_{ij}}{N_i p_j}$	$r(c-1)$

7. PROBLEMS

7.1. Estimate the divergences corresponding to the information components in table 4.1.

7.2. Estimate the divergences corresponding to the information components in table 4.2.

7.3. Estimate the divergences corresponding to the within components in table 6.2.

7.4. Express the information components in table 6.2 in terms of the form $n \log n$.

7.5. Complete the details of the discussion of the analysis of the null hypothesis and the distributions for table 6.3.

7.6. Estimate the divergences corresponding to the information components in table 6.3.

7.7. Express the information components in table 6.3 in terms of the form $n \log n$.

7.8. Estimate the divergences corresponding to the information components in table 6.4.

7.9. Express the within component in table 6.4 in terms of the form $n \log n$.

7.10. Fisher (1956, p. 144) defines a consistent statistic as: "a function of the observed frequencies which takes the exact parametric value when for these frequencies their expectations are substituted." Which of the information statistics in chapter 6 are *Fisher consistent*, that is, consistent in the sense of the foregoing definition? [Cf. Fisher (1922b, p. 316).]

7.11. Are the following six independent multinomial samples homogeneous?

2,	7,	8,	4,	9
8,	10,	8,	13,	4
6,	6,	5,	10,	12
5,	6,	4,	6,	4
5,	7,	4,	2,	14
4,	8,	12,	7,	5

7.12. Are the following four independent multinomial samples homogeneous?

2,	8,	5,	1,	2
8,	2,	5,	2,	1
3,	5,	7,	2,	1
3,	2,	7,	3,	3

7.13. Are the following test results for five manufacturers homogeneous?

	Manufacturer				
	A	*B*	*C*	*D*	*E*
Failed	26	72	61	29	135
Passed	172	169	142	36	542
Total	198	241	203	65	677

7.14. From the analysis in table 4.3 and the properties of the discrimination information, show that for $N \to \infty$, if $x_i/N \to p_i$ with probability 1, then $x_i/N \to \tilde{p}_i$ with probability 1 and $\tilde{p}_i \to p_i$ with probability 1, $i = 1, 2, \cdots, c$. [Cf. Rao (1957).] (See lemma 2.1 of chapter 4.)

7.15. What is the relation, if any, between (4.3) and problem 5.12 in chapter 1?

7.16. From the analysis in table 6.4 and the properties of the discrimination information, show that for $N_i \to \infty$, if $x_{ij}/N_i \to p_j$ with probability 1, then $x_{ij}/N_i \to \tilde{p}_j$ with probability 1, $\tilde{p}_j \to p_j$ with probability 1, and $x_j/N \to \tilde{p}_j$ with probability 1, $i = 1, 2, \cdots, r$; $j = 1, 2, \cdots, c$. (See problem 7.14.)

7.17. Show that $\inf_{p \le p_0} \left(x \log \frac{x}{Np} + (N - x) \log \frac{N - x}{Nq} \right) = x \log \frac{x}{Np_0} + (N - x) \log \frac{N - x}{Nq_0}$, $x > Np_0$. [See (4.18).]

7.18. Find the approximate value of $\hat{J}(H_1, H_2)$ in (6.6) using the procedure for (4.5) and (4.6).

CHAPTER 7

Poisson Populations

1. BACKGROUND

Suppose two simple statistical hypotheses, say H_1 and H_2, specify respectively the Poisson populations

$$(1.1)\quad p(x, m_i) = \frac{e^{-m_i}m_i^x}{x!}, \qquad x = 0, 1, 2, \cdots; \qquad i = 1, 2, m_i > 0.$$

The mean information per observation from the population hypothesized by H_1, for discriminating for H_1 against H_2, is (see section 2 of chapter 1)

$$(1.2)\qquad I(1:2) = \sum_{x=0}^{\infty} p(x, m_1) \log \frac{p(x, m_1)}{p(x, m_2)}$$

$$= m_1 \log \frac{m_1}{m_2} + m_2 - m_1.$$

The divergence between H_1 and H_2, a measure of the difficulty of discriminating between them, is (see section 3 of chapter 1)

$$(1.3)\qquad J(1, 2) = \sum_{x=0}^{\infty} (p(x, m_1) - p(x, m_2)) \log \frac{p(x, m_1)}{p(x, m_2)}$$

$$= (m_1 - m_2) \log \frac{m_1}{m_2}.$$

The mean discrimination information and divergence for a random sample of n independent observations O_n are

$$I(1:2; O_n) = n\left(m_1 \log \frac{m_1}{m_2} + m_2 - m_1\right) = nI(1:2),$$

$$J(1, 2; O_n) = n(m_1 - m_2) \log \frac{m_1}{m_2} = nJ(1, 2).$$

These may be calculated directly or derived from the additivity property (see section 2 in chapter 2).

2. CONJUGATE DISTRIBUTIONS

Suppose that every possible observation from $p^*(x)$, any distribution on the nonnegative integers, is also a possible observation from the Poisson distribution $p(x, m) = e^{-m}m^x/x!$, $x = 0, 1, 2, \cdots$. This is to avoid the contingency that $p^*(x) \neq 0$ and $p(x, m) = 0$. (See section 7 of chapter 2.)

Theorem 2.1 of chapter 3 permits us to assert:

LEMMA 2.1. *The least informative distribution on the nonnegative integers, with given expected value, for discrimination against the Poisson distribution* $p(x, m) = e^{-m}m^x/x!$, *namely, the distribution* $p^*(x)$ *such that* $E^*(x) = \theta$ *and* $\sum_{x=0}^{\infty} p^*(x) \log (p^*(x)/p(x, m))$ *is a minimum, is the distribution*

$$(2.1)\quad p^*(x) = e^{\tau x}p(x, m)/e^{-m+me^\tau} = e^{-me^\tau}(me^\tau)^x/x! = e^{-m^*}(m^*)^x/x!,$$

where $\sum_{x=0}^{\infty} e^{\tau x}p(x, m) = e^{-m+me^\tau}$, $m^* = me^\tau = \theta = \dfrac{d}{d\tau}\log e^{-m+me^\tau}$, *and* τ *is a real parameter.*

Note that the least informative distribution $p^*(x)$ here is a Poisson distribution. [Cf. Sanov (1957, p. 25).]

We illustrate lemma 2.1 with a numerical example (cf. example 2.1 of chapter 3). Table 2.1 gives the negative binomial distribution $p_1^*(x) = (\Gamma(n + x)/x!\Gamma(n))p^xq^{-n-x}$, $n = 2$, $p = 0.5$, $q = 1.5$, mean = 1; the Poisson distribution $p_2^*(x) = e^{-m}m^x/x!$, $m = 1$; and the Poisson distribution $p(x) = e^{-m}m^x/x!$, $m = 1.5$, which is taken as the distribution $p(x)$ of the lemma. The other two are distributions with $E^*(x) = 1$. [The numerical values of the negative binomial are taken from Cochran (1954, p. 419). See example 2.2 in chapter 4.]

TABLE 2.1

x	$p_1^*(x)$	$p_2^*(x)$	$p(x)$	$p_1^*(x) \log \dfrac{p_1^*(x)}{p(x)}$	$p_2^*(x) \log \dfrac{p_2^*(x)}{p(x)}$
0	0.4444	0.3679	0.2231	0.30624	0.18402
1	0.2963	0.3679	0.3347	−0.03611	0.03479
2	0.1482	0.1839	0.2510	−0.07813	−0.05720
3	0.0658	0.0613	0.1255	−0.04249	−0.04392
4+	0.0453	0.0190	0.0657	−0.01678	−0.02357
	1.0000	1.0000	1.0000	0.13273	0.09412

Note that the Poisson distribution does provide the smaller value of $\Sigma p^*(x) \log (p^*(x)/p(x))$, and [see (1.2)] that $1 \log (1/1.5) + 1.5 - 1 = 0.09453$. The difference between 0.09412 in table 2.1 and 0.09453 computed from the formula for $I(1:2)$ is due to the grouping for $x \geqq 4$, illustrating the statement that grouping loses information (see sections 3 and 4 of chapter 2, example 2.2 of chapter 4, and problem 6.6).

The Poisson distribution $p^*(x)$ in (2.1) is the conjugate distribution (see section 1 of chapter 5) of the Poisson distribution $p(x, m)$. We thus have

$$I(p^*:p) = \sum_{x=0}^{\infty} p^*(x) \log \frac{p^*(x)}{p(x, m)} \tag{2.2}$$

$$= \theta\tau + m - me^{\tau} = \theta \log \frac{\theta}{m} + m - \theta,$$

$$J(p^*, p) = \sum_{x=0}^{\infty} (p^*(x) - p(x, m)) \log \frac{p^*(x)}{p(x, m)}$$

$$= \tau(\theta - m) = (\theta - m) \log \frac{\theta}{m}.$$

For applications to testing hypotheses about Poisson distributions, the basic Poisson distribution $p(x, m)$ will be that of the null hypothesis H_2, whereas the conjugate distribution will range over the populations of the alternative hypothesis H_1.

3. r SAMPLES

3.1. Basic Problem

Suppose we have r independent samples of $n_1, n_2, \cdots, n_r$ independent observations from Poisson populations. We want to test the hypotheses:

(3.1) H_1: the Poisson population parameters are $m_1, m_2, \cdots, m_r$,
H_2: the Poisson population parameters are $m_1 = m_2 = \cdots = m_r = m$,

that is, a null hypothesis of homogeneity H_2, the samples are from the same Poisson population.

From the additivity property (see section 2 in chapter 2), or by direct evaluation for the r samples, we have

$$I(H_1:H_2) = \sum_{i=1}^{r} n_i\left(m_i \log \frac{m_i}{m} + m - m_i\right), \tag{3.2}$$

$$J(H_1, H_2) = \sum_{i=1}^{r} n_i(m_i - m) \log \frac{m_i}{m}. \tag{3.3}$$

With the observed sample best unbiased estimates, the respective sample averages, as the θ_i of the conjugate distributions we have

$$(3.4) \qquad \hat{I}(m^*:m) = \sum_{i=1}^{r} (\hat{\tau}_i n_i \bar{x}_i + n_i(m - me^{\hat{\tau}_i})),$$

where [see (2.1)] $\hat{\tau}_i = \log(\bar{x}_i/m)$, $i = 1, 2, \cdots, r$, and [see (2.2)]

$$(3.5) \qquad \hat{I}(m^*:m) = \sum_{i=1}^{r} n_i\left(\bar{x}_i \log \frac{\bar{x}_i}{m} + m - \bar{x}_i\right).$$

The hypothesis H_2 of (3.1) usually does not specify m. We can analyze $\hat{I}(m^*:m)$ into two additive components, one due to the deviation between m and its best unbiased estimate from the pooled samples, the other due to what may be termed error within the samples. Letting $n\bar{x} = n_1\bar{x}_1 + \cdots + n_r\bar{x}_r$, $n = n_1 + n_2 + \cdots + n_r$, we have the analysis summarized in table 3.1. The degrees of freedom are those of the asymptotic χ^2-distributions under the null hypothesis H_2 of (3.1). Note that $\hat{I}(m^*:\hat{m})$ $(\hat{I}(H_1:H_2))$ in table 3.1 is the minimum value of $\hat{I}(m^*:m)$ in (3.5) for variations of $m > 0$, that is, over the populations H_2, and that by the convexity property (see section 3 of chapter 2)

$$\sum_{i=1}^{r}\left(n_i\bar{x}_i \log \frac{\bar{x}_i}{m} + n_i(m - \bar{x}_i)\right) \geqq n\bar{x} \log \frac{\bar{x}}{m} + n(m - \bar{x}).$$

We remark that the analysis in table 3.1 is a reflection of the fact that the hypothesis H_2 in (3.1) is the intersection of the hypotheses $H_2(\cdot)$, the samples are homogeneous, and $H_2(\cdot|m)$, the parameter of the homogeneous samples is m, that is, $H_2 = H_2(\cdot) \cap H_2(\cdot|m)$. The between component in table 3.1, $2\hat{I}(\hat{m}:m)$, is a test for the hypothesis $H_2(\cdot|m)$, and the within component in table 3.1, $2\hat{I}(m^*:\hat{m})$ or $2\hat{I}(H_1:H_2)$, is a conditional test for the hypothesis $H_2(\cdot)$, subject to the observed $n\bar{x} = n_1\bar{x}_1 + \cdots + n_r\bar{x}_r$.

The error component in table 3.1 may be expressed as

$$(3.6) \qquad \hat{I}(H_1:H_2) = \sum_{i=1}^{r} n_i\bar{x}_i \log \bar{x}_i - n\bar{x} \log \bar{x} = \sum_{i=1}^{r} n_i\bar{x}_i \log n_i\bar{x}_i$$
$$- \sum_{i=1}^{r} \bar{x}_i n_i \log n_i - n\bar{x} \log n\bar{x} + \bar{x}n \log n,$$

for computational convenience with the table of $n \log n$, since $n_i\bar{x}_i$, $i = 1, 2, \cdots, r$, and $n\bar{x}$ are integral.

The divergences do not provide a similar additive analysis (with these estimates), but the estimate of the divergence corresponding to the error component is

$$\hat{J}(m^*, \hat{m}) = \hat{J}(H_1, H_2) = \sum_{i=1}^{r} n_i(\bar{x}_i - \bar{x}) \log \frac{\bar{x}_i}{\bar{x}}. \tag{3.7}$$

Note that $\hat{I}(H_1 : H_2) = \hat{I}(m^* : \hat{m})$ in table 3.1 is (3.2) with the substitution of $\bar{x}_i$ for m_i and $\bar{x}$ for m, and that (3.7) is (3.3) with the same substitutions.

TABLE 3.1

Component due to	Information	D.F.
Between $\hat{m} = \bar{x}$ and m, $2\hat{I}(\hat{m} : m)$	$2\left(n\bar{x} \log \frac{\bar{x}}{m} + n(m - \bar{x})\right)$	1
Error, $2\hat{I}(m^* : \hat{m})$ (Within)	$2 \sum_{i=1}^{r} n_i\bar{x}_i \log \frac{\bar{x}_i}{\bar{x}}$	$r - 1$
Total, $2\hat{I}(m^* : m)$	$2 \sum_{i=1}^{r} \left(n_i\bar{x}_i \log \frac{\bar{x}_i}{m} + n_i(m - \bar{x}_i)\right)$	r

Under the null hypothesis H_2 of (3.1) (the samples are from the same population), $2\hat{I}(H_1 : H_2)$ and $\hat{J}(H_1, H_2)$ are asymptotically distributed as χ^2 with $(r - 1)$ degrees of freedom. Under the alternative hypothesis H_1 of (3.1), $2\hat{I}(H_1 : H_2)$ and $\hat{J}(H_1, H_2)$ are asymptotically distributed as noncentral χ^2 with $(r - 1)$ degrees of freedom and respective noncentrality parameters $2 \sum_{i=1}^{r} n_i m_i \log (m_i/m)$ and $\sum_{i=1}^{r} n_i(m_i - m) \log (m_i/m)$, $nm = \sum_{i=1}^{r} n_i m_i$, corresponding to $n\bar{x} = \sum_{i=1}^{r} n_i\bar{x}_i$.

With the approximation used in (4.5) and (4.6) of chapter 6, we find [cf. Cochran (1954), Fisher (1950), Rao and Chakravarti (1956)]

$$2\hat{I}(H_1 : H_2) \approx \sum \frac{n_i(\bar{x}_i - \bar{x})^2}{\bar{x}} = \chi^2, \tag{3.8}$$

$$\hat{J}(H_1, H_2) \approx \frac{1}{2} \sum \frac{n_i(\bar{x}_i - \bar{x})^2}{\bar{x}} + \frac{1}{2} \sum \frac{n_i(\bar{x}_i - \bar{x})^2}{\bar{x}_i}.$$

3.2. Partition

The error component in table 3.1 can be analyzed into $(r - 1)$ comparisons, each of 1 degree of freedom, between each sample and the pooled sample of all its predecessors. (Compare the analysis of the error component in section 6.2 of chapter 6. We leave the comparable details to the reader.) This permits an assessment of each sample as it is added, for changes that may have occurred. [See Cochran (1954), Lancaster (1949).]

To indicate the successive pooling of the samples, we define

$$N_i y_i = n_1\bar{x}_1 + n_2\bar{x}_2 + \cdots + n_i\bar{x}_i, \qquad i = 2, \cdots, r-1,$$

and

$$N_i = n_1 + n_2 + \cdots + n_i.$$

The analysis in table 3.2 is derived in a straightforward fashion from these definitions and the properties of the logarithm. Note that the

TABLE 3.2

Component due to	Information	D.F.
Within samples 1 and 2	$2\sum_{i=1}^{2} n_i\bar{x}_i \log \frac{\bar{x}_i}{y_2}$	1
Between sample 3 and pooled samples 1and 2	$2\left(n_3\bar{x}_3 \log \frac{\bar{x}_3}{y_3} + N_2 y_2 \log \frac{y_2}{y_3}\right)$	1
⋮	⋮	⋮
Within samples 1 to $(r-2)$	$2\sum_{i=1}^{r-2} n_i\bar{x}_i \log \frac{\bar{x}_i}{y_{r-2}}$	$r-3$
Between sample $(r-1)$ and pooled samples 1 to $(r-2)$	$2\left(n_{r-1}\bar{x}_{r-1} \log \frac{\bar{x}_{r-1}}{y_{r-1}} + N_{r-2} y_{r-2} \log \frac{y_{r-2}}{y_{r-1}}\right)$	1
Within samples 1 to $r-1$	$2\sum_{i=1}^{r-1} n_i\bar{x}_i \log \frac{\bar{x}_i}{y_{r-1}}$	$r-2$
Between sample r and pooled samples 1 to $r-1$	$2\left(n_r\bar{x}_r \log \frac{\bar{x}_r}{\bar{x}} + N_{r-1} y_{r-1} \log \frac{y_{r-1}}{\bar{x}}\right)$	1
Error, $2\hat{I}(H_1:H_2)$ (Within r samples)	$2\sum_{i=1}^{r} n_i\bar{x}_i \log \frac{\bar{x}_i}{\bar{x}}$	$r-1$

convexity property (see section 3 of chapter 2) ensures that each between component is the minimum value of the within component below it in table 3.2 for the given pooling. The degrees of freedom are those of the asymptotic χ^2-distributions under the null hypothesis H_2 of (3.1). We leave to the reader the estimation of the divergences, as well as the expression in terms of the form $n \log n$ for computational convenience.

4. "ONE-SIDED" HYPOTHESIS, SINGLE SAMPLE

(Cf. section 4.4 of chapter 6.)

It is also of interest to examine a "one-sided" hypothesis. Suppose we have a random sample of n independent observations from a Poisson population, and we want to test the hypotheses:

$$\begin{aligned} &H_1\text{: the Poisson population parameter is } m_1 > m,\\ &H_2\text{: the Poisson population parameter is equal to } m. \end{aligned} \tag{4.1}$$

The conjugate distribution (2.1) ranges over the Poisson populations H_1 in (4.1) if $m^* = me^\tau > m$. Only values of $\tau > 0$ are therefore admissible. With the value of the observed sample best unbiased estimate (the sample average $\bar{x}$) as $\hat{\theta}$ of the conjugate distribution, we have [cf. (3.4)],

$$\hat{I}(m^*:m) = \hat{\tau} n\bar{x} + n(m - me^{\hat{\tau}}), \qquad \hat{\tau} = \log (\bar{x}/m). \tag{4.2}$$

If $\bar{x} > m$, then $\hat{\tau} > 0$ is admissible. If $\bar{x} \leq m$, then $\hat{\tau} \leq 0$ is not admissible. On the boundary $\hat{\tau} = 0$ of the admissible region, $\hat{I}(m^*:m) = 0$. We thus have:

$$\begin{aligned} \hat{I}(H_1:H_2) &= n\bar{x} \log \frac{\bar{x}}{m} + n(m - \bar{x}), \qquad \bar{x} > m,\\ &= 0, \qquad \bar{x} \leq m. \end{aligned} \tag{4.3}$$

Asymptotically, $2\hat{I}(H_1:H_2)$ has a χ^2-distribution with 1 degree of freedom under the null hypothesis H_2 of (4.1), but the α significance level must be taken from the usual χ^2 tables at the 2α level, since we do not consider values of $\bar{x} < m$ for which $\hat{I}(H_1:H_2)$ is the same as for some value of $\bar{x} > m$.

Instead of the simple null hypothesis H_2 of (4.1), let us consider the composite null hypothesis H_2':

$$\begin{aligned} &H_1\text{: the Poisson population parameter is } m_1 > m_0,\\ &H_2'\text{: the Poisson population parameter is } m \leq m_0. \end{aligned} \tag{4.4}$$

It may be shown that [cf. (4.18) of chapter 6] (see problem 6.7)

$$\text{(4.5)}\quad \inf_{m \leq m_0} \left(n\bar{x} \log \frac{\bar{x}}{m} + n(m - \bar{x}) \right) = n\bar{x} \log \frac{\bar{x}}{m_0} + n(m_0 - \bar{x}), \qquad \bar{x} > m_0,$$

and therefore

$$\text{(4.6)}\qquad \begin{aligned} \hat{I}(H_1 : H_2') &= n\bar{x} \log \frac{\bar{x}}{m_0} + n(m_0 - \bar{x}), \qquad \bar{x} > m_0, \\ &= 0, \qquad \bar{x} \leqq m_0. \end{aligned}$$

Under the null hypothesis H_2' of (4.4), asymptotically, Prob $[2\hat{I}(H_1 : H_2') \geqq \chi_{2\alpha}^2] \leqq \alpha$, where $\chi_{2\alpha}^2$ is the entry in the usual χ^2 tables at the 2α level for 1 degree of freedom.

Similarly, for the hypotheses

$$\text{(4.7)}\qquad \begin{aligned} &H_3\text{: the Poisson population parameter is } m_1 < m_0, \\ &H_2''\text{: the Poisson population parameter is } m \geqq m_0, \end{aligned}$$

we have

$$\text{(4.8)}\qquad \begin{aligned} \hat{I}(H_3 : H_2'') &= n\bar{x} \log \frac{\bar{x}}{m_0} + n(m_0 - \bar{x}), \qquad \bar{x} < m_0, \\ &= 0, \qquad \bar{x} \geqq m_0. \end{aligned}$$

Under the null hypothesis H_2'' of (4.7), asymptotically, Prob $[2\hat{I}(H_3 : H_2'') \geqq \chi_{2\alpha}^2] \leqq \alpha$ where $\chi_{2\alpha}^2$ is as above.

The two-sided hypothesis

$$\text{(4.9)}\qquad \begin{aligned} &H_4\text{: the Poisson population parameter is } m_1 \neq m_0, \\ &H_2\text{: the Poisson population parameter is } m = m_0, \end{aligned}$$

is a special case of section 3.1, and

$$\text{(4.10)}\qquad 2\hat{I}(H_4 : H_2) = 2\left(n\bar{x} \log \frac{\bar{x}}{m_0} + n(m_0 - \bar{x}) \right)$$

is asymptotically distributed as χ^2 with 1 degree of freedom under the null hypothesis H_2 of (4.9).

A $100(1 - \alpha)\ \%$ asymptotic confidence interval for m_0 is given by

$$\text{(4.11)}\qquad 2n\bar{x} \log \frac{\bar{x}}{m_0} + 2n(m_0 - \bar{x}) \leqq \chi^2(\alpha, 1),$$

where $\chi^2(\alpha, 1)$ is the value for which the χ^2-distribution with 1 degree of freedom yields Prob $[\chi^2 \geqq \chi^2(\alpha, 1)] = \alpha$. (Cf. section 5 of chapter 5.) (See problem 6.4.)

Note that H_2, H_2', and H_2'', respectively of (4.9), (4.4), and (4.7), satisfy $H_2 \rightleftharpoons H_2' \cap H_2''$, that is, $(m = m_0)$ if and only if $(m \leqq m_0)$ and $(m \geqq m_0)$; also H_4, H_1, and H_3, respectively of (4.9), (4.4), and (4.7), satisfy $H_4 \rightleftharpoons H_1 \cup H_3$, that is $(m_1 \neq m_0)$ if and only if $(m_1 > m_0)$ or $(m_1 < m_0)$. The region of acceptance common to the hypotheses H_2' and H_2'', $n\bar{x} \log (\bar{x}/m_0) + n(m_0 - \bar{x}) \leqq$ constant, is also the region of acceptance of H_2.

For illustration we take $nm_0 = 50$ and compute for (4.6) some probabilities under H_2' and H_1 from Molina's tables (1942) for the exact Poisson values (see tables 4.1 and 4.2), and the χ^2- and noncentral χ^2-distributions for approximating values (see table 4.3). (Cf. section 4.5 of chapter 6.)

TABLE 4.1

Prob $(n\bar{x} \leqq 50)$, Poisson

nm	H_2'	H_1				
40	0.9474	$nm_1 = 55$	60	65	70	80
45	0.7963					
50	0.5375	0.2768	0.1077	0.0321	0.0075	0.0002

TABLE 4.2

Prob $(n\bar{x} \geqq 63)$, Poisson

nm	H_2'	H_1				
40	0.0005	$nm_1 = 55$	60	65	70	80
45	0.0065					
50	0.0424	0.1559	0.3662	0.6146	0.8140	0.9781

$$2(62 \log \tfrac{62}{50} + 50 - 62) = 2.67381$$
$$2(63 \log \tfrac{63}{50} + 50 - 63) = 3.12007$$
$$2\hat{I}(H_1 : H_2') \text{ (corrected)} = 2.90 \text{ (cf. section 4.5.2 of chapter 6).}$$

The central value in table 4.3 was computed from $\dfrac{1}{\sqrt{2\pi}} \displaystyle\int_{1.70}^{\infty} e^{-x^2/2}\, dx$ and the noncentral values from $\dfrac{1}{\sqrt{2\pi}} \left(\displaystyle\int_{1.70}^{\infty} e^{-(x-\mu)^2/2}\, dx + \int_{1.70}^{\infty} e^{-(x+\mu)^2/2}\, dx \right)$

where $\mu^2 = 2\left(nm_1 \log \frac{nm_1}{50} + 50 - nm_1\right)$. [Cf. section 4.5.2 of chapter 6 and (6.9) in chapter 12.]

TABLE 4.3

Prob ($\chi^2 \geqq 2.90$), Upper Tail Only

Central	Noncentral		
0.0443	nm_1	$2\left(nm_1 \log \frac{nm_1}{50} + 50 - nm_1\right)$	
	55	0.48412	0.1652
	60	1.87859	0.3710
	80	15.20058	0.9860

We summarize comparable values in table 4.4, that is, the exact and approximate probabilities of rejecting H_2' when it is true, and when one of the indicated alternatives is true.

TABLE 4.4

	H_2'	H_1		
		$nm_1 = 55$	60	80
Poisson	0.0424	0.1559	0.3662	0.9781
χ^2	0.0443	0.1652	0.3710	0.9860

5. "ONE-SIDED" HYPOTHESIS, TWO SAMPLES

(Cf. section 5.2 of chapter 6.)

We now test a "one-sided" hypothesis for two samples. Suppose we have two independent samples of n_1 and n_2 independent observations each, from Poisson populations. We want to test the hypotheses:

$$\begin{aligned} &H_1'\text{: the Poisson population parameters are } m_1 > m_2,\\ &H_2\text{: the Poisson population parameters are } m_1 = m_2 = m. \end{aligned} \tag{5.1}$$

The conjugate distributions [cf. (2.1)] range over the Poisson populations of H_1' in (5.1) if

$$m_1^* = me^{\tau_1} > m_2^* = me^{\tau_2}.$$

Only values $\tau_1 > \tau_2$ are therefore admissible. For $r = 2$, we get from (3.4)

$$\hat{I}(m^*:m) = \hat{\tau}_1 n_1 \bar{x}_1 + n_1(m - me^{\hat{\tau}_1}) + \hat{\tau}_2 n_2 \bar{x}_2 + n_2(m - me^{\hat{\tau}_2}), \tag{5.2}$$

$$n_1\bar{x}_1 = n_1 me^{\hat{\tau}_1}, \qquad n_2\bar{x}_2 = n_2 me^{\hat{\tau}_2}, \tag{5.3}$$

or

$$\hat{\tau}_1 = \log\frac{\bar{x}_1}{m}, \qquad \hat{\tau}_2 = \log\frac{\bar{x}_2}{m}. \tag{5.4}$$

If $\bar{x}_1 > \bar{x}_2$, then $\hat{\tau}_1 > \hat{\tau}_2$ are admissible. However, if $\bar{x}_1 \leqq \bar{x}_2$, then $\hat{\tau}_1 \leqq \hat{\tau}_2$ are not admissible, and we must find the value of $\hat{I}(m^*:m)$ along the boundary $\hat{\tau}_2 = \hat{\tau}_1 = \hat{\tau}$ of the admissible area. For $\hat{\tau}_2 = \hat{\tau}_1 = \hat{\tau}$ in (5.2), we have

$$n\bar{x} = nme^{\hat{\tau}}, \quad \text{or} \quad \hat{\tau} = \log\frac{\bar{x}}{m}, \qquad n\bar{x} = n_1\bar{x}_1 + n_2\bar{x}_2, \qquad n = n_1 + n_2, \tag{5.5}$$

and consequently,

$$\hat{I}(m^*:m) = \sum_{i=1}^{2}\left(n_i\bar{x}_i \log\frac{\bar{x}_i}{m} + n_i(m - \bar{x}_i)\right), \qquad \bar{x}_1 > \bar{x}_2, \tag{5.6}$$

$$\hat{I}(m^*:m) = n\bar{x}\log\frac{\bar{x}}{m} + n(m - \bar{x}), \qquad \bar{x}_1 \leqq \bar{x}_2. \tag{5.7}$$

If we examine the analysis in table 3.1 for $r = 2$, corresponding to the two-sided hypothesis

$$\begin{aligned} &H_1\text{: the Poisson population parameters are } m_1 \neq m_2,\\ &H_2\text{: the Poisson population parameters are } m_1 = m_2 = m, \end{aligned} \tag{5.8}$$

we see that $2\hat{I}(m^*:m)$ in (5.6) is the total of table 3.1 when $\bar{x}_1 > \bar{x}_2$, and $2\hat{I}(m^*:m)$ in (5.7) is the between component of table 3.1 when $\bar{x}_1 \leqq \bar{x}_2$.

The hypothesis H_2 of (5.1) usually does not specify the value of m, and we then have

$$\begin{aligned} 2\hat{I}(H_1':H_2) &= \text{error component of table 3.1}, \qquad r = 2, \qquad \bar{x}_1 > \bar{x}_2,\\ &= 0, \qquad \bar{x}_1 \leqq \bar{x}_2. \end{aligned} \tag{5.9}$$

Asymptotically, $2\hat{I}(H_1':H_2)$ in (5.9) has a χ^2-distribution with 1 degree of freedom under the null hypothesis H_2 of (5.1), but the α significance level must be taken from the usual χ^2 tables at the 2α level.

Similarly, for testing the hypotheses

$$\begin{aligned} &H_1''\text{: the Poisson population parameters are } m_1 < m_2,\\ &H_2\text{: the Poisson population parameters are } m_1 = m_2 = m, \end{aligned} \tag{5.10}$$

we have $\hat{I}(m^*:m)$ in (5.6) when $\bar{x}_1 < \bar{x}_2$ and $\hat{I}(m^*:m)$ in (5.7) when $\bar{x}_1 \geqq \bar{x}_2$. The hypothesis H_2 of (5.10) usually does not specify the value of m, and we then have

$$(5.11)\quad 2\hat{I}(H_1'':H_2) = \text{error component of table 3.1}, \quad r = 2, \quad \bar{x}_1 < \bar{x}_2,$$
$$= 0, \quad \bar{x}_1 \geqq \bar{x}_2.$$

Asymptotically, $2\hat{I}(H_1'':H_2)$ in (5.11) is distributed as χ^2 with 1 degree of freedom under the null hypothesis H_2 of (5.10), but the α significance level must be taken from the usual χ^2 tables at the 2α level.

Note that H_1, H_1', and H_1'' of (5.8), (5.1), and (5.10) respectively, satisfy $H_1 \rightleftharpoons H_1' \cup H_1''$, that is, $(m_1 \neq m_2)$ if and only if $(m_1 > m_2)$ or $(m_1 < m_2)$.

We summarize the preceding in table 5.1. (See table 5.3 of chapter 6.)

TABLE 5.1

	$H_1':H_2$ (5.1)	$H_1'':H_2$ (5.10)	$H_1:H_2$ (5.8)
Between $2\left(n\bar{x}\log\frac{\bar{x}}{m_0} + n(m_0 - \bar{x})\right)$	$\bar{x}_1 \leqq \bar{x}_2$, $m = m_0$	$\bar{x}_1 \geqq \bar{x}_2$, $m = m_0$	$m = m_0$
Within $2\sum_{i=1}^{2} n_i\bar{x}_i\log\frac{\bar{x}_i}{\bar{x}}$	$\bar{x}_1 > \bar{x}_2$, $m = \hat{m} = \bar{x}$; 0, $\bar{x}_1 \leqq \bar{x}_2$	$\bar{x}_1 < \bar{x}_2$, $m = \hat{m} = \bar{x}$; 0, $\bar{x}_1 \geqq \bar{x}_2$	$m = \hat{m} = \bar{x}$
Total $2\sum_{i=1}^{2} n_i\left(\bar{x}_i\log\frac{\bar{x}_i}{m_0} + (m_0 - \bar{x}_i)\right)$	$\bar{x}_1 > \bar{x}_2$, $m = m_0$	$\bar{x}_1 < \bar{x}_2$, $m = m_0$	$m = m_0$

6. PROBLEMS

6.1. Complete the details of the analysis in table 3.2.

6.2. Estimate the divergences corresponding to the information components in table 3.2.

6.3. Express the within components in table 3.2 in terms of the form $n \log n$.

6.4. Compute the confidence interval for m_0 from (4.11) for $\bar{x} = 10$, $n = 10, 100$.

6.5. The following represent the totals of successive samples of the same size from Poisson populations: 427, 440, 494, 422, 409, 310, 302 [from Lancaster (1949, p. 127)].

Are the successive samples homogeneous? If not, where does the change in homogeneity occur? (Lancaster gives the data derived from successive plates of bacterial culture mixed with disinfectant.)

6.6. Compute $\Sigma p_3^*(x) \log \dfrac{p_3^*(x)}{p(x)}$ with $p(x)$ given in table 2.1 and $p_3^*(x) = \dfrac{n!}{x!(n-x)!} p^x q^{n-x}$, $n = 10$, $p = 0.1$, $q = 1 - p$, and compare with table 2.1.

6.7. Show that $\inf_{m \leq m_0} \left(n\bar{x} \log \dfrac{\bar{x}}{m} + n(m - \bar{x}) \right) = n\bar{x} \log \dfrac{\bar{x}}{m_0} + n(m_0 - \bar{x})$, $\bar{x} > m_0$. [See (4.5).]

6.8. Show that with the approximation used in (4.5) and (4.6) of chapter 6 the between component and the total in table 3.1 yield:

(*a*) $2\hat{I}(\hat{m}:m) \approx n(\bar{x} - m)^2/m$,

(*b*) $2\hat{I}(m^*:m) \approx \sum_{i=1}^{r} n_i(\bar{x}_i - m)^2/m$.

CHAPTER 8

Contingency Tables

1. INTRODUCTION

A contingency table is essentially a sample from a multivalued population with the various probabilities and partitions of the categories subject to restrictions in addition to those of the multinomial distribution. The analyses of contingency tables in this chapter are therefore closely related to those of multinomial samples in chapter 6. Studies and applications of contingency tables have a long history in statistical theory. See, for example, Pearson (1904), Fisher (1925a, all editions), Yule and Kendall (1937), Kendall (1943), Wilks (1943), Cramér (1946a), Rao (1952), Mitra (1955), Roy and Kastenbaum (1955, 1956), Roy and Mitra (1956), Roy (1957).

McGill (1954) has applied the communication theory measure of transmitted information to the analysis of contingency tables. McGill's approach, though somewhat different from ours, employs closely related concepts, and we derive similar results for contingency tables. Garner and McGill (1954, 1956) have pointed out some of the parallels that exist among the analysis of variance, correlation analysis, and an information measure they call uncertainty, as methods of analyzing component variation [cf. the article by McGill on pp. 56–62 in Quastler (1955)].

We shall study two-way and three-way tables in detail. The extension of the procedures to higher order tables poses no new conceptual problems, only algebraic complexities of detail, and we leave this to the reader.

2. TWO-WAY TABLES

We first study a two-factor or two-way table. Suppose we have N independent observations, each characterized by a value of two classifications, row and column, distributed among r row-categories and c

column-categories. Let x_{ij} be the frequency of occurrence in the ith row and jth column, and

$$x_{i\cdot} = \sum_{j=1}^{c} x_{ij}, \qquad x_{\cdot j} = \sum_{i=1}^{r} x_{ij},$$
$$N = \sum_{i=1}^{r} \sum_{j=1}^{c} x_{ij} = \sum_{i=1}^{r} x_{i\cdot} = \sum_{j=1}^{c} x_{\cdot j}.$$

We denote the probabilities by p's with corresponding subscripts.

We are first concerned with testing a null hypothesis H_2, the row and column classifications are independent, that is,

$$(2.1)\qquad \begin{array}{ll} H_1: p_{ij} \neq p_{i\cdot}p_{\cdot j}, & i = 1, 2, \cdots, r, \quad j = 1, 2, \cdots, c, \quad \text{for at least one } (i, j), \quad \sum_{i=1}^{r}\sum_{j=1}^{c} p_{ij} = 1, \quad p_{ij} > 0, \\ H_2: p_{ij} = p_{i\cdot}p_{\cdot j}, & p_{1\cdot} + p_{2\cdot} + \cdots + p_{r\cdot} = 1 = p_{\cdot 1} + p_{\cdot 2} + \cdots + p_{\cdot c}, \quad p_{i\cdot} > 0, \quad p_{\cdot j} > 0. \end{array}$$

Without repeating the detailed argument (similar to that for a sample of N independent observations from a multinomial population with rc categories), we have [cf. (2.6) and (2.8) of chapter 6]

$$(2.2)\qquad I(H_1:H_2) = N\sum_{i=1}^{r}\sum_{j=1}^{c} p_{ij} \log \frac{p_{ij}}{p_{i\cdot}p_{\cdot j}},$$

$$(2.3)\qquad J(H_1, H_2) = N\sum_{i=1}^{r}\sum_{j=1}^{c} (p_{ij} - p_{i\cdot}p_{\cdot j}) \log \frac{p_{ij}}{p_{i\cdot}p_{\cdot j}}.$$

Note that $I(H_1:H_2)/N$ in (2.2) is a measure of the relation between the row- and column-categories and has also been defined as the mean information in the row-categories about the column-categories or vice versa (see examples 4.3 and 4.4 of chapter 1).

For the conjugate distribution (see section 3 of chapter 6) with parameters the same as the observed sample best unbiased estimates, we have

$$(2.4)\qquad \hat{I}((p)^*:(p)) = \sum_{i=1}^{r}\sum_{j=1}^{c} x_{ij} \log \frac{x_{ij}}{Np_{i\cdot}p_{\cdot j}}.$$

The null hypothesis of independence H_2 in (2.1) usually does not specify $p_{i\cdot}$, $i = 1, 2, \cdots, r$, and $p_{\cdot j}$, $j = 1, 2, \cdots, c$. We can analyze $\hat{I}((p)^*:(p))$ of (2.4) into three additive components: a marginal component due to the deviations between the $p_{i\cdot}$ and their best unbiased estimates from the row totals, a marginal component due to the deviations between the $p_{\cdot j}$ and their best unbiased estimates from the column totals, and a conditional component due to the independence hypothesis. These components correspond respectively to a hypothesis $H_2(R)$ specifying the values of the $p_{i\cdot}$, a hypothesis $H_2(C)$ specifying the values of the $p_{\cdot j}$, and a hypothesis $H_2(R \times C)$ of independence, that is, H_2 in (2.1) is the intersection $H_2(R) \cap H_2(C) \cap H_2(R \times C)$. The analysis summarized in table 2.1 is an analogue of that in table 4.3 of chapter 6. Here there are $(r - 1)$ independent parameters $p_{1\cdot}, p_{2\cdot}, \cdots, p_{(r-1)\cdot}$, and $(c - 1)$ independent parameters $p_{\cdot 1}, p_{\cdot 2}, \cdots, p_{\cdot (c-1)}$. The equations (4.11) of chapter 6 are here

$$\sum_{j=1}^{c}\left(\frac{x_{ij}}{p_{ij}}p_{\cdot j} - \frac{x_{rj}}{p_{rj}}p_{\cdot j}\right) = 0, \qquad i = 1, 2, \cdots, r - 1,$$

$$\sum_{i=1}^{r}\left(\frac{x_{ij}}{p_{ij}}p_{i\cdot} - \frac{x_{ic}}{p_{ic}}p_{i\cdot}\right) = 0, \qquad j = 1, 2, \cdots, c - 1.$$

Since $p_{ij} = p_{i\cdot}p_{\cdot j}$, these equations reduce to

$$\frac{x_{i\cdot}}{p_{i\cdot}} = \frac{x_{r\cdot}}{p_{r\cdot}}, \qquad i = 1, 2, \cdots, r - 1, \qquad \frac{x_{\cdot j}}{p_{\cdot j}} = \frac{x_{\cdot c}}{p_{\cdot c}}, \qquad i = 1, 2, \cdots, c - 1,$$

yielding

$$\tilde{p}_{i\cdot} = \frac{x_{i\cdot}}{N}, \qquad \tilde{p}_{\cdot j} = \frac{x_{\cdot j}}{N}, \qquad \tilde{p}_{ij} = \frac{x_{i\cdot}}{N}\frac{x_{\cdot j}}{N},$$

$$i = 1, 2, \cdots, r, \qquad j = 1, 2, \cdots, c.$$

[Cf. Cramér (1946a, pp. 442–443).] Note that the independence component in table 2.1 is the minimum value of the total for variations of the $p_{i\cdot}$ and $p_{\cdot j}$,

$$\sum_{i=1}^{r} p_{i\cdot} = 1 = \sum_{j=1}^{c} p_{\cdot j},$$

that is, over the populations of H_2 with the given marginal values, and that by the convexity property (see section 3 of chapter 2)

$$\sum_{i=1}^{r}\sum_{j=1}^{c} x_{ij} \log \frac{x_{ij}}{Np_{i\cdot}p_{\cdot j}} \geqq \sum_{i=1}^{r} x_{i\cdot} \log \frac{x_{i\cdot}}{Np_{i\cdot}},$$

$$\sum_{i=1}^{r}\sum_{j=1}^{c} x_{ij} \log \frac{x_{ij}}{Np_{i\cdot}p_{\cdot j}} \geqq \sum_{j=1}^{c} x_{\cdot j} \log \frac{x_{\cdot j}}{Np_{\cdot j}}.$$

The degrees of freedom in table 2.1 are those of the asymptotic χ^2-distributions under the null hypothesis H_2 of (2.1). [Cf. Wilks (1935a).]

The independence component in table 2.1 may also be expressed as

$$(2.5) \quad \hat{I}(H_1:H_2) = \sum_{i=1}^{r}\sum_{j=1}^{c} x_{ij} \log x_{ij} - \sum_{i=1}^{r} x_{i\cdot} \log x_{i\cdot} - \sum_{j=1}^{c} x_{\cdot j} \log x_{\cdot j} + N \log N,$$

for computational convenience with the table of $n \log n$.

The divergences do not provide a similar additive analysis (with these estimates), but the estimate of the divergence corresponding to the independence component in (2.5) is [cf. (4.12) of chapter 6]

$$\hat{J}(H_1, H_2) = N\sum_{i=1}^{r}\sum_{j=1}^{c}\left(\frac{x_{ij}}{N} - \frac{x_{i\cdot}}{N}\frac{x_{\cdot j}}{N}\right)\log\frac{Nx_{ij}}{x_{i\cdot}x_{\cdot j}} \tag{2.6}$$

$$= \sum_{i=1}^{r}\sum_{j=1}^{c}\left(x_{ij} - \frac{x_{i\cdot}x_{\cdot j}}{N}\right)\log\frac{Nx_{ij}}{x_{i\cdot}x_{\cdot j}}.$$

TABLE 2.1

Component due to	Information	D.F.
Rows, $H_2(R)$ $\tilde{p}_{i\cdot} = x_{i\cdot}/N$	$2\sum_{i=1}^{r} x_{i\cdot}\log\frac{x_{i\cdot}}{Np_{i\cdot}}$	$r - 1$
Columns, $H_2(C)$ $\tilde{p}_{\cdot j} = x_{\cdot j}/N$	$2\sum_{j=1}^{c} x_{\cdot j}\log\frac{x_{\cdot j}}{Np_{\cdot j}}$	$c - 1$
Independence, $H_2(R \times C)$ $2\hat{I}(H_1:H_2) = 2\hat{I}((p)^*:(\hat{p}))$	$2\sum_{i=1}^{r}\sum_{j=1}^{c} x_{ij}\log\frac{Nx_{ij}}{x_{i\cdot}x_{\cdot j}}$	$(r - 1)(c - 1)$
Total, $2\hat{I}((p)^*:(p))$	$2\sum_{i=1}^{r}\sum_{j=1}^{c} x_{ij}\log\frac{x_{ij}}{Np_{i\cdot}p_{\cdot j}}$	$rc - 1$

Note that the independence component $\hat{I}(H_1:H_2)$ in table 2.1 is (2.2) with the substitution of x_{ij}/N for p_{ij}, $x_{i\cdot}/N$ for $p_{i\cdot}$, and $x_{\cdot j}/N$ for $p_{\cdot j}$, and that (2.6) is (2.3) with the same substitutions.

If the row and column classifications are independent, $2\hat{I}(H_1:H_2)$ and $\hat{J}(H_1, H_2)$ are asymptotically distributed as χ^2 with $(r - 1)(c - 1)$ degrees of freedom. Under the alternative hypothesis H_1 of (2.1), $2\hat{I}(H_1:H_2)$ and $\hat{J}(H_1, H_2)$ are asymptotically distributed as noncentral χ^2 with $(r - 1)(c - 1)$ degrees of freedom and respective noncentrality parameters $2I(H_1:H_2)$ and $J(H_1, H_2)$ as given by (2.2) and (2.3) with

$$p_{i\cdot} = \sum_{j=1}^{c} p_{ij}, \qquad p_{\cdot j} = \sum_{i=1}^{r} p_{ij}.$$

(See problem 13.11.)

With the approximations in (4.5) and (4.6) of chapter 6, we find that [cf. Cramér (1946a, pp. 441–445), Hsu (1949, pp. 367–369), Roy (1957, p. 128)]

$$(2.7)\quad 2\hat{I}(H_1:H_2) \approx \sum_{i=1}^{r}\sum_{j=1}^{c} \frac{\left(x_{ij} - \dfrac{x_{i\cdot}x_{\cdot j}}{N}\right)^2}{\dfrac{x_{i\cdot}x_{\cdot j}}{N}} = \chi^2,$$

$$(2.8)\quad \hat{J}(H_1, H_2) \approx \frac{1}{2}\sum_{i=1}^{r}\sum_{j=1}^{c} \frac{\left(x_{ij} - \dfrac{x_{i\cdot}x_{\cdot j}}{N}\right)^2}{\dfrac{x_{i\cdot}x_{\cdot j}}{N}} + \frac{1}{2}\sum_{i=1}^{r}\sum_{j=1}^{c} \frac{\left(x_{ij} - \dfrac{x_{i\cdot}x_{\cdot j}}{N}\right)^2}{x_{ij}}.$$

The reader now should express (2.2) in terms of the entropies defined in problem 8.30 of chapter 2.

The test of homogeneity for r samples in section 6.1 of chapter 6 may also be viewed as that for a two-way contingency table with the hypotheses in (6.1) in chapter 6 subject to fixed row totals, that is, for given N_i, $i = 1, \cdots, r$. We leave it to the reader to relate the components in table 2.1 and table 6.1 of chapter 6. [Cf. Good (1950, pp. 97–101).]

3. THREE-WAY TABLES

The possible combinations of hypotheses of interest become more numerous for three-way and higher order contingency tables. We shall examine several cases for a three-way table to illustrate the general procedure. [Cf. McGill (1954), Mitra (1955), Roy and Mitra (1956), Roy (1957, pp. 116–120).]

Suppose we have N independent observations, each characterized by a value of three classifications, row, column, and depth, distributed among r row-categories, c column-categories, and d depth-categories. Let x_{ijk} be the frequency of occurrence in the ith row, jth column, kth depth, and

$$\sum_{i=1}^{r}\sum_{j=1}^{c}\sum_{k=1}^{d} x_{ijk} = N, \qquad x_{i\cdot\cdot} = \sum_{j=1}^{c}\sum_{k=1}^{d} x_{ijk},$$

$$x_{\cdot j\cdot} = \sum_{i=1}^{r}\sum_{k=1}^{d} x_{ijk}, \qquad x_{\cdot\cdot k} = \sum_{i=1}^{r}\sum_{j=1}^{c} x_{ijk}, \qquad x_{ij\cdot} = \sum_{k=1}^{d} x_{ijk},$$

$$x_{i\cdot k} = \sum_{j=1}^{c} x_{ijk}, \qquad x_{\cdot jk} = \sum_{i=1}^{r} x_{ijk}, \qquad \sum_{j=1}^{c}\sum_{k=1}^{d} x_{\cdot jk} = N,$$

$$\sum_{i=1}^{r} x_{i\cdot\cdot} = \sum_{j=1}^{c} x_{\cdot j\cdot} = \sum_{k=1}^{d} x_{\cdot\cdot k} = N.$$

We denote the probabilities by p's with corresponding subscripts.

3.1. Independence of the Three Classifications

Consider a three-way table and the hypotheses:

$$H_1: p_{ijk} \neq p_{i\cdot\cdot}p_{\cdot j\cdot}p_{\cdot\cdot k}, \quad \text{for at least one } (i, j, k), \quad \Sigma\Sigma\Sigma p_{ijk} = 1, \quad p_{ijk} > 0,$$

$$(3.1) \quad H_2: p_{ijk} = p_{i\cdot\cdot}p_{\cdot j\cdot}p_{\cdot\cdot k}, \quad i = 1, 2, \cdots, r, \quad j = 1, 2, \cdots, c, \quad k = 1, 2, \cdots, d, \quad p_{1\cdot\cdot} + p_{2\cdot\cdot} + \cdots + p_{r\cdot\cdot} = p_{\cdot 1\cdot} + p_{\cdot 2\cdot} + \cdots + p_{\cdot c\cdot} = p_{\cdot\cdot 1} + p_{\cdot\cdot 2} + \cdots + p_{\cdot\cdot d} = 1, \quad p_{i\cdot\cdot} > 0, \quad p_{\cdot j\cdot} > 0, \quad p_{\cdot\cdot k} > 0.$$

Without repeating the detailed argument (similar to that for a sample of N independent observations from a multinominal population with rcd categories), we have [cf. (2.6) and (2.8) of chapter 6]

$$(3.2) \qquad I(H_1:H_2) = N\Sigma\Sigma\Sigma p_{ijk} \log \frac{p_{ijk}}{p_{i\cdot\cdot}p_{\cdot j\cdot}p_{\cdot\cdot k}},$$

$$(3.3) \qquad J(H_1, H_2) = N\Sigma\Sigma\Sigma(p_{ijk} - p_{i\cdot\cdot}p_{\cdot j\cdot}p_{\cdot\cdot k}) \log \frac{p_{ijk}}{p_{i\cdot\cdot}p_{\cdot j\cdot}p_{\cdot\cdot k}}.$$

Note that $I(H_1:H_2)/N$ in (3.2) is a measure of the joint relation among the row-, column-, and depth-categories [see the remarks following (2.3)].

For the conjugate distribution (see section 3 of chapter 6) with parameters the same as the observed sample best unbiased estimates, we have

$$(3.4) \qquad \hat{I}((p^*):(p)) = \sum_{i=1}^{r} \sum_{j=1}^{c} \sum_{k=1}^{d} x_{ijk} \log \frac{x_{ijk}}{Np_{i\cdot\cdot}p_{\cdot j\cdot}p_{\cdot\cdot k}}.$$

The null hypothesis of independence H_2 in (3.1) usually does not specify $p_{i\cdot\cdot}, p_{\cdot j\cdot}, p_{\cdot\cdot k}, i = 1, 2, \cdots, r, j = 1, 2, \cdots, c, k = 1, 2, \cdots, d$. We can analyze $\hat{I}((p)^*:(p))$ of (3.4) into several additive components. These components correspond to a hypothesis $H_2(R)$ specifying the values of the $p_{i\cdot\cdot}$, a hypothesis $H_2(C)$ specifying the values of the $p_{\cdot j\cdot}$, a hypothesis $H_2(D)$ specifying the values of the $p_{\cdot\cdot k}$, and a hypothesis $H_2(R \times C \times D)$ of independence, that is, H_2 in (3.1) is the intersection $H_2(R) \cap H_2(C) \cap H_2(D) \cap H_2(R \times C \times D)$. The analysis summarized in table 3.1 is an analogue of that in table 4.3 of chapter 6. Here there are $(r - 1)$ independent parameters $p_{i\cdot\cdot}, i = 1, 2, \cdots, r - 1, (c - 1)$ independent parameters $p_{\cdot j\cdot}, j = 1, 2, \cdots, c - 1$, and $(d - 1)$ independent parameters $p_{\cdot\cdot k}, k = 1, 2, \cdots, d - 1$. The equations (4.11) of chapter 6 are here

$$\sum_{j=1}^{c} \sum_{k=1}^{d} \left(\frac{x_{ijk}}{p_{ijk}} p_{\cdot j\cdot}p_{\cdot\cdot k} - \frac{x_{rjk}}{p_{rjk}} p_{\cdot j\cdot}p_{\cdot\cdot k}\right) = 0, \qquad i = 1, 2, \cdots, r - 1,$$

$$\sum_{i=1}^{r}\sum_{k=1}^{d}\left(\frac{x_{ijk}}{p_{ijk}}p_{i\cdot\cdot}p_{\cdot\cdot k}-\frac{x_{ick}}{p_{ick}}p_{i\cdot\cdot}p_{\cdot\cdot k}\right)=0,\qquad j=1,2,\cdots,c-1,$$

$$\sum_{i=1}^{r}\sum_{j=1}^{c}\left(\frac{x_{ijk}}{p_{ijk}}p_{i\cdot\cdot}p_{\cdot j\cdot}-\frac{x_{ijd}}{p_{ijd}}p_{i\cdot\cdot}p_{\cdot j\cdot}\right)=0,\qquad k=1,2,\cdots,d-1.$$

Since $p_{ijk} = p_{i\cdot\cdot}p_{\cdot j\cdot}p_{\cdot\cdot k}$, these equations reduce to

$$\frac{x_{i\cdot\cdot}}{p_{i\cdot\cdot}}=\frac{x_{r\cdot\cdot}}{p_{r\cdot\cdot}},\qquad i=1,2,\cdots,r-1,\qquad \frac{x_{\cdot j\cdot}}{p_{\cdot j\cdot}}=\frac{x_{\cdot c\cdot}}{p_{\cdot c\cdot}},$$

$$j=1,2,\cdots,c-1,\frac{x_{\cdot\cdot k}}{p_{\cdot\cdot k}}=\frac{x_{\cdot\cdot d}}{p_{\cdot\cdot d}},\qquad k=1,2,\cdots,d-1,$$

yielding

$$\hat{p}_{i\cdot\cdot}=\frac{x_{i\cdot\cdot}}{N},\qquad \hat{p}_{\cdot j\cdot}=\frac{x_{\cdot j\cdot}}{N},\qquad \hat{p}_{\cdot\cdot k}=\frac{x_{\cdot\cdot k}}{N},\qquad \hat{p}_{ijk}=\frac{x_{i\cdot\cdot}}{N}\cdot\frac{x_{\cdot j\cdot}}{N}\cdot\frac{x_{\cdot\cdot k}}{N},$$

$$i=1,2,\cdots,r,\qquad j=1,2,\cdots,c,\qquad k=1,2,\cdots,d.$$

(We write $\hat{p}$ here rather than $\tilde{p}$ because we shall need $\tilde{p}$ for different estimates in the analysis in section 10.) Note that the independence component in table 3.1 is the minimum value of the total for variations of the $p_{i\cdot\cdot}$, $p_{\cdot j\cdot}$, $p_{\cdot\cdot k}$,

$$\sum_{i=1}^{r}p_{i\cdot\cdot}=\sum_{j=1}^{c}p_{\cdot j\cdot}=\sum_{k=1}^{d}p_{\cdot\cdot k}=1,$$

that is, over the populations of H_2 with the given marginal values, and that by the convexity property (see section 3 of chapter 2) the total is not less than the row, column, or depth component. The degrees of freedom in table 3.1 are those of the asymptotic χ^2-distributions under the null hypothesis H_2 of (3.1).

The independence component in table 3.1 may also be expressed as

$$(3.5)\quad \hat{I}(H_1:H_2)=\Sigma\Sigma\Sigma x_{ijk}\log x_{ijk}-\Sigma x_{i\cdot\cdot}\log x_{i\cdot\cdot}-\Sigma x_{\cdot j\cdot}\log x_{\cdot j\cdot}$$
$$-\Sigma x_{\cdot\cdot k}\log x_{\cdot\cdot k}+2N\log N,$$

for computational convenience with the table of $n \log n$.

The divergences do not provide a similar additive analysis (with these estimates), but the estimate of the divergence corresponding to the independence component in table 3.1 is [cf. (4.12) of chapter 6]

$$(3.6)\qquad \hat{J}(H_1,H_2)=N\Sigma\Sigma\Sigma\left(\frac{x_{ijk}}{N}-\frac{x_{i\cdot\cdot}}{N}\frac{x_{\cdot j\cdot}}{N}\frac{x_{\cdot\cdot k}}{N}\right)\log\frac{N^2x_{ijk}}{x_{i\cdot\cdot}x_{\cdot j\cdot}x_{\cdot\cdot k}}.$$

Note that the independence component $\hat{I}(H_1:H_2)$ in table 3.1 is (3.2) with the substitution of x_{ijk}/N for p_{ijk}, $x_{i\cdot\cdot}/N$ for $p_{i\cdot\cdot}$, $x_{\cdot j\cdot}/N$ for $p_{\cdot j\cdot}$, and $x_{\cdot\cdot k}/N$ for $p_{\cdot\cdot k}$ and that (3.6) is (3.3) with the same substitutions.

TABLE 3.1

Component due to	Information	D.F.
Rows, $H_2(R)$ $\hat{p}_{i\cdot\cdot} = x_{i\cdot\cdot}/N$	$2\sum_{i=1}^{r} x_{i\cdot\cdot} \log \frac{x_{i\cdot\cdot}}{Np_{i\cdot\cdot}}$	$r - 1$
Columns, $H_2(C)$ $\hat{p}_{\cdot j\cdot} = x_{\cdot j\cdot}/N$	$2\sum_{j=1}^{c} x_{\cdot j\cdot} \log \frac{x_{\cdot j\cdot}}{Np_{\cdot j\cdot}}$	$c - 1$
Depths, $H_2(D)$ $\hat{p}_{\cdot\cdot k} = x_{\cdot\cdot k}/N$	$2\sum_{k=1}^{d} x_{\cdot\cdot k} \log \frac{x_{\cdot\cdot k}}{Np_{\cdot\cdot k}}$	$d - 1$
Independence, $H_2(R \times C \times D)$ $2\hat{I}(H_1:H_2)$	$2\sum_{i=1}^{r}\sum_{j=1}^{c}\sum_{k=1}^{d} x_{ijk} \log \frac{N^2 x_{ijk}}{x_{i\cdot\cdot}x_{\cdot j\cdot}x_{\cdot\cdot k}}$	$rcd - r - c - d + 2$
Total, $2\hat{I}((p^*):(p))$	$2\sum_{i=1}^{r}\sum_{j=1}^{c}\sum_{k=1}^{d} x_{ijk} \log \frac{x_{ijk}}{Np_{i\cdot\cdot}p_{\cdot j\cdot}p_{\cdot\cdot k}}$	$rcd - 1$

If the row, column, and depth classifications are independent, $2\hat{I}(H_1:H_2)$ and $\hat{J}(H_1, H_2)$ are asymptotically distributed as χ^2 with $(rcd - r - c - d + 2)$ degrees of freedom. Under the alternative hypothesis H_1 of (3.1), $2\hat{I}(H_1:H_2)$ and $\hat{J}(H_1, H_2)$ are asymptotically distributed as noncentral χ^2 with $(rcd - r - c - d + 2)$ degrees of freedom and respective noncentrality parameters $2I(H_1:H_2)$ and $J(H_1, H_2)$ as given by (3.2) and (3.3) with

$$p_{i\cdot\cdot} = \sum_j\sum_k p_{ijk}, \qquad p_{\cdot j\cdot} = \sum_i\sum_k p_{ijk}, \qquad p_{\cdot\cdot k} = \sum_i\sum_j p_{ijk}.$$

(See problem 13.8.)

3.2. Row Classification Independent of the Other Classifications

Consider a three-way table and the hypotheses

$$H_1: p_{ijk} \neq p_{i\cdot\cdot}p_{\cdot jk}, \quad \text{for at least one } (i, jk), \quad \Sigma\Sigma\Sigma p_{ijk} = 1, \quad p_{ijk} > 0,$$

$$(3.7)\quad \begin{aligned} H_2: p_{ijk} = p_{i\cdot\cdot}p_{\cdot jk}, \quad & i = 1, 2, \cdots, r, \quad j = 1, 2, \cdots, c, \\ & k = 1, 2, \cdots, d, \quad p_{1\cdot\cdot} + p_{2\cdot\cdot} + \cdots + p_{r\cdot\cdot} = 1 \\ & = \sum_{j=1}^{c}\sum_{k=1}^{d} p_{\cdot jk}, \quad p_{i\cdot\cdot} > 0, \quad p_{\cdot jk} > 0. \end{aligned}$$

Note that H_2 in (3.7) implies

$$p_{ij\cdot} = \sum_{k=1}^{d} p_{ijk} = p_{i\cdot\cdot} \sum_{k=1}^{d} p_{\cdot jk} = p_{i\cdot\cdot}p_{\cdot j\cdot},$$

and

$$p_{i\cdot k} = \sum_{j=1}^{c} p_{ijk} = p_{i\cdot\cdot} \sum_{j=1}^{c} p_{\cdot jk} = p_{i\cdot\cdot}p_{\cdot\cdot k},$$

that is, the row and column classifications are independent and the row and depth classifications are independent. [Is the converse true? See Feller (1950, pp. 87–88), Kolmogorov (1950, p. 11).]

Without repeating the detailed argument (similar to that already used), we have [cf. (2.6) and (2.8) of chapter 6],

$$I(H_1:H_2) = N\Sigma\Sigma\Sigma p_{ijk} \log \frac{p_{ijk}}{p_{i\cdot\cdot}p_{\cdot jk}}, \tag{3.8}$$

$$J(H_1, H_2) = N\Sigma\Sigma\Sigma(p_{ijk} - p_{i\cdot\cdot}p_{\cdot jk}) \log \frac{p_{ijk}}{p_{i\cdot\cdot}p_{\cdot jk}}. \tag{3.9}$$

Note that $I(H_1:H_2)/N$ in (3.8) is a measure of the relation between the row- and (column, depth)-categories and may be defined as the information in the row-categories about the (column, depth)-categories or vice versa [see the remarks following (2.3)].

For the conjugate distribution (see section 3 of chapter 6) with parameters the same as the observed sample best unbiased estimates, we have

$$\hat{I}((p)^*:(p)) = \sum_{i=1}^{r} \sum_{j=1}^{c} \sum_{k=1}^{d} x_{ijk} \log \frac{x_{ijk}}{Np_{i\cdot\cdot}p_{\cdot jk}}. \tag{3.10}$$

The null hypothesis H_2 of (3.7) usually does not specify $p_{i\cdot\cdot}$, $p_{\cdot jk}$, $i = 1, 2, \cdots, r$, $j = 1, 2, \cdots, c$, $k = 1, 2, \cdots, d$. We can analyze $\hat{I}((p)^*:(p))$ of (3.10) into several additive components. These components correspond to a hypothesis $H_2(R)$ specifying the values of the $p_{i\cdot\cdot}$, a hypothesis $H_2(CD)$ specifying the values of the $p_{\cdot jk}$, and a hypothesis $H_2(R \times CD)$ of independence, that is, H_2 in (3.7) is the intersection $H_2(R) \cap H_2(CD) \cap H_2(R \times CD)$. The analysis summarized in table 3.2 is an analogue of that in table 4.3 of chapter 6. Here there are $(r - 1)$ independent parameters $p_{i\cdot\cdot}$, $i = 1, 2, \cdots, r - 1$, and $(cd - 1)$ independent parameters $p_{\cdot jk}$, $j = 1, 2, \cdots, c$, $k = 1, 2, \cdots, d$, omitting the combination $j = c$ and $k = d$. The equations (4.11) of chapter 6 are here

$$\sum_{j=1}^{c} \sum_{k=1}^{d} \left(\frac{x_{ijk}}{p_{ijk}} p_{\cdot jk} - \frac{x_{rjk}}{p_{rjk}} p_{\cdot jk}\right) = 0, \qquad i = 1, 2, \cdots, r - 1,$$

$$\sum_{i=1}^{r} \left(\frac{x_{ijk}}{p_{ijk}} p_{i\cdot\cdot} - \frac{x_{icd}}{p_{icd}} p_{i\cdot\cdot}\right) = 0, \qquad j = 1, 2, \cdots, c, \quad k = 1, 2, \cdots, d,$$

omitting $j = c$ and $k = d$.

Since $p_{ijk} = p_{i\cdot\cdot}p_{\cdot jk}$, these equations reduce to

$$\frac{x_{i\cdot\cdot}}{p_{i\cdot\cdot}} = \frac{x_{r\cdot\cdot}}{p_{r\cdot\cdot}}, \quad i = 1, 2, \cdots, r-1, \quad \frac{x_{\cdot jk}}{p_{\cdot jk}} = \frac{x_{\cdot cd}}{p_{\cdot cd}}, \quad j = 1, 2, \cdots, c,$$
$$k = 1, 2, \cdots, d, \text{ omitting } j = c \text{ and } k = d,$$

yielding

$$\tilde{p}_{i\cdot\cdot} = \frac{x_{i\cdot\cdot}}{N}, \quad \tilde{p}_{\cdot jk} = \frac{x_{\cdot jk}}{N}, \quad \tilde{p}_{ijk} = \frac{x_{i\cdot\cdot}}{N}\frac{x_{\cdot jk}}{N}, \quad i = 1, 2, \cdots, r,$$
$$j = 1, 2, \cdots, c, \quad k = 1, 2, \cdots, d.$$

Note that the independence component in table 3.2 is the minimum value of the total for variations of the $p_{i\cdot\cdot}$ and $p_{\cdot jk}$,

$$\sum_{i=1}^{r} p_{i\cdot\cdot} = 1 = \sum_{j=1}^{c}\sum_{k=1}^{d} p_{\cdot jk},$$

that is, over the populations of H_2 with the given row and (column, depth) marginal values, and that by the convexity property (see section 3 of chapter 2) the total is not less than the row or (column, depth) component. The degrees of freedom in table 3.2 are those of the asymptotic χ^2-distributions under the null hypothesis H_2 of (3.7).

TABLE 3.2

Component due to	Information	D.F.
Rows, $H_2(R)$ $\tilde{p}_{i\cdot\cdot} = x_{i\cdot\cdot}/N$	$2\sum_{i=1}^{r} x_{i\cdot\cdot}\log\frac{x_{i\cdot\cdot}}{Np_{i\cdot\cdot}}$	$r-1$
Column, depth, $H_2(CD)$ $\tilde{p}_{\cdot jk} = x_{\cdot jk}/N$	$2\sum_{j=1}^{c}\sum_{k=1}^{d} x_{\cdot jk}\log\frac{x_{\cdot jk}}{Np_{\cdot jk}}$	$cd-1$
Rows × (column, depth) Independence, $H_2(R \times CD)$ $2\hat{I}(H_1:H_2)$	$2\sum_{i=1}^{r}\sum_{j=1}^{c}\sum_{k=1}^{d} x_{ijk}\log\frac{Nx_{ijk}}{x_{i\cdot\cdot}x_{\cdot jk}}$	$(r-1)(cd-1)$
Total, $2\hat{I}((p^*):(p))$	$2\sum_{i=1}^{r}\sum_{j=1}^{c}\sum_{k=1}^{d} x_{ijk}\log\frac{x_{ijk}}{Np_{i\cdot\cdot}p_{\cdot jk}}$	$rcd-1$

The independence component in table 3.2 may also be expressed as

$$(3.11)\quad \hat{I}(H_1:H_2) = \Sigma\Sigma\Sigma x_{ijk}\log x_{ijk} - \Sigma x_{i\cdot\cdot}\log x_{i\cdot\cdot} - \Sigma\Sigma x_{\cdot jk}\log x_{\cdot jk} + N\log N$$

for computational convenience with the table of $n \log n$. The divergences do not provide a similar additive analysis (with these estimates), but the estimate of the divergence corresponding to the independence component in table 3.2 is [cf. (4.12) of chapter 6]

$$(3.12) \qquad \hat{J}(H_1, H_2) = N\Sigma\Sigma\Sigma\left(\frac{x_{ijk}}{N} - \frac{x_{i\cdot\cdot}}{N}\frac{x_{\cdot jk}}{N}\right)\log\frac{Nx_{ijk}}{x_{i\cdot\cdot}x_{\cdot jk}}.$$

Note that $\hat{I}(H_1{:}H_2)$ in table 3.2 is (3.8) with the substitution of x_{ijk}/N for p_{ijk}, $x_{i\cdot\cdot}/N$ for $p_{i\cdot\cdot}$, and $x_{\cdot jk}/N$ for $p_{\cdot jk}$ and that (3.12) is (3.9) with the same substitutions.

If the row classification is independent of the other two classifications, $2\hat{I}(H_1{:}H_2)$ of (3.11) and $\hat{J}(H_1, H_2)$ of (3.12) are asymptotically distributed as χ^2 with $(r-1)(cd-1)$ degrees of freedom. Under the alternative hypothesis H_1 of (3.7), $2\hat{I}(H_1{:}H_2)$ and $\hat{J}(H_1, H_2)$ are asymptotically distributed as noncentral χ^2 with $(r-1)(cd-1)$ degrees of freedom and respective noncentrality parameters $2I(H_1{:}H_2)$ and $J(H_1, H_2)$ given by (3.8) and (3.9) with $p_{i\cdot\cdot} = \sum_j\sum_k p_{ijk}$, $p_{\cdot jk} = \sum_i p_{ijk}$. (See problem 13.9.)

Similar analyses are of course possible if the independence hypothesis is that either the column or depth classification is independent of the other two classifications. We leave the details to the reader.

3.3. Independence Hypotheses

The independence component in table 3.1 is analyzed into additive components in table 3.3. This is a reflection of the fact that $H_2(R \times C \times D) \rightleftharpoons H_2(R \times CD) \cap H_2(C \times D)$, that is, the three classifications are independent if and only if the row classification is independent of the (column, depth) classifications and the column and depth classifications

TABLE 3.3

Component due to	Information	D.F.
Column × depth $H_2(C \times D)$	$2\sum_{j=1}^{c}\sum_{k=1}^{d} x_{\cdot jk}\log\frac{Nx_{\cdot jk}}{x_{\cdot j\cdot}x_{\cdot\cdot k}}$	$(c-1)(d-1)$
Row × (column, depth) $H_2(R \times CD)$	$2\sum_{i=1}^{r}\sum_{j=1}^{c}\sum_{k=1}^{d} x_{ijk}\log\frac{Nx_{ijk}}{x_{i\cdot\cdot}x_{\cdot jk}}$	$(r-1)(cd-1)$
Independence, $H_2(R \times C \times D)$	$2\sum_{i=1}^{r}\sum_{j=1}^{c}\sum_{k=1}^{d} x_{ijk}\log\frac{N^2x_{ijk}}{x_{i\cdot\cdot}x_{\cdot j\cdot}x_{\cdot\cdot k}}$	$rcd - r - c - d + 2$

are independent, since $p_{ijk} = p_{i\cdot\cdot}p_{\cdot jk}$ and $p_{\cdot jk} = p_{\cdot j\cdot}p_{\cdot\cdot k}$ imply that $p_{ijk} = p_{i\cdot\cdot}p_{\cdot j\cdot}p_{\cdot\cdot k}$; and $p_{ijk} = p_{i\cdot\cdot}p_{\cdot j\cdot}p_{\cdot\cdot k}$ implies $\sum_{i=1}^{r} p_{ijk} = p_{\cdot jk} = p_{\cdot j\cdot}p_{\cdot\cdot k}$ or $p_{ijk} = p_{i\cdot\cdot}p_{\cdot jk}$. It is of course also true that $H_2(R \times C \times D) \rightleftharpoons H_2(C \times RD) \cap H_2(R \times D)$ and $H_2(R \times C \times D) \rightleftharpoons H_2(RC \times D) \cap H_2(R \times C)$, but we leave the details to the reader. Note that the convexity property (see section 3 of chapter 2) ensures that the $H_2(C \times D)$ component is the minimum value of the $H_2(R \times C \times D)$ component for the given grouping. (See example 12.3 and problem 8.30 in chapter 12.)

3.4. Conditional Independence

Suppose for some category, say the kth, of the depth classification we want to test a null hypothesis that the row and column classifications are independent.

The argument here parallels that for the two-way contingency table. We shall follow it with the notation introduced for the three-way contingency table, our basic problem. We want to test the hypotheses:

$$H_1: p_{ijk} \neq \frac{p_{i\cdot k}p_{\cdot jk}}{p_{\cdot\cdot k}}, \quad \text{for at least one } (i, j), \quad \sum_{i=1}^{r}\sum_{j=1}^{c} p_{ijk} = p_{\cdot\cdot k}, \quad p_{ijk} > 0,$$

(3.13)

$$H_2: p_{ijk} = \frac{p_{i\cdot k}p_{\cdot jk}}{p_{\cdot\cdot k}}, \quad i = 1, 2, \cdots, r, \quad j = 1, 2, \cdots, c, \quad \sum_{i=1}^{r}\sum_{j=1}^{c} p_{ijk}$$
$$= p_{\cdot\cdot k}, \quad \sum_{i=1}^{r} p_{i\cdot k} = p_{\cdot\cdot k} = \sum_{j=1}^{c} p_{\cdot jk}, \quad p_{i\cdot k} > 0, \quad p_{\cdot jk} > 0, \quad p_{\cdot\cdot k} > 0.$$

Note that we are dealing with conditional probabilities $p_{ijk}/p_{\cdot\cdot k}$, $p_{i\cdot k}/p_{\cdot\cdot k}$, and $p_{\cdot jk}/p_{\cdot\cdot k}$. The analysis in table 3.4 is derived from that in table 2.1. We shall denote a conditional hypothesis about the rows for the kth category of the depth classification by $H_2(R|k)$ and the corresponding hypothesis for any category of the depth classification by $H_2(R|D)$; similarly for the columns.

If H_2 in (3.13) is true for all k, that is, the row and column classifications are conditionally independent given the depth classification, the appropriate analysis corresponds to that in table 3.4 with each information component now summed over $k = 1, 2, \cdots, d$, and each degree of freedom multiplied by d. In particular, the information component for a null hypothesis of conditional independence $H_2((R|D) \times (C|D))$ is

$$(3.14) \qquad 2\hat{I}(H_1:H_2) = 2\sum_{i=1}^{r}\sum_{j=1}^{c}\sum_{k=1}^{d} x_{ijk} \log \frac{x_{ijk}}{x_{i\cdot k}x_{\cdot jk}/x_{\cdot\cdot k}},$$

with $d(r-1)(c-1)$ degrees of freedom for the asymptotic χ^2-distribution under the null hypothesis of conditional independence.

TABLE 3.4

Component due to	Information	D.F.
Rows, $H_2(R\|k)$ $\tilde{p}_{i\cdot k}/\tilde{p}_{\cdot\cdot k} = x_{i\cdot k}/x_{\cdot\cdot k}$	$2\sum_{i=1}^{r} x_{i\cdot k} \log \frac{x_{i\cdot k}}{x_{\cdot\cdot k}(p_{i\cdot k}/p_{\cdot\cdot k})}$	$r - 1$
Columns, $H_2(C\|k)$ $\tilde{p}_{\cdot jk}/\tilde{p}_{\cdot\cdot k} = x_{\cdot jk}/x_{\cdot\cdot k}$	$2\sum_{j=1}^{c} x_{\cdot jk} \log \frac{x_{\cdot jk}}{x_{\cdot\cdot k}(p_{\cdot jk}/p_{\cdot\cdot k})}$	$c - 1$
Conditional independence, $H_2((R\|k) \times (C\|k))$	$2\sum_{i=1}^{r}\sum_{j=1}^{c} x_{ijk} \log \frac{x_{ijk}}{x_{i\cdot k}x_{\cdot jk}/x_{\cdot\cdot k}}$	$(r - 1)(c - 1)$
Total, $2\hat{I}((p^*):(p))$	$2\sum_{i=1}^{r}\sum_{j=1}^{c} x_{ijk} \log \frac{x_{ijk}}{x_{\cdot\cdot k}(p_{i\cdot k}/p_{\cdot\cdot k})(p_{\cdot jk}/p_{\cdot\cdot k})}$	$rc - 1$

3.5. Further Analysis

The $H_2(R \times CD)$ component in table 3.3 [the test for the hypothesis that the row and (column, depth) classifications are independent] is analyzed into additive components in table 3.5. This is a reflection of the fact that $H_2(R \times CD) \rightleftharpoons H_2((R|D) \times (C|D)) \cap H_2(R \times D)$, that is, the row classification is independent of the (column, depth) classifications if and only if the row and column classifications are conditionally independent given the depth classification and the row and depth classifications are independent, since $p_{ijk} = p_{i\cdot k}p_{\cdot jk}/p_{\cdot\cdot k}$ and $p_{i\cdot k} = p_{i\cdot\cdot}p_{\cdot\cdot k}$ imply $p_{ijk} = p_{i\cdot\cdot}p_{\cdot jk}$; and $p_{ijk} = p_{i\cdot\cdot}p_{\cdot jk}$ implies $\sum_{j=1}^{c} p_{ijk} = p_{i\cdot k} = p_{i\cdot\cdot}p_{\cdot\cdot k}$ or $p_{ijk} = p_{i\cdot k}p_{\cdot jk}/p_{\cdot\cdot k}$. Note that the convexity property (see section 3 of chapter 2)

TABLE 3.5

Component due to	Information	D.F.
Row × depth $H_2(R \times D)$	$2\sum_{i=1}^{r}\sum_{k=1}^{d} x_{i\cdot k} \log \frac{Nx_{i\cdot k}}{x_{i\cdot\cdot}x_{\cdot\cdot k}}$	$(r - 1)(d - 1)$
(Row, depth) × (column, depth) $H_2((R\|D) \times (C\|D))$	$2\sum_{i=1}^{r}\sum_{j=1}^{c}\sum_{k=1}^{d} x_{ijk} \log \frac{x_{ijk}}{\frac{x_{i\cdot k}x_{\cdot jk}}{x_{\cdot\cdot k}}}$	$d(r - 1)(c - 1)$
Row × (column, depth) $H_2(R \times CD)$	$2\sum_{i=1}^{r}\sum_{j=1}^{c}\sum_{k=1}^{d} x_{ijk} \log \frac{Nx_{ijk}}{x_{i\cdot\cdot}x_{\cdot jk}}$	$(r - 1)(cd - 1)$

ensures that the $H_2(R \times D)$ component is the minimum value of the $H_2(R \times CD)$ component for the given grouping. (See example 12.3 and problem 8.31 in chapter 12.)

Similar analyses follow if the conditional independence hypothesis is applied to other combinations of the classifications. We leave the details to the reader.

4. HOMOGENEITY OF TWO-WAY TABLES

We may treat r independent samples of a c × d table as an r × c × d three-way table with suitable hypotheses and restrictions. Thus, suppose we want to test a null hypothesis that r samples of a c × d table are homogeneous, subject to a fixed total for each c × d table. With the three-way table notation, the hypotheses are [cf. (6.1) of chapter 6]:

$$\begin{aligned} H_1: p_{ijk} &\neq p_{\cdot jk}, \quad \sum_{j=1}^{c}\sum_{k=1}^{d} p_{ijk} = 1, \qquad i = 1, 2, \cdots, r, \\ H_2: p_{ijk} &= p_{\cdot jk}, \quad i = 1, 2, \cdots, r, \qquad j = 1, 2, \cdots, c, \\ & k = 1, 2, \cdots d, \qquad \sum_{j=1}^{c}\sum_{k=1}^{d} p_{\cdot jk} = 1. \end{aligned} \tag{4.1}$$

The analysis in table 4.1 is derived from that in table 6.1 of chapter 6 for the basic problem of the homogeneity of r samples from multinomial populations with cd categories.

TABLE 4.1

Component due to	Information	D.F.
$\hat{p}_{\cdot jk} = x_{\cdot jk}/N$ (Between)	$2\sum_{j=1}^{c}\sum_{k=1}^{d} x_{\cdot jk} \log \frac{x_{\cdot jk}}{Np_{\cdot jk}}$	$cd - 1$
Error, $2\hat{I}(H_1:H_2)$ (Within, homogeneity)	$2\sum_{i=1}^{r}\sum_{j=1}^{c}\sum_{k=1}^{d} x_{ijk} \log \frac{Nx_{ijk}}{x_{i\cdot\cdot}x_{\cdot jk}}$	$(r-1)(cd-1)$
Total, $2\hat{I}((p^*):(p))$	$2\sum_{i=1}^{r}\sum_{j=1}^{c}\sum_{k=1}^{d} x_{ijk} \log \frac{x_{ijk}}{x_{i\cdot\cdot}p_{\cdot jk}}$	$r(cd-1)$

The degrees of freedom in table 4.1 are those of the asymptotic χ^2-distributions under the null hypothesis H_2 of (4.1). Note that the error or within or homogeneity component in table 4.1 is the minimum value of the total for variations of the $p_{\cdot jk}$, $\sum_{j=1}^{c}\sum_{k=1}^{d} p_{\cdot jk} = 1$, given the c × d table totals, that is, over the populations of H_2, and that by the convexity property (see section 3 of chapter 2) the total is not less than the between component. As might be expected, the analysis in table 4.1 is related to that in table 3.2 for the hypothesis of independence of the row classification and the other two classifications. In fact, the total of table 4.1 is the total minus the row component of table 3.2, the between component of table 4.1 is the (column, depth) component of table 3.2, and the within or homogeneity component in table 4.1 is the independence component of table 3.2. (See problem 13.10.)

5. CONDITIONAL HOMOGENEITY

Suppose we have the r samples of section 4, and for some category, say the jth, of the column classification we want to test a null hypothesis that the r samples of the depth classification are homogeneous. The argument here parallels that in section 6.1 of chapter 6 for the basic problem of the homogeneity of r samples. We shall follow it with the notation for the three-way contingency table. We want to test the hypotheses:

$$(5.1)\qquad \begin{aligned} &H_1: p_{ijk} \neq p_{\cdot jk}, \quad \sum_{k=1}^{d} p_{ijk} = p_{ij\cdot}, \quad i = 1, 2, \cdots, r,\\ &H_2: p_{ijk} = p_{\cdot jk}, \quad i = 1, 2, \cdots, r, \quad k = 1, 2, \cdots, d,\\ &\qquad\qquad\qquad \sum_{k=1}^{d} p_{\cdot jk} = p_{\cdot j\cdot}. \end{aligned}$$

The analysis in table 5.1 is derived from that in table 6.1 of chapter 6.

TABLE 5.1

Component due to	Information	D.F.
$\hat{p}_{\cdot jk}/\hat{p}_{\cdot j\cdot} = x_{\cdot jk}/x_{\cdot j\cdot}$ (Between)	$2\sum_{k=1}^{d} x_{\cdot jk}\log\dfrac{x_{\cdot jk}}{x_{\cdot j\cdot}(p_{\cdot jk}/p_{\cdot j\cdot})}$	$d-1$
Error, $2\hat{I}(H_1:H_2)$, (Within)	$2\sum_{i=1}^{r}\sum_{k=1}^{d} x_{ijk}\log\dfrac{x_{ijk}}{(x_{ij\cdot}x_{\cdot jk})/x_{\cdot j\cdot}}$	$(r-1)(d-1)$
Total, $2\hat{I}((p^*):(p))$	$2\sum_{i=1}^{r}\sum_{k=1}^{d} x_{ijk}\log\dfrac{x_{ijk}}{x_{ij\cdot}(p_{\cdot jk}/p_{\cdot j\cdot})}$	$r(d-1)$

If H_2 in (5.1) is true for all j, that is, the depth classification is conditionally homogeneous given the column classification, the appropriate analysis corresponds to that in table 5.1 with each information component now summed over $j = 1, 2, \cdots, c$, and each degree of freedom multiplied by c. In particular, the information component for a null hypothesis of conditional homogeneity $p_{ijk}/p_{ij\cdot} = p_{\cdot jk}/p_{\cdot j\cdot}$, $i = 1, 2, \cdots, r$, $j = 1, 2, \cdots, c$, $k = 1, 2, \cdots, d$, is

$$(5.2) \qquad 2\hat{I}(H_1:H_2) = 2\sum_{i=1}^{r}\sum_{j=1}^{c}\sum_{k=1}^{d} x_{ijk} \log \frac{x_{ijk}}{(x_{ij\cdot}x_{\cdot jk})/x_{\cdot j\cdot}},$$

with $c(r-1)(d-1)$ degrees of freedom for the asymptotic χ^2-distribution under the null hypothesis of conditional homogeneity.

Note that $2\hat{I}(H_1:H_2)$ in (5.2) is similar to the component for the test of a null hypothesis of conditional independence $H_2((R|C) \times (D|C))$ (cf. table 3.5).

6. HOMOGENEITY

The homogeneity component in table 4.1 is analyzed into additive components in table 6.1 (cf. table 3.5).

TABLE 6.1

Component due to	Information	D.F.
(C)-homogeneity	$2\sum_{i=1}^{r}\sum_{j=1}^{c} x_{ij\cdot} \log \frac{Nx_{ij\cdot}}{x_{i\cdot\cdot}x_{\cdot j\cdot}}$	$(r-1)(c-1)$
Conditional homogeneity-$(D\|C)$	$2\sum_{i=1}^{r}\sum_{j=1}^{c}\sum_{k=1}^{d} x_{ijk} \log \frac{x_{ijk}}{(x_{ij\cdot}x_{\cdot jk})/x_{\cdot j\cdot}}$	$c(r-1)(d-1)$
(C, D)-homogeneity	$2\sum_{i=1}^{r}\sum_{j=1}^{c}\sum_{k=1}^{d} x_{ijk} \log \frac{Nx_{ijk}}{x_{i\cdot\cdot}x_{\cdot jk}}$	$(r-1)(cd-1)$

The analysis in table 6.1 is a reflection of the fact that (C, D)-homogeneity $\rightleftharpoons$ conditional homogeneity-$(D|C) \cap (C)$-homogeneity, that is, the two-way (column, depth) tables are homogeneous if and only if the depth classifications are conditionally homogeneous given the column classification, and the column classifications are homogeneous, since $p_{ijk}/p_{ij\cdot} = p_{\cdot jk}/p_{\cdot j\cdot}$, $i = 1, 2, \cdots, r$, $j = 1, 2, \cdots, c$, $k = 1, 2, \cdots, d$, and $p_{ij\cdot} = p_{\cdot j\cdot}$ imply $p_{ijk} = p_{\cdot jk}$, $i = 1, 2, \cdots, r$, $j = 1, 2, \cdots, c$,

$k = 1, 2, \cdots, d$; and $p_{ijk} = p_{\cdot jk}$, $i = 1, 2, \cdots, r$, $j = 1, 2, \cdots, c$, $k = 1, 2, \cdots, d$, implies $p_{ij\cdot} = p_{\cdot j\cdot}$, $i = 1, 2, \cdots, r$, $j = 1, 2, \cdots, c$, and $p_{ijk}/p_{ij\cdot} = p_{\cdot jk}/p_{\cdot j\cdot}$. Note that the convexity property (see section 3 of chapter 2) ensures that the (C)-homogeneity component is the minimum value of the (C, D)-homogeneity component for the given grouping (see examples 12.2 and 12.4).

7. INTERACTION

Since the information component for conditional homogeneity in table 6.1 is a convex function (see section 3 of chapter 2),

$$(7.1) \qquad \sum_{i=1}^{r} \sum_{j=1}^{c} \sum_{k=1}^{d} x_{ijk} \log \frac{x_{ijk}}{\dfrac{x_{ij\cdot}x_{\cdot jk}}{x_{\cdot j\cdot}}} \geqq \sum_{i=1}^{r} \sum_{k=1}^{d} x_{i\cdot k} \log \frac{x_{i\cdot k}}{\sum_{j=1}^{c} \dfrac{x_{ij\cdot}x_{\cdot jk}}{x_{\cdot j\cdot}}},$$

with equality in (7.1) if and only if (cf. example 3.2 of chapter 2)

$$\frac{x_{ijk}}{\dfrac{x_{ij\cdot}x_{\cdot jk}}{x_{\cdot j\cdot}}} = \frac{x_{imk}}{\dfrac{x_{im\cdot}x_{\cdot mk}}{x_{\cdot m\cdot}}}, i = 1, 2, \cdots, r, k = 1, 2, \cdots, d;\ j, m = 1, 2, \cdots, c.$$

We may therefore analyze the conditional homogeneity component in table 6.1 into two additive components as shown in table 7.1, with $y_{i\cdot k} = \sum_{j=1}^{c} x_{ij\cdot}x_{\cdot jk}/x_{\cdot j\cdot}$. Note that $y_{i\cdot\cdot} = \sum_{k=1}^{d} y_{i\cdot k} = x_{i\cdot\cdot}$ and that $y_{\cdot\cdot k} = \sum_{i=1}^{r} y_{i\cdot k} = x_{\cdot\cdot k}$. (See example 12.4.) The analysis in table 7.1 is

TABLE 7.1

Component due to	Information	D.F.
(RD)-interaction	$2\sum_{i=1}^{r}\sum_{k=1}^{d} x_{i\cdot k} \log \dfrac{x_{i\cdot k}}{y_{i\cdot k}}$	$(r-1)(d-1)$
(RD, C)-interaction	$2\sum_{i=1}^{r}\sum_{j=1}^{c}\sum_{k=1}^{d} x_{ijk} \log \dfrac{x_{ijk}}{\dfrac{x_{i\cdot k}x_{ij\cdot}x_{\cdot jk}}{y_{i\cdot k}x_{\cdot j\cdot}}}$	$(r-1)(c-1)(d-1)$
Conditional homogeneity-$(D\|C)$	$2\sum_{i=1}^{r}\sum_{j=1}^{c}\sum_{k=1}^{d} x_{ijk} \log \dfrac{x_{ijk}}{\dfrac{x_{ij\cdot}x_{\cdot jk}}{x_{\cdot j\cdot}}}$	$c(r-1)(d-1)$

a reflection of the fact that $p_{i\cdot k} = \sum_{j=1}^{c} \frac{p_{ij\cdot}p_{\cdot jk}}{p_{\cdot j\cdot}}$ and $p_{ijk} = \frac{p_{i\cdot k}p_{ij\cdot}p_{\cdot jk}}{\left(\sum_{j=1}^{c} \frac{p_{ij\cdot}p_{\cdot jk}}{p_{\cdot j\cdot}}\right)p_{\cdot j\cdot}}$ imply the null hypothesis of conditional homogeneity $p_{ijk}/p_{ij\cdot} = p_{\cdot jk}/p_{\cdot j\cdot}$; and $p_{ijk}/p_{ij\cdot} = p_{\cdot jk}/p_{\cdot j\cdot}$ implies $p_{i\cdot k} = \sum_{j=1}^{c} \frac{p_{ij\cdot}p_{\cdot jk}}{p_{\cdot j\cdot}}$ and $p_{ijk} = \frac{p_{i\cdot k}p_{ij\cdot}p_{\cdot jk}}{\left(\sum_{j=1}^{c} \frac{p_{ij\cdot}p_{\cdot jk}}{p_{\cdot j\cdot}}\right)p_{\cdot j\cdot}}$. The degrees of freedom in table 7.1 are those of the asymptotic χ^2-distributions under the null hypothesis of conditional homogeneity. [Cf. Roy and Kastenbaum (1956).]

8. NEGATIVE INTERACTION

It is true that the conditional homogeneity component of table 7.1 may also be analyzed *algebraically* as shown in table 8.1. However, the (D)-homogeneity component is not necessarily smaller than the conditional homogeneity component. The interaction component in table 8.1 may therefore have to be negative. This is illustrated in example 12.4. The contrary is illustrated in example 12.2.

Note that if $x_{ij\cdot} = x_{i\cdot\cdot}x_{\cdot j\cdot}/N$, that is, the ($C$, D)-homogeneity component is the same as the conditional homogeneity-($D|C$) component, then $y_{i\cdot k} = \sum_{j=1}^{c} \frac{x_{ij\cdot}x_{\cdot jk}}{x_{\cdot j\cdot}} = x_{i\cdot\cdot}x_{\cdot\cdot k}/N$ and the (RD)-interaction component in table 7.1 becomes the (D)-homogeneity component of table 8.1, and the (RD, C)-interaction component in table 7.1 becomes the RCD-interaction component in table 8.1. [Cf. McGill (1954, p. 108), Sakaguchi (1957b, p. 26).]

TABLE 8.1

Component due to	Information
(D)-homogeneity	$2\sum_{i=1}^{r}\sum_{k=1}^{d} x_{i\cdot k}\log\frac{Nx_{i\cdot k}}{x_{i\cdot\cdot}x_{\cdot\cdot k}}$
RCD-interaction	$2\sum_{i=1}^{r}\sum_{j=1}^{c}\sum_{k=1}^{d} x_{ijk}\log\frac{x_{ijk}}{\frac{Nx_{ij\cdot}x_{i\cdot k}x_{\cdot jk}}{x_{i\cdot\cdot}x_{\cdot j\cdot}x_{\cdot\cdot k}}}$
Conditional homogeneity-($D|C$)	$2\sum_{i=1}^{r}\sum_{j=1}^{c}\sum_{k=1}^{d} x_{ijk}\log\frac{x_{ijk}}{\frac{x_{ij\cdot}x_{\cdot jk}}{x_{\cdot j\cdot}}}$

9. PARTITIONS

The independence component in table 2.1 can be analyzed into components depending on partitions of the r × c contingency table [cf. Cochran (1954), Irwin (1949), Kimball (1954), Lancaster (1949)]. The partitionings correspond to possible dependence between subsets of the row and column classifications. See section 3.6 of chapter 12 for the analogous problem for a multivariate normal population. Suppose, for example, we partition a two-way contingency table into four parts by grouping the rows into two sets of r_1, r_2 rows respectively, $r_1 + r_2 = r$, and the columns into two sets of c_1, c_2 columns respectively, $c_1 + c_2 = c$. We supplement the notation by defining

$$N_{\alpha\beta} = \Sigma\Sigma x_{ij}, \qquad \alpha = 1 = \beta \text{ for } i = 1, 2, \cdots, r_1, \quad j = 1, 2, \cdots, c_1,$$
$$\alpha = 2 = \beta \text{ for } i = r_1 + 1, \cdots, r_1 + r_2,$$
$$j = c_1 + 1, \cdots, c_1 + c_2,$$

$$x_{i\cdot}^{\alpha\beta} = \sum_j x_{ij}, \qquad x_{\cdot j}^{\alpha\beta} = \sum_i x_{ij}, \qquad x_{i\cdot}^{\alpha\cdot} = x_{i\cdot}^{\alpha 1} + x_{i\cdot}^{\alpha 2}, \qquad x_{\cdot j}^{\cdot\beta} = x_{\cdot j}^{1\beta} + x_{\cdot j}^{2\beta},$$

$$N_{\alpha\cdot} = N_{\alpha 1} + N_{\alpha 2}, \qquad N_{\cdot\beta} = N_{1\beta} + N_{2\beta}, \qquad \alpha = 1, 2, \qquad \beta = 1, 2,$$

$$N = N_{11} + N_{12} + N_{21} + N_{22} = N_{1\cdot} + N_{2\cdot} = N_{\cdot 1} + N_{\cdot 2}.$$

The components of the analysis are those for the four subcontingency tables, the pair of row subtotals, the pair of column subtotals, and the 2 × 2 table of the partitioned total.

The analysis in table 9.1 follows in a straightforward fashion from the definitions of the notation and the properties of the logarithm. The degrees of freedom are those of the asymptotic χ^2-distributions under the null hypothesis H_2 of (2.1).

The same procedure will apply for any partitioning of the original contingency table into subtables either *ab initio* or by further partitioning of the subtables. This procedure is applicable when there is reason to test for possible dependence between subsets of the row classifications and subsets of the column classifications, after finding a significantly large independence component in table 2.1.

Similarly, partitioning of three-way and higher order contingency tables leads to analysis of the independence components. Thus the independence component in table 3.1 can be further analyzed in addition to the analysis in table 3.3. Suppose, for example, we partition a three-way contingency table into eight parts by grouping the rows into two sets r_1, r_2 respectively, $r_1 + r_2 = r$, the columns into two sets c_1, c_2 respectively, $c_1 + c_2 = c$, and the depth into two sets d_1, d_2 respectively, $d_1 + d_2 = d$.

TABLE 9.1

Component due to	Information	D.F.
Partition totals	$2\sum_{\alpha=1}^{2}\sum_{\beta=1}^{2} N_{\alpha\beta}\log (NN_{\alpha\beta}/N_{\alpha\cdot}N_{\cdot\beta})$	1
Partition column totals	$2\sum_{j=c_1+1}^{c_1+c_2}\left(x_{\cdot j}^{12}\log\frac{N_{\cdot 2}x_{\cdot j}^{12}}{N_{12}x_{\cdot j}^{\cdot 2}} + x_{\cdot j}^{22}\log\frac{N_{\cdot 2}x_{\cdot j}^{22}}{N_{22}x_{\cdot j}^{\cdot 2}}\right)$	$c_2 - 1$
Partition column totals	$2\sum_{j=1}^{c_1}\left(x_{\cdot j}^{11}\log\frac{N_{\cdot 1}x_{\cdot j}^{11}}{N_{11}x_{\cdot j}^{\cdot 1}} + x_{\cdot j}^{21}\log\frac{N_{\cdot 1}x_{\cdot j}^{21}}{N_{21}x_{\cdot j}^{\cdot 1}}\right)$	$c_1 - 1$
Partition row totals	$2\sum_{i=r_1+1}^{r_1+r_2}\left(x_{i\cdot}^{21}\log\frac{N_{2\cdot}x_{i\cdot}^{21}}{N_{21}x_{i\cdot}^{2\cdot}} + x_{i\cdot}^{22}\log\frac{N_{2\cdot}x_{i\cdot}^{22}}{N_{22}x_{i\cdot}^{2\cdot}}\right)$	$r_2 - 1$
Partition row totals	$2\sum_{i=1}^{r_1}\left(x_{i\cdot}^{11}\log\frac{N_{1\cdot}x_{i\cdot}^{11}}{N_{11}x_{i\cdot}^{1\cdot}} + x_{i\cdot}^{12}\log\frac{N_{1\cdot}x_{i\cdot}^{12}}{N_{12}x_{i\cdot}^{1\cdot}}\right)$	$r_1 - 1$
Subcontingency table	$2\sum_{i=r_1+1}^{r_1+r_2}\sum_{j=c_1+1}^{c_1+c_2} x_{ij}\log\frac{N_{22}x_{ij}}{x_{i\cdot}^{22}x_{\cdot j}^{22}}$	$(r_2 - 1)(c_2 - 1)$
Subcontingency table	$2\sum_{i=r_1+1}^{r_1+r_2}\sum_{j=1}^{c_1} x_{ij}\log\frac{N_{21}x_{ij}}{x_{i\cdot}^{21}x_{\cdot j}^{21}}$	$(r_2 - 1)(c_1 - 1)$
Subcontingency table	$2\sum_{i=1}^{r_1}\sum_{j=c_1+1}^{c_1+c_2} x_{ij}\log\frac{N_{12}x_{ij}}{x_{i\cdot}^{12}x_{\cdot j}^{12}}$	$(r_1 - 1)(c_2 - 1)$
Subcontingency table	$2\sum_{i=1}^{r_1}\sum_{j=1}^{c_1} x_{ij}\log\frac{N_{11}x_{ij}}{x_{i\cdot}^{11}x_{\cdot j}^{11}}$	$(r_1 - 1)(c_1 - 1)$
Independence, $H_2(R \times C)$ $2\hat{I}(H_1:H_2)$	$2\sum_{i=1}^{r}\sum_{j=1}^{c} x_{ij}\log\frac{Nx_{ij}}{x_{i\cdot}x_{\cdot j}}$	$(r - 1)(c - 1)$

We supplement the notation by defining

$$N_{\alpha\beta\gamma} = \Sigma\Sigma\Sigma x_{ijk}, \qquad \alpha = \beta = \gamma = 1 \text{ for } i = 1, 2, \cdots, r_1,$$
$$j = 1, 2, \cdots, c_1,\ k = 1, 2, \cdots, d_1,$$
$$\alpha = \beta = \gamma = 2 \text{ for } i = r_1 + 1, \cdots, r_1 + r_2,$$
$$j = c_1 + 1, \cdots, c_1 + c_2,\ k = d_1 + 1, \cdots, d_1 + d_2,$$

$$x_{i\cdot\cdot}^{\alpha\beta\gamma} = \sum_j\sum_k x_{ijk}, \qquad x_{\cdot j\cdot}^{\alpha\beta\gamma} = \sum_i\sum_k x_{ijk}, \qquad x_{\cdot\cdot k}^{\alpha\beta\gamma} = \sum_i\sum_j x_{ijk},$$

$$x_{i\cdot\cdot}^{\alpha\cdot\cdot} = \sum_\beta\sum_\gamma x_{i\cdot\cdot}^{\alpha\beta\gamma}, \qquad x_{\cdot j\cdot}^{\cdot\beta\cdot} = \sum_\alpha\sum_\gamma x_{\cdot j\cdot}^{\alpha\beta\gamma}, \qquad x_{\cdot\cdot k}^{\cdot\cdot\gamma} = \sum_\alpha\sum_\beta x_{\cdot\cdot k}^{\alpha\beta\gamma},$$

$$N_{\alpha\cdot\cdot} = \sum_\beta\sum_\gamma N_{\alpha\beta\gamma}, \qquad N_{\cdot\beta\cdot} = \sum_\alpha\sum_\gamma N_{\alpha\beta\gamma}, \qquad N_{\cdot\cdot\gamma} = \sum_\alpha\sum_\beta N_{\alpha\beta\gamma},$$

$$N = \sum_\alpha\sum_\beta\sum_\gamma N_{\alpha\beta\gamma} = \sum_\alpha N_{\alpha\cdot\cdot} = \sum_\beta N_{\cdot\beta\cdot} = \sum_\gamma N_{\cdot\cdot\gamma}.$$

The components of the analysis are those for the eight three-way subtables, two sets each of row, column, and depth subtotals with four elements per set, and the 2 × 2 × 2 table of the partitioned total.

The analysis in table 9.2 follows in a straightforward fashion from the definitions of the notation and the properties of the logarithm.

TABLE 9.2

Component due to	Information	D.F.
Partition totals	$2\sum_{\alpha=1}^{2}\sum_{\beta=1}^{2}\sum_{\gamma=1}^{2} N_{\alpha\beta\gamma}\log\frac{N^2N_{\alpha\beta\gamma}}{N_{\alpha\cdot\cdot}N_{\cdot\beta\cdot}N_{\cdot\cdot\gamma}}$	4
Two partition depth totals for $\gamma = 1, 2$ and $k = 1, 2, \cdots, d_1$ for $\gamma = 1$, $k = d_1 + 1, \cdots, d_1 + d_2$ for $\gamma = 2$	$2\sum_{\alpha}\sum_{\beta}\sum_{k} x_{\cdot\cdot k}^{\alpha\beta\gamma}\log\frac{N_{\cdot\cdot\gamma}x_{\cdot\cdot k}^{\alpha\beta\gamma}}{N_{\alpha\beta\gamma}x_{\cdot\cdot k}^{\cdot\cdot\gamma}}$	$3(d_1 - 1)$ $3(d_2 - 1)$
Two partition column totals for $\beta = 1, 2$ and $j = 1, 2, \cdots, c_1$ for $\beta = 1$, $j = c_1 + 1, \cdots, c_1 + c_2$ for $\beta = 2$	$2\sum_{\alpha}\sum_{\gamma}\sum_{j} x_{\cdot j\cdot}^{\alpha\beta\gamma}\log\frac{N_{\cdot\beta\cdot}x_{\cdot j\cdot}^{\alpha\beta\gamma}}{N_{\alpha\beta\gamma}x_{\cdot j\cdot}^{\cdot\beta\cdot}}$	$3(c_1 - 1)$ $3(c_2 - 1)$
Two partition row totals for $\alpha = 1, 2$ and $i = 1, 2, \cdots, r_1$ for $\alpha = 1$, $i = r_1 + 1, \cdots, r_1 + r_2$ for $\alpha = 2$	$2\sum_{\beta}\sum_{\gamma}\sum_{i} x_{i\cdot\cdot}^{\alpha\beta\gamma}\log\frac{N_{\alpha\cdot\cdot}x_{i\cdot\cdot}^{\alpha\beta\gamma}}{N_{\alpha\beta\gamma}x_{i\cdot\cdot}^{\alpha\cdot\cdot}}$	$3(r_1 - 1)$ $3(r_2 - 1)$
Eight three-way subcontingency tables for $\alpha, \beta, \gamma = 1, 2$ with $i = 1, 2, \cdots, r_1$ for $\alpha = 1$, $i = r_1 + 1, \cdots, r_1 + r_2$ for $\alpha = 2$, etc.	$2\sum_{i}\sum_{j}\sum_{k} x_{ijk}\log\frac{N_{\alpha\beta\gamma}^2x_{ijk}}{x_{i\cdot\cdot}^{\alpha\beta\gamma}x_{\cdot j\cdot}^{\alpha\beta\gamma}x_{\cdot\cdot k}^{\alpha\beta\gamma}}$	$r_1c_1d_1 - r_1 - c_1 - d_1 + 2$ $r_1c_1d_2 - r_1 - c_1 - d_2 + 2$ $r_1c_2d_1 - r_1 - c_2 - d_1 + 2$ $r_1c_2d_2 - r_1 - c_2 - d_2 + 2$ $r_2c_1d_1 - r_2 - c_1 - d_1 + 2$ $r_2c_1d_2 - r_2 - c_1 - d_2 + 2$ $r_2c_2d_1 - r_2 - c_2 - d_1 + 2$ $r_2c_2d_2 - r_2 - c_2 - d_2 + 2$
Independence, $H_2(R \times C \times D)$	$2\sum_{i=1}^{r}\sum_{j=1}^{c}\sum_{k=1}^{d} x_{ijk}\log\frac{N^2x_{ijk}}{x_{i\cdot\cdot}x_{\cdot j\cdot}x_{\cdot\cdot k}}$	$rcd - r - c - d + 2$

The partitioning procedure can also be applied to any of the subtables, but we leave the details to the reader. The degrees of freedom are those of the asymptotic χ^2-distributions under the null hypothesis H_2 of independence in (3.1). For tables 9.1 and 9.2 we leave to the reader the estimation of the corresponding divergences, as well as the expression in terms of the form $n \log n$ for computational convenience. (See problem 8.26 in chapter 12 for the analogous problem for a multivariate normal sample.)

10. PARAMETRIC CASE

Suppose that in table 3.1, the $p_{i\cdot\cdot}$, $i = 1, 2, \cdots, r$, are known functions of independent parameters $\alpha_1, \alpha_2, \cdots, \alpha_m$, $m < r$, the $p_{\cdot j\cdot}$, $j = 1, 2, \cdots, c$, are known functions of independent parameters $\beta_1, \beta_2, \cdots, \beta_n$, $n < c$, the $p_{\cdot\cdot k}$, $k = 1, 2, \cdots, d$, are known functions of independent parameters $\gamma_1, \gamma_2, \cdots, \gamma_s$, $s < d$. We "fit" the contingency table with estimates (by some procedure to be determined) $\tilde{\alpha}_1, \cdots, \tilde{\alpha}_m$, $\tilde{\beta}_1, \cdots, \tilde{\beta}_n$, $\tilde{\gamma}_1, \cdots, \tilde{\gamma}_s$, of the α's, β's, γ's, letting $\tilde{p}_{i\cdot\cdot} = p_{i\cdot\cdot}(\tilde{\alpha}_1, \cdots, \tilde{\alpha}_m)$, $i = 1, 2, \cdots, r$; $\tilde{p}_{\cdot j\cdot} = p_{\cdot j\cdot}(\tilde{\beta}_1, \cdots, \tilde{\beta}_n)$, $j = 1, 2, \cdots, c$; $\tilde{p}_{\cdot\cdot k} = p_{\cdot\cdot k}(\tilde{\gamma}_1, \cdots, \tilde{\gamma}_s)$, $k = 1, 2, \cdots, d$; $\tilde{p}_{1\cdot\cdot} + \tilde{p}_{2\cdot\cdot} + \cdots + \tilde{p}_{r\cdot\cdot} = \tilde{p}_{\cdot 1\cdot} + \tilde{p}_{\cdot 2\cdot} + \cdots + \tilde{p}_{\cdot c\cdot} = \tilde{p}_{\cdot\cdot 1} + \tilde{p}_{\cdot\cdot 2} + \cdots + \tilde{p}_{\cdot\cdot d} = 1$.

If the $\tilde{p}_{i\cdot\cdot}$, $\tilde{p}_{\cdot j\cdot}$, $\tilde{p}_{\cdot\cdot k}$, or the $\tilde{\alpha}$'s, $\tilde{\beta}$'s, $\tilde{\gamma}$'s, are such that identically in the α's, β's, γ's,

$$(10.1)\quad \sum_{i=1}^{r} \frac{x_{i\cdot\cdot}}{N} \log \frac{\tilde{p}_{i\cdot\cdot}}{p_{i\cdot\cdot}} + \sum_{j=1}^{c} \frac{x_{\cdot j\cdot}}{N} \log \frac{\tilde{p}_{\cdot j\cdot}}{p_{\cdot j\cdot}} + \sum_{k=1}^{d} \frac{x_{\cdot\cdot k}}{N} \log \frac{\tilde{p}_{\cdot\cdot k}}{p_{\cdot\cdot k}}$$
$$= \sum_{i=1}^{r} \tilde{p}_{i\cdot\cdot} \log \frac{\tilde{p}_{i\cdot\cdot}}{p_{i\cdot\cdot}} + \sum_{j=1}^{c} \tilde{p}_{\cdot j\cdot} \log \frac{\tilde{p}_{\cdot j\cdot}}{p_{\cdot j\cdot}} + \sum_{k=1}^{d} \tilde{p}_{\cdot\cdot k} \log \frac{\tilde{p}_{\cdot\cdot k}}{p_{\cdot\cdot k}},$$

we have a further analysis of table 3.1 in table 10.1. We see (cf. (4.8)–(4.11) of chapter 6) that (10.1) implies the $\tilde{\alpha}$'s, $\tilde{\beta}$'s, $\tilde{\gamma}$'s are the solutions of

$$(10.2)\quad \begin{aligned} &\sum_{i=1}^{r} \frac{x_{i\cdot\cdot}}{p_{i\cdot\cdot}} \frac{\partial p_{i\cdot\cdot}}{\partial \alpha_a} = 0, \qquad a = 1, 2, \cdots, m, \\ &\sum_{j=1}^{c} \frac{x_{\cdot j\cdot}}{p_{\cdot j\cdot}} \frac{\partial p_{\cdot j\cdot}}{\partial \beta_b} = 0, \qquad b = 1, 2, \cdots, n, \\ &\sum_{k=1}^{d} \frac{x_{\cdot\cdot k}}{p_{\cdot\cdot k}} \frac{\partial p_{\cdot\cdot k}}{\partial \gamma_g} = 0, \qquad g = 1, 2, \cdots, s. \end{aligned}$$

These are the maximum-likelihood equations for estimating the α's, β's, γ's, or minimizing the total in table 3.1. We leave to the reader the estimation of the divergences, as well as the expression in terms of the form $n \log n$ for computational convenience. The degrees of freedom are those of the asymptotic χ^2-distributions under the null hypothesis of independence H_2 in (3.1), the $p_{i..}$'s, $p_{.j.}$'s, $p_{..k}$'s understood as functions respectively of the α's, β's, γ's. (See problem 13.12.)

TABLE 10.1

Component due to	Information	D.F.
Between $\hat{p}_{i..} = x_{i..}/N$ and $\tilde{p}_{i..}$	$2\sum_{i=1}^{r} x_{i..} \log \frac{x_{i..}}{N\tilde{p}_{i..}}$	$r - m - 1$
Between $\hat{p}_{.j.} = x_{.j.}/N$ and $\tilde{p}_{.j.}$	$2\sum_{j=1}^{c} x_{.j.} \log \frac{x_{.j.}}{N\tilde{p}_{.j.}}$	$c - n - 1$
Between $\hat{p}_{..k} = x_{..k}/N$ and $\tilde{p}_{..k}$	$2\sum_{k=1}^{d} x_{..k} \log \frac{x_{..k}}{N\tilde{p}_{..k}}$	$d - s - 1$
Independence, $H_2(R \times C \times D)$ $2\hat{I}(H_1:H_2)$	$2\sum_{i=1}^{r}\sum_{j=1}^{c}\sum_{k=1}^{d} x_{ijk} \log \frac{N^2 x_{ijk}}{x_{i..}x_{.j.}x_{..k}}$	$rcd - r - c - d + 2$
$2\hat{I}(\tilde{\alpha}, \tilde{\beta}, \tilde{\gamma})$	$2\sum_{i=1}^{r}\sum_{j=1}^{c}\sum_{k=1}^{d} x_{ijk} \log \frac{x_{ijk}}{N\tilde{p}_{i..}\tilde{p}_{.j.}\tilde{p}_{..k}}$	$rcd - m - n - s - 1$
$\tilde{p}_{i..}$'s	$2N\sum_{i=1}^{r} \tilde{p}_{i..} \log \frac{\tilde{p}_{i..}}{p_{i..}}$	m
$\tilde{p}_{.j.}$'s	$2N\sum_{j=1}^{c} \tilde{p}_{.j.} \log \frac{\tilde{p}_{.j.}}{p_{.j.}}$	n
$\tilde{p}_{..k}$'s	$2N\sum_{k=1}^{d} \tilde{p}_{..k} \log \frac{\tilde{p}_{..k}}{p_{..k}}$	s
Total, $2\hat{I}((p^*):(p))$	$2\sum_{i=1}^{r}\sum_{j=1}^{c}\sum_{k=1}^{d} x_{ijk} \log \frac{x_{ijk}}{Np_{i..}p_{.j.}p_{..k}}$	$rcd - 1$

11. SYMMETRY

For two-way contingency tables with the same number of rows and columns arising from related classifications, it is often of interest to test a null hypothesis of symmetry H_2, the events in cells symmetrically situated about the main diagonal have the same probability of occurrence, that is, the hypotheses [see Bowker (1948)],

$$(11.1)\quad \begin{aligned} &H_1: p_{ij} \neq p_{ji}, \quad i = 1, 2, \cdots, c, \quad j = 1, 2, \cdots, c, \quad i \neq j, \text{ for at least one } (i, j),\\ &H_2: p_{ij} = p_{ji}. \end{aligned}$$

For the conjugate distribution (see section 3 of chapter 6) with parameters the same as the observed sample best unbiased estimates, we have

$$(11.2)\qquad \hat{I}(p^*:p) = \sum_{i=1}^{c} \sum_{j=1}^{c} x_{ij} \log \frac{x_{ij}}{Np_{ij}}, \qquad p_{ij} = p_{ji}.$$

The null hypothesis H_2 of (11.1) usually does not specify the p_{ij}, $i = 1, 2, \cdots, c$, $j = 1, 2, \cdots, c$. We analyze $\hat{I}(p^*:p)$ of (11.2) into several additive components in table 11.1. The degrees of freedom are those of the asymptotic χ^2-distributions under the null hypothesis of symmetry H_2 in (11.1). Note that the convexity property (see section 3 of chapter 2) ensures that the component due to $\hat{p}_{ij}$ is the minimum value of the total for the symmetric grouping, and the symmetry component is the sum of all but the diagonal terms of the total with p_{ij} replaced by $\hat{p}_{ij}$.

TABLE 11.1

Component due to	Information	D.F.
Diagonal terms	$2\sum_{i=1}^{c} x_{ii} \log \frac{x_{ii}}{Np_{ii}} + 2(N - \Sigma x_{ii}) \log \frac{N - \Sigma x_{ii}}{N(1 - \Sigma p_{ii})}$	c
$\hat{p}_{ij} = \frac{x_{ij} + x_{ji}}{2N}$	$2\sum\sum_{i<j} (x_{ij} + x_{ji}) \log \frac{(x_{ij} + x_{ji})(1 - \Sigma p_{ii})}{2p_{ij}(N - \Sigma x_{ii})}$	$\frac{c(c-1)}{2} - 1$
Symmetry, $2\hat{I}(H_1:H_2)$	$2\sum\sum_{i\neq j} x_{ij} \log \frac{2x_{ij}}{x_{ij} + x_{ji}}$	$\frac{c(c-1)}{2}$
Total, $2\hat{I}(p^*:p)$	$2\sum_{i=1}^{c}\sum_{j=1}^{c} x_{ij} \log \frac{x_{ij}}{Np_{ij}}, \qquad p_{ij} = p_{ji}$	$c^2 - 1$

The symmetry component in table 11.1 may also be expressed as

$$(11.3)\quad \hat{I}(H_1:H_2) = \sum\sum_{i\neq j} x_{ij} \log x_{ij} - \sum\sum_{i<j} (x_{ij} + x_{ji}) \log (x_{ij} + x_{ji}) + (\log 2) \sum\sum_{i\neq j} x_{ij}$$

for computational convenience with the table of $n \log n$.

The divergences do not provide a similar additive analysis (with these estimates), but the estimate of the divergence corresponding to the symmetry component in (11.3) is

$$(11.4)\qquad \hat{J}(H_1, H_2) = N\sum_{i\neq j}\sum\left(\frac{x_{ij}}{N} - \frac{x_{ij}+x_{ji}}{2N}\right)\log\frac{2x_{ij}}{x_{ij}+x_{ji}}$$

$$= \frac{1}{2}\sum_{i\neq j}\sum(x_{ij}-x_{ji})\log\frac{2x_{ij}}{x_{ij}+x_{ji}}.$$

Under the null hypothesis H_2 of (11.1) (the events in the cells symmetrically situated about the main diagonal have the same probability of occurrence), $2\hat{I}(H_1:H_2)$ and $\hat{J}(H_1, H_2)$ are asymptotically distributed as χ^2 with $c(c-1)/2$ degrees of freedom.

With the approximations used in (4.5) and (4.6) of chapter 6, we find [cf. Bowker (1948, p. 573)]:

$$(11.5)\qquad 2\hat{I}(H_1:H_2) \approx \sum_{i<j}\sum\frac{(x_{ij}-x_{ji})^2}{x_{ij}+x_{ji}} = \chi^2,$$

$$(11.6)\qquad \hat{J}(H_1, H_2) \approx \frac{1}{2}\sum_{i<j}\sum\frac{(x_{ij}-x_{ji})^2}{x_{ij}+x_{ji}} + \frac{1}{2}\sum_{i\neq j}\sum\frac{(x_{ij}-x_{ji})^2}{4x_{ij}}.$$

If $p_{ij} = p_{ji}$, $i = 1, 2, \cdots, c$, $j = 1, 2, \cdots, c$, $i \neq j$, the marginal distributions for the row and column classifications are the same, that is, $p_{i\cdot} = p_{i1} + p_{i2} + \cdots + p_{ic} = p_{\cdot i} = p_{1i} + p_{2i} + \cdots + p_{ci}$, $i = 1, 2, \cdots, c$. The weaker hypothesis of equality of marginal distributions is also of interest, especially in the absence of symmetry. For the test of the weaker hypothesis see Stuart (1955a) and section 7 of chapter 12.

12. EXAMPLES

Example 12.1. As an example of the test for symmetry consider the data in table 12.1 for 3242 men aged 30–39 with unaided distance vision [taken from Stuart (1953, p. 109)]. From (11.3) and the table of $n \log n$ we find

$$\sum_{i\neq j}\sum x_{ij}\log x_{ij} = 4622.580,\qquad \sum_{i<j}\sum(x_{ij}+x_{ji})\log(x_{ij}+x_{ji}) = 5322.353,$$

$$\sum_{i\neq j}\sum x_{ij} = 1013,\qquad 1013\log 2 = 702.158,\qquad \text{and } 2\hat{I}(H_1:H_2) = 4.770,$$

which as a χ^2 with 6 degrees of freedom is not significant. We therefore accept the null hypothesis of symmetry of vision in the left eye and right eye of the population from which the sample was drawn.

TABLE 12.1. 3242 Men Aged 30–39; Unaided Distance Vision

Right Eye \ Left Eye	Highest Grade	Second Grade	Third Grade	Lowest Grade	Total
Highest Grade	821	112	85	35	1053
Second Grade	116	494	145	27	782
Third Grade	72	151	583	87	893
Lowest Grade	43	34	106	331	514
Total	1052	791	919	480	3242

Example 12.2. The data in table 12.2 represent the number of items passing, P, or failing, F, two tests, T_1, T_2, on certain manufactured products from manufacturers A, B, C, D. With tests as the row classification, manufacturers as the column classification, and result as the depth classification, we find

$$\sum_{i=1}^{2}\sum_{j=1}^{4}\sum_{k=1}^{2} x_{ijk}\log x_{ijk} = 2893.819, \qquad \sum_{i=1}^{2}\sum_{j=1}^{4} x_{ij\cdot}\log x_{ij\cdot} = 3215.410,$$

$$\sum_{i=1}^{2}\sum_{k=1}^{2} x_{i\cdot k}\log x_{i\cdot k} = 3829.547, \qquad \sum_{j=1}^{4}\sum_{k=1}^{2} x_{\cdot jk}\log x_{\cdot jk} = 3376.470,$$

$$\sum_{i=1}^{2} x_{i\cdot\cdot}\log x_{i\cdot\cdot} = 4158.008, \qquad \sum_{j=1}^{4} x_{\cdot j\cdot}\log x_{\cdot j\cdot} = 3701.858,$$

$$\sum_{k=1}^{2} x_{\cdot\cdot k}\log x_{\cdot\cdot k} = 4317.737, \qquad N\log N = 4646.210.$$

TABLE 12.2

	T_1			T_2		
	P	F	Total	P	F	Total
A	112	32	144	84	24	108
B	76	20	96	86	10	96
C	87	9	96	58	14	72
D	41	7	48	40	8	48
	316	68	384	268	56	324

These values and the analysis in table 6.1 yield table 12.3 to test the homogeneity of the results and manufacturers over the tests.

TABLE 12.3

Component due to	Information	D.F.
Manufacturer homogeneity	3.508	3
Conditional homogeneity, results given manufacturer	7.594	4
Manufacturer, result homogeneity	11.102	7

Since the 5% values of χ^2 for 3, 4, 7 degrees of freedom are, respectively, 7.81, 9.49, 14.07, we accept the null hypothesis that the results for the different manufacturers over the tests are homogeneous. We also illustrate table 8.1 in table 12.4. In view of the values in table 12.4, we may accept the null hypothesis that the failure rate is the same for the two tests.

TABLE 12.4

Component due to	Information	D.F.
Result homogeneity	0.024	1
Test, manufacturer, result, interaction	7.570	3
Conditional homogeneity, results given manufacturer	7.594	4

Example 12.3. In table 12.5 the 124 failures of table 12.2 are also classified by the defects, D_1, D_2. For the 4 × 2 × 2 table 12.5(a), we test the hypotheses of (3.1) with $i = A, B, C, D, j = T_1, T_2, k = D_1, D_2$, that is, the null hypothesis of independence among manufacturers, tests, and defects. From the data we find

$$\sum_{i=1}^{4}\sum_{j=1}^{2}\sum_{k=1}^{2} x_{ijk}\log x_{ijk} = 280.642, \qquad \sum_{i=1}^{4}\sum_{j=1}^{2} x_{ij\cdot}\log x_{ij\cdot} = 357.097,$$

$$\sum_{i=1}^{4}\sum_{k=1}^{2} x_{i\cdot k}\log x_{i\cdot k} = 359.061, \qquad \sum_{j=1}^{2}\sum_{k=1}^{2} x_{\cdot jk}\log x_{\cdot jk} = 429.705,$$

$$\sum_{i=1}^{4} x_{i\cdot\cdot}\log x_{i\cdot\cdot} = 440.193, \qquad \sum_{j=1}^{2} x_{\cdot j\cdot}\log x_{\cdot j\cdot} = 512.347,$$

$$\sum_{k=1}^{2} x_{\cdot\cdot k}\log x_{\cdot\cdot k} = 512.023, \qquad N\log N = 597.715.$$

These values and the analysis in table 3.3 yield table 12.6.

TABLE 12.5

		T_1	T_2	
A	D_1	24	11	$x_{A\cdot\cdot} = 56$
	D_2	8	13	
B	D_1	7	2	$x_{B\cdot\cdot} = 30$
	D_2	13	8	
C	D_1	7	7	$x_{C\cdot\cdot} = 23$
	D_2	2	7	
D	D_1	5	3	$x_{D\cdot\cdot} = 15$
	D_2	2	5	
		68	56	$N = 124$

(a)

	T_1	T_2	
D_1	43	23	$x_{\cdot\cdot D_1} = 66$
D_2	25	33	$x_{\cdot\cdot D_2} = 58$
	$x_{\cdot T_1\cdot} = 68$	$x_{\cdot T_2\cdot} = 56$	

(b)

TABLE 12.6

Component due to	Information	D.F.
H_2 (Test × defect)	6.100	1
H_2 (Manufacturer × (test, defect))	16.918	9
H_2 (Manufacturer × test × defect)	23.018	10

Since the 5% values of χ^2 for 1, 9, 10 degrees of freedom are, respectively, 3.84, 16.92, 18.31, and the 1% values are 6.63, 21.67, 23.21, we reject the hypothesis of independence between test and defect and also of course the three-way independence and examine further the hypothesis of independence between manufacturer and the pair, test, defect.

The analysis for conditional independence in table 3.5 applied to the H_2 (Manufacturer × (test, defect)) component in table 12.6 yields tables 12.7 and 12.8.

Since the 5% values of χ^2 for 3, 6 degrees of freedom are, respectively, 7.81, 12.59, and the 1% values are 11.34, 16.81, we infer from tables 12.6, 12.7, 12.8 that manufacturer and test are independent but not defect and test, and defect and manufacturer, with the manufacturers and defects conditionally independent given the test.

TABLE 12.7

Component due to	Information	D.F.
H_2 (Manufacturer × test)	4.544	3
Conditional independence, manufacturer and defect given test	12.374	6
H_2 (Manufacturer × (test, defect))	16.918	9

TABLE 12.8

Component due to	Information	D.F.
H_2 (Manufacturer × defect)	9.120	3
Conditional independence, manufacturer and test given defect	7.798	6
H_2 (Manufacturer × (test, defect))	16.918	9

Example 12.4. Table 12.9, taken from Campbell, Snedecor, and Simanton (1939, p. 64), gives the distribution of 1397 houseflies by sex and mortality among 12 successive tests with a standard insecticide [also discussed by Norton (1945)]. The problem here is to test the homogeneity of the sex, mortality results over the 12 successive tests. With level as the row classification, sex as the column classification, and mortality as the depth classification, we find

$$\sum_{i=1}^{12}\sum_{j=1}^{2}\sum_{k=1}^{2} x_{ijk}\log x_{ijk} = 5118.828, \qquad \sum_{i=1}^{12}\sum_{j=1}^{2} x_{ij\cdot}\log x_{ij\cdot} = 5713.331,$$

$$\sum_{i=1}^{12}\sum_{k=1}^{2} x_{i\cdot k}\log x_{i\cdot k} = 5766.322, \qquad \sum_{j=1}^{2}\sum_{k=1}^{2} x_{\cdot jk}\log x_{\cdot jk} = 8554.522,$$

$$\sum_{i=1}^{12} x_{i\cdot\cdot}\log x_{i\cdot\cdot} = 6652.973, \qquad \sum_{j=1}^{2} x_{\cdot j\cdot}\log x_{\cdot j\cdot} = 9159.110,$$

$$\sum_{k=1}^{2} x_{\cdot\cdot k}\log x_{\cdot\cdot k} = 9215.809, \qquad N\log N = 10117.189.$$

These values and the analysis in table 6.1 yield table 12.10.

TABLE 12.9. Mortality of Male and Female Houseflies in 12 Successive Tests of a Standard Insecticide

	Males			Females					
Level	Alive	Dead	Total	Alive	Dead	Total	Level Total	Total Alive	Total Dead
1	17	40	57	46	6	52	109	63	46
2	14	44	58	44	5	49	107	58	49
3	19	42	61	48	5	53	114	67	47
4	21	33	54	41	4	45	99	62	37
5	9	39	48	68	8	76	124	77	47
6	21	38	59	70	5	75	134	91	43
7	19	40	59	56	4	60	119	75	44
8	15	32	47	51	8	59	106	66	40
9	20	35	55	73	9	82	137	93	44
10	15	29	44	78	5	83	127	93	34
11	12	19	31	69	2	71	102	81	21
12	12	29	41	75	3	78	119	87	32
	194	420	614	719	64	783	1397	913	484

TABLE 12.10

Component due to		Information		D.F.	
Homogeneity, sexes		36.874		11	
Conditional homogeneity, mortality given sex		20.170		22	
	Male		8.906		11
	Female		11.264		11
Homogeneity, sex, mortality		57.044		33	
Homogeneity, mortality		29.458		11	
Conditional homogeneity, sex given mortality		27.586		22	
	Alive		21.340		11
	Dead		6.246		11

Campbell, Snedecor, and Simanton (1939), with the classical χ^2, found 8.6 and 10.5, respectively, for the conditional homogeneity for the males and for the females; 36.5 for the homogeneity of sexes; and 28.7 for the homogeneity of mortality. We accept a null hypothesis of conditional homogeneity for mortality given the sex. Since the 1% values of χ^2 for 11, 22, 33 degrees of freedom are, respectively, 24.72, 40.29, approximately 55, we infer that the mortality results are not homogeneous, the results for the sexes are not homogeneous, and the sex, mortality results are not homogeneous, although there is conditional homogeneity for mortality given the sex and for the sex given the mortality.

Note that the homogeneity component for mortality is greater than the conditional homogeneity for mortality given sex, also the homogeneity component for the sexes is greater than the conditional homogeneity for sex given mortality, so that here the analysis in table 8.1 would lead to a negative interaction component. To apply the analysis in table 7.1 we compute

$$y_{ij\cdot} = \sum_{k=1}^{2} \frac{x_{i\cdot k}x_{\cdot jk}}{x_{\cdot\cdot k}}, \qquad i = 1, 2, \cdots, 12, \qquad j = 1, 2,$$

and

$$y_{i\cdot k} = \sum_{j=1}^{2} \frac{x_{ij\cdot}x_{\cdot jk}}{x_{\cdot j\cdot}}, \qquad i = 1, 2, \cdots, 12, \qquad k = 1, 2,$$

getting table 12.11. We also find

$$\sum_{i=1}^{12}\sum_{k=1}^{2} x_{i\cdot k} \log y_{i\cdot k} = 5762.541, \qquad \sum_{i=1}^{12}\sum_{j=1}^{2} x_{ij\cdot} \log y_{ij\cdot} = 5707.284.$$

TABLE 12.11

	$y_{ij\cdot}$		$y_{i\cdot k}$	
	j		k	
i	1	2	1	2
1	53.30	55.70	65.76	43.24
2	54.84	52.16	63.32	43.68
3	55.02	58.98	67.94	46.06
4	45.28	53.72	58.38	40.62
5	57.15	66.85	84.95	39.05
6	56.65	77.35	87.51	46.49
7	54.12	64.88	73.74	45.26
8	48.73	57.27	69.03	36.97
9	57.94	79.06	92.68	44.32
10	49.27	77.73	90.12	36.88
11	35.43	66.57	74.99	27.01
12	46.25	72.75	84.58	34.42

The analysis of the conditional homogeneity terms in table 7.1 yields tables 12.12 and 12.13. We infer that table 12.9 is homogeneous, those factors that differed from level to level affected the two sexes similarly as to mortality. [Cf. Norton (1945), who makes the same inference by a different approach and interaction.]

TABLE 12.12

Component due to	Information	D.F.
(Level, mortality)-interaction	7.562	11
((Level, mortality), sex)-interaction	12.608	11
Conditional homogeneity, mortality given sex	20.170	22

TABLE 12.13

Component due to	Information	D.F.
(Level, sex)-interaction	12.094	11
((Level, sex), mortality)-interaction	15.492	11
Conditional homogeneity, sex given mortality	27.586	22

13. PROBLEMS

13.1. Relate the components in table 2.1 and table 6.1 of chapter 6.

13.2. Derive the equivalent of table 3.5 for the null hypothesis that the column and (row, depth) classifications are independent.

13.3. Estimate the divergences corresponding to the information components in tables 9.1 and 9.2.

13.4. Express the information components in tables 9.1 and 9.2 in terms of the form $n \log n$.

13.5. Are the two sets of data given in table 13.1 homogeneous?

TABLE 13.1

	Process	
	A	*B*
Failed	68	38
Passed	450	413

	Process	
	A	*B*
Failed	76	17
Passed	365	82

13.6. Table 13.2, from Cochran (1954, Table 8, p. 442), gives the distribution of mothers of children in the Baltimore schools who had been referred by their teachers as presenting behavior problems, and mothers of a comparable group of control children who had not been so referred. For each mother it was recorded whether she had suffered any infant losses (for example, stillbirths) previous to the birth of the child in the study. The data are further classified into three birth-order classes. The comparison is part of a study of possible associations between behavior problems in children and complications of pregnancy of the mother. Analyze the data.

TABLE 13.2

	Problems			Controls		
Birth Order	Losses	None	Total	Losses	None	Total
2	20	82	102	10	54	64
3–4	26	41	67	16	30	46
5+	27	22	49	14	23	37
	73	145	218	40	107	147

13.7. Table 13.3 [from Bartlett (1935, p. 249), who refers to data from Hoblyn and Palmer] is the result of an experiment designed to investigate the propagation of plum root stocks from root cuttings. There were 240 cuttings for each of the four treatments. Analyze the data.

TABLE 13.3

	Alive			Dead		
	Time of Planting			Time of Planting		
Length of Cutting	At Once	In Spring	Total	At Once	In Spring	Total
Long	156	84	240	84	156	240
Short	107	31	138	133	209	342
Total	263	115	378	217	365	582

13.8. From the analysis in table 3.1, and the properties of the discrimination information, show that for $N \to \infty$, if $x_{ijk}/N \to p_{i\cdot\cdot}p_{\cdot j\cdot}p_{\cdot\cdot k}$ with probability 1, then $\frac{x_{ijk}}{N} \to \frac{x_{i\cdot\cdot}}{N}\frac{x_{\cdot j\cdot}}{N}\frac{x_{\cdot\cdot k}}{N}$, $\hat{p}_{i\cdot\cdot} \to p_{i\cdot\cdot}$, $\hat{p}_{\cdot j\cdot} \to p_{\cdot j\cdot}$, $\hat{p}_{\cdot\cdot k} \to p_{\cdot\cdot k}$ with probability 1,

$i = 1, 2, \cdots, r$, $j = 1, 2, \cdots, c$, $k = 1, 2, \cdots, d$. (See problems 7.14 and 7.16 in chapter 6.)

13.9. From the analysis in table 3.2, and the properties of the discrimination information, show that for $N \to \infty$, if $x_{ijk}/N \to p_{i\cdot\cdot}p_{\cdot jk}$ with probability 1, then $\frac{x_{ijk}}{N} \to \frac{x_{i\cdot\cdot}}{N}\frac{x_{\cdot jk}}{N}$, $\tilde{p}_{i\cdot\cdot} \to p_{i\cdot\cdot}$, $\tilde{p}_{\cdot jk} \to p_{\cdot jk}$ with probability 1, $i = 1, 2, \cdots, r$, $j = 1, 2, \cdots, c$, $k = 1, 2, \cdots, d$. (See problem 13.8 above and problems 7.14 and 7.16 in chapter 6.)

13.10. Brownlee, in Quastler (1955, p. 63), gives the data shown in table 13.4 on the numbers of defective fertilizer drums of two different types in two different locations. Show that quality × type is not homogeneous over the location (see section 4) and should therefore not be pooled over location. (Brownlee raises the question of pooling over location because "it is usually assumed that pooling is permissible when second-order interaction is absent" [he refers to Snedecor (1946)]. Absence of second-order interaction is defined as equality of the ratios of the products of the diagonal terms. Here there is no second-order interaction in this sense because $\frac{72 \times 180}{48 \times 420} = \frac{18 \times 720}{42 \times 480}$.

TABLE 13.4

Quality	Location A: Type of Drum I	II	Total	Location B: Type of Drum I	II	Total
Defective	72	48	120	18	42	60
Acceptable	420	180	600	480	720	1200
Total	492	228	720	498	762	1260

13.11. From the analysis in table 2.1, and the properties of the discrimination information, show that for $N \to \infty$, if $x_{ij}/N \to p_{ij}$, $x_{i\cdot}/N \to p_{i\cdot}$, $x_{\cdot j}/N \to p_{\cdot j}$ with probability 1, then $2\hat{I}(H_1\!:\!H_2)/N \to 2I(H_1\!:\!H_2)/N$, with probability 1, $i = 1, 2, \cdots, r$, $j = 1, 2, \cdots, c$, where $I(H_1\!:\!H_2)$ is given in (2.2). (See problems 13.8 and 13.9 above.)

13.12. From the analysis in table 10.1, and the properties of the discrimination information, show that for $N \to \infty$, if $x_{ijk}/N \to p_{i\cdot\cdot}p_{\cdot j\cdot}p_{\cdot\cdot k}$ with probability 1, then $\tilde{p}_{i\cdot\cdot} \to p_{i\cdot\cdot}$, $\tilde{p}_{\cdot j\cdot} \to p_{\cdot j\cdot}$, $\tilde{p}_{\cdot\cdot k} \to p_{\cdot\cdot k}$, $N^2x_{ijk}/x_{i\cdot\cdot}x_{\cdot j\cdot}x_{\cdot\cdot k} \to 1$, $x_{i\cdot\cdot}/N\tilde{p}_{i\cdot\cdot} \to 1$, $x_{\cdot j\cdot}/N\tilde{p}_{\cdot j\cdot} \to 1$, $x_{\cdot\cdot k}/N\tilde{p}_{\cdot\cdot k} \to 1$, with probability 1. On the other hand, what do you infer if $x_{ijk}/N \to p_{ijk}$, $x_{i\cdot\cdot}/N \to p_{i\cdot\cdot}$, $x_{\cdot j\cdot}/N \to p_{\cdot j\cdot}$, $x_{\cdot\cdot k}/N \to p_{\cdot\cdot k}$? (See problems 13.8, 13.9, and 13.11 above.)

CHAPTER 9

Multivariate Normal Populations

1. INTRODUCTION

We continue in the spirit of the preceding chapters, especially 6, 7, and 8, and now take up the analysis of one or more samples from multivariate normal populations for tests of statistical hypotheses. Before we consider questions of estimation, distribution, and testing, it will be helpful to derive in this chapter certain values as parameters of the populations. Matrix notation and theory are used. Matrices are denoted by upper case boldface type, for example, $\mathbf{A} = (a_{ij})$, $\mathbf{X}_1 = (x_{1ij})$, etc., $i = 1, 2, \cdots, m$; $j = 1, 2, \cdots, n$. One-row or one-column matrices (vectors) are denoted by lower case boldface type, for example, $\mathbf{x}' = (x_1, x_2, \cdots, x_k)$, $\boldsymbol{\mu}_1' = (\mu_{11}, \mu_{12}, \cdots, \mu_{1p})$, etc. ($\mathbf{x}'$ is the transpose of the one-column matrix $\mathbf{x}$, etc.)

Suppose we have two k-variate normal populations $N(\boldsymbol{\mu}_i, \boldsymbol{\Sigma}_i)$, with $\boldsymbol{\mu}_i' = (\mu_{i1}, \mu_{i2}, \cdots, \mu_{ik})$, $i = 1, 2$, the one-row matrices (vectors) of mean values, and $\boldsymbol{\Sigma}_i = (\sigma_{irs})$, $i = 1, 2$; $r, s = 1, 2, \cdots, k$, the covariance matrices. Denoting the respective population densities by [cf. Anderson (1958, p. 17), Roy (1957, p. 15)]

$$f_i(x_1, x_2, \cdots, x_k) = \frac{1}{|2\pi\boldsymbol{\Sigma}_i|^{1/2}} \exp\left(-\tfrac{1}{2}(\mathbf{x} - \boldsymbol{\mu}_i)'\boldsymbol{\Sigma}_i^{-1}(\mathbf{x} - \boldsymbol{\mu}_i)\right)$$

we find (see problem 10.1)

$$(1.1)\quad \log \frac{f_1(x_1, x_2, \cdots, x_k)}{f_2(x_1, x_2, \cdots, x_k)} = \tfrac{1}{2} \log \frac{|\boldsymbol{\Sigma}_2|}{|\boldsymbol{\Sigma}_1|} - \tfrac{1}{2} \operatorname{tr} \boldsymbol{\Sigma}_1^{-1}(\mathbf{x} - \boldsymbol{\mu}_1)(\mathbf{x} - \boldsymbol{\mu}_1)' + \tfrac{1}{2} \operatorname{tr} \boldsymbol{\Sigma}_2^{-1}(\mathbf{x} - \boldsymbol{\mu}_2)(\mathbf{x} - \boldsymbol{\mu}_2)',$$

from which we get

$$(1.2)\quad I(1:2) = \int f_1(x_1, \cdots, x_k) \log \frac{f_1(x_1, \cdots, x_k)}{f_2(x_1, \cdots, x_k)}\, dx_1 \cdots dx_k$$
$$= \tfrac{1}{2} \log \frac{|\boldsymbol{\Sigma}_2|}{|\boldsymbol{\Sigma}_1|} + \tfrac{1}{2} \operatorname{tr} \boldsymbol{\Sigma}_1(\boldsymbol{\Sigma}_2^{-1} - \boldsymbol{\Sigma}_1^{-1}) + \tfrac{1}{2} \operatorname{tr} \boldsymbol{\Sigma}_2^{-1}(\boldsymbol{\mu}_1 - \boldsymbol{\mu}_2)(\boldsymbol{\mu}_1 - \boldsymbol{\mu}_2)',$$

$$(1.3)\quad J(1, 2) = \int (f_1(x_1, \cdots, x_k) - f_2(x_1, \cdots, x_k)) \log \frac{f_1(x_1, \cdots, x_k)}{f_2(x_1, \cdots, x_k)} dx_1 \cdots dx_k$$

$$= \tfrac{1}{2} \operatorname{tr} (\mathbf{\Sigma}_1 - \mathbf{\Sigma}_2)(\mathbf{\Sigma}_2^{-1} - \mathbf{\Sigma}_1^{-1}) + \tfrac{1}{2} \operatorname{tr} (\mathbf{\Sigma}_1^{-1} + \mathbf{\Sigma}_2^{-1}) (\boldsymbol{\mu}_1 - \boldsymbol{\mu}_2)(\boldsymbol{\mu}_1 - \boldsymbol{\mu}_2)'.$$

Assuming equal population covariance matrices, $\mathbf{\Sigma}_1 = \mathbf{\Sigma}_2 = \mathbf{\Sigma}$, (1.2) and (1.3) become, respectively,

$$(1.4)\quad I(1:2; \boldsymbol{\mu}) = \tfrac{1}{2} \operatorname{tr} \mathbf{\Sigma}^{-1}(\boldsymbol{\mu}_1 - \boldsymbol{\mu}_2)(\boldsymbol{\mu}_1 - \boldsymbol{\mu}_2)' = \tfrac{1}{2} \operatorname{tr} \mathbf{\Sigma}^{-1}\boldsymbol{\delta}\boldsymbol{\delta}' = \tfrac{1}{2}\boldsymbol{\delta}'\mathbf{\Sigma}^{-1}\boldsymbol{\delta},$$

$$(1.5)\quad J(1, 2; \boldsymbol{\mu}) = \operatorname{tr} \mathbf{\Sigma}^{-1}(\boldsymbol{\mu}_1 - \boldsymbol{\mu}_2)(\boldsymbol{\mu}_1 - \boldsymbol{\mu}_2)' = \boldsymbol{\delta}'\mathbf{\Sigma}^{-1}\boldsymbol{\delta},$$

where $\boldsymbol{\delta} = \boldsymbol{\mu}_1 - \boldsymbol{\mu}_2$. Mahalanobis' generalized distance is $k\boldsymbol{\delta}'\mathbf{\Sigma}^{-1}\boldsymbol{\delta}$ [Mahalanobis (1936)]. [See section 3 of chapter 1 and Anderson (1958, p. 135).]

Assuming equal population means, $\boldsymbol{\mu}_1 = \boldsymbol{\mu}_2$, $\boldsymbol{\delta} = \boldsymbol{\mu}_1 - \boldsymbol{\mu}_2 = \mathbf{0}$ (or variables centered at their respective means), (1.2) and (1.3) become, respectively,

$$(1.6)\quad I(1:2; \mathbf{\Sigma}) = \tfrac{1}{2} \log \frac{|\mathbf{\Sigma}_2|}{|\mathbf{\Sigma}_1|} + \tfrac{1}{2} \operatorname{tr} \mathbf{\Sigma}_1(\mathbf{\Sigma}_2^{-1} - \mathbf{\Sigma}_1^{-1}) = \tfrac{1}{2} \log \frac{|\mathbf{\Sigma}_2|}{|\mathbf{\Sigma}_1|} - \frac{k}{2} + \tfrac{1}{2} \operatorname{tr} \mathbf{\Sigma}_1\mathbf{\Sigma}_2^{-1},$$

$$(1.7)\quad J(1, 2; \mathbf{\Sigma}) = \tfrac{1}{2} \operatorname{tr} (\mathbf{\Sigma}_1 - \mathbf{\Sigma}_2)(\mathbf{\Sigma}_2^{-1} - \mathbf{\Sigma}_1^{-1}) = \tfrac{1}{2} \operatorname{tr} \mathbf{\Sigma}_1\mathbf{\Sigma}_2^{-1} + \tfrac{1}{2} \operatorname{tr} \mathbf{\Sigma}_2\mathbf{\Sigma}_1^{-1} - k.$$

The corresponding values for

$$I(2:1) = \int f_2(x_1, \cdots, x_k) \log \frac{f_2(x_1, \cdots, x_k)}{f_1(x_1, \cdots, x_k)} dx_1 \cdots dx_k$$

are easily derived from the fact that $I(1:2) + I(2:1) = J(1, 2)$. Note that the general values for the mean discrimination information and divergence in (1.2) and (1.3) are expressible as the sum of two components, one due to the difference in means, the other due to the difference in variances and covariances; these may also be characterized respectively as differences in size and shape. For single-variate normal populations, $k = 1$, corresponding to (1.4)–(1.7) respectively, we have

$$I(1:2;\mu) = \frac{1}{2}\frac{\delta^2}{\sigma^2} = \frac{1}{2}\frac{(\mu_1 - \mu_2)^2}{\sigma^2}, \tag{1.8}$$

$$J(1,2;\mu) = \frac{\delta^2}{\sigma^2}, \tag{1.9}$$

$$I(1:2;\sigma^2) = \frac{1}{2}\log\frac{\sigma_2^{\,2}}{\sigma_1^{\,2}} - \frac{1}{2} + \frac{1}{2}\frac{\sigma_1^{\,2}}{\sigma_2^{\,2}}, \tag{1.10}$$

$$J(1,2;\sigma^2) = \frac{1}{2}\frac{\sigma_1^{\,2}}{\sigma_2^{\,2}} + \frac{1}{2}\frac{\sigma_2^{\,2}}{\sigma_1^{\,2}} - 1. \tag{1.11}$$

2. COMPONENTS OF INFORMATION

Since $I(1:2)$ and $J(1,2)$ are additive for independent random variables, we have for a random sample of n observations, O_n, $I(1:2;O_n) = nI(1:2)$ and $J(1,2;O_n) = nJ(1,2)$ where $I(1:2)$ and $J(1,2)$ are respectively (1.2) and (1.3). (See sections 2 and 5 of chapter 2.)

The averages and the variances and covariances in a sample O_n from a multivariate normal population, $N(\boldsymbol{\mu}, \boldsymbol{\Sigma})$, are independently distributed. The averages are normally distributed, $N(\boldsymbol{\mu}, (1/n)\boldsymbol{\Sigma})$, and the variances and covariances are distributed according to the Wishart distribution. [See Anderson (1958, pp. 53, 154), Kendall (1946, pp. 330–335), Rao (1952, pp. 66–74), Wilks (1943, pp. 120, 226–233).] Since the averages are normally distributed (1.2) and (1.3) yield

$$I(1:2;\bar{\mathbf{x}}) = \tfrac{1}{2}\log\frac{|\boldsymbol{\Sigma}_2|}{|\boldsymbol{\Sigma}_1|} + \tfrac{1}{2}\operatorname{tr}\boldsymbol{\Sigma}_1(\boldsymbol{\Sigma}_2^{-1} - \boldsymbol{\Sigma}_1^{-1}) + \frac{n}{2}\operatorname{tr}\boldsymbol{\Sigma}_2^{-1}\boldsymbol{\delta}\boldsymbol{\delta}', \tag{2.1}$$

$$J(1:2;\bar{\mathbf{x}}) = \tfrac{1}{2}\operatorname{tr}(\boldsymbol{\Sigma}_1 - \boldsymbol{\Sigma}_2)(\boldsymbol{\Sigma}_2^{-1} - \boldsymbol{\Sigma}_1^{-1}) + \frac{n}{2}\operatorname{tr}(\boldsymbol{\Sigma}_1^{-1} + \boldsymbol{\Sigma}_2^{-1})\boldsymbol{\delta}\boldsymbol{\delta}'. \tag{2.2}$$

Note that the sample size appears in (2.1) and (2.2) as a factor only in the components due to the difference in means.

Designating the density of the Wishart distribution by

$$W(s_{11}, \cdots, s_{kk}) = \frac{(\frac{N}{2})^{kN/2}|\mathbf{S}|^{(N-k-1)/2}\exp(-\frac{1}{2}\operatorname{tr}N\mathbf{S}\boldsymbol{\Sigma}^{-1})}{\pi^{k(k-1)/4}|\boldsymbol{\Sigma}|^{N/2}\prod_{\alpha=1}^{k}\Gamma(N+1-\alpha)/2},$$

we find (see problem 10.5)

$$(2.3)\quad \log \frac{W_1(s_{11}, \cdots, s_{kk})}{W_2(s_{11}, \cdots, s_{kk})} = \frac{N}{2} \log \frac{|\mathbf{\Sigma}_2|}{|\mathbf{\Sigma}_1|} - \frac{N}{2} \operatorname{tr} \mathbf{\Sigma}_1^{-1}\mathbf{S} + \frac{N}{2} \operatorname{tr} \mathbf{\Sigma}_2^{-1}\mathbf{S},$$

$$(2.4)\qquad I(1:2; \mathbf{S}) = \frac{N}{2}\left(\log \frac{|\mathbf{\Sigma}_2|}{|\mathbf{\Sigma}_1|} + \operatorname{tr} \mathbf{\Sigma}_1(\mathbf{\Sigma}_2^{-1} - \mathbf{\Sigma}_1^{-1})\right),$$

$$(2.5)\qquad J(1, 2; \mathbf{S}) = \frac{N}{2} \operatorname{tr} (\mathbf{\Sigma}_1 - \mathbf{\Sigma}_2)(\mathbf{\Sigma}_2^{-1} - \mathbf{\Sigma}_1^{-1}),$$

where $\mathbf{S}$ is the sample covariance matrix of unbiased estimates and $N = n - 1$ degrees of freedom.

We thus see from the preceding, and theorems 2.1 and 5.1 of chapter 2, that

$$(2.6)\quad I(1:2; O_n) = nI(1:2) = I(1:2; \bar{\mathbf{x}}) + I(1:2; \mathbf{S}) = I(1:2; \bar{\mathbf{x}}, \mathbf{S}),$$

$$(2.7)\quad J(1, 2; O_n) = nJ(1, 2) = J(1, 2; \bar{\mathbf{x}}) + J(1, 2; \mathbf{S}) = J(1, 2; \bar{\mathbf{x}}, \mathbf{S}).$$

Assuming that the population covariance matrices differ only in the values of the correlation coefficients, that is, $\mathbf{\Sigma}_1 = \mathbf{D}_\sigma\mathbf{P}_1\mathbf{D}_\sigma$, $\mathbf{\Sigma}_2 = \mathbf{D}_\sigma\mathbf{P}_2\mathbf{D}_\sigma$, where $\mathbf{P}_1$ and $\mathbf{P}_2$ are matrices of correlation coefficients and

$$\mathbf{D}_\sigma = \begin{pmatrix} \sigma_1 & \cdots & 0 \\ \cdot & \cdot & \cdot \\ \cdot & \cdot & \cdot \\ \cdot & & \cdot\ \cdot \\ 0 & \cdots & \sigma_k \end{pmatrix}$$ is a diagonal matrix of standard deviations, (2.4)

and (2.5) become respectively,

$$(2.8)\qquad I(1:2; \mathbf{S}) = \frac{N}{2}\left(\log \frac{|\mathbf{P}_2|}{|\mathbf{P}_1|} + \operatorname{tr} \mathbf{P}_1(\mathbf{P}_2^{-1} - \mathbf{P}_1^{-1})\right),$$

$$(2.9)\qquad J(1, 2; \mathbf{S}) = \frac{N}{2} \operatorname{tr} (\mathbf{P}_1 - \mathbf{P}_2)(\mathbf{P}_2^{-1} - \mathbf{P}_1^{-1}).$$

We now deal with several samples. Suppose we have r independent samples, respectively, of $n_1, n_2, \cdots, n_r$ independent observations each, with $n = n_1 + n_2 + \cdots + n_r$. We may treat the r samples as one large sample from populations with means and covariance matrices given by (the n_i indicate the number of occurrences of the corresponding term):

$$(2.10)\qquad \boldsymbol{\mu}_i' = (\underset{n_1}{\boldsymbol{\mu}_{i1}'\cdots}, \underset{n_2}{\boldsymbol{\mu}_{i2}'\cdots}, \cdots, \underset{n_r}{\boldsymbol{\mu}_{ir}'\cdots}), \qquad i = 1, 2,$$

$$(2.11)\qquad \boldsymbol{\Sigma}_i = \begin{pmatrix} \boldsymbol{\Sigma}_{i1} \cdots \mathbf{0} \cdots \mathbf{0} \cdots \\ \cdot \\ \mathbf{0} \cdots \boldsymbol{\Sigma}_{i2} \cdots \mathbf{0} \cdots \\ \cdot \quad \cdot \quad \cdot \\ \cdot \quad \cdot \quad \cdot \\ \cdot \quad \cdot \quad \cdot \\ \mathbf{0} \cdots \mathbf{0} \cdots \boldsymbol{\Sigma}_{ir} \cdots \\ \cdots\cdots\cdots\cdots \end{pmatrix} \begin{matrix} n_1 \\ \\ n_2 \\ \cdot \\ \cdot \\ \cdot \\ n_r \end{matrix}, \qquad i = 1, 2,$$

$$(2.12)\qquad \boldsymbol{\delta}' = \boldsymbol{\mu}_1' - \boldsymbol{\mu}_2' = (\underset{n_1}{\boldsymbol{\delta}_1'\cdots}, \underset{n_2}{\boldsymbol{\delta}_2'\cdots}, \cdots, \underset{n_r}{\boldsymbol{\delta}_r'\cdots}),$$

$$\boldsymbol{\delta}_j' = \boldsymbol{\mu}_{1j}' - \boldsymbol{\mu}_{2j}'.$$

With the preceding (or from the additivity property), we find for the r samples,

$$(2.13)\quad I(1:2; O_n) = \sum_{j=1}^{r} \frac{n_j}{2}\left(\log\frac{|\boldsymbol{\Sigma}_{2j}|}{|\boldsymbol{\Sigma}_{1j}|} + \operatorname{tr}\boldsymbol{\Sigma}_{1j}(\boldsymbol{\Sigma}_{2j}^{-1} - \boldsymbol{\Sigma}_{1j}^{-1}) + \operatorname{tr}\boldsymbol{\Sigma}_{2j}^{-1}\boldsymbol{\delta}_j\boldsymbol{\delta}_j'\right),$$

$$(2.14)\quad J(1, 2; O_n) = \sum_{j=1}^{r} \frac{n_j}{2}(\operatorname{tr}(\boldsymbol{\Sigma}_{1j} - \boldsymbol{\Sigma}_{2j})(\boldsymbol{\Sigma}_{2j}^{-1} - \boldsymbol{\Sigma}_{1j}^{-1}) + \operatorname{tr}(\boldsymbol{\Sigma}_{1j}^{-1} + \boldsymbol{\Sigma}_{2j}^{-1})\boldsymbol{\delta}_j\boldsymbol{\delta}_j'),$$

$$(2.15)\quad I(1:2; O_n) = \sum_{j=1}^{r} I(1:2; \bar{\mathbf{x}}_j) + \sum_{j=1}^{r} I(1:2; \mathbf{S}_j),$$

$$(2.16)\quad J(1, 2; O_n) = \sum_{j=1}^{r} J(1, 2; \bar{\mathbf{x}}_j) + \sum_{j=1}^{r} J(1, 2; \mathbf{S}_j),$$

where $I(1:2; \bar{\mathbf{x}}_j)$, $I(1:2; \mathbf{S}_j)$, $J(1, 2; \bar{\mathbf{x}}_j)$, $J(1, 2; \mathbf{S}_j)$ are (2.1), (2.4), (2.2), (2.5), respectively, for the jth sample.

When the r samples are from populations with common covariance matrices, $\boldsymbol{\Sigma}_{ij} = \boldsymbol{\Sigma}$, $i = 1, 2, j = 1, 2, \cdots, r$, we find

$$(2.17)\quad I(1:2; O_n) = \sum_{j=1}^{r} I(1:2; \bar{\mathbf{x}}_j) = \tfrac{1}{2}\operatorname{tr}\boldsymbol{\Sigma}^{-1}(n_1\boldsymbol{\delta}_1\boldsymbol{\delta}_1' + \cdots + n_r\boldsymbol{\delta}_r\boldsymbol{\delta}_r')$$
$$= \tfrac{1}{2}\operatorname{tr}\boldsymbol{\Sigma}^{-1}\boldsymbol{\Sigma}^* = \tfrac{1}{2}\sum_{j=1}^{r} J(1, 2; \bar{\mathbf{x}}_j),$$

where $\boldsymbol{\Sigma}^* = n_1\boldsymbol{\delta}_1\boldsymbol{\delta}_1' + \cdots + n_r\boldsymbol{\delta}_r\boldsymbol{\delta}_r'$. [Cf. Hotelling (1951).]

When the r samples are from populations with common means (or the variables are centered at their respective means), $\boldsymbol{\delta}_j = \mathbf{0}$, $j = 1, 2, \cdots, r$, we find

$$(2.18)\quad I(1:2; O_n) = \sum_{j=1}^{r} I(1:2; \mathbf{S}_j)$$

$$= \sum_{j=1}^{r} \frac{N_j}{2}\left(\log \frac{|\boldsymbol{\Sigma}_{2j}|}{|\boldsymbol{\Sigma}_{1j}|} + \operatorname{tr} \boldsymbol{\Sigma}_{1j}(\boldsymbol{\Sigma}_{2j}^{-1} - \boldsymbol{\Sigma}_{1j}^{-1})\right),$$

$$(2.19)\quad J(1, 2; O_n) = \sum_{j=1}^{r} J(1, 2; \mathbf{S}_j) = \sum_{j=1}^{r} \frac{N_j}{2} \operatorname{tr} (\boldsymbol{\Sigma}_{1j} - \boldsymbol{\Sigma}_{2j})(\boldsymbol{\Sigma}_{2j}^{-1} - \boldsymbol{\Sigma}_{1j}^{-1}),$$

where N_j is the number of degrees of freedom in the jth sample for the estimates $\mathbf{S}_j$.

3. CANONICAL FORM

$I(1:2)$ and $J(1, 2)$ are functions of the population parameters under H_1 and H_2. According to corollary 4.1 in chapter 2, $I(1:2)$ and $J(1, 2)$ are invariant for nonsingular transformations of the random variables, and therefore in particular for nonsingular linear transformations. An important connection exists between the invariant properties and linear discriminant functions, and we now examine this in some detail. (This will also reflect itself in invariant properties of the subsequent tests.)

If the random matrix $\mathbf{x}$ is subjected to the nonsingular linear transformation $\mathbf{y} = \mathbf{A}\mathbf{x}$, the means and covariance matrix of the y's are respectively $\boldsymbol{\mu}_y = \mathbf{A}\boldsymbol{\mu}_x$, $\boldsymbol{\Sigma}_y = \mathbf{A}\boldsymbol{\Sigma}\mathbf{A}'$. If the x's are normally distributed, the y's are normally distributed and (see Anderson (1958, pp. 19–27), problems 10.5, 10.10)

$$(3.1)\quad I(1:2; \mathbf{y}) = \tfrac{1}{2} \log \frac{|\mathbf{A}\boldsymbol{\Sigma}_2\mathbf{A}'|}{|\mathbf{A}\boldsymbol{\Sigma}_1\mathbf{A}'|} + \tfrac{1}{2} \operatorname{tr} \mathbf{A}\boldsymbol{\Sigma}_1\mathbf{A}'(\mathbf{A}'^{-1}\boldsymbol{\Sigma}_2^{-1}\mathbf{A}^{-1} - \mathbf{A}'^{-1}\boldsymbol{\Sigma}_1^{-1}\mathbf{A}^{-1}) + \tfrac{1}{2} \operatorname{tr} \mathbf{A}'^{-1}\boldsymbol{\Sigma}_2^{-1}\mathbf{A}^{-1}\mathbf{A}\boldsymbol{\delta}\boldsymbol{\delta}'\mathbf{A}'$$

$$= \tfrac{1}{2} \log \frac{|\boldsymbol{\Sigma}_2|}{|\boldsymbol{\Sigma}_1|} + \tfrac{1}{2} \operatorname{tr} \boldsymbol{\Sigma}_1(\boldsymbol{\Sigma}_2^{-1} - \boldsymbol{\Sigma}_1^{-1}) + \tfrac{1}{2} \operatorname{tr} \boldsymbol{\Sigma}_2^{-1}\boldsymbol{\delta}\boldsymbol{\delta}'$$

$$= I(1:2; \mathbf{x}),$$

$$(3.2)\quad J(1, 2; \mathbf{y}) = \tfrac{1}{2} \operatorname{tr} (\mathbf{A}\boldsymbol{\Sigma}_1\mathbf{A}' - \mathbf{A}\boldsymbol{\Sigma}_2\mathbf{A}')(\mathbf{A}'^{-1}\boldsymbol{\Sigma}_2^{-1}\mathbf{A}^{-1} - \mathbf{A}'^{-1}\boldsymbol{\Sigma}_1^{-1}\mathbf{A}^{-1}) + \tfrac{1}{2} \operatorname{tr} (\mathbf{A}'^{-1}\boldsymbol{\Sigma}_1^{-1}\mathbf{A}^{-1} + \mathbf{A}'^{-1}\boldsymbol{\Sigma}_2^{-1}\mathbf{A}^{-1})\mathbf{A}\boldsymbol{\delta}\boldsymbol{\delta}'\mathbf{A}'$$

$$= \tfrac{1}{2} \operatorname{tr} (\boldsymbol{\Sigma}_1 - \boldsymbol{\Sigma}_2)(\boldsymbol{\Sigma}_2^{-1} - \boldsymbol{\Sigma}_1^{-1}) + \tfrac{1}{2} \operatorname{tr} (\boldsymbol{\Sigma}_1^{-1} + \boldsymbol{\Sigma}_2^{-1})\boldsymbol{\delta}\boldsymbol{\delta}'$$

$$= J(1, 2; \mathbf{x}).$$

Since $\boldsymbol{\Sigma}_1$ and $\boldsymbol{\Sigma}_2$ are positive definite, there exists a real nonsingular matrix $\mathbf{A}$ such that [see Anderson (1958, pp. 337–341), Ferrar (1941, pp. 151–153), Rao (1952, pp. 25–27)]

$$\mathbf{A}\mathbf{\Sigma}_1\mathbf{A}' = \mathbf{\Lambda}, \qquad \mathbf{A}\mathbf{\Sigma}_2\mathbf{A}' = \mathbf{I}, \tag{3.3}$$

where $\mathbf{\Lambda}$ is the diagonal matrix with real and positive elements $\lambda_1, \lambda_2, \cdots, \lambda_k$, and $\mathbf{I}$ is the identity matrix; in fact, the λ's are the roots of the determinantal equation

$$|\mathbf{\Sigma}_1 - \lambda\mathbf{\Sigma}_2| = 0. \tag{3.4}$$

The matrix $\mathbf{A}$ in (3.3) defines a linear transformation of the x's such that the y's are independent with variances $\lambda_1, \lambda_2, \cdots, \lambda_k$ in the population under H_1 and unit variances in the population under H_2. Letting $\mathbf{A}' = (\boldsymbol{\alpha}_1, \boldsymbol{\alpha}_2, \cdots, \boldsymbol{\alpha}_k)$, that is, the one-row matrix (vector) $\boldsymbol{\alpha}_i'$ is the ith row of the matrix $\mathbf{A}$, (3.3) and (3.4) yield

$$\begin{aligned} &\boldsymbol{\alpha}_i'\mathbf{\Sigma}_1\boldsymbol{\alpha}_i = \lambda_i, \qquad \boldsymbol{\alpha}_i'\mathbf{\Sigma}_2\boldsymbol{\alpha}_i = 1, \qquad i = 1, 2, \cdots, k,\\ &\boldsymbol{\alpha}_i'\mathbf{\Sigma}_1\boldsymbol{\alpha}_j = 0, \qquad \boldsymbol{\alpha}_i'\mathbf{\Sigma}_2\boldsymbol{\alpha}_j = 0, \qquad i \neq j,\\ &\mathbf{\Sigma}_1\boldsymbol{\alpha}_i = \lambda_i\mathbf{\Sigma}_2\boldsymbol{\alpha}_i, \qquad i = 1, 2, \cdots, k,\\ &|\mathbf{\Sigma}_1\mathbf{\Sigma}_2^{-1}| = |\mathbf{\Sigma}_1|/|\mathbf{\Sigma}_2| = \lambda_1\lambda_2\cdots\lambda_k,\\ &\operatorname{tr} \mathbf{\Sigma}_1\mathbf{\Sigma}_2^{-1} = \lambda_1 + \lambda_2 + \cdots + \lambda_k,\\ &\operatorname{tr} \mathbf{\Sigma}_2\mathbf{\Sigma}_1^{-1} = \frac{1}{\lambda_1} + \frac{1}{\lambda_2} + \cdots + \frac{1}{\lambda_k},\\ &\boldsymbol{\delta}'\mathbf{A}' = (\boldsymbol{\delta}'\boldsymbol{\alpha}_1, \boldsymbol{\delta}'\boldsymbol{\alpha}_2, \cdots, \boldsymbol{\delta}'\boldsymbol{\alpha}_k). \end{aligned} \tag{3.5}$$

In terms of the *characteristic roots*, the λ's, and the *characteristic vectors*, the $\boldsymbol{\alpha}_i$, we have:

$$\begin{aligned} I(1:2) &= -\tfrac{1}{2}\log \lambda_1\lambda_2\cdots\lambda_k + \tfrac{1}{2}(\lambda_1 + \lambda_2 + \cdots + \lambda_k) - k/2\\ &\qquad + \tfrac{1}{2}[(\boldsymbol{\alpha}_1'\boldsymbol{\delta})^2 + \cdots + (\boldsymbol{\alpha}_k'\boldsymbol{\delta})^2],\\ &= \sum_{i=1}^{k} \tfrac{1}{2}[-\log \lambda_i + \lambda_i - 1 + (\boldsymbol{\alpha}_i'\boldsymbol{\delta})^2], \end{aligned} \tag{3.6}$$

$$\begin{aligned} J(1, 2) &= \tfrac{1}{2}(\lambda_1 + \cdots + \lambda_k) + \frac{1}{2}\left(\frac{1}{\lambda_1} + \frac{1}{\lambda_2} + \cdots + \frac{1}{\lambda_k}\right)\\ &\qquad - k + \frac{1}{2}\left(\frac{1}{\lambda_1} + 1\right)(\boldsymbol{\alpha}_1'\boldsymbol{\delta})^2 + \cdots + \frac{1}{2}\left(\frac{1}{\lambda_k} + 1\right)(\boldsymbol{\alpha}_k'\boldsymbol{\delta})^2\\ &= \sum_{i=1}^{k} \frac{1}{2}\left[\lambda_i + \frac{1}{\lambda_i} - 2 + \left(\frac{1}{\lambda_i} + 1\right)(\boldsymbol{\alpha}_i'\boldsymbol{\delta})^2\right]. \end{aligned} \tag{3.7}$$

(See sections 5 and 6 of chapter 3.)

4. LINEAR DISCRIMINANT FUNCTIONS

The right-hand side of (1.1) is an optimum, or sufficient discriminant function for assigning an observation to one of two multivariate normal populations. This in general is quadratic. [Cf. Neyman and Pearson (1933), Welch (1939).] However, we may prefer to work with one or more linear functions for the convenience they offer. How do we find the best linear function? Which properties of the linear function do we optimize? For the present we shall examine the consequences of *maximizing* the discrimination information or divergence for the linear function. A more detailed discussion and application will take place later.

Consider the linear discriminant function

$$(4.1) \qquad y = \alpha_1 x_1 + \cdots + \alpha_k x_k = \boldsymbol{\alpha}'\mathbf{x},$$

where the x's are k-variate normal $N(\boldsymbol{\mu}_i, \boldsymbol{\Sigma}_i)$, $i = 1, 2$. The linear function y is consequently normally distributed, with parameters

$$(4.2) \qquad E_1(y) = \boldsymbol{\alpha}'\boldsymbol{\mu}_1, \quad E_2(y) = \boldsymbol{\alpha}'\boldsymbol{\mu}_2, \quad \text{var}_1(y) = \boldsymbol{\alpha}'\boldsymbol{\Sigma}_1\boldsymbol{\alpha}, \quad \text{var}_2(y) = \boldsymbol{\alpha}'\boldsymbol{\Sigma}_2\boldsymbol{\alpha}.$$

We consider how to determine $\boldsymbol{\alpha}$ under certain assumptions about the populations.

5. EQUAL COVARIANCE MATRICES

When $\boldsymbol{\Sigma}_1 = \boldsymbol{\Sigma}_2 = \boldsymbol{\Sigma}$, (1.4) and (1.5) yield

$$(5.1) \qquad 2I(1:2) = J(1, 2) = \text{tr}\, \boldsymbol{\Sigma}^{-1}\boldsymbol{\delta}\boldsymbol{\delta}', \qquad \boldsymbol{\delta} = \boldsymbol{\mu}_1 - \boldsymbol{\mu}_2.$$

For the linear discriminant function $y = \boldsymbol{\alpha}'\mathbf{x}$,

$$(5.2) \qquad 2I(1:2; y) = J(1, 2; y) = \boldsymbol{\alpha}'\boldsymbol{\delta}\boldsymbol{\delta}'\boldsymbol{\alpha}/\boldsymbol{\alpha}'\boldsymbol{\Sigma}\boldsymbol{\alpha}.$$

The value of $\boldsymbol{\alpha}$ for which $\lambda = \boldsymbol{\alpha}'\boldsymbol{\delta}\boldsymbol{\delta}'\boldsymbol{\alpha}/\boldsymbol{\alpha}'\boldsymbol{\Sigma}\boldsymbol{\alpha}$ is a maximum satisfies (by the usual calculus procedures, see problems 10.2, 10.4) $\boldsymbol{\delta}\boldsymbol{\delta}'\boldsymbol{\alpha} = \lambda\boldsymbol{\Sigma}\boldsymbol{\alpha}$, where λ is the largest root of $|\boldsymbol{\delta}\boldsymbol{\delta}' - \lambda\boldsymbol{\Sigma}| = 0$. Here, since $\boldsymbol{\delta}\boldsymbol{\delta}'$ is of rank 1, there is only one nonzero root, $\lambda = \boldsymbol{\delta}'\boldsymbol{\Sigma}^{-1}\boldsymbol{\delta} = \text{tr}\, \boldsymbol{\Sigma}^{-1}\boldsymbol{\delta}\boldsymbol{\delta}'$. The linear discriminant function with $\boldsymbol{\Sigma}\boldsymbol{\alpha} = \boldsymbol{\delta}$, or $\boldsymbol{\alpha} = \boldsymbol{\Sigma}^{-1}\boldsymbol{\delta}$, is sufficient, since,

$$(5.3) \qquad 2I(1:2; y) = J(1, 2; y) = \frac{\boldsymbol{\alpha}'\boldsymbol{\delta}\boldsymbol{\delta}'\boldsymbol{\alpha}}{\boldsymbol{\alpha}'\boldsymbol{\Sigma}\boldsymbol{\alpha}} = \frac{\boldsymbol{\delta}'\boldsymbol{\Sigma}^{-1}\boldsymbol{\delta}\boldsymbol{\delta}'\boldsymbol{\Sigma}^{-1}\boldsymbol{\delta}}{\boldsymbol{\delta}'\boldsymbol{\Sigma}^{-1}\boldsymbol{\Sigma}\boldsymbol{\Sigma}^{-1}\boldsymbol{\delta}}$$

$$= \boldsymbol{\delta}'\boldsymbol{\Sigma}^{-1}\boldsymbol{\delta} = \text{tr}\, \boldsymbol{\Sigma}^{-1}\boldsymbol{\delta}\boldsymbol{\delta}'$$

$$= 2I(1:2) = J(1, 2).$$

For r samples from populations with common covariance matrices, but different means, (2.17) is

$$(5.4) \quad 2I(1:2; O_n) = J(1, 2; O_n) = \operatorname{tr} \boldsymbol{\Sigma}^{-1}\boldsymbol{\Sigma}^* = \operatorname{tr} \boldsymbol{\Sigma}^{-1}(n_1\boldsymbol{\delta}_1\boldsymbol{\delta}_1' + \cdots + n_r\boldsymbol{\delta}_r\boldsymbol{\delta}_r').$$

If we propose to use the same linear discriminant function, $y = \boldsymbol{\alpha}'\mathbf{x}$, for all the samples, (5.4) yields for the linear discriminant function:

$$(5.5) \quad 2I(1:2; O_n, y) = J(1, 2; O_n, y) = \frac{n_1(\boldsymbol{\alpha}'\boldsymbol{\delta}_1\boldsymbol{\delta}_1'\boldsymbol{\alpha}) + \cdots + n_r(\boldsymbol{\alpha}'\boldsymbol{\delta}_r\boldsymbol{\delta}_r'\boldsymbol{\alpha})}{\boldsymbol{\alpha}'\boldsymbol{\Sigma}\boldsymbol{\alpha}} = \frac{\boldsymbol{\alpha}'\boldsymbol{\Sigma}^*\boldsymbol{\alpha}}{\boldsymbol{\alpha}'\boldsymbol{\Sigma}\boldsymbol{\alpha}}.$$

The value of $\boldsymbol{\alpha}$ for which $\lambda = \boldsymbol{\alpha}'\boldsymbol{\Sigma}^*\boldsymbol{\alpha}/\boldsymbol{\alpha}'\boldsymbol{\Sigma}\boldsymbol{\alpha}$ is a maximum satisfies (by the usual calculus procedures) $\boldsymbol{\Sigma}^*\boldsymbol{\alpha} = \lambda\boldsymbol{\Sigma}\boldsymbol{\alpha}$, where λ is the largest root of $|\boldsymbol{\Sigma}^* - \lambda\boldsymbol{\Sigma}| = 0$. From its definition, the rank of $\boldsymbol{\Sigma}^*$ is not greater than r. The determinantal equation has $p \leq \min(k, r)$ nonzero roots, designated in descending order as $\lambda_1, \lambda_2, \cdots, \lambda_p$. Each root λ_i is associated with a one-column matrix (vector) $\boldsymbol{\alpha}_i$, $\boldsymbol{\Sigma}^*\boldsymbol{\alpha}_i = \lambda_i\boldsymbol{\Sigma}\boldsymbol{\alpha}_i$, and a linear discriminant function $y_i = \boldsymbol{\alpha}_i'\mathbf{x}$. Since $\operatorname{tr} \boldsymbol{\Sigma}^*\boldsymbol{\Sigma}^{-1} = \lambda_1 + \lambda_2 + \cdots + \lambda_p$, (5.4) and (5.5) yield

$$(5.6) \quad J(1, 2; O_n) = \operatorname{tr} \boldsymbol{\Sigma}^*\boldsymbol{\Sigma}^{-1} = J(1, 2; O_n, y_1) + J(1, 2; O_n, y_2) + \cdots + J(1, 2; O_n, y_p).$$

The discrimination efficiency of the linear discriminant function y_1 can be measured by the ratio $\lambda_1/(\lambda_1 + \cdots + \lambda_p)$ or $J(1, 2; O_n, y_1)/J(1, 2; O_n)$; the discrimination efficiency of the pair of linear discriminant functions y_1 and y_2 can be measured by the ratio $(\lambda_1 + \lambda_2)/(\lambda_1 + \lambda_2 + \cdots + \lambda_p)$ or $[J(1, 2; O_n, y_1) + J(1, 2; O_n, y_2)]/J(1, 2; O_n)$; etc. (See section 6 of chapter 3.)

The vectors $\boldsymbol{\alpha}_i$ associated with different roots λ_i have the property that $\boldsymbol{\alpha}_i'\boldsymbol{\Sigma}^*\boldsymbol{\alpha}_j = 0 = \boldsymbol{\alpha}_i'\boldsymbol{\Sigma}\boldsymbol{\alpha}_j$, $i \neq j$, and the corresponding linear discriminant functions y_i are independent, with a diagonal covariance matrix of elements $\boldsymbol{\alpha}_i'\boldsymbol{\Sigma}\boldsymbol{\alpha}_i$. There will be one, two, etc., distinct λ_i, and corresponding distinct linear discriminant functions according as the population means are collinear, coplanar, etc. [Cf. Williams (1952, 1955).]

6. PRINCIPAL COMPONENTS

Assuming that the k-variate normal populations are centered at their means, or that $\boldsymbol{\delta} = \mathbf{0}$, the linear discriminant function $y = \boldsymbol{\alpha}'\mathbf{x}$ is normally distributed, and

(6.1) $E_1(y) - E_2(y) = 0, \qquad \text{var}_1(y) = \boldsymbol{\alpha}'\boldsymbol{\Sigma}_1\boldsymbol{\alpha}, \qquad \text{var}_2(y) = \boldsymbol{\alpha}'\boldsymbol{\Sigma}_2\boldsymbol{\alpha},$

(6.2) $$I(1:2; y) = \frac{1}{2}\log\frac{\boldsymbol{\alpha}'\boldsymbol{\Sigma}_2\boldsymbol{\alpha}}{\boldsymbol{\alpha}'\boldsymbol{\Sigma}_1\boldsymbol{\alpha}} - \frac{1}{2} + \frac{\boldsymbol{\alpha}'\boldsymbol{\Sigma}_1\boldsymbol{\alpha}}{2\boldsymbol{\alpha}'\boldsymbol{\Sigma}_2\boldsymbol{\alpha}},$$

(6.3) $$J(1, 2; y) = \frac{1}{2}\frac{\boldsymbol{\alpha}'\boldsymbol{\Sigma}_1\boldsymbol{\alpha}}{\boldsymbol{\alpha}'\boldsymbol{\Sigma}_2\boldsymbol{\alpha}} + \frac{1}{2}\frac{\boldsymbol{\alpha}'\boldsymbol{\Sigma}_2\boldsymbol{\alpha}}{\boldsymbol{\alpha}'\boldsymbol{\Sigma}_1\boldsymbol{\alpha}} - 1.$$

The value of $\boldsymbol{\alpha}$ for which $I(1:2; y)$ is a maximum satisfies (by the usual calculus procedures)

(6.4) $$\boldsymbol{\Sigma}_1\boldsymbol{\alpha} = \lambda\boldsymbol{\Sigma}_2\boldsymbol{\alpha},$$

where λ is a root of the determinantal equation

(6.5) $$|\boldsymbol{\Sigma}_1 - \lambda\boldsymbol{\Sigma}_2| = 0,$$

all roots of which are real and positive. Designate these roots in ascending order as $\lambda_1, \lambda_2, \cdots, \lambda_k$. Seeking $\boldsymbol{\alpha}$ for which $J(1, 2; y)$ is a maximum, we find the same conditions, (6.4) and (6.5), as for maximizing $I(1:2; y)$. Each root λ_i is associated with a vector $\boldsymbol{\alpha}_i$ and the linear discriminant function $y_i = \boldsymbol{\alpha}_i'\mathbf{x}$. We thus have for the linear discriminant function y_i,

(6.6) $$I(1:2; y_i) = -\frac{1}{2}\log\lambda_i - \frac{1}{2} + \frac{\lambda_i}{2},$$

(6.7) $$J(1, 2; y_i) = \frac{\lambda_i}{2} + \frac{1}{2\lambda_i} - 1,$$

and from (3.6) and (3.7), with $\boldsymbol{\delta} = \mathbf{0}$,

(6.8) $$I(1:2) = I(1:2; y_1) + I(1:2; y_2) + \cdots + I(1:2; y_k),$$

(6.9) $$J(1, 2) = J(1, 2; y_1) + J(1, 2; y_2) + \cdots + J(1, 2; y_k).$$

We determine the value of λ_i for which (6.6) is a maximum (the most informative linear discriminant function) as follows. Since the function $g(\lambda) = -\frac{1}{2}\log\lambda - \frac{1}{2} + (\lambda/2)$ is convex [see problem 8.31(*a*) in chapter 2], nonnegative, and equal to zero for $\lambda = 1$, the maximum of (6.6) occurs for λ_1 or λ_k according as $g(\lambda_1) > g(\lambda_k)$, or $g(\lambda_1) < g(\lambda_k)$, that is,

(6.10) $$\log\frac{\lambda_k}{\lambda_1} > \lambda_k - \lambda_1 \qquad \text{or} \qquad \log\frac{\lambda_k}{\lambda_1} < \lambda_k - \lambda_1.$$

We determine the value of λ_i for which (6.7) is a maximum (the most divergent linear discriminant function) as follows. Since the function $f(\lambda) = (\lambda/2) + (1/2\lambda) - 1$, $\lambda > 0$, is convex [see problem 8.31(*a*) in chapter 2], nonnegative, equal to zero for $\lambda = 1$, and $f(\lambda) = f(1/\lambda)$, the maximum of (6.7) occurs for λ_1 or λ_k according as

$$\lambda_1\lambda_k < 1 \qquad \text{or} \qquad \lambda_1\lambda_k > 1. \tag{6.11}$$

Note that the linear discriminant functions of this section define the transformation with matrix **A** in section 3. The "best" lincar discriminant function is not necessarily associated with the largest λ.

Assuming that $\mathbf{\Sigma}_1 = \mathbf{D}_\sigma \mathbf{P} \mathbf{D}_\sigma$, $\mathbf{\Sigma}_2 = \mathbf{D}_\sigma \mathbf{D}_\sigma$, where $\mathbf{P}$ is a matrix of correlation coefficients and $\mathbf{D}_\sigma$ a diagonal matrix of standard deviations, $\mathbf{A\Sigma}_1\mathbf{A}' = \mathbf{\Lambda} = \mathbf{AD}_\sigma\mathbf{PD}_\sigma\mathbf{A}' = \mathbf{BPB}'$, and $\mathbf{A\Sigma}_2\mathbf{A}' = \mathbf{I} = \mathbf{AD}_\sigma\mathbf{D}_\sigma\mathbf{A}' = \mathbf{BB}'$. $\mathbf{B} = \mathbf{AD}_\sigma$ is an orthogonal matrix, (6.5) becomes $|\mathbf{P} - \lambda\mathbf{I}| = 0$, and (6.4) becomes $\mathbf{PD}_\sigma\boldsymbol{\alpha} = \lambda\mathbf{D}_\sigma\boldsymbol{\alpha}$, or $\mathbf{P}\boldsymbol{\beta} = \lambda\boldsymbol{\beta}$, with $\boldsymbol{\beta} = \mathbf{D}_\sigma\boldsymbol{\alpha}$, that is, $\mathbf{B}' = (\boldsymbol{\beta}_1, \boldsymbol{\beta}_2, \cdots, \boldsymbol{\beta}_k) = \mathbf{D}_\sigma\mathbf{A}' = (\mathbf{D}_\sigma\boldsymbol{\alpha}_1, \mathbf{D}_\sigma\boldsymbol{\alpha}_2, \cdots, \mathbf{D}_\sigma\boldsymbol{\alpha}_k)$. The linear discriminant functions $y_1, y_2, \cdots, y_k$ such that $\mathbf{y} = \mathbf{Bx}$ are called principal components by Hotelling (1933) [cf. Anderson (1958, pp. 272–279), Girshick (1936)]. Since $\operatorname{tr} \mathbf{P} = \lambda_1 + \cdots + \lambda_k = k$, here (see problem 10.7)

$$\begin{aligned} I(1:2) &= -\tfrac{1}{2}\log|\mathbf{P}| \\ &= -\tfrac{1}{2}\log(1-\rho_{1\cdot 23\cdots k}^2)(1-\rho_{2\cdot 3\cdots k}^2)\cdots(1-\rho_{k-1,k}^2) \\ &= -\tfrac{1}{2}\log\lambda_1 - \tfrac{1}{2}\log\lambda_2 - \cdots - \tfrac{1}{2}\log\lambda_k, \end{aligned} \tag{6.12}$$

$$\begin{aligned} J(1,2) &= \frac{1}{2}\operatorname{tr}\mathbf{P}^{-1} - \frac{k}{2} = \frac{1}{2}\sum_{i=1}^{k}\frac{\rho_{i\cdot 12\cdots(i-1)(i+1)\cdots k}^2}{1-\rho_{i\cdot 12\cdots(i-1)(i+1)\cdots k}^2} \\ &= \frac{1-\lambda_1}{2\lambda_1} + \cdots + \frac{1-\lambda_k}{2\lambda_k}, \end{aligned} \tag{6.13}$$

where $\rho_{i\cdot 12\cdots(i-1)(i+1)\cdots k}$, $i = 1, 2, \cdots, k$, $\rho_{j\cdot j+1\cdots k}$, $j = 1, 2, \cdots, k-1$, are multiple correlation coefficients in the population under H_1, and the λ's are the roots of $|\mathbf{P} - \lambda\mathbf{I}| = 0$.

Note that $I(1:2)$ in (6.12) is a measure of the joint relation among the k variates (see the remarks following (3.3) in chapter 8).

For bivariate populations in particular, we have

$$\mathbf{P} = \begin{pmatrix} 1 & \rho \\ \rho & 1 \end{pmatrix}, \tag{6.14}$$

$$I(1:2) = -\tfrac{1}{2}\log(1-\rho^2), \qquad J(1,2) = \rho^2/(1-\rho^2), \tag{6.15}$$

$$|\mathbf{P} - \lambda\mathbf{I}| = \lambda^2 - 2\lambda + 1 - \rho^2 = 0, \tag{6.16}$$

$$\lambda_1 = 1 - \rho, \qquad \lambda_2 = 1 + \rho, \qquad \rho > 0, \tag{6.17}$$

$$(6.18)\quad \boldsymbol{\beta}_1' = \left(\frac{1}{\sqrt{2}}, -\frac{1}{\sqrt{2}}\right), \quad \boldsymbol{\beta}_2' = \left(\frac{1}{\sqrt{2}}, \frac{1}{\sqrt{2}}\right), \quad \mathbf{B} = \begin{pmatrix} \frac{1}{\sqrt{2}} & -\frac{1}{\sqrt{2}} \\ \frac{1}{\sqrt{2}} & \frac{1}{\sqrt{2}} \end{pmatrix},$$

$$(6.19)\quad y_1 = (x_1 - x_2)/\sqrt{2}, \qquad y_2 = (x_1 + x_2)/\sqrt{2},$$

$$(6.20)\quad \begin{aligned} I(1:2; y_1) &= -\tfrac{1}{2}\log(1-\rho) - (\rho/2), \\ I(1:2; y_2) &= -\tfrac{1}{2}\log(1+\rho) + (\rho/2), \end{aligned}$$

$$(6.21)\quad J(1, 2; y_1) = \rho^2/2(1-\rho), \qquad J(1, 2; y_2) = \rho^2/2(1+\rho).$$

Note that for $\rho > 0$, the most informative and most divergent linear discriminant function is $y_1 = (x_1 - x_2)/\sqrt{2}$, since $\log[(1+\rho)/(1-\rho)] > 2\rho$ and $\lambda_1\lambda_2 = 1 - \rho^2 < 1$ [or see (6.20) and (6.21)].

7. CANONICAL CORRELATION

[Cf. Anderson (1958, pp. 288–298).] We now want to examine a partitioning of the k variates into two sets, $\mathbf{x}' = (\mathbf{x}_1', \mathbf{x}_2')$, $\mathbf{x}_1' = (x_1, x_2, \cdots, x_{k_1})$, $\mathbf{x}_2' = (x_{k_1+1}, x_{k_1+2}, \cdots, x_{k_1+k_2})$. For a partitioning into more than two sets see problem 10.13 and section 3.6 of chapter 12. Assume that the populations are centered at their means, or $\boldsymbol{\delta} = \mathbf{0}$, and that

$$(7.1)\quad \boldsymbol{\Sigma}_1 = \begin{pmatrix} \boldsymbol{\Sigma}_{11} & \boldsymbol{\Sigma}_{12} \\ \boldsymbol{\Sigma}_{21} & \boldsymbol{\Sigma}_{22} \end{pmatrix}, \qquad \boldsymbol{\Sigma}_2 = \begin{pmatrix} \boldsymbol{\Sigma}_{11} & \mathbf{0} \\ \mathbf{0} & \boldsymbol{\Sigma}_{22} \end{pmatrix},$$

where $\boldsymbol{\Sigma}_{11} = (\sigma_{ij})$, $i, j = 1, 2, \cdots, k_1$,
$\boldsymbol{\Sigma}_{22} = (\sigma_{rs})$, $r, s = k_1 + 1, \cdots, k_1 + k_2 = k$,
$\boldsymbol{\Sigma}_{12} = (\sigma_{is})$, $\boldsymbol{\Sigma}_{21} = \boldsymbol{\Sigma}_{12}'$,

that is, the two sets are independent in the population under H_2.

Since, as may be verified ($\mathbf{I}_{k_1}$ is the identity matrix of order k_1, etc.),

$$(7.2)\quad \begin{pmatrix} \mathbf{I}_{k_1} & \mathbf{0} \\ -\boldsymbol{\Sigma}_{21}\boldsymbol{\Sigma}_{11}^{-1} & \mathbf{I}_{k_2} \end{pmatrix} \begin{pmatrix} \boldsymbol{\Sigma}_{11} & \boldsymbol{\Sigma}_{12} \\ \boldsymbol{\Sigma}_{21} & \boldsymbol{\Sigma}_{22} \end{pmatrix} \begin{pmatrix} \mathbf{I}_{k_1} & -\boldsymbol{\Sigma}_{11}^{-1}\boldsymbol{\Sigma}_{12} \\ \mathbf{0} & \mathbf{I}_{k_2} \end{pmatrix} = \begin{pmatrix} \boldsymbol{\Sigma}_{11} & \mathbf{0} \\ \mathbf{0} & \boldsymbol{\Sigma}_{22\cdot 1} \end{pmatrix},$$

where $\boldsymbol{\Sigma}_{22\cdot 1} = \boldsymbol{\Sigma}_{22} - \boldsymbol{\Sigma}_{21}\boldsymbol{\Sigma}_{11}^{-1}\boldsymbol{\Sigma}_{12}$, we have

$$(7.3)\quad \begin{aligned} \begin{pmatrix} \boldsymbol{\Sigma}_{11} & \boldsymbol{\Sigma}_{12} \\ \boldsymbol{\Sigma}_{21} & \boldsymbol{\Sigma}_{22} \end{pmatrix}^{-1} &= \begin{pmatrix} \mathbf{I}_{k_1} & -\boldsymbol{\Sigma}_{11}^{-1}\boldsymbol{\Sigma}_{12} \\ \mathbf{0} & \mathbf{I}_{k_2} \end{pmatrix} \begin{pmatrix} \boldsymbol{\Sigma}_{11}^{-1} & \mathbf{0} \\ \mathbf{0} & \boldsymbol{\Sigma}_{22\cdot 1}^{-1} \end{pmatrix} \begin{pmatrix} \mathbf{I}_{k_1} & \mathbf{0} \\ -\boldsymbol{\Sigma}_{21}\boldsymbol{\Sigma}_{11}^{-1} & \mathbf{I}_{k_2} \end{pmatrix} \\ &= \begin{pmatrix} \boldsymbol{\Sigma}_{11}^{-1} + \boldsymbol{\Sigma}_{11}^{-1}\boldsymbol{\Sigma}_{12}\boldsymbol{\Sigma}_{22\cdot 1}^{-1}\boldsymbol{\Sigma}_{21}\boldsymbol{\Sigma}_{11}^{-1} & -\boldsymbol{\Sigma}_{11}^{-1}\boldsymbol{\Sigma}_{12}\boldsymbol{\Sigma}_{22\cdot 1}^{-1} \\ -\boldsymbol{\Sigma}_{22\cdot 1}^{-1}\boldsymbol{\Sigma}_{21}\boldsymbol{\Sigma}_{11}^{-1} & \boldsymbol{\Sigma}_{22\cdot 1}^{-1} \end{pmatrix}. \end{aligned}$$

Note that the matrix $\begin{pmatrix} \mathbf{I}_{k_1} & \mathbf{0} \\ -\mathbf{\Sigma}_{21}\mathbf{\Sigma}_{11}^{-1} & \mathbf{I}_{k_2} \end{pmatrix}$ in (7.2) is that of a nonsingular linear transformation, and (7.2) implies that in the population under H_1, $\mathbf{x}_1$ and $\mathbf{x}_2 - \mathbf{\Sigma}_{21}\mathbf{\Sigma}_{11}^{-1}\mathbf{x}_1$ are independent with covariance matrix the right-hand side of (7.2) (see section 3). We thus have (see problems 10.6 and 10.11)

$$(7.4)\quad I(1:2) = \frac{1}{2}\log\frac{\begin{vmatrix} \mathbf{\Sigma}_{11} & \mathbf{0} \\ \mathbf{0} & \mathbf{\Sigma}_{22} \end{vmatrix}}{\begin{vmatrix} \mathbf{\Sigma}_{11} & \mathbf{\Sigma}_{12} \\ \mathbf{\Sigma}_{21} & \mathbf{\Sigma}_{22} \end{vmatrix}} + \frac{1}{2}\operatorname{tr}\begin{pmatrix} \mathbf{\Sigma}_{11} & \mathbf{\Sigma}_{12} \\ \mathbf{\Sigma}_{21} & \mathbf{\Sigma}_{22} \end{pmatrix}\left[\begin{pmatrix} \mathbf{\Sigma}_{11}^{-1} & \mathbf{0} \\ \mathbf{0} & \mathbf{\Sigma}_{22}^{-1} \end{pmatrix} - \begin{pmatrix} \mathbf{\Sigma}_{11} & \mathbf{\Sigma}_{12} \\ \mathbf{\Sigma}_{21} & \mathbf{\Sigma}_{22} \end{pmatrix}^{-1}\right]$$

$$= \frac{1}{2}\log\frac{|\mathbf{\Sigma}_{11}|\,|\mathbf{\Sigma}_{22}|}{\begin{vmatrix} \mathbf{\Sigma}_{11} & \mathbf{\Sigma}_{12} \\ \mathbf{\Sigma}_{21} & \mathbf{\Sigma}_{22} \end{vmatrix}} = \frac{1}{2}\log\frac{|\mathbf{\Sigma}_{22}|}{|\mathbf{\Sigma}_{22\cdot 1}|},$$

a measure of the relation between the sets $\mathbf{x}_1'$ and $\mathbf{x}_2'$, or the mean information in $\mathbf{x}_1'$ about $\mathbf{x}_2'$, or in $\mathbf{x}_2'$ about $\mathbf{x}_1'$ (see example 4.3 of chapter 1),

$$(7.5)\quad J(1, 2)$$

$$= \frac{1}{2}\operatorname{tr}\left[\begin{pmatrix} \mathbf{\Sigma}_{11} & \mathbf{\Sigma}_{12} \\ \mathbf{\Sigma}_{21} & \mathbf{\Sigma}_{22} \end{pmatrix} - \begin{pmatrix} \mathbf{\Sigma}_{11} & \mathbf{0} \\ \mathbf{0} & \mathbf{\Sigma}_{22} \end{pmatrix}\right]\left[\begin{pmatrix} \mathbf{\Sigma}_{11}^{-1} & \mathbf{0} \\ \mathbf{0} & \mathbf{\Sigma}_{22}^{-1} \end{pmatrix} - \begin{pmatrix} \mathbf{\Sigma}_{11} & \mathbf{\Sigma}_{12} \\ \mathbf{\Sigma}_{21} & \mathbf{\Sigma}_{22} \end{pmatrix}^{-1}\right]$$

$$= \frac{1}{2}\operatorname{tr}\begin{pmatrix} \mathbf{0} & \mathbf{\Sigma}_{12} \\ \mathbf{\Sigma}_{21} & \mathbf{0} \end{pmatrix}\begin{pmatrix} -\mathbf{\Sigma}_{11}^{-1}\mathbf{\Sigma}_{12}\mathbf{\Sigma}_{22\cdot 1}^{-1}\mathbf{\Sigma}_{21}\mathbf{\Sigma}_{11}^{-1} & \mathbf{\Sigma}_{11}^{-1}\mathbf{\Sigma}_{12}\mathbf{\Sigma}_{22\cdot 1}^{-1} \\ \mathbf{\Sigma}_{22\cdot 1}^{-1}\mathbf{\Sigma}_{21}\mathbf{\Sigma}_{11}^{-1} & \mathbf{\Sigma}_{22}^{-1} - \mathbf{\Sigma}_{22\cdot 1}^{-1} \end{pmatrix}$$

$$= \frac{1}{2}\operatorname{tr}\begin{pmatrix} \mathbf{\Sigma}_{12}\mathbf{\Sigma}_{22\cdot 1}^{-1}\mathbf{\Sigma}_{21}\mathbf{\Sigma}_{11}^{-1} & (\cdot) \\ (\cdot) & \mathbf{\Sigma}_{21}\mathbf{\Sigma}_{11}^{-1}\mathbf{\Sigma}_{12}\mathbf{\Sigma}_{22\cdot 1}^{-1} \end{pmatrix}$$

$$= \operatorname{tr}\mathbf{\Sigma}_{21}\mathbf{\Sigma}_{11}^{-1}\mathbf{\Sigma}_{12}\mathbf{\Sigma}_{22\cdot 1}^{-1} = \operatorname{tr}\mathbf{\Sigma}_{22}\mathbf{\Sigma}_{22\cdot 1}^{-1} - k_2,$$

where $(\cdot)$ indicates matrices whose values are not needed.

To highlight the partition of the variates, we write the linear discriminant function $y = \boldsymbol{\alpha}'\mathbf{x}$ as

$$(7.6)\quad \begin{aligned} y &= \beta_1 x_1 + \cdots + \beta_{k_1} x_{k_1} + \gamma_1 x_{k_1+1} + \cdots + \gamma_{k_2} x_{k_1+k_2} \\ &= \boldsymbol{\beta}'\mathbf{x}_1 + \boldsymbol{\gamma}'\mathbf{x}_2, \end{aligned}$$

where $\boldsymbol{\beta}$ and $\boldsymbol{\gamma}$ are respectively the one-column matrices of $\beta_1, \cdots, \beta_{k_1}$, and $\gamma_1, \cdots, \gamma_{k_2}$, $\boldsymbol{\alpha}' = (\boldsymbol{\beta}', \boldsymbol{\gamma}')$, $\mathbf{x}' = (\mathbf{x}_1', \mathbf{x}_2')$. Now, (6.4) and (6.5) are

$$(7.7)\quad \begin{pmatrix} \mathbf{\Sigma}_{11} & \mathbf{\Sigma}_{12} \\ \mathbf{\Sigma}_{21} & \mathbf{\Sigma}_{22} \end{pmatrix}\begin{pmatrix} \boldsymbol{\beta} \\ \boldsymbol{\gamma} \end{pmatrix} = \lambda\begin{pmatrix} \mathbf{\Sigma}_{11} & \mathbf{0} \\ \mathbf{0} & \mathbf{\Sigma}_{22} \end{pmatrix}\begin{pmatrix} \boldsymbol{\beta} \\ \boldsymbol{\gamma} \end{pmatrix},$$

$$(7.8)\qquad \begin{vmatrix} (1-\lambda)\mathbf{\Sigma}_{11} & \mathbf{\Sigma}_{12} \\ \mathbf{\Sigma}_{21} & (1-\lambda)\mathbf{\Sigma}_{22} \end{vmatrix} = 0.$$

Since (7.7) is equivalent to

$$(7.9)\qquad \begin{aligned} \mathbf{\Sigma}_{11}\boldsymbol{\beta} + \mathbf{\Sigma}_{12}\boldsymbol{\gamma} &= \lambda\mathbf{\Sigma}_{11}\boldsymbol{\beta} \\ \mathbf{\Sigma}_{21}\boldsymbol{\beta} + \mathbf{\Sigma}_{22}\boldsymbol{\gamma} &= \lambda\mathbf{\Sigma}_{22}\boldsymbol{\gamma}, \end{aligned}$$

or

$$(7.10)\qquad \begin{aligned} \boldsymbol{\beta} &= -\frac{1}{1-\lambda}\mathbf{\Sigma}_{11}^{-1}\mathbf{\Sigma}_{12}\boldsymbol{\gamma} \\ \mathbf{0} &= -\frac{1}{1-\lambda}\mathbf{\Sigma}_{21}\mathbf{\Sigma}_{11}^{-1}\mathbf{\Sigma}_{12}\boldsymbol{\gamma} + (1-\lambda)\mathbf{\Sigma}_{22}\boldsymbol{\gamma}, \end{aligned}$$

(7.8) is equivalent to

$$(7.11)\qquad |\mathbf{\Sigma}_{21}\mathbf{\Sigma}_{11}^{-1}\mathbf{\Sigma}_{12} - \rho^2\mathbf{\Sigma}_{22}| = 0,$$

where $\rho^2 = (1-\lambda)^2$. The roots of (7.8) [see (6.5)] are real and positive. If $k_2 \le k_1$, since $k = k_1 + k_2$, and the determinant of (7.11) is of order k_2,

$$(7.12)\qquad \begin{aligned} &\lambda_i = 1 - \rho_i, \qquad \lambda_{k_1+i} = 1 + \rho_{k_2+1-i}, \qquad i = 1, 2, \cdots, k_2, \\ &\lambda_{k_2+1} = \cdots = \lambda_{k_2+(k_1-k_2)} = 1, \end{aligned}$$

where $\rho_1 \ge \rho_2 \ge \cdots \ge \rho_{k_2}$. Note that $-1 \le \rho_i \le 1$ since the λ's cannot be negative. Hotelling (1936) called the ρ_i canonical correlations. For the associated linear discriminant functions [see (6.6) and (6.7)], we now have

$$(7.13)\qquad \begin{aligned} &I(1:2; y_i) = -\tfrac{1}{2}\log(1-\rho_i) - \frac{\rho_i}{2}, \\ &I(1:2; y_{k_1+i}) = -\tfrac{1}{2}\log(1+\rho_{k_2+1-i}) + \frac{\rho_{k_2+1-i}}{2}, \\ &I(1:2; y_j) = 0, \qquad i = 1, 2, \cdots, k_2, \qquad j = k_2 + 1, \cdots, \\ &\qquad\qquad\qquad\qquad\qquad\qquad\qquad k_2 + (k_1 - k_2), \end{aligned}$$

$$(7.14)\qquad \begin{aligned} &J(1, 2; y_i) = \frac{1-\rho_i}{2} + \frac{1}{2(1-\rho_i)} - 1 = \frac{1}{2}\frac{\rho_i^2}{1-\rho_i}, \\ &J(1, 2; y_{k_1+i}) = \tfrac{1}{2}(1+\rho_{k_2+1-i}) + \frac{1}{2(1+\rho_{k_2+1-i})} - 1 = \frac{\rho_{k_2+1-i}^2}{2(1+\rho_{k_2+1-i})}, \\ &J(1, 2; y_j) = \tfrac{1}{2} + \tfrac{1}{2} - 1 = 0, \qquad i = 1, 2, \cdots, k_2, \\ &\qquad\qquad\qquad j = k_2 + 1, \cdots, k_2 + (k_1 - k_2), \end{aligned}$$

from which we see that

$$(7.15)\quad I(1:2;y_i) + I(1:2;y_{k+1-i}) = -\tfrac{1}{2}\log(1-\rho_i^2), \quad i = 1, 2, \cdots, k_2,$$
$$J(1,2;y_i) + J(1,2;y_{k+1-i}) = \rho_i^2/(1-\rho_i^2),$$

and

$$(7.16)\quad I(1:2) = \frac{1}{2}\log\frac{|\mathbf{\Sigma}_{11}||\mathbf{\Sigma}_{22}|}{\begin{vmatrix}\mathbf{\Sigma}_{11} & \mathbf{\Sigma}_{12}\\ \mathbf{\Sigma}_{21} & \mathbf{\Sigma}_{22}\end{vmatrix}} = -\tfrac{1}{2}\log(1-\rho_1^2)(1-\rho_2^2)\cdots(1-\rho_{k_2}^2),$$

$$J(1,2) = \operatorname{tr}\mathbf{\Sigma}_{22}\mathbf{\Sigma}_{22\cdot 1}^{-1} - k_2 = \frac{\rho_1^2}{1-\rho_1^2} + \frac{\rho_2^2}{1-\rho_2^2} + \cdots + \frac{\rho_{k_2}^2}{1-\rho_{k_2}^2}.$$

Since $\log[(1+\rho_1)/(1-\rho_1)] > 2\rho_1$ and $\lambda_1\lambda_k = (1-\rho_1)(1+\rho_1) = 1-\rho_1^2 < 1$, the most informative and most divergent linear discriminant function (7.6) is associated with the root λ_1 or the largest canonical correlation.

Note that for bivariate populations, $k = 2$, $k_1 = k_2 = 1$, (7.11) becomes $(\sigma_{21}\sigma_{12}/\sigma_{11} - \rho^2\sigma_{22}) = 0$, or the canonical correlation is the simple correlation between the variates, and [see (6.15)]

$$(7.17)\quad I(1:2) = -\tfrac{1}{2}\log(1-\rho^2),$$
$$J(1,2) = \rho^2/(1-\rho^2).$$

For k-variate populations with $k_1 = k-1$, $k_2 = 1$, (7.11) yields the canonical correlation $\rho^2 = \mathbf{\Sigma}_{21}\mathbf{\Sigma}_{11}^{-1}\mathbf{\Sigma}_{12}/\sigma_{kk}$. But now $\mathbf{\Sigma}_{21} = (\sigma_{k1}, \sigma_{k2}, \cdots, \sigma_{k\,k-1})$ or $\rho^2 = 1 - \dfrac{|\mathbf{\Sigma}_1|}{\sigma_{kk}|\mathbf{\Sigma}_{11}|}$ [see (7.1)], and thus the canonical correlation is the multiple correlation between x_k and the other variates [cf. Cramér (1946a), pp. 109, 308)], and

$$(7.18)\quad I(1:2) = -\tfrac{1}{2}\log(1-\rho_{k\cdot 12\cdots(k-1)}^2),$$
$$J(1,2) = \rho_{k\cdot 12\cdots(k-1)}^2/(1-\rho_{k\cdot 12\cdots(k-1)}^2).$$

Instead of a single linear discriminant function, suppose we examine the pair of linear discriminant functions

$$(7.19)\quad \begin{cases} u = \beta_1 x_1 + \cdots + \beta_{k_1}x_{k_1} = \boldsymbol{\beta}'\mathbf{x}_1\\ v = \gamma_1 x_{k_1+1} + \cdots + \gamma_{k_2}x_{k_1+k_2} = \boldsymbol{\gamma}'\mathbf{x}_2.\end{cases}$$

We have

$$\operatorname{var}_i(u) = \boldsymbol{\beta}'\mathbf{\Sigma}_{11}\boldsymbol{\beta}, \qquad \operatorname{var}_i(v) = \boldsymbol{\gamma}'\mathbf{\Sigma}_{22}\boldsymbol{\gamma}, \qquad i = 1, 2,$$
$$\operatorname{cov}_1(u,v) = \boldsymbol{\beta}'\mathbf{\Sigma}_{12}\boldsymbol{\gamma}, \qquad \operatorname{cov}_2(u,v) = 0,$$

$$(7.20)\qquad I(1:2;u,v) = \frac{1}{2}\log\frac{\boldsymbol{\beta}'\boldsymbol{\Sigma}_{11}\boldsymbol{\beta}\,\boldsymbol{\gamma}'\boldsymbol{\Sigma}_{22}\boldsymbol{\gamma}}{\begin{vmatrix}\boldsymbol{\beta}'\boldsymbol{\Sigma}_{11}\boldsymbol{\beta} & \boldsymbol{\beta}'\boldsymbol{\Sigma}_{12}\boldsymbol{\gamma}\\ \boldsymbol{\gamma}'\boldsymbol{\Sigma}_{21}\boldsymbol{\beta} & \boldsymbol{\gamma}'\boldsymbol{\Sigma}_{22}\boldsymbol{\gamma}\end{vmatrix}} = -\tfrac{1}{2}\log(1-\rho_{uv}^2),$$

$$J(1,2;u,v) = \frac{(\boldsymbol{\beta}'\boldsymbol{\Sigma}_{12}\boldsymbol{\gamma})^2}{(\boldsymbol{\beta}'\boldsymbol{\Sigma}_{11}\boldsymbol{\beta})(\boldsymbol{\gamma}'\boldsymbol{\Sigma}_{22}\boldsymbol{\gamma}) - (\boldsymbol{\beta}'\boldsymbol{\Sigma}_{12}\boldsymbol{\gamma})^2} = \frac{\rho_{uv}^2}{1-\rho_{uv}^2}.$$

The values of $\boldsymbol{\beta}$ and $\boldsymbol{\gamma}$ which maximize $I(1:2;u,v)$ [or $J(1,2;u,v)$] in (7.20) satisfy (by the usual calculus procedures) (7.9), where $(1-\lambda)^2 = \rho_{uv}^2$. The canonical correlations are thus the correlations of the pair of linear discriminant functions (7.19). From (7.15) we have

$$(7.21)\qquad I(1:2;u_i,v_i) = I(1:2;y_i) + I(1:2;y_{k+1-i}), \qquad i = 1, 2, \cdots, k_2,$$

$$J(1,2;u_i,v_i) = J(1,2;y_i) + J(1,2;y_{k+1-i}).$$

The discrimination information and divergence for the pairs of linear discriminant functions thus define an ordering according to the values of the canonical correlations.

8. COVARIANCE VARIATES

Two k-variate normal populations with the same covariance matrices may differ only in the means of the last k_2 variates. The first $k - k_2 = k_1$ variates are then called the covariance variates, and we shall now find the discrimination information provided by using the covariance variates also, compared with that provided by using only the last k_2 variates [cf. Cochran and Bliss (1948)].

We have the partition $\mathbf{x}' = (\mathbf{x}_1', \mathbf{x}_2')$, $\boldsymbol{\mu}_1' - \boldsymbol{\mu}_2' = \boldsymbol{\delta}' = (\boldsymbol{\delta}_1', \boldsymbol{\delta}_2')$, with $\boldsymbol{\delta}_1 = \mathbf{0}$, and $\boldsymbol{\Sigma}_1 = \boldsymbol{\Sigma}_2 = \boldsymbol{\Sigma} = \begin{pmatrix}\boldsymbol{\Sigma}_{11} & \boldsymbol{\Sigma}_{12}\\ \boldsymbol{\Sigma}_{21} & \boldsymbol{\Sigma}_{22}\end{pmatrix}$.

From (5.1) and (7.3) we now have

$$(8.1)\qquad 2I(1:2;\mathbf{x}') = J(1,2;\mathbf{x}') = \operatorname{tr}\boldsymbol{\Sigma}^{-1}\boldsymbol{\delta}\boldsymbol{\delta}'$$

$$= \operatorname{tr}\begin{pmatrix}\boldsymbol{\Sigma}_{11} & \boldsymbol{\Sigma}_{12}\\ \boldsymbol{\Sigma}_{21} & \boldsymbol{\Sigma}_{22}\end{pmatrix}^{-1}\begin{pmatrix}\mathbf{0}\\ \boldsymbol{\delta}_2\end{pmatrix}(\mathbf{0}\quad \boldsymbol{\delta}_2')$$

$$= \operatorname{tr}\boldsymbol{\Sigma}_{22\cdot 1}^{-1}\boldsymbol{\delta}_2\boldsymbol{\delta}_2' = \boldsymbol{\delta}_2'\boldsymbol{\Sigma}_{22\cdot 1}^{-1}\boldsymbol{\delta}_2.$$

On the other hand, with the last k_2 variates only:

$$(8.2)\qquad 2I(1:2;\mathbf{x}_2') = J(1,2;\mathbf{x}_2') = \operatorname{tr}\boldsymbol{\Sigma}_{22}^{-1}\boldsymbol{\delta}_2\boldsymbol{\delta}_2' = \boldsymbol{\delta}_2'\boldsymbol{\Sigma}_{22}^{-1}\boldsymbol{\delta}_2.$$

Since $I(1:2; \mathbf{x}') > I(1:2; \mathbf{x}_2')$ (see sections 3 and 4 of chapter 2), the contribution of the covariance variates is

$$\boldsymbol{\delta}_2'\boldsymbol{\Sigma}_{22\cdot1}^{-1}\boldsymbol{\delta}_2 - \boldsymbol{\delta}_2'\boldsymbol{\Sigma}_{22}^{-1}\boldsymbol{\delta}_2, \tag{8.3}$$

and the gain ratio is

$$\lambda = \frac{\boldsymbol{\delta}_2'\boldsymbol{\Sigma}_{22\cdot1}^{-1}\boldsymbol{\delta}_2}{\boldsymbol{\delta}_2'\boldsymbol{\Sigma}_{22}^{-1}\boldsymbol{\delta}_2}, \tag{8.4}$$

where λ lies between the smallest and largest root of the determinantal equation

$$|\boldsymbol{\Sigma}_{22\cdot1}^{-1} - \lambda\boldsymbol{\Sigma}_{22}^{-1}| = 0 = |\boldsymbol{\Sigma}_{22} - \lambda\boldsymbol{\Sigma}_{22\cdot1}|, \tag{8.5}$$

$$|\boldsymbol{\Sigma}_{21}\boldsymbol{\Sigma}_{11}^{-1}\boldsymbol{\Sigma}_{12} - \rho^2\boldsymbol{\Sigma}_{22}| = 0, \tag{8.6}$$

where $\rho^2 = (\lambda - 1)/\lambda$. The roots of (8.6) are the canonical correlations [see (7.11)]; hence the largest value of λ in (8.4) cannot exceed $1/(1 - \rho_1^2)$, where ρ_1 is the largest canonical correlation.

We now study the linear discriminant functions with and without the covariance variates. Since the covariance matrices in the populations are equal, a unique sufficient linear discriminant function exists (see section 5). For all the k variates with the partitioning of the coefficients of the linear discriminant function as in (7.6), $\boldsymbol{\alpha} = \boldsymbol{\Sigma}^{-1}\boldsymbol{\delta}$ becomes [see (7.3)]

$$\begin{pmatrix}\boldsymbol{\beta}\\ \boldsymbol{\gamma}\end{pmatrix} = \begin{pmatrix}\boldsymbol{\Sigma}_{11} & \boldsymbol{\Sigma}_{12}\\ \boldsymbol{\Sigma}_{21} & \boldsymbol{\Sigma}_{22}\end{pmatrix}^{-1}\begin{pmatrix}\mathbf{0}\\ \boldsymbol{\delta}_2\end{pmatrix} = \begin{pmatrix}-\boldsymbol{\Sigma}_{11}^{-1}\boldsymbol{\Sigma}_{12}\boldsymbol{\Sigma}_{22\cdot1}^{-1}\boldsymbol{\delta}_2\\ \boldsymbol{\Sigma}_{22\cdot1}^{-1}\boldsymbol{\delta}_2\end{pmatrix}. \tag{8.7}$$

If the covariance variates are ignored, the coefficients of the linear discriminant function are

$$\boldsymbol{\beta} = \mathbf{0}, \qquad \boldsymbol{\gamma} = \boldsymbol{\Sigma}_{22}^{-1}\boldsymbol{\delta}_2. \tag{8.8}$$

For bivariate populations, $k = 2$, $k_1 = k_2 = 1$, the canonical correlation is the simple correlation between the variates, $\boldsymbol{\Sigma}_{22\cdot1} = \sigma_2^2(1 - \rho^2)$, and (8.4) becomes $\lambda = 1/(1 - \rho^2)$. For k-variate populations with $k_2 = 1$, $k_1 = k - 1$, there is only one canonical correlation (the multiple correlation of x_k with $x_1, x_2, \cdots, x_{k-1}$), $\boldsymbol{\Sigma}_{22\cdot1} = \sigma_k^2(1 - \rho_{k\cdot12\cdots(k-1)}^2)$, and (8.4) becomes $\lambda = 1/(1 - \rho_{k\cdot12\cdots(k-1)}^2)$. (See problem 10.9.)

9. GENERAL CASE

[Cf. Greenhouse (1954).] With no restrictive assumptions about the means and covariance matrices of the k-variate normal populations under H_1 and H_2, the parameters of the normal distributions of the linear discriminant function $y = \boldsymbol{\alpha}'\mathbf{x}$ are:

$$E_1(y) = \boldsymbol{\alpha}'\boldsymbol{\mu}_1, \quad E_2(y) = \boldsymbol{\alpha}'\boldsymbol{\mu}_2, \quad \text{var}_1(y) = \boldsymbol{\alpha}'\boldsymbol{\Sigma}_1\boldsymbol{\alpha}, \quad \text{var}_2(y) = \boldsymbol{\alpha}'\boldsymbol{\Sigma}_2\boldsymbol{\alpha}, \tag{9.1}$$

and

$$(9.2)\quad I(1:2;y) = \frac{1}{2}\log\frac{\boldsymbol{\alpha}'\boldsymbol{\Sigma}_2\boldsymbol{\alpha}}{\boldsymbol{\alpha}'\boldsymbol{\Sigma}_1\boldsymbol{\alpha}} - \frac{1}{2} + \frac{1}{2}\frac{\boldsymbol{\alpha}'\boldsymbol{\Sigma}_1\boldsymbol{\alpha}}{\boldsymbol{\alpha}'\boldsymbol{\Sigma}_2\boldsymbol{\alpha}} + \frac{1}{2}\frac{\boldsymbol{\alpha}'\boldsymbol{\delta}\boldsymbol{\delta}'\boldsymbol{\alpha}}{\boldsymbol{\alpha}'\boldsymbol{\Sigma}_2\boldsymbol{\alpha}},$$

$$(9.3)\quad I(2:1;y) = \frac{1}{2}\log\frac{\boldsymbol{\alpha}'\boldsymbol{\Sigma}_1\boldsymbol{\alpha}}{\boldsymbol{\alpha}'\boldsymbol{\Sigma}_2\boldsymbol{\alpha}} - \frac{1}{2} + \frac{1}{2}\frac{\boldsymbol{\alpha}'\boldsymbol{\Sigma}_2\boldsymbol{\alpha}}{\boldsymbol{\alpha}'\boldsymbol{\Sigma}_1\boldsymbol{\alpha}} + \frac{1}{2}\frac{\boldsymbol{\alpha}'\boldsymbol{\delta}\boldsymbol{\delta}'\boldsymbol{\alpha}}{\boldsymbol{\alpha}'\boldsymbol{\Sigma}_1\boldsymbol{\alpha}},$$

$$(9.4)\quad J(1,2;y) = \frac{1}{2}\frac{\boldsymbol{\alpha}'\boldsymbol{\Sigma}_2\boldsymbol{\alpha}}{\boldsymbol{\alpha}'\boldsymbol{\Sigma}_1\boldsymbol{\alpha}} + \frac{1}{2}\frac{\boldsymbol{\alpha}'\boldsymbol{\Sigma}_1\boldsymbol{\alpha}}{\boldsymbol{\alpha}'\boldsymbol{\Sigma}_2\boldsymbol{\alpha}} - 1 + \frac{1}{2}\left(\frac{1}{\boldsymbol{\alpha}'\boldsymbol{\Sigma}_1\boldsymbol{\alpha}} + \frac{1}{\boldsymbol{\alpha}'\boldsymbol{\Sigma}_2\boldsymbol{\alpha}}\right)\boldsymbol{\alpha}'\boldsymbol{\delta}\boldsymbol{\delta}'\boldsymbol{\alpha}.$$

For a given y it is true that $J(1,2;y) = I(1:2;y) + I(2:1;y)$. It is not true, however, that the same y will yield the maximum value for $I(1:2;y)$, $I(2:1;y)$, $J(1,2;y)$.

The value of $\boldsymbol{\alpha}$ for which $I(1:2;y)$ in (9.2) is a maximum satisfies (by the usual calculus procedures)

$$(9.5)\quad \boldsymbol{\Sigma}_1\boldsymbol{\alpha} - \lambda\boldsymbol{\Sigma}_2\boldsymbol{\alpha} = \gamma\boldsymbol{\delta},$$

where

$$\lambda = \frac{\boldsymbol{\alpha}'\boldsymbol{\Sigma}_1\boldsymbol{\alpha}}{\boldsymbol{\alpha}'\boldsymbol{\Sigma}_2\boldsymbol{\alpha}}\left(1 - \frac{(\boldsymbol{\alpha}'\boldsymbol{\delta})^2}{\boldsymbol{\alpha}'\boldsymbol{\Sigma}_2\boldsymbol{\alpha} - \boldsymbol{\alpha}'\boldsymbol{\Sigma}_1\boldsymbol{\alpha}}\right), \qquad \gamma = \frac{(\boldsymbol{\alpha}'\boldsymbol{\delta})(\boldsymbol{\alpha}'\boldsymbol{\Sigma}_1\boldsymbol{\alpha})}{\boldsymbol{\alpha}'\boldsymbol{\Sigma}_2\boldsymbol{\alpha} - \boldsymbol{\alpha}'\boldsymbol{\Sigma}_1\boldsymbol{\alpha}}.$$

Since γ is a proportionality factor, we may set $\gamma = 1$, and $\boldsymbol{\alpha}$ satisfies

$$(9.6)\quad \boldsymbol{\Sigma}_1\boldsymbol{\alpha} - \lambda\boldsymbol{\Sigma}_2\boldsymbol{\alpha} = \boldsymbol{\delta},$$

where λ, given in (9.5), must not be a root of $|\boldsymbol{\Sigma}_1 - \lambda\boldsymbol{\Sigma}_2| = 0$.

The value of $\boldsymbol{\alpha}$ for which $I(2:1;y)$ in (9.3) is a maximum satisfies (by the usual calculus procedures) an equation of the same form as (9.5) but with

$$(9.7)\quad \lambda = \frac{\boldsymbol{\alpha}'\boldsymbol{\Sigma}_1\boldsymbol{\alpha}(\boldsymbol{\alpha}'\boldsymbol{\Sigma}_1\boldsymbol{\alpha} - \boldsymbol{\alpha}'\boldsymbol{\Sigma}_2\boldsymbol{\alpha})}{\boldsymbol{\alpha}'\boldsymbol{\Sigma}_2\boldsymbol{\alpha}(\boldsymbol{\alpha}'\boldsymbol{\Sigma}_1\boldsymbol{\alpha} - \boldsymbol{\alpha}'\boldsymbol{\Sigma}_2\boldsymbol{\alpha} - (\boldsymbol{\alpha}'\boldsymbol{\delta})^2)}, \qquad \gamma = \frac{(\boldsymbol{\alpha}'\boldsymbol{\Sigma}_1\boldsymbol{\alpha})(\boldsymbol{\alpha}'\boldsymbol{\delta})}{\boldsymbol{\alpha}'\boldsymbol{\Sigma}_2\boldsymbol{\alpha} - \boldsymbol{\alpha}'\boldsymbol{\Sigma}_1\boldsymbol{\alpha} + (\boldsymbol{\alpha}'\boldsymbol{\delta})^2}.$$

Again setting the proportionality factor $\gamma = 1$, $\boldsymbol{\alpha}$ must satisfy an equation of the form (9.6) where λ, given in (9.7), must not be a root of $|\boldsymbol{\Sigma}_1 - \lambda\boldsymbol{\Sigma}_2| = 0$.

The value of $\boldsymbol{\alpha}$ for which $J(1,2;y)$ in (9.4) is a maximum satisfies (by the usual calculus procedures) an equation of the same form as (9.5) but with

$$(9.8)\quad \lambda = \frac{\boldsymbol{\alpha}'\boldsymbol{\Sigma}_1\boldsymbol{\alpha}((\boldsymbol{\alpha}'\boldsymbol{\Sigma}_2\boldsymbol{\alpha})^2 - (\boldsymbol{\alpha}'\boldsymbol{\Sigma}_1\boldsymbol{\alpha})^2 - (\boldsymbol{\alpha}'\boldsymbol{\delta})^2(\boldsymbol{\alpha}'\boldsymbol{\Sigma}_1\boldsymbol{\alpha}))}{\boldsymbol{\alpha}'\boldsymbol{\Sigma}_2\boldsymbol{\alpha}((\boldsymbol{\alpha}'\boldsymbol{\Sigma}_2\boldsymbol{\alpha})^2 - (\boldsymbol{\alpha}'\boldsymbol{\Sigma}_1\boldsymbol{\alpha})^2 + (\boldsymbol{\alpha}'\boldsymbol{\delta})^2(\boldsymbol{\alpha}'\boldsymbol{\Sigma}_2\boldsymbol{\alpha}))},$$

$$\gamma = \frac{(\boldsymbol{\alpha}'\boldsymbol{\delta})(\boldsymbol{\alpha}'\boldsymbol{\Sigma}_1\boldsymbol{\alpha})(\boldsymbol{\alpha}'\boldsymbol{\Sigma}_1\boldsymbol{\alpha} + \boldsymbol{\alpha}'\boldsymbol{\Sigma}_2\boldsymbol{\alpha})}{(\boldsymbol{\alpha}'\boldsymbol{\Sigma}_2\boldsymbol{\alpha})^2 - (\boldsymbol{\alpha}'\boldsymbol{\Sigma}_1\boldsymbol{\alpha})^2 + (\boldsymbol{\alpha}'\boldsymbol{\delta})^2(\boldsymbol{\alpha}'\boldsymbol{\Sigma}_2\boldsymbol{\alpha})}.$$

Again setting the proportionality factor $\gamma = 1$, $\boldsymbol{\alpha}$ must satisfy an equation of the form (9.6) where λ, given in (9.8), must not be a root of $|\boldsymbol{\Sigma}_1 - \lambda\boldsymbol{\Sigma}_2| = 0$.

Note that here we find three types of linear discriminant functions. Since λ depends on $\boldsymbol{\alpha}$, an iterative procedure must be employed to solve for $\boldsymbol{\alpha}$. This is studied in chapter 13.

10. PROBLEMS

10.1. Show that $\mathbf{x}'\boldsymbol{\Sigma}^{-1}\mathbf{x} = \operatorname{tr} \boldsymbol{\Sigma}^{-1}\mathbf{x}\mathbf{x}'$.

10.2. If $\dfrac{\mathbf{d}}{\mathbf{d}\boldsymbol{\alpha}} = \begin{pmatrix} \dfrac{\partial}{\partial \alpha_1} \\ \cdot \\ \cdot \\ \cdot \\ \dfrac{\partial}{\partial \alpha_k} \end{pmatrix}$, show that $\dfrac{\mathbf{d}}{\mathbf{d}\boldsymbol{\alpha}}\boldsymbol{\alpha}'\boldsymbol{\Sigma}\boldsymbol{\alpha} = 2\boldsymbol{\Sigma}\boldsymbol{\alpha}$, where $\boldsymbol{\Sigma}$ is a symmetric $k \times k$ matrix and $\boldsymbol{\alpha}' = (\alpha_1, \alpha_2, \cdots, \alpha_k)$. [Cf. Anderson (1958, p. 347).]

10.3. If $d\mathbf{A}$ denotes the matrix each element of which is the differential of the corresponding element of the matrix $\mathbf{A}$, show that

(*a*) $d \operatorname{tr} \boldsymbol{\Sigma} = \operatorname{tr} d\boldsymbol{\Sigma}$.

(*b*) $d\boldsymbol{\Sigma}^{-1} = -\boldsymbol{\Sigma}^{-1}\, d\boldsymbol{\Sigma}\boldsymbol{\Sigma}^{-1}$.

(*c*) $d \log |\boldsymbol{\Sigma}| = \operatorname{tr} \boldsymbol{\Sigma}^{-1}\, d\boldsymbol{\Sigma}$.

[Cf. Dwyer and MacPhail (1948).]

10.4. Show that (see section 5)

$$\left|1 - \frac{1}{\lambda}\boldsymbol{\delta}'\boldsymbol{\Sigma}^{-1}\boldsymbol{\delta}\right| \cdot |\lambda\boldsymbol{\Sigma}| = \begin{vmatrix} 1 & \boldsymbol{\delta}' \\ \boldsymbol{\delta} & \lambda\boldsymbol{\Sigma} \end{vmatrix} = |\lambda\boldsymbol{\Sigma} - \boldsymbol{\delta}\boldsymbol{\delta}'|.$$

10.5. Show that $\operatorname{tr} \mathbf{A}\mathbf{B}' = \operatorname{tr} \mathbf{B}'\mathbf{A} = \sum_{i=1}^{k}\sum_{j=1}^{k} a_{ij}b_{ij}$, $\mathbf{A} = (a_{ij})$, $\mathbf{B} = (b_{ij})$, $i, j = 1, 2, \cdots, k$.

10.6. Show that (see section 7)

$$\begin{vmatrix} \boldsymbol{\Sigma}_{11} & \boldsymbol{\Sigma}_{12} \\ \boldsymbol{\Sigma}_{21} & \boldsymbol{\Sigma}_{22} \end{vmatrix} = |\boldsymbol{\Sigma}_{11}| \cdot |\boldsymbol{\Sigma}_{22} - \boldsymbol{\Sigma}_{21}\boldsymbol{\Sigma}_{11}^{-1}\boldsymbol{\Sigma}_{12}|.$$

10.7. (*a*) Show that $\rho^2_{1\cdot 23\cdots k} = 1 - 1/\rho^{11}$, where $\mathbf{P}$ is the matrix of correlation coefficients, $\mathbf{P}^{-1} = (\rho^{ij})$, and $\rho_{1\cdot 23\cdots k}$ is the multiple correlation coefficient of x_1 with $x_2, x_3, \cdots, x_k$.

(*b*) Show that $|\mathbf{P}| = (1 - \rho^2_{1\cdot 23\cdots k})(1 - \rho^2_{2\cdot 3\cdots k}) \cdots (1 - \rho^2_{k-1,k})$, where $\rho_{j\cdot j+1\cdots k}$ is the multiple correlation coefficient of x_j with $x_{j+1}, x_{j+2}, \cdots, x_k$, $j = 1, 2, \cdots, k - 1$.

10.8. Show that a necessary and sufficient condition for the independence of the k variates of a multivariate normal population is that the k multiple correlation coefficients of each x with the other x's are all zero, or that $\rho^{11} = \rho^{22} = \cdots = \rho^{kk} = 1$.

10.9. Suppose that in section 8 [cf. Cochran and Bliss (1948, p. 157)]

$$\mathbf{\Sigma} = \begin{pmatrix} \mathbf{\Sigma}_{11} & \mathbf{\Sigma}_{12} \\ \mathbf{\Sigma}_{21} & \mathbf{\Sigma}_{22} \end{pmatrix} = \left(\begin{array}{c|cc} 2351 & 1259 & 1340 \\ \hline 1259 & 3223 & 1200 \\ 1340 & 1200 & 3137 \end{array}\right),$$

$$\boldsymbol{\delta}' = (0, \boldsymbol{\delta}_2') = (0, -1197.2, -844.3).$$

Verify that:

(*a*) $\rho_{1\cdot 23}^2 = 0.33$.

(*b*) $\operatorname{tr} \mathbf{\Sigma}^{-1}\boldsymbol{\delta}\boldsymbol{\delta}' = 729.556$.

(*c*) $\operatorname{tr} \mathbf{\Sigma}_{22}^{-1}\boldsymbol{\delta}_2\boldsymbol{\delta}_2' = 503.845$.

(*d*) The canonical correlation is $\rho_1^2 = 0.33$.

(*e*) The gain ratio does not exceed $1/(1 - \rho_1^2)$.

10.10. Let $x_1, x_2, \cdots, x_n$ be distributed with the multivariate normal density $\frac{1}{|2\pi\mathbf{\Sigma}|^{1/2}} \exp\left(-\frac{1}{2}\mathbf{x}'\mathbf{\Sigma}^{-1}\mathbf{x}\right)$, where $\mathbf{x}' = (x_1, x_2, \cdots, x_n)$. If $\mathbf{y} = \mathbf{A}\mathbf{x}$, $\mathbf{y}' = (y_1, y_2, \cdots, y_m)$, $\mathbf{A} = (a_{ij})$, $i = 1, 2, \cdots, m$, $j = 1, 2, \cdots, n$, $m < n$, $\mathbf{A}$ of rank m, show that the y's are normally distributed $N(\mathbf{0}, \mathbf{A}\mathbf{\Sigma}\mathbf{A}')$, that is, with zero means and covariance matrix $\mathbf{A}\mathbf{\Sigma}\mathbf{A}'$ of rank $m < n$.

10.11. Let $\mathbf{y} = \mathbf{A}\mathbf{x}$, $\mathbf{z} = \mathbf{B}\mathbf{x}$, where $\mathbf{x}$, $\mathbf{y}$, $\mathbf{A}$ are defined in problem 10.10 and $\mathbf{z}' = (z_1, z_2, \cdots, z_{n-m})$, $\mathbf{B} = (b_{ij})$, $i = 1, 2, \cdots, n - m$, $j = 1, 2, \cdots, n$, $\mathbf{B}$ of rank $n - m$. Show that the set of y's is independent of the set of z's if $\mathbf{A}\mathbf{\Sigma}\mathbf{B}' = \mathbf{0} = \mathbf{B}\mathbf{\Sigma}\mathbf{A}'$.

10.12. Show that a necessary and sufficient condition for the independence of the k variates of a multivariate normal population is that $\rho_{1\cdot 23\cdots k}^2 = \rho_{2\cdot 3\cdots k}^2 = \cdots = \rho_{k-1\cdot k}^2 = 0$, where $\rho_{j\cdot j+1\cdots k}$ is the multiple correlation coefficient of x_j with $x_{j+1}, \cdots, x_k$.

10.13. Partition k variates into $m \le k$ sets, $\mathbf{x}' = (\mathbf{x}_1', \mathbf{x}_2', \cdots, \mathbf{x}_m')$, $\mathbf{x}_i' = (x_{k_1+k_2+\cdots+k_{i-1}+1}, \cdots, x_{k_1+k_2+\cdots+k_i})$. Assume that the multivariate normal populations under H_1 and H_2 (see section 7) are centered at their means, or $\boldsymbol{\delta} = \mathbf{0}$, and

$$\mathbf{\Sigma}_1 = \begin{pmatrix} \mathbf{\Sigma}_{11} & \mathbf{\Sigma}_{12} & \cdots & \mathbf{\Sigma}_{1m} \\ \mathbf{\Sigma}_{21} & \mathbf{\Sigma}_{22} & \cdots & \mathbf{\Sigma}_{2m} \\ \cdot & \cdot & \cdots & \cdot \\ \mathbf{\Sigma}_{m1} & \mathbf{\Sigma}_{m2} & \cdots & \mathbf{\Sigma}_{mm} \end{pmatrix}, \qquad \mathbf{\Sigma}_2 = \begin{pmatrix} \mathbf{\Sigma}_{11} & 0 & \cdots & 0 \\ 0 & \mathbf{\Sigma}_{22} & \cdots & 0 \\ \cdot & \cdot & \cdots & \cdot \\ 0 & 0 & \cdots & \mathbf{\Sigma}_{mm} \end{pmatrix},$$

where $\mathbf{\Sigma}_{ii} = (\sigma_{\alpha\beta})$, $\alpha, \beta = k_1 + k_2 + \cdots + k_{i-1} + 1, \cdots, k_1 + k_2 + \cdots + k_i$, and $\mathbf{\Sigma}_{ji}' = \mathbf{\Sigma}_{ij} = (\sigma_{rs})$, $r = k_1 + k_2 + \cdots + k_{i-1} + 1, \cdots, k_1 + k_2 + \cdots + k_i$, $s = k_1 + k_2 + \cdots + k_{j-1} + 1, \cdots, k_1 + k_2 + \cdots + k_j$, $k_1 + k_2 + \cdots + k_m = k$. Show that

$$I(1:2) = \frac{1}{2} \log \frac{|\mathbf{\Sigma}_{11}| \cdot |\mathbf{\Sigma}_{22}| \cdots |\mathbf{\Sigma}_{mm}|}{\begin{vmatrix} \mathbf{\Sigma}_{11} & \mathbf{\Sigma}_{12} & \cdots & \mathbf{\Sigma}_{1m} \\ \mathbf{\Sigma}_{21} & \mathbf{\Sigma}_{22} & \cdots & \mathbf{\Sigma}_{2m} \\ \cdot & \cdot & \cdots & \cdot \\ \mathbf{\Sigma}_{m1} & \mathbf{\Sigma}_{m2} & \cdots & \mathbf{\Sigma}_{mm} \end{vmatrix}}.$$

10.14. Show that (see problem 10.6)

$$\begin{vmatrix} \mathbf{\Sigma}_{11} & \mathbf{\Sigma}_{12} & \cdots & \mathbf{\Sigma}_{1m} \\ \mathbf{\Sigma}_{21} & \mathbf{\Sigma}_{22} & \cdots & \mathbf{\Sigma}_{2m} \\ \cdot & \cdot & \cdots & \cdot \\ \mathbf{\Sigma}_{m1} & \mathbf{\Sigma}_{m2} & \cdots & \mathbf{\Sigma}_{mm} \end{vmatrix} = |\mathbf{\Sigma}_{11}| \cdot |\mathbf{\Sigma}_{22\cdot 1}| \cdot |\mathbf{\Sigma}_{33\cdot 12}| \cdot |\mathbf{\Sigma}_{44\cdot 123}| \cdots |\mathbf{\Sigma}_{mm\cdot 12\cdots m-1}|,$$

where

$$\mathbf{\Sigma}_{ij\cdot 1} = \mathbf{\Sigma}_{ij} - \mathbf{\Sigma}_{i1}\mathbf{\Sigma}_{11}^{-1}\mathbf{\Sigma}_{1j}, \quad \mathbf{\Sigma}_{ij\cdot 12} = \mathbf{\Sigma}_{ij\cdot 1} - \mathbf{\Sigma}_{i2\cdot 1}\mathbf{\Sigma}_{22\cdot 1}^{-1}\mathbf{\Sigma}_{2j\cdot 1},$$

$$\mathbf{\Sigma}_{ij\cdot 123} = \mathbf{\Sigma}_{ij\cdot 12} - \mathbf{\Sigma}_{i3\cdot 12}\mathbf{\Sigma}_{33\cdot 12}^{-1}\mathbf{\Sigma}_{3j\cdot 12}, \qquad \mathbf{\Sigma}_{mm\cdot 12\cdots m-1} = \mathbf{\Sigma}_{mm\cdot 12\cdots m-2}$$
$$- \mathbf{\Sigma}_{mm-1\cdot 12\cdots m-2}\mathbf{\Sigma}_{m-1\,m-1\cdot 12\cdots m-2}^{-1}\mathbf{\Sigma}_{m-1\,m\cdot 12\cdots m-2}.$$

10.15. Suppose the k variates of a multivariate normal population have been partitioned into the $m \le k$ sets of problem 10.13. Show that a necessary and sufficient condition for the sets to be mutually independent is that $|\mathbf{\Sigma}_{22}| = |\mathbf{\Sigma}_{22\cdot 1}|$, $|\mathbf{\Sigma}_{33}| = |\mathbf{\Sigma}_{33\cdot 12}|$, $|\mathbf{\Sigma}_{44}| = |\mathbf{\Sigma}_{44\cdot 123}|, \cdots, |\mathbf{\Sigma}_{mm}| = |\mathbf{\Sigma}_{mm\cdot 12\cdots m-1}|$, where the matrices are defined in problems 10.13 and 10.14 above.

10.16. Show that [cf. (7.4) and problem 10.17],

$$\frac{\begin{vmatrix} \mathbf{\Sigma}_{11} & \mathbf{\Sigma}_{12} & \cdots & \mathbf{\Sigma}_{1r-1} \\ \mathbf{\Sigma}_{21} & \mathbf{\Sigma}_{22} & \cdots & \mathbf{\Sigma}_{2r-1} \\ \cdot & \cdot & \cdots & \cdot \\ \mathbf{\Sigma}_{r-11} & \mathbf{\Sigma}_{r-12} & \cdots & \mathbf{\Sigma}_{r-1\,r-1} \end{vmatrix} \cdot |\mathbf{\Sigma}_{rr}|}{\begin{vmatrix} \mathbf{\Sigma}_{11} & \mathbf{\Sigma}_{12} & \cdots & \mathbf{\Sigma}_{1r} \\ \mathbf{\Sigma}_{21} & \mathbf{\Sigma}_{22} & \cdots & \mathbf{\Sigma}_{2r} \\ \cdot & \cdot & \cdots & \cdot \\ \mathbf{\Sigma}_{r1} & \mathbf{\Sigma}_{r2} & \cdots & \mathbf{\Sigma}_{rr} \end{vmatrix}} = \frac{|\mathbf{\Sigma}_{rr}|}{|\mathbf{\Sigma}_{rr\cdot 12\cdots r-1}|},$$

where the matrices are defined in problems 10.13 and 10.14 above.

10.17. Partition k variates into the $m \le k$ sets of problem 10.13. Assume that the multivariate normal populations under H_1 and H_2 (see section 7) are centered at their means, or $\boldsymbol{\delta} = \mathbf{0}$, and

$$\mathbf{\Sigma}_1 = \begin{pmatrix} \mathbf{\Sigma}_{11} & \mathbf{\Sigma}_{12} & \cdots & \mathbf{\Sigma}_{1m} \\ \mathbf{\Sigma}_{21} & \mathbf{\Sigma}_{22} & \cdots & \mathbf{\Sigma}_{2m} \\ \cdot & \cdot & \cdots & \cdot \\ \mathbf{\Sigma}_{m1} & \mathbf{\Sigma}_{m2} & \cdots & \mathbf{\Sigma}_{mm} \end{pmatrix}, \quad \mathbf{\Sigma}_2 = \begin{pmatrix} \mathbf{\Sigma}_{11} & \mathbf{\Sigma}_{12} & \cdots & \mathbf{\Sigma}_{1m-1} & \mathbf{0} \\ \mathbf{\Sigma}_{21} & \mathbf{\Sigma}_{22} & \cdots & \mathbf{\Sigma}_{2m-1} & \mathbf{0} \\ \cdot & \cdot & \cdots & \cdot & \cdot \\ \mathbf{\Sigma}_{m-11} & \mathbf{\Sigma}_{m-12} & \cdots & \mathbf{\Sigma}_{m-1\,m-1} & \mathbf{0} \\ \mathbf{0} & \mathbf{0} & \cdots & \mathbf{0} & \mathbf{\Sigma}_{mm} \end{pmatrix},$$

where the matrices are defined in problem 10.13 above. Show that

$$I(1:2) = \frac{1}{2} \log \frac{|\mathbf{\Sigma}_{mm}|}{|\mathbf{\Sigma}_{mm\cdot 12\cdots m-1}|},$$

and that for $k_m = 1$, $I(1:2)$ is given in (7.18).

10.18. Suppose the k variates of a multivariate normal population have been partitioned into the $m \le k$ sets of problem 10.13. Show that a necessary and sufficient condition that the mth set be independent of the preceding $m - 1$ sets is that $|\mathbf{\Sigma}_{mm}| = |\mathbf{\Sigma}_{mm\cdot 12\cdots m-1}|$, where the matrices are defined in problems 10.13 and 10.14 above.

10.19. Show that the $I(1:2)$'s in (7.4), problem 10.13, and problem 10.17 are unchanged when the covariance matrices are replaced by the corresponding correlation matrices. Show that the equalities in problems 10.14, 10.15, 10.16, and 10.18 are also unchanged when the covariance matrices are replaced by the corresponding correlation matrices.

10.20. Partition k variates into the $m \le k$ sets of problem 10.13. Assume that the multivariate normal populations under H_1 and H_2 are centered at their means, or $\boldsymbol{\delta} = \mathbf{0}$, and

$$\boldsymbol{\Sigma}_1 = \begin{pmatrix} \boldsymbol{\Sigma}_{m-1\,m-1\cdot 12\cdots m-2} & \boldsymbol{\Sigma}_{m-1\,m\cdot 12\cdots m-2} \\ \boldsymbol{\Sigma}_{mm-1\cdot 12\cdots m-2} & \boldsymbol{\Sigma}_{mm\cdot 12\cdots m-2} \end{pmatrix},$$

$$\boldsymbol{\Sigma}_2 = \begin{pmatrix} \boldsymbol{\Sigma}_{m-1\,m-1\cdot 12\cdots m-2} & \mathbf{0} \\ \mathbf{0} & \boldsymbol{\Sigma}_{mm\cdot 12\cdots m-2} \end{pmatrix},$$

where the matrices are defined in problem 10.14 above. Show that

$$I(1:2) = \frac{1}{2}\log\frac{|\boldsymbol{\Sigma}_{mm\cdot 12\cdots m-2}|}{|\boldsymbol{\Sigma}_{mm\cdot 12\cdots m-1}|} = \frac{1}{2}\log\frac{|\mathbf{P}_{mm\cdot 12\cdots m-2}|}{|\mathbf{P}_{mm\cdot 12\cdots m-1}|}$$

and that for $k_m = k_{m-1} = 1$, $I(1:2) = -\frac{1}{2}\log(1 - \rho^2_{mm-1\cdot 12\cdots m-2})$, where $\rho_{mm-1\cdot 12\cdots m-2}$ is a partial correlation coefficient.

10.21. Show that the characteristic function of the distribution of $y = \mathbf{x}'\boldsymbol{\Sigma}^{-1}\mathbf{x}$, where $\mathbf{x}$ is k-variate normal $N(\mathbf{0}, \boldsymbol{\Sigma})$, is $E(\exp it\mathbf{x}'\boldsymbol{\Sigma}^{-1}\mathbf{x}) = (1 - 2it)^{-k/2}$, the characteristic function of the χ^2-distribution with k degrees of freedom.

10.22. Show that when $\mathbf{x}$ in problem 10.21 is k-variate normal $N(\boldsymbol{\mu}, \boldsymbol{\Sigma})$, $E(\exp it\mathbf{x}'\boldsymbol{\Sigma}^{-1}\mathbf{x}) = \exp[it\boldsymbol{\mu}'\boldsymbol{\Sigma}^{-1}\boldsymbol{\mu}/(1 - 2it)](1 - 2it)^{-k/2}$, the characteristic function of the noncentral χ^2-distribution with k degrees of freedom and noncentrality parameter $\boldsymbol{\mu}'\boldsymbol{\Sigma}^{-1}\boldsymbol{\mu}$. (See section 6.1 in chapter 12.)

CHAPTER 10

The Linear Hypothesis

1. INTRODUCTION

In this chapter we pick up again the general line of reasoning in chapters 6, 7, and 8 to examine the analysis of samples from normal populations in order to test the general linear hypothesis [Kolodziejczyk (1935)]. The analyses of this chapter may be derived as special cases of those on the multivariate linear hypothesis in chapter 11. Nevertheless, the development and study of the linear hypothesis first is thought to be worth while for its own sake as well as an aid in the exposition. The treatment is not intended to be exhaustive, and has wider applicability than to the specific cases considered.

2. BACKGROUND*

Suppose two simple statistical hypotheses, say H_1 and H_2, specify respectively the n-variate normal populations $N(\boldsymbol{\mu}_i, \boldsymbol{\Sigma})$, $i = 1, 2$, where $\boldsymbol{\mu}_i' = (\mu_{i1}, \mu_{i2}, \cdots, \mu_{in})$, $i = 1, 2$, are the one-row matrices (vectors) of means, and $\boldsymbol{\Sigma} = (\sigma_{rs})$, $r, s = 1, 2, \cdots, n$, is the common matrix of variances and covariances, so that [see (1.4) and (1.5) in chapter 9]:

$$(2.1) \qquad 2I(1:2) = J(1, 2) = (\boldsymbol{\mu}_1 - \boldsymbol{\mu}_2)'\boldsymbol{\Sigma}^{-1}(\boldsymbol{\mu}_1 - \boldsymbol{\mu}_2).$$

If the variates are independent, $\sigma_{rs} = 0$, $r \neq s$, $\boldsymbol{\Sigma}^{-1} = (\sigma^{rs})$, with $\sigma^{rs} = 0$, $r \neq s$, $\sigma^{rr} = 1/\sigma_{rr}$, $r = 1, 2, \cdots, n$, and (2.1) becomes, writing $\sigma_{rr} = \sigma_r^2$,

$$(2.2) \quad 2I(1:2) = J(1, 2) = \frac{(\mu_{11} - \mu_{21})^2}{\sigma_1^2} + \frac{(\mu_{12} - \mu_{22})^2}{\sigma_2^2} + \cdots + \frac{(\mu_{1n} - \mu_{2n})^2}{\sigma_n^2}.$$

If the variates are also identically distributed, as well as independent, that

* Sections 2–8 are mainly taken from an article by Kullback and Rosenblatt, which appeared in *Biometrika*, Vol. 44 (1957), pp. 67–83, and are reprinted with the permission of the editor.

is, $\mu_{ij} = \mu_i$, $i = 1, 2$, $j = 1, 2, \cdots, n$, and $\sigma_r^2 = \sigma^2$, $r = 1, 2, \cdots, n$, then (cf. example 4.2 of chapter 3)

$$(2.3) \qquad J(1, 2) = n(\mu_1 - \mu_2)^2/\sigma^2 = 2I(1:2).$$

3. THE LINEAR HYPOTHESIS

We now consider

$$(3.1) \qquad \mathbf{z} = \mathbf{y} - \mathbf{X}\boldsymbol{\beta},$$

where $\mathbf{z}' = (z_1, z_2, \cdots, z_n)$, $\mathbf{y}' = (y_1, y_2, \cdots, y_n)$, $\boldsymbol{\beta}' = (\beta_1, \beta_2, \cdots, \beta_p)$, $\mathbf{X} = (x_{ir})$, $i = 1, 2, \cdots, n$, $r = 1, 2, \cdots, p$; $p < n$, such that:

(*a*) the z's are independent, normally distributed random variables with zero means and common variance σ^2,

(*b*) the x_{ir}'s are assumed to be known,

(*c*) $\mathbf{X}$ is of rank p,

(*d*) $\boldsymbol{\beta} = \boldsymbol{\beta}^1$ and $\boldsymbol{\beta} = \boldsymbol{\beta}^2$ are one-column parameter matrices (vectors) specified respectively by the hypotheses H_1 and H_2, and

(*e*) $E_1(\mathbf{y}) = \mathbf{X}\boldsymbol{\beta}^1$ and $E_2(\mathbf{y}) = \mathbf{X}\boldsymbol{\beta}^2$.

We find that (2.1) yields here

$$(3.2) \qquad \begin{aligned} J(1, 2) &= (\mathbf{X}\boldsymbol{\beta}^1 - \mathbf{X}\boldsymbol{\beta}^2)'(\sigma^2\mathbf{I})^{-1}(\mathbf{X}\boldsymbol{\beta}^1 - \mathbf{X}\boldsymbol{\beta}^2) \\ &= (\boldsymbol{\beta}^1 - \boldsymbol{\beta}^2)'\mathbf{X}'\mathbf{X}(\boldsymbol{\beta}^1 - \boldsymbol{\beta}^2)/\sigma^2 \\ &= (\boldsymbol{\beta}^1 - \boldsymbol{\beta}^2)'\mathbf{S}(\boldsymbol{\beta}^1 - \boldsymbol{\beta}^2)/\sigma^2, \end{aligned}$$

where $\mathbf{S} = \mathbf{X}'\mathbf{X}$ is a $p \times p$ matrix of rank p and $\mathbf{I}$ is the $n \times n$ identity matrix.

We remark that $J(1, 2)$ $[2I(1:2)]$ in (3.2) is equivalent to the divergence between two multivariate normal populations with respective means $\boldsymbol{\beta}^1$, $\boldsymbol{\beta}^2$ and common covariance matrix $\sigma^2\mathbf{S}^{-1}$.

Suitable specification of the matrices $\mathbf{X}$ and $\boldsymbol{\beta}$ provides the appropriate model for many statistical problems of interest. [Cf. Kolodziejczyk (1935), Rao (1952, p. 119), Tocher (1952), Wilks (1938b; 1943, pp. 176–199), Zelen (1957, p. 312).]

4. THE MINIMUM DISCRIMINATION INFORMATION STATISTIC

We first state some facts about estimates of the parameters $\boldsymbol{\beta}$ and σ^2 of section 3. The classical least squares procedure of minimizing $\mathbf{z}'\mathbf{z} = (\mathbf{y} - \mathbf{X}\boldsymbol{\beta})'(\mathbf{y} - \mathbf{X}\boldsymbol{\beta})$ leads to the normal equations

$$(4.1) \qquad \mathbf{S}\hat{\boldsymbol{\beta}} = \mathbf{X}'\mathbf{y}.$$

It is shown in section 9 that the $\hat{\beta}_i$'s [solutions of (4.1)] are minimum variance, unbiased, sufficient estimates of the β_i's. [Cf. Durbin and Kendall (1951), Kempthorne (1952), Kolodziejczyk (1935), Plackett (1949), Rao (1952).]

It is a known result in regression theory that the components of $\hat{\boldsymbol{\beta}}$ (linear functions of the z's) are normally distributed with covariance matrix $\sigma^2\mathbf{S}^{-1}$. An unbiased estimate of σ^2 with $(n - p)$ degrees of freedom is obtained from $(n - p)\hat{\sigma}^2 = \hat{\mathbf{z}}'\hat{\mathbf{z}} = (\mathbf{y} - \mathbf{X}\hat{\boldsymbol{\beta}})'(\mathbf{y} - \mathbf{X}\hat{\boldsymbol{\beta}}) = \mathbf{y}'\mathbf{y} - \hat{\boldsymbol{\beta}}'\mathbf{S}\hat{\boldsymbol{\beta}}$. [Cf. Kempthorne (1952, pp. 54–59), Rao (1952, pp. 58–62).] (See problems 4.1–4.6 at the end of this section.)

In accordance with chapter 5, and as illustrated in the analyses in chapters 6, 7, and 8, the minimum discrimination information statistic is obtained by replacing the population parameters in $I(1:2)$ by the best unbiased estimates under the hypotheses. (See examples 4.1 and 4.2 in chapter 5 for the analysis of the conjugate distribution for single-variate normal populations. The multivariate normal generalizations of these examples are in sections 2 and 3.1 of chapter 12.)

The remark at the end of section 3 and the behavior of the least squares estimates imply that the analyses are essentially dependent on the implications of the hypotheses for the distributions of the estimates of $\boldsymbol{\beta}$.

Suppose the hypothesis H_1 imposes no restriction on $\boldsymbol{\beta}$ and the null hypothesis H_2 specifies $\boldsymbol{\beta} = \boldsymbol{\beta}^2$. Writing $\hat{\boldsymbol{\beta}}^1$ to indicate the solution of (4.1) under H_1, we have (cf. example 4.2 of chapter 5, section 3.1 of chapter 12)

$$2\hat{I}(H_1:H_2) = \hat{J}(H_1, H_2) = (\hat{\boldsymbol{\beta}}^1 - \boldsymbol{\beta}^2)'\mathbf{S}(\hat{\boldsymbol{\beta}}^1 - \boldsymbol{\beta}^2)/\hat{\sigma}^2. \tag{4.2}$$

In particular, for the common null hypothesis $H_2:\boldsymbol{\beta}^2 = \mathbf{0}$, (4.2) becomes (hereafter we shall just use J)

$$\hat{J}(H_1, H_2) = \hat{\boldsymbol{\beta}}^{1\prime}\mathbf{S}\hat{\boldsymbol{\beta}}^1/\hat{\sigma}^2. \tag{4.3}$$

Note that under the null hypothesis $H_2:\boldsymbol{\beta}^2 = \mathbf{0}$, $\hat{J}(H_1, H_2)$ in (4.3) is the quadratic form in the exponent of the multivariate normal distribution of the $\hat{\beta}_i$'s with the covariance matrix replaced by an unbiased estimate with $(n - p)$ degrees of freedom. $\hat{J}(H_1, H_2)$ is therefore Hotelling's generalized Student ratio (Hotelling's T^2) and

$$\hat{J}(H_1, H_2) = pF, \tag{4.4}$$

where F has the analysis of variance distribution (the F-distribution) with $n_1 = p$ and $n_2 = n - p$ degrees of freedom. [Cf. Anderson (1958, pp. 101–107), Hotelling (1951, p. 25), Hsu (1938), Kendall (1946, pp. 335–337), Rao (1952, p. 73), Simaika (1941), Wijsman (1957), Wilks (1943, p. 238).]

This approach, in contrast to the now classic method of derivation as a ratio of independent χ^2's divided by their degrees of freedom, is especially important for the generalizations in chapter 11. [See section 4 of chapter 11, particularly (4.5).] We need not appeal here to the general asymptotic distribution theory which is consistent with the conclusions above. We summarize in the usual analysis of variance table 4.1, where $\hat{\boldsymbol{\beta}}^{1\prime}\mathbf{S}\hat{\boldsymbol{\beta}}^1 = \hat{\boldsymbol{\beta}}^{1\prime}\mathbf{X}'\mathbf{y} = \mathbf{y}'\mathbf{X}\mathbf{S}^{-1}\mathbf{X}'\mathbf{y}$ [cf. Kempthorne (1952, p. 42), Rao (1952, p. 105)].

TABLE 4.1

Variation due to	Sum of Squares	D.F.
Linear regression	$\hat{\boldsymbol{\beta}}^{1\prime}\mathbf{S}\hat{\boldsymbol{\beta}}^1 = \mathbf{y}'\mathbf{X}\mathbf{S}^{-1}\mathbf{X}'\mathbf{y} = \hat{\sigma}^2\hat{J}(H_1, H_2)$	p
Difference	$\mathbf{y}'\mathbf{y} - \hat{\boldsymbol{\beta}}^{1\prime}\mathbf{S}\hat{\boldsymbol{\beta}}^1 = \mathbf{y}'(\mathbf{I} - \mathbf{X}\mathbf{S}^{-1}\mathbf{X}')\mathbf{y} = (n-p)\hat{\sigma}^2$	$n - p$
Total	$\mathbf{y}'\mathbf{y}$	n

For the null hypothesis H_2: $\boldsymbol{\beta} = \boldsymbol{\beta}^2 \neq \mathbf{0}$, (4.4) still holds, with $\hat{J}(H_1, H_2)$ given by (4.2).

Problem 4.1. Show that $\hat{\boldsymbol{\beta}} = \mathbf{S}^{-1}\mathbf{X}'\mathbf{z} + \boldsymbol{\beta}$ and therefore $E_1(\hat{\boldsymbol{\beta}}) = \boldsymbol{\beta}^1$, $E_2(\hat{\boldsymbol{\beta}}) = \boldsymbol{\beta}^2$.

Problem 4.2. Show that $E_1(\hat{\boldsymbol{\beta}}^1 - \boldsymbol{\beta}^1)(\hat{\boldsymbol{\beta}}^1 - \boldsymbol{\beta}^1)' = \sigma^2\mathbf{S}^{-1}$.

Problem 4.3. Show that $(\mathbf{I} - \mathbf{X}\mathbf{S}^{-1}\mathbf{X}')(\mathbf{X}\mathbf{S}^{-1}\mathbf{X}') = \mathbf{0}$. What does this imply about the quadratic forms $\mathbf{y}'\mathbf{X}\mathbf{S}^{-1}\mathbf{X}'\mathbf{y}$ and $\mathbf{y}'(\mathbf{I} - \mathbf{X}\mathbf{S}^{-1}\mathbf{X}')\mathbf{y}$?

Problem 4.4. Show that $J(1, 2; \hat{\boldsymbol{\beta}}) = J(1, 2)$ given by (3.2). Why does this imply that $\hat{\boldsymbol{\beta}}$ is sufficient?

Problem 4.5. Use lemma 5.3 of chapter 3 to show that $\mathbf{y}'\mathbf{y} \geq \mathbf{y}'\mathbf{X}\mathbf{S}^{-1}\mathbf{X}'\mathbf{y}$.

Problem 4.6. Show that $(n-p)\hat{\sigma}^2 = \dfrac{\begin{vmatrix} \mathbf{y}'\mathbf{y} & \mathbf{y}'\mathbf{X} \\ \mathbf{X}'\mathbf{y} & \mathbf{X}'\mathbf{X} \end{vmatrix}}{|\mathbf{X}'\mathbf{X}|}$.

5. SUBHYPOTHESES

5.1. Two-Partition Subhypothesis

[See Grundy (1951), Kempthorne (1952).] Suppose we partition the parameters into two sets, and instead of (3.1) we now consider

$$\mathbf{z} = \mathbf{y} - (\mathbf{X}_1, \mathbf{X}_2)\begin{pmatrix} \boldsymbol{\beta}_1 \\ \boldsymbol{\beta}_2 \end{pmatrix}, \tag{5.1}$$

where

$$\mathbf{X} = (\mathbf{X}_1, \mathbf{X}_2) = \left(\begin{array}{c|c} x_{11}, \cdots, x_{1q} & x_{1q+1}, \cdots, x_{1p} \\ \cdot & \cdot \\ \cdot & \cdot \\ \cdot & \cdot \\ x_{n1}, \cdots, x_{nq} & x_{nq+1}, \cdots, x_{np} \end{array}\right), \qquad \boldsymbol{\beta} = \begin{pmatrix} \boldsymbol{\beta}_1 \\ \boldsymbol{\beta}_2 \end{pmatrix}$$

with $\mathbf{X}_1$ and $\mathbf{X}_2$ respectively of ranks q and $p-q$, and $\boldsymbol{\beta}_1' = (\beta_1, \beta_2, \cdots, \beta_q)$, $\boldsymbol{\beta}_2' = (\beta_{q+1}, \cdots, \beta_p)$. The z's are still assumed to be independent, normally distributed random variables with zero means and common variance σ^2, and under H_1 and H_2,

$$E_1(\mathbf{y}) = \mathbf{X}_1\boldsymbol{\beta}_1^1 + \mathbf{X}_2\boldsymbol{\beta}_2^1 \tag{5.2}$$
$$E_2(\mathbf{y}) = \mathbf{X}_1\boldsymbol{\beta}_1^2 + \mathbf{X}_2\boldsymbol{\beta}_2^2.$$

We also write

$$\mathbf{S} = \mathbf{X}'\mathbf{X} = \begin{pmatrix} \mathbf{S}_{11} & \mathbf{S}_{12} \\ \mathbf{S}_{21} & \mathbf{S}_{22} \end{pmatrix}, \tag{5.3}$$

where

$$\mathbf{S}_{11} = \mathbf{X}_1'\mathbf{X}_1, \qquad \mathbf{S}_{12} = \mathbf{X}_1'\mathbf{X}_2 = \mathbf{S}_{21}', \qquad \mathbf{S}_{22} = \mathbf{X}_2'\mathbf{X}_2.$$

Now (3.2) becomes

$$J(1,2) = (\boldsymbol{\beta}_1^1 - \boldsymbol{\beta}_1^2, \boldsymbol{\beta}_2^1 - \boldsymbol{\beta}_2^2)' \begin{pmatrix} \mathbf{S}_{11} & \mathbf{S}_{12} \\ \mathbf{S}_{21} & \mathbf{S}_{22} \end{pmatrix} \begin{pmatrix} \boldsymbol{\beta}_1^1 - \boldsymbol{\beta}_1^2 \\ \boldsymbol{\beta}_2^1 - \boldsymbol{\beta}_2^2 \end{pmatrix} \Big/ \sigma^2. \tag{5.4}$$

The normal equations (4.1) under H_1 become

$$\begin{pmatrix} \mathbf{S}_{11} & \mathbf{S}_{12} \\ \mathbf{S}_{21} & \mathbf{S}_{22} \end{pmatrix} \begin{pmatrix} \hat{\boldsymbol{\beta}}_1^1 \\ \hat{\boldsymbol{\beta}}_2^1 \end{pmatrix} = \begin{pmatrix} \mathbf{X}_1' \\ \mathbf{X}_2' \end{pmatrix} \mathbf{y}, \tag{5.5}$$

or

$$\mathbf{S}_{11}\hat{\boldsymbol{\beta}}_1^1 + \mathbf{S}_{12}\hat{\boldsymbol{\beta}}_2^1 = \mathbf{X}_1'\mathbf{y} \tag{5.6}$$
$$\mathbf{S}_{21}\hat{\boldsymbol{\beta}}_1^1 + \mathbf{S}_{22}\hat{\boldsymbol{\beta}}_2^1 = \mathbf{X}_2'\mathbf{y},$$

and

$$(n-p)\hat{\sigma}^2 = \mathbf{y}'\mathbf{y} - (\hat{\boldsymbol{\beta}}_1^{1\prime}, \hat{\boldsymbol{\beta}}_2^{1\prime}) \begin{pmatrix} \mathbf{S}_{11} & \mathbf{S}_{12} \\ \mathbf{S}_{21} & \mathbf{S}_{22} \end{pmatrix} \begin{pmatrix} \hat{\boldsymbol{\beta}}_1^1 \\ \hat{\boldsymbol{\beta}}_2^1 \end{pmatrix}.$$

Letting

$$\mathbf{S}_{22\cdot1} = \mathbf{S}_{22} - \mathbf{S}_{21}\mathbf{S}_{11}^{-1}\mathbf{S}_{12}, \qquad \mathbf{X}_{2\cdot1}' = \mathbf{X}_2' - \mathbf{S}_{21}\mathbf{S}_{11}^{-1}\mathbf{X}_1',$$
$$\mathbf{S}_{11\cdot2} = \mathbf{S}_{11} - \mathbf{S}_{12}\mathbf{S}_{22}^{-1}\mathbf{S}_{21}, \qquad \mathbf{X}_{1\cdot2}' = \mathbf{X}_1' - \mathbf{S}_{12}\mathbf{S}_{22}^{-1}\mathbf{X}_2',$$

(5.6) yields

$$\hat{\boldsymbol{\beta}}_2^1 = \mathbf{S}_{22\cdot1}^{-1}\mathbf{X}_{2\cdot1}'\mathbf{y}, \tag{5.7}$$

$$\hat{\boldsymbol{\beta}}_1^1 = \mathbf{S}_{11}^{-1}\mathbf{X}_1'\mathbf{y} - \mathbf{S}_{11}^{-1}\mathbf{S}_{12}\hat{\boldsymbol{\beta}}_2^1 = \mathbf{S}_{11\cdot2}^{-1}\mathbf{X}_{1\cdot2}'\mathbf{y}. \tag{5.8}$$

It is useful to note [see, for example, Frazer, Duncan, and Collar (1938, para. 4.9); also section 7 of chapter 9] that

$$\mathbf{S}^{-1} = \begin{pmatrix} \mathbf{S}_{11} & \mathbf{S}_{12} \\ \mathbf{S}_{21} & \mathbf{S}_{22} \end{pmatrix}^{-1} = \begin{pmatrix} \mathbf{S}_{11\cdot2}^{-1} & \mathbf{M} \\ \mathbf{M}' & \mathbf{S}_{22\cdot1}^{-1} \end{pmatrix},$$

where the $q \times (p-q)$ matrix

$$\mathbf{M} = -\mathbf{S}_{11}^{-1}\mathbf{S}_{12}\mathbf{S}_{22\cdot1}^{-1} = -\mathbf{S}_{11\cdot2}^{-1}\mathbf{S}_{12}\mathbf{S}_{22}^{-1},$$

so that in the applications the elements of the matrix $\mathbf{S}_{11\cdot2}^{-1}$ or $\mathbf{S}_{22\cdot1}^{-1}$ are available once the matrix $\mathbf{S}^{-1}$ is obtained.

Suppose now that in particular we want to test the null hypothesis $H_2: \boldsymbol{\beta} = \boldsymbol{\beta}^2 = \begin{pmatrix} \boldsymbol{\beta}_1^2 \\ \mathbf{0} \end{pmatrix}$, that is, $\boldsymbol{\beta}_2^2 = \mathbf{0}$, with no restrictions on $\boldsymbol{\beta}_1^2$, against the alternative hypothesis $H_1: \boldsymbol{\beta} = \boldsymbol{\beta}^1 = \begin{pmatrix} \boldsymbol{\beta}_1^1 \\ \boldsymbol{\beta}_2^1 \end{pmatrix}$ with no restrictions on the parameters. Again we estimate $J(1, 2)$ by replacing the parameters by the best unbiased estimates under the hypotheses. Under H_1 we have (5.8), (5.7), (5.6) for $\hat{\boldsymbol{\beta}}_1^1$, $\hat{\boldsymbol{\beta}}_2^1$, and $\hat{\sigma}^2$. Under H_2 the normal equations (4.1) now yield

$$\hat{\boldsymbol{\beta}}_1^2 = \mathbf{S}_{11}^{-1}\mathbf{X}_1'\mathbf{y}. \tag{5.9}$$

From (5.4), (5.8), and (5.9) we have

$$\hat{\sigma}^2 \hat{J}(H_1, H_2) = (-\hat{\boldsymbol{\beta}}_2^{1'}\mathbf{S}_{21}\mathbf{S}_{11}^{-1}, \hat{\boldsymbol{\beta}}_2^{1'}) \begin{pmatrix} \mathbf{S}_{11} & \mathbf{S}_{12} \\ \mathbf{S}_{21} & \mathbf{S}_{22} \end{pmatrix} \begin{pmatrix} -\mathbf{S}_{11}^{-1}\mathbf{S}_{12}\hat{\boldsymbol{\beta}}_2^1 \\ \hat{\boldsymbol{\beta}}_2^1 \end{pmatrix} = \hat{\boldsymbol{\beta}}_2^{1'}\mathbf{S}_{22\cdot1}\hat{\boldsymbol{\beta}}_2^1. \tag{5.10}$$

It may be verified that

$$\mathbf{X}\mathbf{S}^{-1}\mathbf{X}' = \mathbf{X}_{2\cdot1}\mathbf{S}_{22\cdot1}^{-1}\mathbf{X}_{2\cdot1}' + \mathbf{X}_1\mathbf{S}_{11}^{-1}\mathbf{X}_1', \tag{5.11}$$

that is

$$\hat{\boldsymbol{\beta}}^{1'}\mathbf{S}\hat{\boldsymbol{\beta}}^1 = \hat{\boldsymbol{\beta}}_2^{1'}\mathbf{S}_{22\cdot1}\hat{\boldsymbol{\beta}}_2^1 + \hat{\boldsymbol{\beta}}_1^{2'}\mathbf{S}_{11}\hat{\boldsymbol{\beta}}_1^2, \tag{5.12}$$

or

$$\hat{\boldsymbol{\beta}}^{1'}\mathbf{X}'\mathbf{y} = \hat{\boldsymbol{\beta}}_2^{1'}\mathbf{X}_{2\cdot1}'\mathbf{y} + \hat{\boldsymbol{\beta}}_1^{2'}\mathbf{X}_1'\mathbf{y},$$

and

$$\hat{\sigma}^2 \hat{J}(H_1, H_2) = \hat{\boldsymbol{\beta}}^{1'}\mathbf{S}\hat{\boldsymbol{\beta}}^1 - \hat{\boldsymbol{\beta}}_1^{2'}\mathbf{S}_{11}\hat{\boldsymbol{\beta}}_1^2. \tag{5.13}$$

The foregoing is summarized in the analysis of variance table 5.1. $\hat{J}(H_1, H_2) = (\hat{\boldsymbol{\beta}}_2^{1'}\mathbf{S}_{22\cdot1}\hat{\boldsymbol{\beta}}_2^1)/\hat{\sigma}^2 = (p - q)F$, where F has the analysis of

TABLE 5.1

Variation due to	Sum of Squares	D.F.
$H_2: \boldsymbol{\beta}^{2'} = (\boldsymbol{\beta}_1^{2'}, \mathbf{0}')$	$\hat{\boldsymbol{\beta}}_1^{2'}\mathbf{S}_{11}\hat{\boldsymbol{\beta}}_1^2 = \mathbf{y}'\mathbf{X}_1\mathbf{S}_{11}^{-1}\mathbf{X}_1'\mathbf{y}$	q
Difference	$\hat{\boldsymbol{\beta}}_2^{1'}\mathbf{S}_{22\cdot1}\hat{\boldsymbol{\beta}}_2^1 = \mathbf{y}'\mathbf{X}_{2\cdot1}\mathbf{S}_{22\cdot1}^{-1}\mathbf{X}_{2\cdot1}'\mathbf{y} = \hat{\sigma}^2\hat{J}(H_1, H_2)$	$p - q$
$H_1: \boldsymbol{\beta}^{1'} = (\boldsymbol{\beta}_1^{1'}, \boldsymbol{\beta}_2^{1'})$	$\hat{\boldsymbol{\beta}}^{1'}\mathbf{S}\hat{\boldsymbol{\beta}}^1 = \mathbf{y}'\mathbf{X}\mathbf{S}^{-1}\mathbf{X}'\mathbf{y}$	p
Difference	$\mathbf{y}'\mathbf{y} - \hat{\boldsymbol{\beta}}^{1'}\mathbf{S}\hat{\boldsymbol{\beta}}^1 = \mathbf{y}'(\mathbf{I} - \mathbf{X}\mathbf{S}^{-1}\mathbf{X}')\mathbf{y} = (n - p)\hat{\sigma}^2$	$n - p$
Total	$\mathbf{y}'\mathbf{y}$	n

variance distribution with $n_1 = p - q$ and $n_2 = n - p$ degrees of freedom, under the null hypothesis H_2: $\boldsymbol{\beta}_2^2 = \mathbf{0}$.

We may center the y's about constants (their means) by letting

$$\text{(5.14)} \qquad \mathbf{X} = (\mathbf{X}_1, \mathbf{X}_2) = \begin{pmatrix} 1 & x_{12} \cdots x_{1p} \\ \cdot & \cdot \quad\quad \cdot \\ \cdot & \cdot \quad\quad \cdot \\ \cdot & \cdot \quad\quad \cdot \\ 1 & x_{n2} \cdots x_{np} \end{pmatrix}.$$

It may be verified that

$$\text{(5.15)} \qquad \mathbf{X}'\mathbf{X} = \mathbf{S} = \begin{pmatrix} \mathbf{S}_{11} & \mathbf{S}_{12} \\ \mathbf{S}_{21} & \mathbf{S}_{22} \end{pmatrix} = \begin{pmatrix} n & n\bar{x}_2 \cdots n\bar{x}_p \\ n\bar{x}_2 & \\ \cdot & \\ \cdot & \mathbf{S}_{22} \\ \cdot & \\ n\bar{x}_p & \end{pmatrix},$$

$$\text{(5.16)} \quad \mathbf{S}_{22\cdot 1} = \mathbf{S}_{22} - \begin{pmatrix} n\bar{x}_2 \\ \cdot \\ \cdot \\ \cdot \\ n\bar{x}_p \end{pmatrix} \frac{1}{n} (n\bar{x}_2 \cdots n\bar{x}_p) = \mathbf{S}_{22} - \begin{pmatrix} n\bar{x}_2\bar{x}_2 \cdots n\bar{x}_2\bar{x}_p \\ \cdot \quad\quad \cdot \\ \cdot \quad\quad \cdot \\ \cdot \quad\quad \cdot \\ n\bar{x}_p\bar{x}_2 \cdots n\bar{x}_p\bar{x}_p \end{pmatrix}$$

$$= \left(\sum_{i=1}^{n} (x_{ij} - \bar{x}_j)(x_{ik} - \bar{x}_k)\right), \qquad j, k = 2, \cdots, p,$$

$$\text{(5.17)} \quad \mathbf{X}_{2\cdot 1} = \mathbf{X}_2 - \mathbf{X}_1\mathbf{S}_{11}^{-1}\mathbf{S}_{12} = \mathbf{X}_2 - \begin{pmatrix} 1 \\ \cdot \\ \cdot \\ \cdot \\ 1 \end{pmatrix} \frac{1}{n} (n\bar{x}_2 \cdots n\bar{x}_p)$$

$$= \mathbf{X}_2 - \begin{pmatrix} \bar{x}_2 \cdots \bar{x}_p \\ \cdot \quad\quad \cdot \\ \cdot \quad\quad \cdot \\ \cdot \quad\quad \cdot \\ \bar{x}_2 \cdots \bar{x}_p \end{pmatrix} = \left(x_{ij} - \bar{x}_j\right), \qquad \begin{matrix} i = 1, 2, \cdots, n, \\ j = 2, \cdots, p, \end{matrix}$$

$$\text{(5.18)} \qquad \mathbf{X}_1'\mathbf{y} = n\bar{y},$$

$$\text{(5.19)} \qquad \hat{\boldsymbol{\beta}}_1^2 = \bar{y}.$$

The analysis of variance table 5.1 now becomes table 5.2.

TABLE 5.2

Variation due to	Sum of Squares	D.F.
$H_2:\boldsymbol{\beta}^{2'} = (\boldsymbol{\beta}_1^{2'}, \mathbf{0}')$	$n\bar{y}^2$	1
Difference (linear regression)	$\mathbf{y}'\mathbf{X}_{2\cdot 1}\mathbf{S}_{22\cdot 1}^{-1}\mathbf{X}_{2\cdot 1}'\mathbf{y} = \hat{\sigma}^2\hat{J}(H_1, H_2)$	$p - 1$
$H_1:\boldsymbol{\beta}^{1'} = (\boldsymbol{\beta}_1^{1'}, \boldsymbol{\beta}_2^{1'})$	$\mathbf{y}'\mathbf{X}\mathbf{S}^{-1}\mathbf{X}'\mathbf{y}$	p
Difference	$\mathbf{y}'\mathbf{y} - \mathbf{y}'\mathbf{X}\mathbf{S}^{-1}\mathbf{X}'\mathbf{y} = (n - p)\hat{\sigma}^2$	$n - p$
Total	$\mathbf{y}'\mathbf{y}$	n

Problem 5.1. Show that $\mathbf{X}_{2\cdot 1}'\mathbf{X}_{2\cdot 1} = \mathbf{S}_{22\cdot 1}$.
Problem 5.2. Show that $\hat{\boldsymbol{\beta}}_2^1 = \mathbf{S}_{22\cdot 1}^{-1}\mathbf{X}_{2\cdot 1}'\mathbf{z} + \boldsymbol{\beta}_2^1$.
Problem 5.3. Show that $E_1(\hat{\boldsymbol{\beta}}_2^1 - \boldsymbol{\beta}_2^1)(\hat{\boldsymbol{\beta}}_2^1 - \boldsymbol{\beta}_2^1)' = \sigma^2\mathbf{S}_{22\cdot 1}^{-1}$.
Problem 5.4. Show that $\mathbf{X}_{2\cdot 1}'\mathbf{X}_1 = \mathbf{0}$.
Problem 5.5. Show that $\mathbf{X}_{2\cdot 1}'\mathbf{X}_2 = \mathbf{S}_{22\cdot 1}$.
Problem 5.6. Show that $E_1(\hat{\boldsymbol{\beta}}_1^1 - \boldsymbol{\beta}_1^1)(\hat{\boldsymbol{\beta}}_1^1 - \boldsymbol{\beta}_1^1)' = \sigma^2\mathbf{S}_{11\cdot 2}^{-1}$.
Problem 5.7. Show that $\hat{\boldsymbol{\beta}}_1^2 = \mathbf{S}_{11}^{-1}\mathbf{X}_1'\mathbf{z} + \boldsymbol{\beta}_1^2$.
Problem 5.8. Show that $E_2(\hat{\boldsymbol{\beta}}_1^2 - \boldsymbol{\beta}_1^2)(\hat{\boldsymbol{\beta}}_1^2 - \boldsymbol{\beta}_1^2)' = \sigma^2\mathbf{S}_{11}^{-1}$.
Problem 5.9. Show that $\mathbf{S}_{11\cdot 2}^{-1} = \mathbf{S}_{11}^{-1} + \mathbf{S}_{11}^{-1}\mathbf{S}_{12}\mathbf{S}_{22\cdot 1}^{-1}\mathbf{S}_{21}\mathbf{S}_{11}^{-1}$.
Problem 5.10. Show that $\mathbf{X}_{2\cdot 1}\mathbf{S}_{22\cdot 1}^{-1}\mathbf{X}_{2\cdot 1}'(\mathbf{I} - \mathbf{X}\mathbf{S}^{-1}\mathbf{X}') = \mathbf{0}$.

5.2. Three-Partition Subhypothesis

If the subhypothesis requires partitioning the matrices $\boldsymbol{\beta}$ and $\mathbf{X}$ into three submatrices

$$\boldsymbol{\beta}' = (\boldsymbol{\beta}_1', \boldsymbol{\beta}_2', \boldsymbol{\beta}_3') \qquad \text{and} \qquad \mathbf{X} = (\mathbf{X}_1, \mathbf{X}_2, \mathbf{X}_3),$$

we obtain from $\mathbf{S}\hat{\boldsymbol{\beta}} = \mathbf{X}'\mathbf{y}$ the solutions

$$\begin{aligned} \hat{\boldsymbol{\beta}}_3 &= \mathbf{S}_{33\cdot 12}^{-1}\mathbf{X}_{3\cdot 12}'\mathbf{y}, \\ \hat{\boldsymbol{\beta}}_2 &= \mathbf{S}_{22\cdot 1}^{-1}(\mathbf{X}_{2\cdot 1}'\mathbf{y} - \mathbf{S}_{23\cdot 1}\hat{\boldsymbol{\beta}}_3), \\ \hat{\boldsymbol{\beta}}_1 &= \mathbf{S}_{11}^{-1}(\mathbf{X}_1'\mathbf{y} - \mathbf{S}_{12}\hat{\boldsymbol{\beta}}_2 - \mathbf{S}_{13}\hat{\boldsymbol{\beta}}_3), \end{aligned} \tag{5.20}$$

where

$$\mathbf{S} = \begin{pmatrix} \mathbf{S}_{11} & \mathbf{S}_{12} & \mathbf{S}_{13} \\ \mathbf{S}_{21} & \mathbf{S}_{22} & \mathbf{S}_{23} \\ \mathbf{S}_{31} & \mathbf{S}_{32} & \mathbf{S}_{33} \end{pmatrix}, \qquad \mathbf{S}_{tu} = \mathbf{X}_t'\mathbf{X}_u, \qquad t, u = 1, 2, 3,$$

and

$$\mathbf{S}_{33\cdot12} = \mathbf{S}_{33\cdot1} - \mathbf{S}_{32\cdot1}\mathbf{S}_{22\cdot1}^{-1}\,\mathbf{S}_{23\cdot1},$$

$$\mathbf{S}_{33\cdot1} = \mathbf{S}_{33} - \mathbf{S}_{31}\mathbf{S}_{11}^{-1}\mathbf{S}_{13}, \qquad \mathbf{S}_{32\cdot1} = \mathbf{S}_{32} - \mathbf{S}_{31}\mathbf{S}_{11}^{-1}\mathbf{S}_{12} = \mathbf{S}_{23\cdot1}',$$

$$\mathbf{S}_{22\cdot1} = \mathbf{S}_{22} - \mathbf{S}_{21}\mathbf{S}_{11}^{-1}\mathbf{S}_{12}, \qquad \mathbf{X}_{2\cdot1}'\mathbf{y} = (\mathbf{X}_2' - \mathbf{S}_{21}\mathbf{S}_{11}^{-1}\mathbf{X}_1')\mathbf{y},$$

$$\mathbf{X}_{3\cdot12}'\mathbf{y} = (\mathbf{X}_{3\cdot1}' - \mathbf{S}_{32\cdot1}\mathbf{S}_{22\cdot1}^{-1}\mathbf{X}_{2\cdot1}')\mathbf{y}, \qquad \mathbf{X}_{3\cdot1}'\mathbf{y} = (\mathbf{X}_3' - \mathbf{S}_{31}\mathbf{S}_{11}^{-1}\mathbf{X}_1')\mathbf{y};$$

also [cf. (5.12)]

$$(5.21) \qquad \hat{\boldsymbol{\beta}}'\mathbf{S}\hat{\boldsymbol{\beta}} = \mathbf{y}'\mathbf{X}_1\mathbf{S}_{11}^{-1}\mathbf{X}_1'\mathbf{y} + \mathbf{y}'\mathbf{X}_{2\cdot1}\mathbf{S}_{22\cdot1}^{-1}\mathbf{X}_{2\cdot1}'\mathbf{y} + \hat{\boldsymbol{\beta}}_3'\mathbf{S}_{33\cdot12}\hat{\boldsymbol{\beta}}_3.$$

Using (5.20) and collecting terms, we obtain other useful forms of (5.21), for example,

$$(5.22) \qquad \hat{\boldsymbol{\beta}}'\mathbf{S}\hat{\boldsymbol{\beta}} = \mathbf{y}'\mathbf{X}_1\mathbf{S}_{11}^{-1}\mathbf{X}_1'\mathbf{y} + \hat{\boldsymbol{\beta}}_2'\mathbf{X}_{2\cdot1}'\mathbf{y} + \hat{\boldsymbol{\beta}}_3'\mathbf{X}_{3\cdot1}'\mathbf{y}.$$

This is convenient when the data are raw observations and $x_{i1} = 1$ for all i, so that the first partition includes the x_{i1} and β_{i1} and obtains deviations about averages for basically a two-partition problem, and

$$(5.23) \qquad \hat{\boldsymbol{\beta}}'\mathbf{S}\hat{\boldsymbol{\beta}} = \hat{\boldsymbol{\beta}}_1'\mathbf{X}_1'\mathbf{y} + \hat{\boldsymbol{\beta}}_2'\mathbf{X}_2'\mathbf{y} + \hat{\boldsymbol{\beta}}_3'\mathbf{X}_3'\mathbf{y}$$

for a three-partition problem where the variables are already centered about their averages.

The above can be extended by induction to any number of partitions as required.

6. ANALYSIS OF REGRESSION: ONE-WAY CLASSIFICATION, k CATEGORIES

For $p = 1$, also identified as the analysis of covariance, see Federer (1955, p. 485), Kempthorne (1952, p. 48), Kendall (1946, p. 237), Smith (1957), Welch (1935). For p general, see also Kullback and Rosenblatt (1957), Rosenblatt (1953).

Suppose we have k categories each with n_j observations on $(y, x_1, \cdots, x_p)$ for which the general linear regression for each category is

$$(6.1) \qquad z_{ji} = y_{ji} - (\beta_{j1}x_{ji1} + \cdots + \beta_{jr}x_{jir} + \cdots + \beta_{jp}x_{jip}),$$

where $j = 1, 2, \cdots, k$ categories,

$i = 1, 2, \cdots, n_j$ observations for category j,

$r = 1, 2, \cdots, p$ independent variables ($p < n_j$),

the z_{ji} are independent, normally distributed random variables with zero means and common variance σ^2, and the x_{jir} are known.

The linear regressions for each category can be written as

$$(6.2) \qquad \mathbf{z}_j = \mathbf{y}_j - \mathbf{X}_j\boldsymbol{\beta}_j,$$

where for $j = 1, 2, \cdots, k$,

$$\mathbf{z}_j' = (z_{j1}, z_{j2}, \cdots, z_{jn_j}), \qquad \mathbf{y}_j' = (y_{j1}, y_{j2}, \cdots, y_{jn_j}),$$
$$\mathbf{X}_j = (\mathbf{x}_{j1}, \mathbf{x}_{j2}, \cdots, \mathbf{x}_{jp}), \qquad \mathbf{x}_{jr}' = (x_{j1r}, x_{j2r}, \cdots, x_{jn_jr}),$$

and

$$\boldsymbol{\beta}_j' = (\beta_{j1}, \beta_{j2}, \cdots, \beta_{jp}).$$

We may write the k sets of regression equations (6.2) for k categories combined as

$$\text{(6.3)} \qquad \mathbf{z} = \mathbf{y} - \mathbf{X}\boldsymbol{\beta}$$

by defining

$$\mathbf{X} = \begin{pmatrix} \mathbf{X}_1 & \cdot & \cdot & \cdot & \mathbf{0} \\ & \mathbf{X}_2 & & & \\ \cdot & & \cdot & & \cdot \\ \cdot & & & \cdot & \cdot \\ \cdot & & & \cdot & \cdot \\ \mathbf{0} & \cdot & \cdot & \cdot & \mathbf{X}_k \end{pmatrix}, \qquad \boldsymbol{\beta}' = (\boldsymbol{\beta}_1', \boldsymbol{\beta}_2', \cdots, \boldsymbol{\beta}_k'),$$

$$\mathbf{z}' = (\mathbf{z}_1', \mathbf{z}_2', \cdots, \mathbf{z}_k'), \qquad \mathbf{y}' = (\mathbf{y}_1', \mathbf{y}_2', \cdots, \mathbf{y}_k').$$

By the preceding definitions we treat $\boldsymbol{\beta}$ in (6.3) as a parameter matrix of all kp regression coefficients β_{jr} whether or not any of them are equal, or have a particular value including zero, under any hypothesis.

Suppose we specify a null hypothesis with regard to certain groups or sets of the kp parameters β_{jr} among the k categories, and wish to estimate the parameters and test the null hypothesis against some alternative. To distinguish between matrices or parameter vectors under various hypotheses H_α, $\alpha = 1, 2, \cdots$, we shall use, where desirable for clarity or emphasis, the notation $\mathbf{X}^\alpha$, $\boldsymbol{\beta}^\alpha$, and $\mathbf{S}^\alpha = \mathbf{X}^{\alpha'}\mathbf{X}^\alpha$. Where this notation is not used, the applicable hypothesis and definition of the matrices should be clear from the context. For any hypothesis H_α, we shall represent the linear regressions for the k categories combined, under H_α, as

$$\text{(6.4)} \qquad \mathbf{z} = \mathbf{y} - \mathbf{X}^\alpha\boldsymbol{\beta}^\alpha,$$

where $\mathbf{z}$ and $\mathbf{y}$ are defined in (6.3). However, we now define $\boldsymbol{\beta}^\alpha$ as the matrix of distinct regression coefficients specified by the hypothesis H_α, and $\mathbf{X}^\alpha$ as the matrix of x_{jir} with distinct regression effects, specified according to the regression model defined by the hypothesis $\mathbf{H}_\alpha$ for the k categories combined.

With the representation (6.4) of the k-category regression under H_α, the normal equations (4.1) become

$$\text{(6.5)} \qquad \begin{aligned} \mathbf{S}^\alpha\hat{\boldsymbol{\beta}}^\alpha &= \mathbf{X}^{\alpha'}\mathbf{y} \\ \hat{\boldsymbol{\beta}}^\alpha &= \mathbf{S}^{\alpha^{-1}}\mathbf{X}^{\alpha'}\mathbf{y}, \end{aligned}$$

where the elements of $\mathbf{S}^\alpha = \mathbf{X}^{\alpha\prime}\mathbf{X}^\alpha$ will, of course, depend on the particular specification of the matrix $\mathbf{X}^\alpha$.

Also, equivalent to (4.2) and (5.13) we have, for a null hypothesis H_2, and an alternative hypothesis H_1 [cf. (4.7) in chapter 5],

$$(6.6)\quad \hat{J}(H_1, H_2) = (\hat{\boldsymbol{\beta}}^1 - \hat{\boldsymbol{\beta}}^2)'\mathbf{S}(\hat{\boldsymbol{\beta}}^1 - \hat{\boldsymbol{\beta}}^2)/\hat{\sigma}^2 = (\hat{\boldsymbol{\beta}}^{1\prime}\mathbf{S}^1\hat{\boldsymbol{\beta}}^1 - \hat{\boldsymbol{\beta}}^{2\prime}\mathbf{S}^2\hat{\boldsymbol{\beta}}^2)/\hat{\sigma}^2,$$

where

$$(6.7)\qquad (n - pk)\hat{\sigma}^2 = \mathbf{y}'\mathbf{y} - \hat{\boldsymbol{\beta}}^{1\prime}\mathbf{S}^1\hat{\boldsymbol{\beta}}^1,$$
$$n = n_1 + n_2 + \cdots + n_k,$$

and $\mathbf{S} = \mathbf{X}'\mathbf{X} = \mathbf{S}^1$ for $\mathbf{X}$ defined in (6.3).

Thus, for any particular hypothesis on the sets of regression coefficients in k-category regression, the estimates of the coefficients and the test of the hypothesis are readily obtained solely by proper specification of the matrices $\mathbf{X}^\alpha$ and $\boldsymbol{\beta}^\alpha$ in (6.4).

Consider the two hypotheses

$$(6.8)\qquad H_1: \beta_{jr} = \beta_{jr}, \qquad j = 1, 2, \cdots, k, \qquad r = 1, 2, \cdots, p,$$

that is, the β_{jr} are different for all categories and for each $r = 1, 2, \cdots, p$, and the null hypothesis of homogeneity

$$(6.9)\qquad H_2: \beta_{jr} = \beta_{\cdot r}, \qquad j = 1, 2, \cdots, k, \qquad r = 1, 2, \cdots, p,$$

or equivalently, $\boldsymbol{\beta}_j' = \boldsymbol{\beta}_\cdot' = (\beta_{\cdot 1}, \beta_{\cdot 2}, \cdots, \beta_{\cdot p})$, $j = 1, 2, \cdots, k$, that is, the regression coefficients are the same for the different categories for each $r = 1, 2, \cdots, p$.

Under H_1 in (6.8) the best unbiased estimate of $\boldsymbol{\beta}$ is derived from (6.5), where $\boldsymbol{\beta}^\alpha$ and $\mathbf{X}^\alpha$ in (6.4), defining the k-category regression model, are the same as $\boldsymbol{\beta}$ and $\mathbf{X}$ in (6.3), or

$$(6.10)\qquad \begin{pmatrix} \mathbf{S}_1 & \cdots & \mathbf{0} \\ \cdot & \cdot & \cdot \\ \cdot & \cdot & \cdot \\ \cdot & \cdot & \cdot \\ \mathbf{0} & \cdots & \mathbf{S}_k \end{pmatrix} \begin{pmatrix} \hat{\boldsymbol{\beta}}_1 \\ \cdot \\ \cdot \\ \cdot \\ \hat{\boldsymbol{\beta}}_k \end{pmatrix} = \begin{pmatrix} \mathbf{X}_1'\mathbf{y}_1 \\ \cdot \\ \cdot \\ \cdot \\ \mathbf{X}_k'\mathbf{y}_k \end{pmatrix}, \qquad \mathbf{S}_j = \mathbf{X}_j'\mathbf{X}_j.$$

This yields k sets of normal equations

$$(6.11)\qquad \mathbf{S}_j\hat{\boldsymbol{\beta}}_j = \mathbf{X}_j'\mathbf{y}_j, \qquad j = 1, 2, \cdots, k,$$

from which

$$\hat{\boldsymbol{\beta}}_j = \mathbf{S}_j^{-1}\mathbf{X}_j'\mathbf{y}_j.$$

Under H_2 in (6.9), however, the matrices $\mathbf{X}^2$ and $\boldsymbol{\beta}^2$ of (6.4), defining the k-category regression model, are

$$\mathbf{X}^{2'} = (\mathbf{X}_1', \cdots, \mathbf{X}_k'), \qquad \boldsymbol{\beta}^{2'}_{\cdot} = (\beta_{\cdot 1}, \beta_{\cdot 2}, \cdots, \beta_{\cdot p}).$$

Thus,

$$\mathbf{S}^2 = \mathbf{X}^{2'}\mathbf{X}^2 = \sum_{j=1}^{k} \mathbf{X}_j'\mathbf{X}_j = \sum_{j=1}^{k} \mathbf{S}_j,$$

$$\mathbf{X}^{2'}\mathbf{y} = \sum_{j=1}^{k} \mathbf{X}_j'\mathbf{y}_j,$$

and the best unbiased estimate of $\boldsymbol{\beta}$ under H_2 is derived from (6.5) as

$$\hat{\boldsymbol{\beta}}^2_{\cdot} = \mathbf{S}^{2^{-1}}\mathbf{X}^{2'}\mathbf{y}. \tag{6.12}$$

We also have, under H_1, corresponding to (6.7),

$$(n - pk)\hat{\sigma}^2 = \mathbf{y}'\mathbf{y} - \hat{\boldsymbol{\beta}}^{1'}\mathbf{S}^1\hat{\boldsymbol{\beta}}^1 = \sum_{j=1}^{k} (\mathbf{y}_j'\mathbf{y}_j - \hat{\boldsymbol{\beta}}_j'\mathbf{S}_j\hat{\boldsymbol{\beta}}_j). \tag{6.13}$$

Corresponding to (6.6), we therefore have

$$\hat{\sigma}^2\hat{J}(H_1, H_2) = \hat{\boldsymbol{\beta}}^{1'}\mathbf{S}^1\hat{\boldsymbol{\beta}}^1 - \hat{\boldsymbol{\beta}}^{2'}\mathbf{S}^2\hat{\boldsymbol{\beta}}^2 = \sum_{j=1}^{k} \hat{\boldsymbol{\beta}}_j'\mathbf{S}_j\hat{\boldsymbol{\beta}}_j - \hat{\boldsymbol{\beta}}^{2'}_{\cdot}\mathbf{S}^2\hat{\boldsymbol{\beta}}^2_{\cdot}, \tag{6.14}$$

a direct generalization of S_2 in para. 24.30 of Kendall (1946).

TABLE 6.1

Variation due to	Sum of Squares	D.F.
$H_2: \boldsymbol{\beta}^2 = \boldsymbol{\beta}^2_{\cdot}$	$\hat{\boldsymbol{\beta}}^{2'}_{\cdot}\mathbf{S}^2\hat{\boldsymbol{\beta}}^2_{\cdot}$	p
Difference	$\hat{\boldsymbol{\beta}}^{1'}\mathbf{S}^1\hat{\boldsymbol{\beta}}^1 - \hat{\boldsymbol{\beta}}^{2'}\mathbf{S}^2\hat{\boldsymbol{\beta}}^2 = \hat{\sigma}^2\hat{J}(H_1, H_2)$	$p(k-1)$
$H_1: \boldsymbol{\beta}^1 = \boldsymbol{\beta}^1$	$\hat{\boldsymbol{\beta}}^{1'}\mathbf{S}^1\hat{\boldsymbol{\beta}}^1 = \sum_{j=1}^{k} \hat{\boldsymbol{\beta}}_j'\mathbf{S}_j\hat{\boldsymbol{\beta}}_j$	pk
Difference	$\mathbf{y}'\mathbf{y} - \hat{\boldsymbol{\beta}}^{1'}\mathbf{S}^1\hat{\boldsymbol{\beta}}^1 = (n - pk)\hat{\sigma}^2$	$n - pk$
Total	$\mathbf{y}'\mathbf{y}$	$n = \Sigma n_j$

We summarize in the analysis of variance table 6.1. $\hat{J}(H_1, H_2) = p(k-1)F$, where F has the analysis of variance distribution with $p(k-1)$ and $(n - pk)$ degrees of freedom under the null hypothesis H_2 of (6.9).

In particular, for testing a null hypothesis of homogeneity H_2, the means of k samples are the same, $p = 1$, $x_{ji1} = 1$, and for the alternative hypothesis H_1, the population means are different,

$$(6.15) \qquad \mathbf{X}^1 = \begin{pmatrix} 1 & 0 & 0\cdots 0 \\ \cdot & \cdot & \cdot \quad \cdot \\ \cdot & \cdot & \cdot \quad \cdot \\ \cdot & \cdot & \cdot \quad \cdot \\ 1 & 0 & 0\cdots 0 \\ 0 & 1 & 0\cdots 0 \\ \cdot & \cdot & \cdot \quad \cdot \\ \cdot & \cdot & \cdot \quad \cdot \\ \cdot & \cdot & \cdot \quad \cdot \\ 0 & 1 & 0\cdots 0 \\ \cdot & \cdot & \cdot \quad \cdot \\ 0 & 0 & 0\cdots 1 \\ \cdot & \cdot & \cdot \quad \cdot \\ \cdot & \cdot & \cdot \quad \cdot \\ \cdot & \cdot & \cdot \quad \cdot \\ 0 & 0 & 0\cdots 1 \end{pmatrix} \begin{matrix} \\ \\ n_1 \\ \\ \\ \\ \\ n_2, \\ \\ \cdot \\ \cdot \\ \cdot \\ \\ n_k \\ \\ \\ \end{matrix} \qquad \mathbf{X}_j' = (\underbrace{1, \cdots, 1}_{n_j}),$$

$$(6.16) \qquad \mathbf{X}^{1\prime}\mathbf{X}^1 = \mathbf{S}^1 = \begin{pmatrix} n_1 & 0 & \cdots & 0 \\ 0 & n_2 & \cdots & 0 \\ \cdot & \cdot & \cdots & \cdot \\ 0 & \cdot & \cdots & n_k \end{pmatrix}, \qquad \mathbf{S}_j = n_j,$$

$$(6.17) \qquad \mathbf{X}_j'\mathbf{y}_j = y_{j1} + y_{j2} + \cdots + y_{jn_j} = n_j\bar{y}_j,$$

and (6.11) yields as the estimates of the population means under H_1,

$$(6.18) \qquad \hat{\boldsymbol{\beta}}_j = \bar{y}_j.$$

For the null hypothesis of homogeneity H_2,

$$(6.19) \qquad \mathbf{X}^{2\prime} = (\underbrace{1, \cdots, 1}_{n_1}, \underbrace{1, \cdots, 1}_{n_2}, \cdots, \underbrace{1, \cdots, 1}_{n_k}),$$

$$(6.20) \qquad \mathbf{X}^{2\prime}\mathbf{X}^2 = \mathbf{S}^2 = n_1 + n_2 + \cdots + n_k = n,$$

$$(6.21) \qquad \mathbf{X}^{2\prime}\mathbf{y} = n_1\bar{y}_1 + \cdots + n_k\bar{y}_k = n\bar{y},$$

and (6.12) yields as the estimate of the population mean under H_2

$$(6.22) \qquad \hat{\boldsymbol{\beta}}. = \bar{y}.$$

From (6.13) and (6.14) we find that

$$(6.23)\qquad (n-k)\hat{\sigma}^2 = \sum_{j=1}^{k}\left(\sum_{i=1}^{n_j} y_{ij}^2 - n_j\bar{y}_j^2\right) = \sum_{j=1}^{k}\sum_{i=1}^{n_j}(y_{ij}-\bar{y}_j)^2,$$

$$(6.24)\qquad \hat{\sigma}^2\hat{J}(H_1, H_2) = \sum_{j=1}^{k} n_j\bar{y}_j^2 - n\bar{y}^2 = \sum_{j=1}^{k} n_j(\bar{y}_j - \bar{y})^2.$$

The analysis of variance table 6.1 becomes now the analysis of variance table 6.2.

TABLE 6.2

Variation due to	Sum of Squares	D.F.
H_2: Homogeneity	$n\bar{y}^2$	1
Difference	$\sum_{j=1}^{k} n_j(\bar{y}_j - \bar{y})^2 = \hat{\sigma}^2\hat{J}(H_1, H_2)$	$k-1$
H_1: Heterogeneity	$\sum_{j=1}^{k} n_j\bar{y}_j^2$	k
Difference	$\sum_{j=1}^{k}\sum_{i=1}^{n_j}(y_{ij}-\bar{y}_j)^2 = (n-k)\hat{\sigma}^2$	$n-k$
Total	$\sum_{j=1}^{k}\sum_{i=1}^{n_j} y_{ij}^2$	$n = n_1 + n_2 + \cdots + n_k$

The analysis in table 6.2 is more commonly found as in table 6.3. $\hat{J}(H_1, H_2) = (k-1)F$, where F has the analysis of variance distribution with $(k-1)$ and $(n-k)$ degrees of freedom under the null hypothesis of homogeneity.

TABLE 6.3

Variation due to	Sum of Squares	D.F.
Between samples	$\sum_{j=1}^{k} n_j(\bar{y}_j - \bar{y})^2 = \hat{\sigma}^2\hat{J}(H_1, H_2)$	$k-1$
Within samples	$\sum_{j=1}^{k}\sum_{i=1}^{n_j}(y_{ij}-\bar{y}_j)^2 = (n-k)\hat{\sigma}^2$	$n-k$
Total	$\sum_{j=1}^{k}\sum_{i=1}^{n_j}(y_{ij}-\bar{y})^2$	$n-1$

7. TWO-PARTITION SUBHYPOTHESIS

7.1. One-Way Classification, k Categories

Partition the parameters of the matrix $\boldsymbol{\beta}_j$, for each category $j = 1, 2, \cdots, k$, into two sets [see (6.2)]

$$\boldsymbol{\beta}_{j1}' = (\beta_{j1}, \cdots, \beta_{jq}) \qquad \text{and} \qquad \boldsymbol{\beta}_{j2}' = (\beta_{jq+1}, \cdots, \beta_{jp})$$

of q and $p - q$ parameters respectively, $q < p$, so that $\boldsymbol{\beta}_j' = (\boldsymbol{\beta}_{j1}', \boldsymbol{\beta}_{j2}')$.

Consider a null subhypothesis H_2, for $j = 1, 2, \cdots, k$, the β_{jr} are different for $r = 1, 2, \cdots, q$, but for $r = q + 1, q + 2, \cdots, p$, there is a common value $\beta_{\cdot r}$ for the β_{jr}, that is,

$$\text{(7.1)} \qquad \begin{aligned} H_2: \beta_{jr} &= \beta_{jr}, \quad j = 1, 2, \cdots, k, \quad r = 1, 2, \cdots, q, \\ \beta_{jr} &= \beta_{\cdot r}, \quad j = 1, 2, \cdots, k, \quad r = q + 1, q + 2, \cdots, p, \end{aligned}$$

or equivalently

$$\begin{aligned} H_2: \boldsymbol{\beta}_{j1}' &= \boldsymbol{\beta}_{j1}' = (\beta_{j1}, \cdots, \beta_{jq}) \\ \boldsymbol{\beta}_{j2}' &= \boldsymbol{\beta}_{\cdot 2}' = (\beta_{\cdot q+1}, \cdots, \beta_{\cdot p}). \end{aligned}$$

Let H_1 remain as in (6.8), that is, the β_{jr} are different for all j and r. Under H_1 we have the same matrix definitions and results as in section 6. However, for H_2 in (7.1), the matrices $\mathbf{X}^2$ and $\boldsymbol{\beta}^2$ for the k-category regression model are

$$\boldsymbol{\beta}^{2'} = (\boldsymbol{\beta}_1', \boldsymbol{\beta}_2'), \qquad \mathbf{X}^2 = (\mathbf{X}_1, \mathbf{X}_2),$$

where

$$\boldsymbol{\beta}_1' = (\boldsymbol{\beta}_{11}', \boldsymbol{\beta}_{21}', \cdots, \boldsymbol{\beta}_{k1}'), \qquad \boldsymbol{\beta}_2' = (\beta_{\cdot q+1}, \beta_{\cdot q+2}, \cdots, \beta_{\cdot p}) = \boldsymbol{\beta}_{\cdot 2}',$$

$$\mathbf{X}_1 = \begin{pmatrix} \mathbf{X}_{11} & \cdots & \mathbf{0} \\ & \ddots & \\ \mathbf{0} & \cdots & \mathbf{X}_{k1} \end{pmatrix}, \qquad \mathbf{X}_2 = \begin{pmatrix} \mathbf{X}_{12} \\ \cdot \\ \cdot \\ \cdot \\ \mathbf{X}_{k2} \end{pmatrix},$$

$$\begin{aligned} \mathbf{X}_{j1} &= (\mathbf{x}_{j1}, \mathbf{x}_{j2}, \cdots, \mathbf{x}_{jq}), \\ \mathbf{X}_{j2} &= (\mathbf{x}_{jq+1}, \mathbf{x}_{jq+2}, \cdots, \mathbf{x}_{jp}), \qquad j = 1, 2, \cdots, k, \qquad r = 1, 2, \cdots, p. \\ \mathbf{x}_{jr}' &= (x_{j1r}, x_{j2r}, \cdots, x_{jn_jr}), \end{aligned}$$

Thus under H_2,

$$\mathbf{S}^2 = \mathbf{X}^{2'}\mathbf{X}^2 = \begin{pmatrix} \mathbf{X}_1' \\ \mathbf{X}_2' \end{pmatrix} (\mathbf{X}_1, \mathbf{X}_2) = \begin{pmatrix} \mathbf{S}_{11} & \mathbf{S}_{12} \\ \mathbf{S}_{21} & \mathbf{S}_{22} \end{pmatrix},$$

where

$$\mathbf{S}_{11} = \mathbf{X}_1'\mathbf{X}_1, \qquad \mathbf{S}_{12} = \mathbf{X}_1'\mathbf{X}_2, \qquad \mathbf{S}_{21} = \mathbf{S}_{12}' = \mathbf{X}_2'\mathbf{X}_1, \qquad \mathbf{S}_{22} = \mathbf{X}_2'\mathbf{X}_2,$$

$$\mathbf{S}_{11} = \begin{pmatrix} \mathbf{S}_{111} & \cdots & \mathbf{0} \\ & \cdot & \\ & \cdot & \\ & \cdot & \\ \mathbf{0} & \cdots & \mathbf{S}_{k11} \end{pmatrix}, \quad \mathbf{S}_{12} = \begin{pmatrix} \mathbf{S}_{112} \\ \cdot \\ \cdot \\ \cdot \\ \mathbf{S}_{k12} \end{pmatrix}, \quad \mathbf{S}_{22} = (\mathbf{S}_{122} + \cdots + \mathbf{S}_{k22}),$$

$$\mathbf{S}_{j11} = \mathbf{X}_{j1}'\mathbf{X}_{j1}, \qquad \mathbf{S}_{j12} = \mathbf{X}_{j1}'\mathbf{X}_{j2} = \mathbf{S}_{j21}', \qquad \mathbf{S}_{j22} = \mathbf{X}_{j2}'\mathbf{X}_{j2}.$$

From the normal equations (6.5) we now obtain

$$\begin{aligned} \mathbf{S}_{11}\hat{\boldsymbol{\beta}}_1 + \mathbf{S}_{12}\hat{\boldsymbol{\beta}}_2 &= \mathbf{X}_1'\mathbf{y} \\ \mathbf{S}_{21}\hat{\boldsymbol{\beta}}_1 + \mathbf{S}_{22}\hat{\boldsymbol{\beta}}_2 &= \mathbf{X}_2'\mathbf{y}, \end{aligned} \tag{7.2}$$

so that [see (5.7)]

$$\hat{\boldsymbol{\beta}}_2 = \mathbf{S}_{22\cdot1}^{-1}\mathbf{X}_{2\cdot1}'\mathbf{y}, \tag{7.3}$$

where

$$\mathbf{S}_{22\cdot1} = (\mathbf{S}_{22} - \mathbf{S}_{21}\mathbf{S}_{11}^{-1}\mathbf{S}_{12}) = \sum_{j=1}^{k}(\mathbf{S}_{j22} - \mathbf{S}_{j21}\mathbf{S}_{j11}^{-1}\mathbf{S}_{j12}) = \sum_{j=1}^{k}\mathbf{S}_{j22\cdot1},$$

$$\mathbf{X}_{2\cdot1}' = (\mathbf{X}_{12\cdot1}', \cdots, \mathbf{X}_{k2\cdot1}'), \qquad \mathbf{X}_{2\cdot1}'\mathbf{y} = \sum_{j=1}^{k}\mathbf{X}_{j2\cdot1}'\mathbf{y}_j,$$

$$\mathbf{X}_{2\cdot1}' = \mathbf{X}_2' - \mathbf{S}_{21}\mathbf{S}_{11}^{-1}\mathbf{X}_1', \qquad \mathbf{X}_{j2\cdot1}' = \mathbf{X}_{j2}' - \mathbf{S}_{j21}\mathbf{S}_{j11}^{-1}\mathbf{X}_{j1}'.$$

From the definition of the matrices under H_2 we have [see (5.8)]

$$\begin{pmatrix} \hat{\boldsymbol{\beta}}_{11} \\ \cdot \\ \cdot \\ \cdot \\ \hat{\boldsymbol{\beta}}_{k1} \end{pmatrix} = \begin{pmatrix} \mathbf{S}_{111}^{-1} & \cdots & \mathbf{0} \\ & \cdot & \\ & \cdot & \\ & \cdot & \\ \mathbf{0} & \cdots & \mathbf{S}_{k11}^{-1} \end{pmatrix} \left[\begin{pmatrix} \mathbf{X}_{11}' & \cdots & \mathbf{0} \\ & \cdot & \\ & \cdot & \\ & \cdot & \\ \mathbf{0} & \cdots & \mathbf{X}_{k1}' \end{pmatrix} \begin{pmatrix} \mathbf{y}_1 \\ \cdot \\ \cdot \\ \cdot \\ \mathbf{y}_k \end{pmatrix} - \begin{pmatrix} \mathbf{S}_{112} \\ \cdot \\ \cdot \\ \cdot \\ \mathbf{S}_{k12} \end{pmatrix} \hat{\boldsymbol{\beta}}_{\cdot2} \right] \tag{7.4}$$

$$= \begin{pmatrix} \mathbf{S}_{111}^{-1}\mathbf{X}_{11}'\mathbf{y}_1 \\ \cdot \\ \cdot \\ \cdot \\ \mathbf{S}_{k11}^{-1}\mathbf{X}_{k1}'\mathbf{y}_k \end{pmatrix} - \begin{pmatrix} \mathbf{S}_{111}^{-1}\mathbf{S}_{112}\hat{\boldsymbol{\beta}}_{\cdot2} \\ \cdot \\ \cdot \\ \cdot \\ \mathbf{S}_{k11}^{-1}\mathbf{S}_{k12}\hat{\boldsymbol{\beta}}_{\cdot2} \end{pmatrix}.$$

Thus under H_2 of (7.1), we have the following estimates of the regression coefficients:

$$\begin{aligned} \hat{\boldsymbol{\beta}}_{j1} &= \mathbf{S}_{j11}^{-1}(\mathbf{X}_{j1}'\mathbf{y}_j - \mathbf{S}_{j12}\hat{\boldsymbol{\beta}}_{\cdot2}) = \hat{\boldsymbol{\beta}}_{j1}^2, \qquad j = 1, 2, \cdots, k, \\ \hat{\boldsymbol{\beta}}_{\cdot2} &= \left(\sum_{j=1}^{k}\mathbf{S}_{j22\cdot1}\right)^{-1}\sum_{j=1}^{k}\mathbf{X}_{j2\cdot1}'\mathbf{y}_j = \hat{\boldsymbol{\beta}}_{\cdot2}^2, \end{aligned} \tag{7.5}$$

where

$$\hat{\boldsymbol{\beta}}_j' = (\hat{\boldsymbol{\beta}}_{j1}^{2'}, \hat{\boldsymbol{\beta}}_{\cdot 2}^{2'}) = \hat{\boldsymbol{\beta}}_j^{2'}.$$

If under H_1 of (6.8) we also define $\boldsymbol{\beta}_j' = (\boldsymbol{\beta}_{j1}', \boldsymbol{\beta}_{j2}')$ but rearrange and partition the submatrices of $\boldsymbol{\beta}$ and $\mathbf{X}$ so that

$$\boldsymbol{\beta}' = (\boldsymbol{\beta}_1', \boldsymbol{\beta}_2'), \qquad \mathbf{X} = (\mathbf{X}_1, \mathbf{X}_2),$$

where

$$\boldsymbol{\beta}_1' = (\boldsymbol{\beta}_{11}', \boldsymbol{\beta}_{21}', \cdots, \boldsymbol{\beta}_{k1}'), \qquad \boldsymbol{\beta}_{j1}' = (\beta_{j1}, \beta_{j2}, \cdots, \beta_{jq}),$$
$$\boldsymbol{\beta}_2' = (\boldsymbol{\beta}_{12}', \boldsymbol{\beta}_{22}', \cdots, \boldsymbol{\beta}_{k2}'), \qquad \boldsymbol{\beta}_{j2}' = (\beta_{jq+1}, \beta_{jq+2}, \cdots, \beta_{jp}),$$

$$\mathbf{X}_1 = \begin{pmatrix} \mathbf{X}_{11} & \cdots & \mathbf{0} \\ & \cdot & \\ & \cdot & \\ & \cdot & \\ \mathbf{0} & \cdots & \mathbf{X}_{k1} \end{pmatrix}, \qquad \mathbf{X}_2 = \begin{pmatrix} \mathbf{X}_{12} & \cdots & \mathbf{0} \\ & \cdot & \\ & \cdot & \\ & \cdot & \\ \mathbf{0} & \cdots & \mathbf{X}_{k2} \end{pmatrix},$$

$\mathbf{X}_{j1} = (\mathbf{x}_{j1}, \mathbf{x}_{j2}, \cdots, \mathbf{x}_{jq})$, $\mathbf{X}_{j2} = (\mathbf{x}_{jq+1}, \mathbf{x}_{jq+2}, \cdots, \mathbf{x}_{jp})$, $j = 1, 2, \cdots, k$, then

$$\mathbf{X}_1'\mathbf{X}_1 = \mathbf{S}_{11} = \begin{pmatrix} \mathbf{S}_{111} & \cdots & \mathbf{0} \\ & \cdot & \\ & \cdot & \\ & \cdot & \\ \mathbf{0} & \cdots & \mathbf{S}_{k11} \end{pmatrix}, \qquad \mathbf{X}_1'\mathbf{X}_2 = \mathbf{S}_{12} = \begin{pmatrix} \mathbf{S}_{112} & \cdots & \mathbf{0} \\ & \cdot & \\ & \cdot & \\ & \cdot & \\ \mathbf{0} & \cdots & \mathbf{S}_{k12} \end{pmatrix} = \mathbf{S}_{21}',$$

$$\mathbf{X}_2'\mathbf{X}_2 = \mathbf{S}_{22} = \begin{pmatrix} \mathbf{S}_{122} & \cdots & \mathbf{0} \\ & \cdot & \\ & \cdot & \\ & \cdot & \\ \mathbf{0} & \cdots & \mathbf{S}_{k22} \end{pmatrix}, \qquad \mathbf{X}'\mathbf{X} = \mathbf{S} = \begin{pmatrix} \mathbf{S}_{11} & \mathbf{S}_{12} \\ \mathbf{S}_{21} & \mathbf{S}_{22} \end{pmatrix},$$

$$\mathbf{S}_{j11} = \mathbf{X}_{j1}'\mathbf{X}_{j1}, \qquad \mathbf{S}_{j12} = \mathbf{X}_{j1}'\mathbf{X}_{j2} = \mathbf{S}_{j21}',$$
$$\mathbf{S}_{j22} = \mathbf{X}_{j2}'\mathbf{X}_{j2}, \qquad j = 1, 2, \cdots, k.$$

We then obtain the same estimate of $\boldsymbol{\beta}_j$, $j = 1, 2, \cdots, k$, as in section 6, by the procedure of section 5; that is, from (6.5) we have (see problem 11.10),

(7.6)
$$\mathbf{S}_{11}\hat{\boldsymbol{\beta}}_1 + \mathbf{S}_{12}\hat{\boldsymbol{\beta}}_2 = \mathbf{X}_1'\mathbf{y},$$
$$\mathbf{S}_{21}\hat{\boldsymbol{\beta}}_1 + \mathbf{S}_{22}\hat{\boldsymbol{\beta}}_2 = \mathbf{X}_2'\mathbf{y},$$

(7.7)
$$\hat{\boldsymbol{\beta}}_1 = \mathbf{S}_{11}^{-1}(\mathbf{X}_1'\mathbf{y} - \mathbf{S}_{12}\hat{\boldsymbol{\beta}}_2),$$

(7.8)
$$\hat{\boldsymbol{\beta}}_2 = \mathbf{S}_{22\cdot 1}^{-1}\mathbf{X}_{2\cdot 1}'\mathbf{y},$$

where

$$\mathbf{S}_{22\cdot1} = (\mathbf{S}_{22} - \mathbf{S}_{21}\mathbf{S}_{11}^{-1}\mathbf{S}_{12}) = \begin{pmatrix} \mathbf{S}_{122\cdot1} & \cdots & \mathbf{0} \\ & \cdot & \\ & \cdot & \\ & \cdot & \\ \mathbf{0} & \cdots & \mathbf{S}_{k22\cdot1} \end{pmatrix},$$

$$\mathbf{S}_{j22\cdot1} = (\mathbf{S}_{j22} - \mathbf{S}_{j21}\mathbf{S}_{j11}^{-1}\mathbf{S}_{j12}), \qquad j = 1, 2, \cdots, k,$$

$$\mathbf{X}_{2\cdot1}' = \mathbf{X}_2' - \mathbf{S}_{21}\mathbf{S}_{11}^{-1}\mathbf{X}_1' = \begin{pmatrix} \mathbf{X}_{12\cdot1}' & \cdots & \mathbf{0} \\ & \cdot & \\ & \cdot & \\ & \cdot & \\ \mathbf{0} & \cdots & \mathbf{X}_{k2\cdot1}' \end{pmatrix},$$

$$\mathbf{X}_{j2\cdot1}' = \mathbf{X}_{j2}' - \mathbf{S}_{j21}\mathbf{S}_{j11}^{-1}\mathbf{X}_{j1}', \qquad j = 1, 2, \cdots, k.$$

From (7.7) we obtain under H_1 for each category $j = 1, 2, \cdots, k,$

$$\hat{\boldsymbol{\beta}}_{j1} = \mathbf{S}_{j11}^{-1}(\mathbf{X}_{j1}'\mathbf{y}_j - \mathbf{S}_{j12}\hat{\boldsymbol{\beta}}_{j2}) = \hat{\boldsymbol{\beta}}_{j1}^1, \tag{7.9}$$

$$\hat{\boldsymbol{\beta}}_{j2} = \mathbf{S}_{j22\cdot1}^{-1}\mathbf{X}_{j2\cdot1}'\mathbf{y}_j = \hat{\boldsymbol{\beta}}_{j2}^1, \tag{7.10}$$

and

$$\hat{\boldsymbol{\beta}}_j' = (\hat{\boldsymbol{\beta}}_{j1}', \hat{\boldsymbol{\beta}}_{j2}') = \hat{\boldsymbol{\beta}}_j^{1'}.$$

With these estimates of the parameters under H_1 of (6.8) and H_2 of (7.1) and noting after some reduction that [cf. (5.11), (5.12)]

$$\hat{\boldsymbol{\beta}}^{\alpha'}\mathbf{S}^{\alpha}\hat{\boldsymbol{\beta}}^{\alpha} = \mathbf{y}'\mathbf{X}_1\mathbf{S}_{11}^{-1}\mathbf{X}_1'\mathbf{y} + \hat{\boldsymbol{\beta}}_2^{\alpha'}\mathbf{S}_{22\cdot1}^{\alpha}\hat{\boldsymbol{\beta}}_2^{\alpha}, \qquad \alpha = 1, 2, \tag{7.11}$$

we obtain [cf. (5.13)]

$$\hat{\sigma}^2\hat{J}(H_1, H_2) = \hat{\boldsymbol{\beta}}_2^{1'}\mathbf{S}_{22\cdot1}^1\hat{\boldsymbol{\beta}}_2^1 - \hat{\boldsymbol{\beta}}_2^{2'}\mathbf{S}_{22\cdot1}^2\hat{\boldsymbol{\beta}}_2^2 = \sum_{j=1}^{k}\hat{\boldsymbol{\beta}}_{j2}^{1'}\mathbf{S}_{j22\cdot1}\hat{\boldsymbol{\beta}}_{j2}^1 - \hat{\boldsymbol{\beta}}_2^{2'}\mathbf{S}_{22\cdot1}^2\hat{\boldsymbol{\beta}}_2^2, \tag{7.12}$$

where for computational convenience we may write

$$\hat{\boldsymbol{\beta}}_2^{1'}\mathbf{S}_{22\cdot1}^1\hat{\boldsymbol{\beta}}_2^1 = \hat{\boldsymbol{\beta}}_2^{1'}\mathbf{X}_{2\cdot1}^{1'}\mathbf{y} = \sum_{j=1}^{k}\hat{\boldsymbol{\beta}}_{j2}^{1'}\mathbf{X}_{j2\cdot1}'\mathbf{y}, \tag{7.13}$$

$$\hat{\boldsymbol{\beta}}_2^{2'}\mathbf{S}_{22\cdot1}^2\hat{\boldsymbol{\beta}}_2^2 = \hat{\boldsymbol{\beta}}_2^{2'}\mathbf{X}_{2\cdot1}^{2'}\mathbf{y}. \tag{7.14}$$

We summarize in the analysis of variance table 7.1. $\hat{J}(H_1, H_2) = (p - q)(k - 1)F$, where F has the analysis of variance distribution with $(p - q)(k - 1)$ and $n - pk$ degrees of freedom under the null hypothesis H_2 of (7.1).

TABLE 7.1

Variation due to	Sum of Squares	D.F.
$H_2: \boldsymbol{\beta}^2_{j1}, \boldsymbol{\beta}^2_{\cdot 2}$	$\hat{\boldsymbol{\beta}}^{2\prime}\mathbf{S}^2\hat{\boldsymbol{\beta}}^2$	$qk + p - q$
Difference	$\hat{\boldsymbol{\beta}}^{1\prime}\mathbf{S}^1\hat{\boldsymbol{\beta}}^1 - \hat{\boldsymbol{\beta}}^{2\prime}\mathbf{S}^2\hat{\boldsymbol{\beta}}^2$	
	$= \sum_{j=1}^{k} \hat{\boldsymbol{\beta}}^{1\prime}_{j2}\mathbf{S}_{j22\cdot 1}\hat{\boldsymbol{\beta}}^1_{j2} - \hat{\boldsymbol{\beta}}^{2\prime}_2\mathbf{S}^2_{22\cdot 1}\hat{\boldsymbol{\beta}}^2_2$	
	$= \hat{\sigma}^2 \hat{J}(H_1, H_2)$	$(p - q)(k - 1)$
$H_1: \boldsymbol{\beta}^1_{j1}, \boldsymbol{\beta}^1_{j2}$	$\hat{\boldsymbol{\beta}}^{1\prime}\mathbf{S}^1\hat{\boldsymbol{\beta}}^1$	pk
Difference	$\mathbf{y}'\mathbf{y} - \hat{\boldsymbol{\beta}}^{1\prime}\mathbf{S}^1\hat{\boldsymbol{\beta}}^1 = (n - pk)\hat{\sigma}^2$	$n - pk$
Total	$\mathbf{y}'\mathbf{y}$	$n = n_1 + n_2 + \cdots + n_k$

7.2. Carter's Regression Case

Carter (1949) considers the case of a correlation effect among the ith observations $i = 1, 2, \cdots, n$ in each of k samples. His regression model can be written as

$$z_{ji} = y_{ji} - \sum_{r=1}^{q} \beta_{jr} x_{jir} - \alpha_i, \tag{7.15}$$

where the correlation effect among samples is due to α_i, an element common to the ith observation in each sample, $j = 1, 2, \cdots, k$. Stochastic dependence among categories is included in the multivariate linear hypothesis in chapter 11.

It can be seen that this model is a particular case of the subhypothesis analysis where the matrices $\boldsymbol{\beta}$ and $\mathbf{X}$ are

$$\boldsymbol{\beta}' = (\boldsymbol{\beta}_1', \boldsymbol{\beta}_2'), \qquad \mathbf{X} = (\mathbf{X}_1, \mathbf{X}_2),$$

and the submatrices are

$$\boldsymbol{\beta}_1' = (\boldsymbol{\beta}_{11}', \boldsymbol{\beta}_{21}', \cdots, \boldsymbol{\beta}_{k1}'), \qquad \boldsymbol{\beta}_{j1}' = (\beta_{j1}, \beta_{j2}, \cdots, \beta_{jq}),$$

$$\boldsymbol{\beta}_2' = (\alpha_1, \alpha_2, \cdots, \alpha_n),$$

$$\mathbf{X}_1 = \begin{pmatrix} \mathbf{X}_{11} & \cdots & \mathbf{0} \\ & \cdot & \\ & \cdot & \\ & \cdot & \\ \mathbf{0} & \cdots & \mathbf{X}_{k1} \end{pmatrix}, \qquad \mathbf{X}_{j1} = \begin{pmatrix} x_{j11} & \cdots & x_{j1q} \\ \cdot & & \cdot \\ \cdot & & \cdot \\ \cdot & & \cdot \\ x_{jn1} & \cdots & x_{jnq} \end{pmatrix},$$

$$\mathbf{X}_2 = \begin{pmatrix} \mathbf{I} \\ \cdot \\ \cdot \\ \cdot \\ \mathbf{I} \end{pmatrix}, \qquad \mathbf{I} = \begin{pmatrix} 1 & \cdots & 0 \\ \cdot & \cdot & \cdot \\ \cdot & \cdot & \cdot \\ \cdot & \cdot & \cdot \\ 0 & \cdots & 1 \end{pmatrix},$$

where $\mathbf{X}_2$ is a $k \times 1$ matrix of submatrices $\mathbf{I}$, the identity matrix of order $n \times n$. With these definitions of $\boldsymbol{\beta}$ and $\mathbf{X}$, the normal equations for estimating the β's given by Carter [1949, eq. (3.3)] follow directly from the normal equations (7.2) by obtaining

$$\mathbf{S}_{11\cdot 2}\hat{\boldsymbol{\beta}}_1 = \mathbf{X}'_{1\cdot 2}\mathbf{y}, \tag{7.16}$$

where

$$\mathbf{S}_{11\cdot 2} = \mathbf{S}_{11} - \mathbf{S}_{12}\mathbf{S}_{22}^{-1}\mathbf{S}_{21}.$$

Here we obtain

$$\mathbf{S}_{11\cdot 2} = \begin{pmatrix} \left(1-\frac{1}{k}\right)\mathbf{X}'_{11}\mathbf{X}_{11} & -\frac{1}{k}\mathbf{X}'_{11}\mathbf{X}_{21} & \cdots & -\frac{1}{k}\mathbf{X}'_{11}\mathbf{X}_{k1} \\ -\frac{1}{k}\mathbf{X}'_{21}\mathbf{X}_{11} & \left(1-\frac{1}{k}\right)\mathbf{X}'_{21}\mathbf{X}_{21} & \cdots & -\frac{1}{k}\mathbf{X}'_{21}\mathbf{X}_{k1} \\ \cdot & \cdot & & \cdot \\ \cdot & \cdot & & \cdot \\ \cdot & \cdot & & \cdot \\ -\frac{1}{k}\mathbf{X}'_{k1}\mathbf{X}_{11} & -\frac{1}{k}\mathbf{X}'_{k1}\mathbf{X}_{21} & \cdots & \left(1-\frac{1}{k}\right)\mathbf{X}'_{k1}\mathbf{X}_{k1} \end{pmatrix},$$

and

$$\mathbf{X}'_{1\cdot 2} = \begin{pmatrix} \left(1-\frac{1}{k}\right)\mathbf{X}'_{11} & -\frac{1}{k}\mathbf{X}'_{11} & \cdots & -\frac{1}{k}\mathbf{X}'_{11} \\ -\frac{1}{k}\mathbf{X}'_{21} & \left(1-\frac{1}{k}\right)\mathbf{X}'_{21} & \cdots & -\frac{1}{k}\mathbf{X}'_{21} \\ \cdot & \cdot & & \cdot \\ \cdot & \cdot & & \cdot \\ \cdot & \cdot & & \cdot \\ -\frac{1}{k}\mathbf{X}'_{k1} & -\frac{1}{k}\mathbf{X}'_{k1} & \cdots & \left(1-\frac{1}{k}\right)\mathbf{X}'_{k1} \end{pmatrix}.$$

As before,

$$\mathbf{S}_{11} = \mathbf{X}'_1\mathbf{X}_1, \qquad \mathbf{S}_{12} = \mathbf{X}'_1\mathbf{X}_2 = \mathbf{S}'_{21}, \qquad \mathbf{S}_{22} = \mathbf{X}'_2\mathbf{X}_2.$$

The estimates of the correlation effects α_i are not given specifically by Carter (1949). The solution

$$\hat{\alpha}_i = \bar{z}_i, \qquad i = 1, 2, \cdots, n, \tag{7.17}$$

where

$$\bar{z}_i = \frac{1}{k}\sum_{j=1}^{k}\hat{z}_{ji} = \frac{1}{k}\sum_{j=1}^{k}\left(y_{ji} - \sum_{r=1}^{q}\hat{\beta}_{jr}x_{jir}\right),$$

follows directly from

$$\mathbf{S}_{22}\hat{\boldsymbol{\beta}}_2 = \mathbf{X}'_2\mathbf{y} - \mathbf{S}_{21}\hat{\boldsymbol{\beta}}_1. \tag{7.18}$$

8. EXAMPLE

[See Kullback and Rosenblatt (1957).] As an example of sections 5, 6, and 7, we examine the performance data of a manufactured product tested under three environmental conditions (categories) each involving three independent variables. In the equation

$$z_{ji} = y_{ji} - \beta_{j1}x_{ji1} - \beta_{j2}x_{ji2} - \beta_{j3}x_{ji3} - \beta_{j4}x_{ji4}, \tag{8.1}$$

the data y_{ji} and x_{jir}, $r = 2, 3, 4$, are raw observations so that $x_{ji1} = 1$ for all $j = 1, 2, 3$, and $i = 1, 2, \cdots, n_j$. In this example $k = 3$, $p = 4$, $n_1 = 16$, $n_2 = 15$, and $n_3 = 16$. The matrices $\mathbf{S}_j$ and $\mathbf{X}_j'\mathbf{y}_j$, $j = 1, 2, 3$, of the computed sums of squares and products about the origin are

$$\mathbf{S}_1 = \begin{pmatrix} 16.0 & 286.8 & 139.0 & 4{,}835.0 \\ 286.8 & 5{,}340.4 & 2{,}452.2 & 86{,}849.0 \\ 139.0 & 2{,}452.2 & 1{,}307.0 & 41{,}990.0 \\ 4{,}835.0 & 86{,}849.0 & 41{,}990.0 & 1{,}465{,}575.0 \end{pmatrix}, \quad \mathbf{X}_1'\mathbf{y}_1 = \begin{pmatrix} 97{,}500 \\ 1{,}788{,}052 \\ 838{,}010 \\ 29{,}484{,}809 \end{pmatrix},$$

$$\mathbf{S}_2 = \begin{pmatrix} 15.0 & 244.6 & 236.0 & 4{,}625.0 \\ 244.6 & 4{,}181.6 & 3{,}869.0 & 75{,}318.0 \\ 236.0 & 3{,}869.0 & 3{,}824.0 & 72{,}500.0 \\ 4{,}625.0 & 75{,}318.0 & 72{,}500.0 & 1{,}427{,}425.0 \end{pmatrix}, \quad \mathbf{X}_2'\mathbf{y}_2 = \begin{pmatrix} 83{,}470 \\ 1{,}404{,}814 \\ 1{,}320{,}100 \\ 25{,}727{,}050 \end{pmatrix},$$

$$\mathbf{S}_3 = \begin{pmatrix} 16.0 & 256.0 & 97.0 & 2{,}995.0 \\ 256.0 & 4{,}221.7 & 1{,}619.2 & 47{,}897.0 \\ 97.0 & 1{,}619.2 & 785.0 & 17{,}840.0 \\ 2{,}995.0 & 47{,}897.0 & 17{,}840.0 & 580{,}475.0 \end{pmatrix}, \quad \mathbf{X}_3'\mathbf{y}_3 = \begin{pmatrix} 89{,}280 \\ 1{,}456{,}596 \\ 554{,}650 \\ 16{,}743{,}450 \end{pmatrix},$$

where

$$\mathbf{S}_j = (s_{jrt}), \qquad s_{jrt} = \sum_{i=1}^{n_j} x_{jir}x_{jit}, \qquad r, t = 1, 2, 3, 4,$$

$$\mathbf{X}_j'\mathbf{y}_j = (s_{jyx_r}), \qquad s_{jyx_r} = \sum_{i=1}^{n_j} y_{ji}x_{jir}.$$

Note that above the element $s_{111} = n_1$, $s_{211} = n_2$, and $s_{311} = n_3$. The multiple regression equation for all three categories combined is given by (6.4) where, it will be remembered, specification of the matrices $\mathbf{X}^\alpha$ and $\boldsymbol{\beta}^\alpha$ depends on the model prescribed by hypothesis. The data in the above matrices can be suitably arranged for analysis according to hypothesis.

To illustrate the statistical method seven hypotheses are considered and tested. The hypothesis H_1 imposes no restriction on the β's, so that

$$H_1: \beta_{j1} = \beta_{j1}, \qquad \beta_{jr} = \beta_{jr}, \qquad r = 2, 3, 4.$$

TABLE 8.1. Analysis of Variance Table for Tests of Various Null Hypotheses H_α, $\alpha = 2, 3, 4, 5, 6, 7$, Against an Alternative Hypothesis H_1

Variation due to	Sum of Squares		D.F.		$\hat{J}(H_1, H_\alpha)$	F
$H_2: \beta_{j1} = \beta_{j1}$		1,556,805,752	$k =$	3		
$\beta_{jr} = 0, r = 2, 3, 4$						
$\boldsymbol{\beta}^2$						
Diff.: H_1, H_2		24,993,036	$pk - k =$	9	870.47	96.7*
$H_3: \beta_{j1} = \beta_{\cdot 1}$		1,553,937,500	1	1		
$\beta_{jr} = \beta_{\cdot r}, r = 2, 3, 4$		27,030,350	$p - 1$	3		
$\boldsymbol{\beta}^3$		1,580,967,850	p	4		
Diff.: H_1, H_3		830,938	$pk - p$	8	28.94	3.62*
$H_4: \beta_{j1} = \beta_{j1}$		1,556,805,752	k	3		
$\beta_{jr} = \beta_{\cdot r}, r = 2, 3, 4$		24,328,284	$p - 1$	3		
$\boldsymbol{\beta}^4$		1,581,134,036	$p + k - 1$	6		
Diff.: H_1, H_4		664,752	$(p - 1)(k - 1)$	6	23.15	3.86*
$H_5: \beta_{j1} = \beta_{j1}$	$\mathbf{y}'\mathbf{X}_1\mathbf{S}_{11}^{-1}\mathbf{X}_1'\mathbf{y}$	1,556,805,752	k	3		
$\beta_{jr} = \beta_{jr}, r = 2$	$\hat{\boldsymbol{\beta}}_2'\mathbf{X}_{2\cdot 1}'\mathbf{y}$	24,333,415	k	3		
$= \beta_{\cdot r}, r = 3, 4$	$\hat{\boldsymbol{\beta}}_3'\mathbf{X}_{3\cdot 1}'\mathbf{y}$	65,381	$p - 2$	2		
$\boldsymbol{\beta}^5$	$\hat{\boldsymbol{\beta}}^{5\prime}\mathbf{S}^5\hat{\boldsymbol{\beta}}^5$	1,581,204,548	$p + 2(k - 1)$	8		
Diff.: H_1, H_5	$\hat{\boldsymbol{\beta}}^{1\prime}\mathbf{S}^1\hat{\boldsymbol{\beta}}^1 - \hat{\boldsymbol{\beta}}^{5\prime}\mathbf{S}^5\hat{\boldsymbol{\beta}}^5 = \hat{\sigma}^2\hat{J}(H_1, H_5)$	594,240	$(p - 2)(k - 1)$	4	20.70	5.17*

TABLE 8.1 (continued)

$H_6: \beta_{j1} = \beta_{j1}$		1,556,805,752	k	3		
$\beta_{jr} = \beta_{jr}, r = 2$		24,352,124	k	3		
$= 0, \quad r = 3, 4$						
$\boldsymbol{\beta}^6$		1,581,157,876	$2k$	6		
Diff.: H_1, H_6		640,912	$(p-2)k$	6	22.32	3.72*
$H_7: \beta_{j1} = \beta_{j1}$		1,556,805,752	k	3		
$\beta_{jr} = \beta_{\cdot r}, r = 2$		22,464,483	1	1		
$= \beta_{jr}, r = 3, 4$		2,450,665	$2(p-1)$	6		
$\boldsymbol{\beta}^7$		1,581,720,900	$2p + k - 1$	10		
Diff.: H_1, H_7		77,888	$p(k-2) - (k-1)$	2	2.71	1.36
$H_1: \beta_{j1} = \beta_{j1}$	$\mathbf{y}'\mathbf{X}_1\mathbf{S}_{11}^{-1}\mathbf{X}_1'\mathbf{y}$	1,556,805,752	k	3		
$\beta_{jr} = \beta_{jr}, r = 2, 3, 4$	$\hat{\boldsymbol{\beta}}_2'\mathbf{X}_{2\cdot 1}'\mathbf{y}$	24,993,036	$pk - k$	9		
$\boldsymbol{\beta}^1$	$\hat{\boldsymbol{\beta}}^{1\prime}\mathbf{S}^1\hat{\boldsymbol{\beta}}^1$	1,581,798,788	pk	12		
Difference	$\mathbf{y}'\mathbf{y} - \hat{\boldsymbol{\beta}}^{1\prime}\mathbf{S}^1\hat{\boldsymbol{\beta}}^1 = (n - pk)\hat{\sigma}^2$	1,004,912	$n - pk$	35	$\hat{\sigma}^2 = 28{,}712$	
Total	$\mathbf{y}'\mathbf{y}$	1,582,803,700	$n = n_1 + n_2 + n_3$	47		

* Significance at 0.01 probability level.

All other hypotheses, suggested by the nature of the data, are compared as null hypotheses against H_1:

$$
\begin{array}{lllll}
H_2:\beta_{j1} = \beta_{j1}, & \beta_{jr} = 0, & r = 2, 3, 4, & & \\
H_3:\beta_{j1} = \beta_{\cdot 1}, & \beta_{jr} = \beta_{\cdot r}, & r = 2, 3, 4, & & \\
H_4:\beta_{j1} = \beta_{j1}, & \beta_{jr} = \beta_{\cdot r}, & r = 2, 3, 4, & & \\
H_5:\beta_{j1} = \beta_{j1}, & \beta_{jr} = \beta_{jr}, & r = 2, & \beta_{jr} = \beta_{\cdot r}, & r = 3, 4, \\
H_6:\beta_{j1} = \beta_{j1}, & \beta_{jr} = \beta_{jr}, & r = 2, & \beta_{jr} = 0, & r = 3, 4, \\
H_7:\beta_{j1} = \beta_{j1}, & \beta_{jr} = \beta_{\cdot r}, & r = 2, & \beta_{jr} = \beta_{jr}, & r = 3, 4.
\end{array}
$$

The statements above of the various hypotheses all apply for $j = 1, 2, 3$. In stating these hypotheses we have specified β_{j1} separately, for convenience, since in this example it represents the constant term which depends on the mean values. Table 8.1 presents the complete summary of the analysis of variance data and the tests of significance of the various hypotheses. Table 8.2 presents the estimated regression coefficients under the various hypotheses. (The computations were carried out by H. M. Rosenblatt, Fred Okano, and the computing staff at the Naval Proving Ground, Dahlgren, Va.) The specification of the matrices $\mathbf{X}^\alpha$ and $\boldsymbol{\beta}^\alpha$ for H_1 and H_5 is also given, following table 8.2; those for the other hypotheses follow on the same lines. (These are left to the reader.)

Using the 0.01 probability level for significance, and the 0.05 probability level for caution, it is concluded, from table 8.1, that:

1. The regression is real; reject H_2.
2. One set of regression coefficients, including equality of means, cannot adequately represent all three categories; reject H_3.
3. One set of regression coefficients is not adequate even after allowing for differences in the mean value for each category; reject H_4.
4. One set of regression coefficients for variables x_3 and x_4 for all three categories cannot be used; reject H_5.
5. The regression coefficients for x_3 and x_4 cannot be ignored; reject H_6.

However,

6. the use of one regression coefficient for the variable x_2 and different ones for x_3 and x_4 and for the constant term is adequate; accept H_7.

For the hypotheses H_1 and H_5 considered in the example, the matrix of parameters $\boldsymbol{\beta}$ and the matrix of observations $\mathbf{X}$ are given below. Note that, since we are dealing with raw observations in the example, the regression coefficients β_{j1} of $\boldsymbol{\beta}$ and the matrix (vector) $\mathbf{x}_{j1}$ of $\mathbf{X}$, $j = 1, 2, 3$,

TABLE 8.2. Estimates of Regression Coefficients Under Various Hypotheses

Hypothesis	$\hat{\beta}_{j1}$	$\hat{\beta}_{j2}$	$\hat{\beta}_{j3}$	$\hat{\beta}_{j4}$
H_1				
$j = 1$	3,587	203.4	−10.69	−3.46
$j = 2$	−7,186	231.1	79.02	25.10
$j = 3$	1,654	227.7	−6.93	1.73
H_2				
$j = 1$	6,094			
$j = 2$	5,565			
$j = 3$	5,580			
H_3				
$j = 1, 2, 3$	2,009	219.8	−11.19	0.646
H_4				
$j = 1$	1,803	216.0	3.70	1.28
$j = 2$	1,589	216.0	3.70	1.28
$j = 3$	1,862	216.0	3.70	1.28
H_5				
$j = 1$	2,071	201.2	.563	1.36
$j = 2$	1,432	227.1	.563	1.36
$j = 3$	1,743	223.7	.563	1.36
H_6				
$j = 1$	2,467	202.3		
$j = 2$	1,872	226.5		
$j = 3$	2,001	223.7		
H_7				
$j = 1$	3,431	219.8	−4.27	−4.10
$j = 2$	−6,768	219.8	79.26	24.33
$i = 3$	1,758	219.8	−4.16	1.77

have been partitioned for every hypothesis. This provides for the usual practice of obtaining sums of squares and products of deviations about average values to simplify further calculations by reducing by one the rank of the matrix **S** (of sums of squares and products) whose inverse must be obtained.

$$H_1:\beta_{j1} = \beta_{j1}, \quad \beta_{jr} = \beta_{jr}, \quad r = 2, 3, 4, \quad j = 1, 2, 3,$$

$$\boldsymbol{\beta}' = (\boldsymbol{\beta}_1', \boldsymbol{\beta}_2'), \quad \mathbf{X} = (\mathbf{X}_1, \mathbf{X}_2),$$

$$\boldsymbol{\beta}_1' = (\beta_{11}, \beta_{21}, \beta_{31}), \quad \boldsymbol{\beta}_2' = (\boldsymbol{\beta}_{12}', \boldsymbol{\beta}_{22}', \boldsymbol{\beta}_{32}'), \quad \boldsymbol{\beta}_{j2}' = (\beta_{j2}, \beta_{j3}, \beta_{j4}),$$

$$\mathbf{X}_1 = \begin{pmatrix} \mathbf{x}_{11} & \cdots & \mathbf{0} \\ \vdots & \mathbf{x}_{21} & \vdots \\ \mathbf{0} & \cdots & \mathbf{x}_{31} \end{pmatrix}, \quad \mathbf{X}_2 = \begin{pmatrix} \mathbf{X}_{12} & \cdots & \mathbf{0} \\ \vdots & \mathbf{X}_{22} & \vdots \\ \mathbf{0} & \cdots & \mathbf{X}_{32} \end{pmatrix},$$

$$\mathbf{x}_{j1}' = (1, 1, \cdots, 1), \quad \mathbf{X}_{j2} = (\mathbf{x}_{j2}, \mathbf{x}_{j3}, \mathbf{x}_{j4}),$$

$$\text{order } 1 \times n_j \quad \mathbf{x}_{jr}' = (x_{j1r}, x_{j2r}, \cdots, x_{jn_jr}).$$

$$H_5:\beta_{j1} = \beta_{j1}, \quad \beta_{jr} = \beta_{jr}, \quad r = 2, \quad \beta_{jr} = \beta_{\cdot r}, \quad r = 3, 4, \quad j = 1, 2, 3,$$

$$\boldsymbol{\beta}' = (\boldsymbol{\beta}_1', \boldsymbol{\beta}_2', \boldsymbol{\beta}_3'), \quad \mathbf{X} = (\mathbf{X}_1, \mathbf{X}_2, \mathbf{X}_3),$$

$$\boldsymbol{\beta}_1' = (\beta_{11}, \beta_{21}, \beta_{31}), \quad \boldsymbol{\beta}_2' = (\beta_{12}, \beta_{22}, \beta_{32}), \quad \boldsymbol{\beta}_3' = (\beta_{\cdot 3}, \beta_{\cdot 4}),$$

$$\mathbf{X}_1 = \begin{pmatrix} \mathbf{x}_{11} & \cdots & \mathbf{0} \\ \vdots & \mathbf{x}_{21} & \vdots \\ \mathbf{0} & \cdots & \mathbf{x}_{31} \end{pmatrix}, \quad \mathbf{X}_2 = \begin{pmatrix} \mathbf{x}_{12} & \cdots & \mathbf{0} \\ \vdots & \mathbf{x}_{22} & \vdots \\ \mathbf{0} & \cdots & \mathbf{x}_{32} \end{pmatrix}, \quad \begin{matrix} \mathbf{X}_3' = (\mathbf{X}_{13}', \mathbf{X}_{23}', \mathbf{X}_{33}'), \\ \mathbf{X}_{j3} = (\mathbf{x}_{j3}, \mathbf{x}_{j4}), \end{matrix}$$

$\mathbf{x}_{j1}$ and $\mathbf{x}_{jr}$, $r = 2, 3, 4$, are defined as under H_1.

In the foregoing example each hypothesis on the parameters applies to all categories, $j = 1, 2, 3$. It should be clear, however, that this need not be the case, for the theory and method are equally applicable for any assertion of the hypotheses about the parameters. For example, we might have an analysis where part of the hypothesis concerned equality of the parameters for certain of the categories but not for all, for example,

$$\begin{aligned} H_8:\beta_{jr} &= \beta_{\cdot r}, & j &= 1, 3, & r &= 1, \\ \beta_{jr} &= \beta_{jr}, & j &= 2, & r &= 1, \\ \beta_{jr} &= \beta_{\cdot r}, & j &= 1, 2, 3, & r &= 2, \\ \beta_{jr} &= \beta_{jr}, & j &= 1, 2, 3, & r &= 3, 4, \end{aligned}$$

and analysis by the three-partition subhypothesis procedure of section 5 would apply.

9. REPARAMETRIZATION

9.1. Hypotheses Not of Full Rank

[Cf. Kempthorne (1952, section 6.2).] Suppose that the components of $\boldsymbol{\beta}$ in (3.1) are not linearly independent, but satisfy $(p - r)$ linear relations. This implies that the matrix $\mathbf{X}$ in (3.1) is of rank $r < p$ (and conversely), and that

$$\boldsymbol{\beta} = \mathbf{G}\boldsymbol{\gamma}, \tag{9.1}$$

where $\boldsymbol{\gamma}' = (\gamma_1, \gamma_2, \cdots, \gamma_r)$, $\mathbf{G} = (g_{ij})$, $i = 1, \cdots, p$, $j = 1, 2, \cdots, r$, and $\mathbf{G}$ is of rank $r < p$. The matrix $\mathbf{S} = \mathbf{X}'\mathbf{X}$ is now a positive (not positive definite) matrix of rank r, is therefore singular and has no inverse, so that we must re-examine the solution of (4.1) for $\hat{\boldsymbol{\beta}}$. We may, however, write (3.1) as

$$\mathbf{z} = \mathbf{y} - \mathbf{X}\mathbf{G}\boldsymbol{\gamma} = \mathbf{y} - \mathbf{A}\boldsymbol{\gamma}, \tag{9.2}$$

where $\mathbf{A} = \mathbf{X}\mathbf{G}$ is an $n \times r$ matrix of rank r. The least squares estimate of $\boldsymbol{\gamma}$ is derived from the normal equations

$$\mathbf{A}'\mathbf{A}\hat{\boldsymbol{\gamma}} = \mathbf{A}'\mathbf{y} \quad \text{or} \quad \mathbf{G}'\mathbf{S}\mathbf{G}\hat{\boldsymbol{\gamma}} = \mathbf{G}'\mathbf{X}'\mathbf{y}. \tag{9.3}$$

The estimate of $\boldsymbol{\beta}$ is obtained from $\hat{\boldsymbol{\beta}} = \mathbf{G}\hat{\boldsymbol{\gamma}}$, or

$$\hat{\boldsymbol{\beta}} = \mathbf{G}(\mathbf{G}'\mathbf{S}\mathbf{G})^{-1}\,\mathbf{G}'\mathbf{X}'\mathbf{y}. \tag{9.4}$$

As in section 4, $\hat{\boldsymbol{\gamma}}$ is a minimum variance, unbiased, sufficient estimate of $\boldsymbol{\gamma}$ and the components of $\hat{\boldsymbol{\gamma}}$ are normally distributed with covariance matrix $\sigma^2(\mathbf{A}'\mathbf{A})^{-1} = \sigma^2(\mathbf{G}'\mathbf{S}\mathbf{G})^{-1}$. Also, $\hat{\boldsymbol{\beta}} = \mathbf{G}\hat{\boldsymbol{\gamma}}$ is an unbiased estimate of $\boldsymbol{\beta}$ and the components of $\hat{\boldsymbol{\beta}}$ are normally distributed with covariance matrix $\sigma^2\mathbf{G}(\mathbf{G}'\mathbf{S}\mathbf{G})^{-1}\mathbf{G}'$. Corresponding to (4.2) we have

$$\hat{J}(H_1, H_2) = \frac{(\hat{\boldsymbol{\gamma}}^1 - \boldsymbol{\gamma}^2)'\mathbf{A}'\mathbf{A}(\hat{\boldsymbol{\gamma}}^1 - \boldsymbol{\gamma}^2)}{\hat{\sigma}^2} = \frac{(\hat{\boldsymbol{\gamma}}^1 - \boldsymbol{\gamma}^2)'\mathbf{G}'\mathbf{S}\mathbf{G}(\hat{\boldsymbol{\gamma}}^1 - \boldsymbol{\gamma}^2)}{\hat{\sigma}^2} = \frac{(\hat{\boldsymbol{\beta}}^1 - \boldsymbol{\beta}^2)'\mathbf{S}(\hat{\boldsymbol{\beta}}^1 - \boldsymbol{\beta}^2)}{\hat{\sigma}^2}, \tag{9.5}$$

where

$$(n - r)\hat{\sigma}^2 = \mathbf{y}'\mathbf{y} - \hat{\boldsymbol{\gamma}}^{1\prime}\mathbf{A}'\mathbf{A}\hat{\boldsymbol{\gamma}}^1 = \mathbf{y}'\mathbf{y} - \hat{\boldsymbol{\gamma}}^{1\prime}\mathbf{G}'\mathbf{S}\mathbf{G}\hat{\boldsymbol{\gamma}}^1 = \mathbf{y}'\mathbf{y} - \hat{\boldsymbol{\beta}}^{1\prime}\mathbf{S}\hat{\boldsymbol{\beta}}^1.$$

Note that $\mathbf{G}'\mathbf{S}\hat{\boldsymbol{\beta}} = \mathbf{G}'\mathbf{X}'\mathbf{y}$ [see (9.3)] represents r linear functions of the y's that are also linear functions of the $\hat{\beta}$'s. These are unbiased estimates of the same linear functions of the β's. Since $\mathbf{G}'\mathbf{S}\hat{\boldsymbol{\beta}} = \mathbf{G}'\mathbf{X}'\mathbf{y} = \mathbf{G}'\mathbf{S}\mathbf{G}\hat{\boldsymbol{\gamma}}$, we may make similar statements about the γ's and their estimates. Consider now any other set of r linear functions of the y's, say $\mathbf{L}\mathbf{y}$, where $\mathbf{L}$ is an $r \times n$ matrix of rank r. Since

$$E(\mathbf{L}\mathbf{y}) = E(\mathbf{L}(\mathbf{X}\boldsymbol{\beta} + \mathbf{z})) = \mathbf{L}\mathbf{X}\boldsymbol{\beta} = \mathbf{L}\mathbf{X}\mathbf{G}\boldsymbol{\gamma}, \tag{9.6}$$

$\mathbf{L}\mathbf{y}$ is an unbiased estimate of $\boldsymbol{\gamma}$ if $\mathbf{L}\mathbf{X}\mathbf{G} = \mathbf{I}_r$, the $r \times r$ identity matrix. The covariance matrix of the components of $\mathbf{L}\mathbf{y}$ is seen to be $\sigma^2\mathbf{L}\mathbf{L}'$.

From lemma 5.4 of chapter 3 with $k = n$, $\mathbf{B} = \sigma^2\mathbf{I}_n$, where $\mathbf{I}_n$ is the $n \times n$ identity matrix, $\mathbf{L} = \mathbf{C}'$, $\mathbf{U}' = \mathbf{XG}$, $\mathbf{C}'\mathbf{U}' = \mathbf{LXG} = \mathbf{I}_r$,

$$\sigma^2\mathbf{LL}' \geqq \sigma^2(\mathbf{G}'\mathbf{SG})^{-1}, \tag{9.7}$$

where (9.7) means that any quadratic form with matrix $\sigma^2\mathbf{LL}'$ is greater than or equal to the quadratic form with matrix $\sigma^2(\mathbf{G}'\mathbf{SG})^{-1}$. Since the covariance matrix of the components of $\hat{\boldsymbol{\gamma}}$ is $\sigma^2(\mathbf{G}'\mathbf{SG})^{-1}$, we confirm by (9.7) and lemma 5.1(c) of chapter 3 the statement that the variances of the components of $\hat{\boldsymbol{\gamma}}$ are the smallest among all linear functions of the y's that are unbiased estimates of $\boldsymbol{\gamma}$. Similarly, $\mathbf{GLy}$ is an unbiased estimate of $\boldsymbol{\beta}$ if $\mathbf{LXG} = \mathbf{I}_r$. From lemma 5.4 of chapter 3 we may conclude that

$$\sigma^2\mathbf{GLL}'\mathbf{G}' \geqq \sigma^2\mathbf{G}(\mathbf{G}'\mathbf{SG})^{-1}\mathbf{G}', \tag{9.8}$$

from which we infer that the variances of the components of $\hat{\boldsymbol{\beta}}$ are the smallest among all linear functions of the y's that are unbiased estimates of $\boldsymbol{\beta}$.

The value of $J(1, 2)$ and its estimate is the same for any reparametrization as is indicated in (9.5). Since there are only r linearly independent linear functions of the β's, any one set of r linearly independent functions of the β's may be derived from any other such set by a nonsingular linear transformation. The information functions are invariant under nonsingular transformations (see section 4 of chapter 2, also section 3 of chapter 9), hence our conclusion. [Cf. Kempthorne (1952).]

Examples of the application of this procedure to the two-way classification with no replication and with no interaction; the two-way classification with missing observations; the two-way classification with replication and interaction; the two-way classification with replication (unequal cell frequencies), interaction, and missing observations; the latin square; and the latin square with missing observations may be found in McCall (1957). See also Anderson and Bancroft (1952), Kempthorne (1952).

9.2. Partition

When the hypotheses call for a partitioning of the parameters into two sets, for example as in (5.1), it is possible that linear relations may exist among the parameters in one of the partitioned sets only. Here it is necessary to apply the procedures of section 9.1 only to the partitioned set not of full rank. Thus, suppose that in (5.1) the $n \times q$ matrix $\mathbf{X}_1$ is of rank $m < q$. This implies [cf. (9.1)] that

$$\boldsymbol{\beta}_1 = \mathbf{G}_1\boldsymbol{\gamma}_1, \tag{9.9}$$

where $\boldsymbol{\gamma}_1' = (\gamma_1, \gamma_2, \cdots, \gamma_m)$, $\mathbf{G}_1 = (g_{ij})$, $i = 1, 2, \cdots, q$, $j = 1, 2, \cdots, m$, and $\mathbf{G}_1$ is of rank $m < q$. The results of section 5.1 are applicable if $\boldsymbol{\beta}_1$ and $\hat{\boldsymbol{\beta}}_1$ are replaced in the various formulas by $\boldsymbol{\gamma}_1$ and $\hat{\boldsymbol{\gamma}}_1$ respectively, $\mathbf{X}_1$ by $\mathbf{X}_1\mathbf{G}_1$, and $(n - q)$ degrees of freedom by $(n - m)$ degrees of freedom. The estimate $\hat{\boldsymbol{\beta}}_1$ is obtained from $\hat{\boldsymbol{\beta}}_1 = \mathbf{G}_1\hat{\boldsymbol{\gamma}}_1$. Thus, for example, $\mathbf{S}_{11}$ in (5.3) is to be replaced by $\mathbf{G}_1'\mathbf{S}_{11}\mathbf{G}_1$, where $\mathbf{S}_{11} = \mathbf{X}_1'\mathbf{X}_1$, and $\mathbf{S}_{12}$ by $\mathbf{G}_1'\mathbf{S}_{12}$, where $\mathbf{S}_{12} = \mathbf{X}_1'\mathbf{X}_2$.

Similar remarks also apply for a partitioning into three sets as in section 5.2, when one of the sets may not be of full rank.

10. ANALYSIS OF REGRESSION, TWO-WAY CLASSIFICATION

To illustrate section 9, and for its own interest, suppose we have a two-way classification with r row-categories and c column-categories, with one observation per cell and no interaction. Suppose furthermore that there are p independent variables $x_1, x_2, \cdots, x_p$. We want to test a null hypothesis that there are no column effects (the column classification is not significant), against an alternative hypothesis that the column classification is significant. For $p = 1$, also identified as the analysis of covariance, see Federer (1955, p. 487), Kempthorne (1952, p. 98). For $p = 2$, also designated as multiple covariance, see Snedecor (1946, section 13.7). For p general, see Anderson and Bancroft (1952, section 21.4).

The general linear regression model for each cell is

$$(10.1) \qquad z_{ij} = y_{ij} - \mu - \rho_i - \tau_j - \beta_1 x_{ij1} - \beta_2 x_{ij2} - \cdots - \beta_p x_{ijp},$$

where $i = 1, 2, \cdots, r$ row-categories,

$j = 1, 2, \cdots, c$ column-categories,

ρ_i is the ith row effect,

τ_j is the jth column effect,

μ is the over-all mean,

the z_{ij} are independent, normally distributed random variables with zero means and common variance σ^2, and the x_{ijk}, $i = 1, 2, \cdots, r$, $j = 1, 2, \cdots, c$, $k = 1, 2, \cdots, p$, are known.

Enumerating the cells from left to right and top to bottom, the linear regressions may be written as

$$(10.2) \qquad \mathbf{z} = \mathbf{y} - \mathbf{X}\boldsymbol{\beta} = \mathbf{y} - \mathbf{X}_1\boldsymbol{\beta}_1 - \mathbf{X}_2\boldsymbol{\beta}_2 - \mathbf{X}_3\boldsymbol{\beta}_3,$$

where

$$\mathbf{z}' = (z_{11}, z_{12}, \cdots, z_{rc}), \qquad \mathbf{y}' = (y_{11}, y_{12}, \cdots, y_{rc}),$$
$$\mathbf{X} = (\mathbf{X}_1, \mathbf{X}_2, \mathbf{X}_3), \qquad \boldsymbol{\beta}' = (\boldsymbol{\beta}_1', \boldsymbol{\beta}_2', \boldsymbol{\beta}_3'),$$

$$\mathbf{X}_1 = \begin{pmatrix} 1 \\ 1 \\ \cdot \\ \cdot \\ \cdot \\ 1 \end{pmatrix}, \qquad \mathbf{X}_2 = \begin{pmatrix} 1 & 0 \cdots 0 & 1 & 0 \cdots 0 \\ 1 & 0 \cdots 0 & 0 & 1 \cdots 0 \\ \cdot & \cdot & \cdot & \cdot \\ 1 & 0 \cdots 0 & 0 & 0 \cdots 1 \\ \cdot & \cdot & \cdot & \cdot \\ 0 & 0 \cdots 1 & 1 & 0 \cdots 0 \\ 0 & 0 \cdots 1 & 0 & 1 \cdots 0 \\ \cdot & \cdot & \cdot & \cdot \\ 0 & 0 \cdots 1 & 0 & 0 \cdots 1 \end{pmatrix},$$

$$\mathbf{X}_3 = \begin{pmatrix} \mathbf{x}_{11}' \\ \mathbf{x}_{12}' \\ \cdot \\ \cdot \\ \cdot \\ \mathbf{x}_{rc}' \end{pmatrix}, \qquad \mathbf{x}_{ij}' = (x_{ij1}, x_{ij2}, \cdots, x_{ijp}),$$

$$\boldsymbol{\beta}_1' = (\mu), \qquad \boldsymbol{\beta}_2' = (\boldsymbol{\rho}', \boldsymbol{\tau}'), \qquad \boldsymbol{\rho}' = (\rho_1, \rho_2, \cdots, \rho_r),$$
$$\boldsymbol{\tau}' = (\tau_1, \tau_2, \cdots, \tau_c), \qquad \boldsymbol{\beta}_3' = (\beta_1, \beta_2, \cdots, \beta_p),$$

that is, $\mathbf{X}_1$ is $rc \times 1$, $\mathbf{X}_2$ is $rc \times (r + c)$, $\mathbf{X}_3$ is $rc \times p$, $\mathbf{x}_{ij}'$ is $1 \times p$, $\boldsymbol{\beta}_1'$ is 1×1, $\boldsymbol{\beta}_2'$ is $1 \times (r + c)$, $\boldsymbol{\rho}'$ is $1 \times r$, $\boldsymbol{\tau}'$ is $1 \times c$, $\boldsymbol{\beta}_3'$ is $1 \times p$.

We want to test the hypothesis

$$(10.3) \qquad H_1: \boldsymbol{\beta}' = \boldsymbol{\beta}^{1'} = (\boldsymbol{\beta}_1^{1'}, \boldsymbol{\beta}_2^{1'}, \boldsymbol{\beta}_3^{1'}), \qquad \boldsymbol{\beta}_2^{1'} = (\boldsymbol{\rho}^{1'}, \boldsymbol{\tau}^{1'}),$$

that is, no restrictions on the parameters, and the null hypothesis

$$(10.4) \quad H_2: \boldsymbol{\beta}' = \boldsymbol{\beta}^{2'} = (\boldsymbol{\beta}_1^{2'}, \boldsymbol{\beta}_2^{2'}, \boldsymbol{\beta}_3^{2'}), \qquad \boldsymbol{\beta}_2^{2'} = (\boldsymbol{\rho}^{2'}, \mathbf{0}), \quad \text{or} \quad \boldsymbol{\tau}^{2'} = \mathbf{0},$$

that is there are no column effects.

Note that the $rc \times (r + c)$ matrix $\mathbf{X}_2$ is of rank $r + c - 2$, since the row and column effects are essentially restricted to satisfy [cf. Anderson and Bancroft (1952), Kempthorne (1952)]

$$(10.5) \quad \rho_1 + \rho_2 + \cdots + \rho_r = 0, \qquad \tau_1 + \tau_2 + \cdots + \tau_c = 0.$$

The new parameters for the second set of the partition, taking (10.5) into account, are given by

$$(10.6) \qquad \boldsymbol{\beta}_2 = \begin{pmatrix} \boldsymbol{\rho} \\ \boldsymbol{\tau} \end{pmatrix} = \mathbf{G}\boldsymbol{\gamma} = \begin{pmatrix} \mathbf{G}_1 & \mathbf{0} \\ \mathbf{0} & \mathbf{G}_2 \end{pmatrix} \begin{pmatrix} \boldsymbol{\gamma}_1 \\ \boldsymbol{\gamma}_2 \end{pmatrix},$$

where $\gamma_1' = (\gamma_{11}, \cdots, \gamma_{1(r-1)})$, $\gamma_2' = (\gamma_{21}, \cdots, \gamma_{2(c-1)})$, and $\mathbf{G}_1$ and $\mathbf{G}_2$ are respectively the $r \times (r-1)$ and $c \times (c-1)$ matrices

$$\mathbf{G}_1 = \begin{pmatrix} 1 & 0 \cdots & 0 \\ 0 & 1 \cdots & 0 \\ \cdot & \cdot \cdot \cdot \cdot & \cdot \\ 0 & 0 \cdots & 1 \\ -1 & -1 \cdots & -1 \end{pmatrix}, \qquad \mathbf{G}_2 = \begin{pmatrix} 1 & 0 \cdots & 0 \\ 0 & 1 \cdots & 0 \\ \cdot & \cdot \cdot \cdot \cdot & \cdot \\ 0 & 0 \cdots & 1 \\ -1 & -1 \cdots & -1 \end{pmatrix}.$$

For the second set of the partition, we find

$$(10.7) \quad \mathbf{X}_2\mathbf{G} = \begin{pmatrix} 1 & 0 \cdots & 0 & 1 & 0 \cdots & 0 \\ 1 & 0 \cdots & 0 & 0 & 1 \cdots & 0 \\ \cdot & \cdot \cdot \cdot & \cdot & \cdot & \cdot \cdot \cdot & \cdot \\ 1 & 0 \cdots & 0 & -1 & -1 \cdots -1 \\ 0 & 1 \cdots & 0 & 1 & 0 \cdots & 0 \\ 0 & 1 \cdots & 0 & 0 & 1 \cdots & 0 \\ \cdot & \cdot \cdot \cdot & \cdot & \cdot & \cdot \cdot \cdot & \cdot \\ 0 & 1 \cdots & 0 & -1 & -1 \cdots -1 \\ \cdot & \cdot \cdot \cdot & \cdot & \cdot & \cdot \cdot \cdot & \cdot \\ -1 & -1 \cdots -1 & 1 & 0 \cdots & 0 \\ -1 & -1 \cdots -1 & 0 & 1 \cdots & 0 \\ \cdot & \cdot \cdot \cdot & \cdot & \cdot & \cdot \cdot \cdot & \cdot \\ -1 & -1 \cdots -1 & -1 & -1 \cdots -1 \end{pmatrix}, \begin{matrix} \\ \\ c \\ \\ \\ \\ c \\ \\ \\ \\ c \\ \\ \end{matrix}$$

$$\underbrace{\qquad\qquad}_{r-1} \quad \underbrace{\qquad\qquad}_{c-1}$$

where $\mathbf{X}_2\mathbf{G}$ has rc rows and $r - 1 + c - 1$ columns,

$$(10.8) \quad \mathbf{G}'\mathbf{X}_2'\mathbf{X}_2\mathbf{G} = \mathbf{G}'\mathbf{S}_{22}\mathbf{G} = \begin{pmatrix} 2c & c \cdots & c & 0 & 0 \cdots & 0 \\ c & 2c \cdots & c & 0 & 0 \cdots & 0 \\ \cdot & \cdot \cdot \cdot & \cdot & \cdot & \cdot \cdot \cdot & \cdot \\ c & c \cdots & 2c & 0 & 0 \cdots & 0 \\ 0 & 0 \cdots & 0 & 2r & r \cdots & r \\ 0 & 0 \cdots & 0 & r & 2r \cdots & r \\ \cdot & \cdot \cdot \cdot & \cdot & \cdot & \cdot \cdot \cdot & \cdot \\ 0 & 0 \cdots & 0 & r & r \cdots & 2r \end{pmatrix} \begin{matrix} \Big\} r-1 \\ \\ \\ \Big\} c-1, \\ \\ \end{matrix}$$

$$\underbrace{\qquad\qquad}_{r-1} \quad \underbrace{\qquad\qquad}_{c-1}$$

$$(10.9) \quad \mathbf{X}_1'\mathbf{X}_2\mathbf{G} = \mathbf{S}_{12}\mathbf{G} = (0, 0, \cdots, 0), \qquad 1 \times (r - 1 + c - 1),$$

$$(10.10) \quad \mathbf{X}_3'\mathbf{X}_2\mathbf{G} = \mathbf{S}_{32}\mathbf{G} = (\mathbf{x}_{1\cdot} - \mathbf{x}_{r\cdot}, \mathbf{x}_{2\cdot} - \mathbf{x}_{r\cdot}, \cdots, \mathbf{x}_{r-1\cdot} - \mathbf{x}_{r\cdot},$$
$$\mathbf{x}_{\cdot 1} - \mathbf{x}_{\cdot c}, \mathbf{x}_{\cdot 2} - \mathbf{x}_{\cdot c}, \cdots, \mathbf{x}_{\cdot c-1} - \mathbf{x}_{\cdot c}),$$

where $\mathbf{x}_{i\cdot} = \mathbf{x}_{i1} + \mathbf{x}_{i2} + \cdots + \mathbf{x}_{ic}$, $\mathbf{x}_{\cdot j} = \mathbf{x}_{1j} + \mathbf{x}_{2j} + \cdots + \mathbf{x}_{rj}$, and $\mathbf{S}_{32}\mathbf{G}$ is a $p \times (r - 1 + c - 1)$ matrix.

We also find that

$$(10.11)\qquad \mathbf{X}_1'\mathbf{X}_1 = \mathbf{S}_{11} = rc, \qquad \mathbf{X}_1'\mathbf{X}_3 = \mathbf{S}_{13} = \mathbf{x}_{..}',$$

where $\mathbf{x}_{..}' = \mathbf{x}_{11}' + \mathbf{x}_{12}' + \cdots + \mathbf{x}_{ij}' + \cdots + \mathbf{x}_{rc}'$, and $\mathbf{x}_{..}'$ is a $1 \times p$ matrix,

$$(10.12)\qquad \mathbf{X}_3'\mathbf{X}_3 = \mathbf{S}_{33} = \sum_{i=1}^{r}\sum_{j=1}^{c} \mathbf{x}_{ij}\mathbf{x}_{ij}',$$

where $\mathbf{S}_{33}$ is a $p \times p$ matrix,

$$(10.13)\qquad \mathbf{X}_1'\mathbf{y} = y_{..},$$

where $y_{..} = y_{11} + y_{12} + \cdots + y_{ij} + \cdots + y_{rc}$,

$$(10.14)\qquad \mathbf{G}'\mathbf{X}_2'\mathbf{y} = \begin{pmatrix} y_{1\cdot} - y_{r\cdot} \\ y_{2\cdot} - y_{r\cdot} \\ \cdot \\ \cdot \\ \cdot \\ y_{r-1\cdot} - y_{r\cdot} \\ y_{\cdot 1} - y_{\cdot c} \\ y_{\cdot 2} - y_{\cdot c} \\ \cdot \\ \cdot \\ \cdot \\ y_{\cdot c-1} - y_{\cdot c} \end{pmatrix} = \begin{pmatrix} \mathbf{y}_{\text{row}} \\ \mathbf{y}_{\text{col}} \end{pmatrix},$$

where $y_{i\cdot} = y_{i1} + y_{i2} + \cdots + y_{ic}$, $y_{\cdot j} = y_{1j} + y_{2j} + \cdots + y_{rj}$,
$\mathbf{y}_{\text{row}}' = (y_{1\cdot} - y_{r\cdot} \cdots y_{r-1\cdot} - y_{r\cdot})$, $\mathbf{y}_{\text{col}}' = (y_{\cdot 1} - y_{\cdot c} \cdots y_{\cdot c-1} - y_{\cdot c})$,

$$(10.15)\qquad \mathbf{X}_3'\mathbf{y} = \mathbf{x}_{11}y_{11} + \mathbf{x}_{12}y_{12} + \cdots + \mathbf{x}_{ij}y_{ij} + \cdots + \mathbf{x}_{rc}y_{rc},$$

where $\mathbf{X}_3'\mathbf{y}$ is a $p \times 1$ matrix.

Since under H_1 the estimates of the parameters are given in (5.20), we proceed to find the other matrices needed:

$$(10.16)\qquad \mathbf{S}_{22\cdot 1} = \mathbf{G}'\mathbf{S}_{22}\mathbf{G} - \mathbf{G}'\mathbf{S}_{21}\mathbf{S}_{11}^{-1}\mathbf{S}_{12}\mathbf{G} = \begin{pmatrix} \mathbf{C} & \mathbf{0} \\ \mathbf{0} & \mathbf{R} \end{pmatrix},$$

where $\mathbf{C}$ is the $(r-1) \times (r-1)$ matrix $\begin{pmatrix} 2c & c & \cdots & c \\ c & 2c & \cdots & c \\ \cdot & \cdot & \cdot & \cdot \\ c & c & \cdots & 2c \end{pmatrix}$ and $\mathbf{R}$ is the $(c-1) \times (c-1)$ matrix $\begin{pmatrix} 2r & r & \cdots & r \\ r & 2r & \cdots & r \\ \cdot & \cdot & \cdot & \cdot \\ r & r & \cdots & 2r \end{pmatrix}$,

$$(10.17)\quad \mathbf{S}_{33\cdot 1} = \mathbf{S}_{33} - \mathbf{S}_{31}\mathbf{S}_{11}^{-1}\mathbf{S}_{13} = \sum_{i=1}^{r}\sum_{j=1}^{c}\mathbf{x}_{ij}\mathbf{x}'_{ij} - \mathbf{x}_{..}\frac{1}{rc}\mathbf{x}'_{..}$$

$$= \sum_{i=1}^{r}\sum_{j=1}^{c}(\mathbf{x}_{ij} - \bar{\mathbf{x}})(\mathbf{x}_{ij} - \bar{\mathbf{x}})',$$

where $\bar{\mathbf{x}} = \frac{1}{rc}\mathbf{x}_{..}$,

$$(10.18)\quad \mathbf{S}_{32\cdot 1} = \mathbf{S}_{32}\mathbf{G} - \mathbf{S}_{31}\mathbf{S}_{11}^{-1}\mathbf{S}_{12}\mathbf{G} = (\mathbf{x}_{1\cdot} - \mathbf{x}_{r\cdot}, \mathbf{x}_{2\cdot} - \mathbf{x}_{r\cdot}, \cdots, \mathbf{x}_{r-1\cdot} - \mathbf{x}_{r\cdot}, \mathbf{x}_{\cdot 1} - \mathbf{x}_{\cdot c}, \mathbf{x}_{\cdot 2} - \mathbf{x}_{\cdot c}, \cdots, \mathbf{x}_{\cdot c-1} - \mathbf{x}_{\cdot c}),$$

$$(10.19)\quad \mathbf{S}_{33\cdot 12} = \mathbf{S}_{33\cdot 1} - \mathbf{S}_{32\cdot 1}\mathbf{S}_{22\cdot 1}^{-1}\mathbf{S}_{23\cdot 1}$$

$$= \sum_{i=1}^{r}\sum_{j=1}^{c}\mathbf{d}_{ij}\mathbf{d}'_{ij} - (\mathbf{d}_{1\cdot}\cdots\mathbf{d}_{r-1\cdot}\,\mathbf{d}_{\cdot 1}\cdots\mathbf{d}_{\cdot c-1})\begin{pmatrix}\mathbf{C}^{-1} & \mathbf{0}\\ \mathbf{0} & \mathbf{R}^{-1}\end{pmatrix}\begin{pmatrix}\mathbf{d}'_{1\cdot}\\ \cdot\\ \cdot\\ \cdot\\ \mathbf{d}'_{r-1\cdot}\\ \mathbf{d}'_{\cdot 1}\\ \cdot\\ \cdot\\ \cdot\\ \mathbf{d}'_{\cdot c-1}\end{pmatrix}$$

$$= \sum_{i=1}^{r}\sum_{j=1}^{c}\mathbf{d}_{ij}\mathbf{d}'_{ij} - (\mathbf{d}_{1\cdot}\cdots\mathbf{d}_{r-1\cdot})\mathbf{C}^{-1}\begin{pmatrix}\mathbf{d}'_{1\cdot}\\ \cdot\\ \cdot\\ \cdot\\ \mathbf{d}'_{r-1\cdot}\end{pmatrix} - (\mathbf{d}_{\cdot 1}\cdots\mathbf{d}_{\cdot c-1})\mathbf{R}^{-1}\begin{pmatrix}\mathbf{d}'_{\cdot 1}\\ \cdot\\ \cdot\\ \cdot\\ \mathbf{d}'_{\cdot c-1}\end{pmatrix}$$

$$= \sum_{i=1}^{r}\sum_{j=1}^{c}\mathbf{d}_{ij}\mathbf{d}'_{ij} - \mathbf{X}'_{\text{row}}\mathbf{C}^{-1}\mathbf{X}_{\text{row}} - \mathbf{X}'_{\text{col}}\mathbf{R}^{-1}\mathbf{X}_{\text{col}},$$

where $\mathbf{d}_{ij} = \mathbf{x}_{ij} - \bar{\mathbf{x}}$, $\mathbf{d}_{i\cdot} = \mathbf{x}_{i\cdot} - \mathbf{x}_{r\cdot}$, $\mathbf{d}_{\cdot j} = \mathbf{x}_{\cdot j} - \mathbf{x}_{\cdot c}$,
$\mathbf{X}'_{\text{row}} = (\mathbf{d}_{1\cdot}\cdots\mathbf{d}_{r-1\cdot})$, $\mathbf{X}'_{\text{col}} = (\mathbf{d}_{\cdot 1}\cdots\mathbf{d}_{\cdot c-1})$,

$$(10.20)\quad \mathbf{X}'_{2\cdot 1}\mathbf{y} = \mathbf{G}'\mathbf{X}'_{2}\mathbf{y} - \mathbf{G}'\mathbf{S}_{21}\mathbf{S}_{11}^{-1}\mathbf{X}'_{1}\mathbf{y} = \begin{pmatrix} y_{1\cdot} - y_{r\cdot}\\ \cdot\\ \cdot\\ \cdot\\ y_{r-1\cdot} - y_{r\cdot}\\ y_{\cdot 1} - y_{\cdot c}\\ \cdot\\ \cdot\\ \cdot\\ y_{\cdot c-1} - y_{\cdot c}\end{pmatrix} = \begin{pmatrix}\mathbf{y}_{\text{row}}\\ \mathbf{y}_{\text{col}}\end{pmatrix},$$

(10.21) $$\mathbf{X}'_{3\cdot1}\mathbf{y} = \mathbf{X}'_3\mathbf{y} - \mathbf{S}_{31}\mathbf{S}_{11}^{-1}\mathbf{X}'_1\mathbf{y}$$

$$= \sum_{i=1}^{r}\sum_{j=1}^{c}\mathbf{x}_{ij}y_{ij} - \mathbf{x}_{..}\frac{1}{rc}y_{..} = \sum_{i=1}^{r}\sum_{j=1}^{c}\mathbf{x}_{ij}(y_{ij} - \bar{y}),$$

where $rc\bar{y} = y_{..}$,

(10.22) $$\mathbf{X}'_{3\cdot12}\mathbf{y} = \mathbf{X}'_{3\cdot1}\mathbf{y} - \mathbf{S}_{32\cdot1}\mathbf{S}_{22\cdot1}^{-1}\mathbf{X}'_{2\cdot1}\mathbf{y} = \sum_{i=1}^{r}\sum_{j=1}^{c}\mathbf{x}_{ij}(y_{ij} - \bar{y})$$

$$- (\mathbf{d}_{1\cdot} \cdots \mathbf{d}_{r-1\cdot}\ \mathbf{d}_{\cdot1} \cdots \mathbf{d}_{\cdot c-1})\begin{pmatrix}\mathbf{C}^{-1} & \mathbf{0} \\ \mathbf{0} & \mathbf{R}^{-1}\end{pmatrix}\begin{pmatrix} y_{1\cdot} - y_{r\cdot} \\ \cdot \\ \cdot \\ \cdot \\ y_{r-1\cdot} - y_{r\cdot} \\ y_{\cdot1} - y_{\cdot c} \\ \cdot \\ \cdot \\ \cdot \\ y_{\cdot c-1} - y_{\cdot c}\end{pmatrix}$$

$$= \sum_{i=1}^{r}\sum_{j=1}^{c}\mathbf{x}_{ij}(y_{ij} - \bar{y}) - (\mathbf{d}_{1\cdot} \cdots \mathbf{d}_{r-1\cdot})\mathbf{C}^{-1}\begin{pmatrix} y_{1\cdot} - y_{r\cdot} \\ \cdot \\ \cdot \\ \cdot \\ y_{r-1\cdot} - y_{r\cdot}\end{pmatrix}$$

$$- (\mathbf{d}_{\cdot1} \cdots \mathbf{d}_{\cdot c-1})\mathbf{R}^{-1}\begin{pmatrix} y_{\cdot1} - y_{\cdot c} \\ \cdot \\ \cdot \\ \cdot \\ y_{\cdot c-1} - y_{\cdot c}\end{pmatrix}$$

$$= \sum_{i=1}^{r}\sum_{j=1}^{c}\mathbf{x}_{ij}(y_{ij} - \bar{y}) - \mathbf{X}'_{\text{row}}\mathbf{C}^{-1}\mathbf{y}_{\text{row}} - \mathbf{X}'_{\text{col}}\mathbf{R}^{-1}\mathbf{y}_{\text{col}}.$$

It may be shown that (see problem 11.4)

(10.23) $$\mathbf{C}^{-1} = \begin{pmatrix} \frac{r-1}{cr} & -\frac{1}{cr} & \cdots & -\frac{1}{cr} \\ -\frac{1}{cr} & \frac{r-1}{cr} & \cdots & -\frac{1}{cr} \\ \cdot & \cdot & \cdots & \cdot \\ -\frac{1}{cr} & -\frac{1}{cr} & \cdots & \frac{r-1}{cr} \end{pmatrix},$$

$$\mathbf{R}^{-1} = \begin{pmatrix} \dfrac{c-1}{cr} & -\dfrac{1}{cr} & \cdots & -\dfrac{1}{cr} \\ -\dfrac{1}{cr} & \dfrac{c-1}{cr} & \cdots & -\dfrac{1}{cr} \\ \cdot & \cdot & \cdots & \cdot \\ -\dfrac{1}{cr} & -\dfrac{1}{cr} & \cdots & \dfrac{c-1}{cr} \end{pmatrix},$$

$$\mathbf{X}'_{\text{row}}\mathbf{C}^{-1}\mathbf{X}_{\text{row}} = \frac{1}{c}\sum_{i=1}^{r}\mathbf{x}_{i\cdot}\mathbf{x}'_{i\cdot} - \frac{1}{rc}\mathbf{x}_{\cdot\cdot}\mathbf{x}'_{\cdot\cdot},$$

$$\mathbf{X}'_{\text{col}}\mathbf{R}^{-1}\mathbf{X}_{\text{col}} = \frac{1}{r}\sum_{j=1}^{c}\mathbf{x}_{\cdot j}\mathbf{x}'_{\cdot j} - \frac{1}{rc}\mathbf{x}_{\cdot\cdot}\mathbf{x}'_{\cdot\cdot},$$

$$\mathbf{X}'_{\text{row}}\mathbf{C}^{-1}\mathbf{y}_{\text{row}} = \frac{1}{c}\sum_{i=1}^{r}\mathbf{x}_{i\cdot}y_{i\cdot} - \frac{1}{rc}\mathbf{x}_{\cdot\cdot}y_{\cdot\cdot},$$

$$\mathbf{X}'_{\text{col}}\mathbf{R}^{-1}\mathbf{y}_{\text{col}} = \frac{1}{r}\sum_{j=1}^{c}\mathbf{x}_{\cdot j}y_{\cdot j} - \frac{1}{rc}\mathbf{x}_{\cdot\cdot}y_{\cdot\cdot},$$

where $\mathbf{C}^{-1}$ and $\mathbf{R}^{-1}$ are respectively $(r-1)\times(r-1)$ and $(c-1)\times(c-1)$ matrices.

Thus, under H_1, we get from (5.20) the estimates

$$\hat{\boldsymbol{\beta}}_3 = \mathbf{S}_{33\cdot 12}^{-1}\mathbf{X}'_{3\cdot 12}\mathbf{y}, \tag{10.24}$$

where $\mathbf{S}_{33\cdot 12}$ is given in (10.19) and $\mathbf{X}'_{3\cdot 12}\mathbf{y}$ is given in (10.22),

$$\hat{\boldsymbol{\gamma}} = \begin{pmatrix} \mathbf{C}^{-1} & \mathbf{0} \\ \mathbf{0} & \mathbf{R}^{-1} \end{pmatrix}\left[\begin{pmatrix} y_{1\cdot} - y_{r\cdot} \\ \cdot \\ \cdot \\ \cdot \\ y_{r-1\cdot} - y_{r\cdot} \\ y_{\cdot 1} - y_{\cdot c} \\ \cdot \\ \cdot \\ \cdot \\ y_{\cdot c-1} - y_{\cdot c} \end{pmatrix} - \begin{pmatrix} \mathbf{d}'_{1\cdot} \\ \cdot \\ \cdot \\ \cdot \\ \mathbf{d}'_{r-1\cdot} \\ \mathbf{d}'_{\cdot 1} \\ \cdot \\ \cdot \\ \cdot \\ \mathbf{d}'_{\cdot c-1} \end{pmatrix}\hat{\boldsymbol{\beta}}_3\right], \tag{10.25}$$

that is

$$\hat{\boldsymbol{\gamma}}_1 = \mathbf{C}^{-1}\left[\begin{pmatrix} y_{1\cdot} - y_{r\cdot} \\ \cdot \\ \cdot \\ \cdot \\ y_{r-1\cdot} - y_{r\cdot} \end{pmatrix} - \begin{pmatrix} \mathbf{d}'_{1\cdot} \\ \cdot \\ \cdot \\ \cdot \\ \mathbf{d}'_{r-1\cdot} \end{pmatrix}\hat{\boldsymbol{\beta}}_3\right],$$

$$\hat{\boldsymbol{\gamma}}_2 = \mathbf{R}^{-1}\left[\begin{pmatrix} y_{\cdot 1} - y_{\cdot c} \\ \cdot \\ \cdot \\ \cdot \\ y_{\cdot c-1} - y_{\cdot c}\end{pmatrix} - \begin{pmatrix} \mathbf{d}'_{\cdot 1} \\ \cdot \\ \cdot \\ \cdot \\ \mathbf{d}'_{\cdot c-1}\end{pmatrix}\hat{\boldsymbol{\beta}}_3\right],$$

$$\hat{\boldsymbol{\gamma}}_1 = \mathbf{C}^{-1}(\mathbf{y}_{\text{row}} - \mathbf{X}_{\text{row}}\hat{\boldsymbol{\beta}}_3), \qquad \hat{\boldsymbol{\gamma}}_2 = \mathbf{R}^{-1}(\mathbf{y}_{\text{col}} - \mathbf{X}_{\text{col}}\hat{\boldsymbol{\beta}}_3),$$

(10.26)
$$\hat{\boldsymbol{\beta}}_1 = \frac{1}{rc}(y_{\cdot\cdot} - \mathbf{x}'_{\cdot\cdot}\hat{\boldsymbol{\beta}}_3).$$

We now have [see (5.21)]

(10.27) $\hat{\boldsymbol{\beta}}^{1\prime}\mathbf{S}^1\hat{\boldsymbol{\beta}}^1 = \mathbf{y}'\mathbf{X}_1\mathbf{S}_{11}^{-1}\mathbf{X}_1'\mathbf{y} + \mathbf{y}'\mathbf{X}_{2\cdot 1}\mathbf{S}_{22\cdot 1}^{-1}\mathbf{X}_{2\cdot 1}'\mathbf{y} + \hat{\boldsymbol{\beta}}_3^{1\prime}\mathbf{S}_{33\cdot 12}\,\hat{\boldsymbol{\beta}}_3^1,$

where $\mathbf{y}'\mathbf{X}_1\mathbf{S}_{11}^{-1}\mathbf{X}_1'\mathbf{y} = (y_{\cdot\cdot})^2/rc$,

$$\mathbf{y}'\mathbf{X}_{2\cdot 1}\mathbf{S}_{22\cdot 1}^{-1}\mathbf{X}_{2\cdot 1}'\mathbf{y} = (y_{1\cdot} - y_{r\cdot} \cdots y_{r-1\cdot} - y_{r\cdot})\mathbf{C}^{-1}\begin{pmatrix} y_{1\cdot} - y_{r\cdot} \\ \cdot \\ \cdot \\ \cdot \\ y_{r-1\cdot} - y_{r\cdot}\end{pmatrix}$$

$$+ (y_{\cdot 1} - y_{\cdot c} \cdots y_{\cdot c-1} - y_{\cdot c})\mathbf{R}^{-1}\begin{pmatrix} y_{\cdot 1} - y_{\cdot c} \\ \cdot \\ \cdot \\ \cdot \\ y_{\cdot c-1} - y_{\cdot c}\end{pmatrix}$$

$$= \mathbf{y}_{\text{row}}'\mathbf{C}^{-1}\mathbf{y}_{\text{row}} + \mathbf{y}_{\text{col}}'\mathbf{R}^{-1}\mathbf{y}_{\text{col}},$$

and $\hat{\boldsymbol{\beta}}_3^1$ is given by (10.24) and $\mathbf{S}_{33\cdot 12}$ by (10.19). [See (10.46).]

The original row and column effect parameters are estimated by

(10.28)
$$\hat{\boldsymbol{\rho}}^1 = \mathbf{G}_1\hat{\boldsymbol{\gamma}}_1, \qquad \hat{\boldsymbol{\tau}}^1 = \mathbf{G}_2\hat{\boldsymbol{\gamma}}_2,$$

where $\mathbf{G}_1$ and $\mathbf{G}_2$ are defined in (10.6) and $\hat{\boldsymbol{\gamma}}_1$ and $\hat{\boldsymbol{\gamma}}_2$ in (10.25).

Under H_2, instead of the matrix $\mathbf{G}$ in (10.6) we have only the matrix $\mathbf{G}_1$, and the matrix $\mathbf{X}_2$ is now given by the $rc \times r$ matrix

(10.29)
$$\mathbf{X}_2^2 = \begin{pmatrix} 1 & 0 \cdots 0 \\ 1 & 0 \cdots 0 \\ \cdot & \cdot \quad \cdot \quad \cdot \\ 1 & 0 \cdots 0 \\ 0 & 1 \cdots 0 \\ 0 & 1 \cdots 0 \\ \cdot & \cdot \quad \cdot \quad \cdot \\ 0 & 1 \cdots 0 \\ \cdot & \cdot \quad \cdot \quad \cdot \\ 0 & 0 \cdots 1 \\ 0 & 0 \cdots 1 \\ \cdot & \cdot \quad \cdot \quad \cdot \\ 0 & 0 \cdots 1 \end{pmatrix}.$$

Instead of (10.7), under H_2 we have

$$\mathbf{X}_2^2\mathbf{G}_1 = \begin{pmatrix} 1 & 0 \cdots & 0 \\ \cdot & \cdot \quad \cdot & \cdot \\ 1 & 0 \cdots & 0 \\ 0 & 1 \cdots & 0 \\ \cdot & \cdot \quad \cdot & \cdot \\ 0 & 1 \cdots & 0 \\ \cdot & \cdot \quad \cdot & \cdot \\ -1 & -1 \cdots & -1 \\ & \vdots & \\ -1 & -1 \cdots & -1 \end{pmatrix}, \tag{10.30}$$

where $\mathbf{X}_2^2\mathbf{G}_1$ has rc rows and $(r-1)$ columns; instead of (10.8), under H_2 we have

$$\mathbf{G}_1'\mathbf{X}_2^{2\prime}\mathbf{X}_2^2\mathbf{G}_1 = \mathbf{G}_1'\mathbf{S}_{22}^2\mathbf{G}_1 = \underbrace{\begin{pmatrix} 2c & c \cdots & c \\ c & 2c \cdots & c \\ \cdot & \cdot \quad \cdot & \cdot \\ c & c \cdots & 2c \end{pmatrix}}_{r-1} \quad r-1; \tag{10.31}$$

instead of (10.9), under H_2 we have

$$\mathbf{X}_1'\mathbf{X}_2^2\mathbf{G}_1 = \mathbf{S}_{12}^2\mathbf{G}_1 = (0, 0, \cdots, 0), \qquad 1 \times (r-1); \tag{10.32}$$

instead of (10.10), under H_2 we have

$$\mathbf{X}_3'\mathbf{X}_2^2\mathbf{G}_1 = \mathbf{S}_{32}^2\mathbf{G}_1 = (\mathbf{x}_{1\cdot} - \mathbf{x}_{r\cdot} \cdots \mathbf{x}_{r-1\cdot} - \mathbf{x}_{r\cdot}) = \mathbf{X}'_{\text{row}}, \tag{10.33}$$

where $\mathbf{S}_{32}^2\mathbf{G}_1$ is a $p \times (r-1)$ matrix; instead of (10.14), under H_2 we have

$$\mathbf{G}_1'\mathbf{X}_2^{2\prime}\mathbf{y} = \begin{pmatrix} y_{1\cdot} - y_{r\cdot} \\ \cdot \\ \cdot \\ \cdot \\ y_{r-1\cdot} - y_{r\cdot} \end{pmatrix} = \mathbf{y}_{\text{row}}; \tag{10.34}$$

instead of (10.16), under H_2 we have

$$\mathbf{S}_{22\cdot 1}^2 = \mathbf{C}; \tag{10.35}$$

instead of (10.18), under H_2 we have

$$\mathbf{S}_{32\cdot 1}^2 = (\mathbf{x}_{1\cdot} - \mathbf{x}_{r\cdot} \cdots \mathbf{x}_{r-1\cdot} - \mathbf{x}_{r\cdot}) = \mathbf{X}'_{\text{row}}; \tag{10.36}$$

instead of (10.19), under H_2 we have

$$(10.37)\qquad \mathbf{S}^2_{33\cdot 12} = \sum_{i=1}^{r}\sum_{j=1}^{c} \mathbf{d}_{ij}\mathbf{d}'_{ij} - (\mathbf{d}_{1\cdot}\cdots\mathbf{d}_{r-1\cdot})\mathbf{C}^{-1}\begin{pmatrix}\mathbf{d}'_{1\cdot}\\ \cdot\\ \cdot\\ \cdot\\ \mathbf{d}'_{r-1\cdot}\end{pmatrix}$$

$$= \sum_{i=1}^{r}\sum_{j=1}^{c} \mathbf{d}_{ij}\mathbf{d}'_{ij} - \mathbf{X}'_{\mathrm{row}}\mathbf{C}^{-1}\mathbf{X}_{\mathrm{row}};$$

instead of (10.20), under H_2 we have

$$(10.38)\qquad \mathbf{X}^{2'}_{2\cdot 1}\mathbf{y} = \begin{pmatrix} y_{1\cdot} - y_{r\cdot}\\ \cdot\\ \cdot\\ \cdot\\ y_{r-1\cdot} - y_{r\cdot}\end{pmatrix} = \mathbf{y}_{\mathrm{row}};$$

instead of (10.22), under H_2 we have

$$(10.39)\qquad \mathbf{X}^{2'}_{3\cdot 12}\mathbf{y} = \sum_{i=1}^{r}\sum_{j=1}^{c} \mathbf{x}_{ij}(y_{ij} - \bar{y}) - (\mathbf{d}_{1\cdot}\cdots\mathbf{d}_{r-1\cdot})\mathbf{C}^{-1}\begin{pmatrix} y_{1\cdot} - y_{r\cdot}\\ \cdot\\ \cdot\\ \cdot\\ y_{r-1\cdot} - y_{r\cdot}\end{pmatrix}$$

$$= \sum_{i=1}^{r}\sum_{j=1}^{c} \mathbf{x}_{ij}(y_{ij} - \bar{y}) - \mathbf{X}'_{\mathrm{row}}\mathbf{C}^{-1}\mathbf{y}_{\mathrm{row}}.$$

Thus, under H_2, we get as the estimates of the parameters, instead of (10.24),

$$(10.40)\qquad \hat{\boldsymbol{\beta}}^2_3 = \mathbf{S}^{2^{-1}}_{33\cdot 12}\mathbf{X}^{2'}_{3\cdot 12}\mathbf{y},$$

where $\mathbf{S}^2_{33\cdot 12}$ is given in (10.37) and $\mathbf{X}^{2'}_{3\cdot 12}\mathbf{y}$ in (10.39); instead of (10.25),

$$(10.41)\qquad \hat{\boldsymbol{\gamma}}^2 = \begin{pmatrix}\hat{\boldsymbol{\gamma}}^2_1\\ \mathbf{0}\end{pmatrix},$$

where $\hat{\boldsymbol{\gamma}}^2_1 = \mathbf{C}^{-1}\left[\begin{pmatrix} y_{1\cdot} - y_{r\cdot}\\ \cdot\\ \cdot\\ \cdot\\ y_{r-1\cdot} - y_{r\cdot}\end{pmatrix} - \begin{pmatrix}\mathbf{d}'_{1\cdot}\\ \cdot\\ \cdot\\ \cdot\\ \mathbf{d}'_{r-1\cdot}\end{pmatrix}\hat{\boldsymbol{\beta}}^2_3\right] = \mathbf{C}^{-1}(\mathbf{y}_{\mathrm{row}} - \mathbf{X}_{\mathrm{row}}\hat{\boldsymbol{\beta}}^2_3);$

instead of (10.26),

$$(10.42)\qquad \hat{\boldsymbol{\beta}}^2_1 = \frac{1}{rc}(y_{\cdot\cdot} - \mathbf{x}'_{\cdot\cdot}\hat{\boldsymbol{\beta}}^2_3);$$

instead of (10.27),

$$\hat{\boldsymbol{\beta}}^{2\prime}\mathbf{S}^2\hat{\boldsymbol{\beta}}^2 = \mathbf{y}'\mathbf{X}_1^2\mathbf{S}_{11}^{2-1}\mathbf{X}_1^{2\prime}\mathbf{y} + \mathbf{y}'\mathbf{X}_{2\cdot 1}^2\mathbf{S}_{22\cdot 1}^{2-1}\mathbf{X}_{2\cdot 1}^{2\prime}\mathbf{y} + \hat{\boldsymbol{\beta}}_3^{2\prime}\mathbf{S}_{33\cdot 12}^2\hat{\boldsymbol{\beta}}_3^2, \tag{10.43}$$

where

$$\mathbf{y}'\mathbf{X}_1^2\mathbf{S}_{11}^{2-1}\mathbf{X}_1^{2\prime}\mathbf{y} = (y_{..})^2/rc,$$

$$\mathbf{y}'\mathbf{X}_{2\cdot 1}^2\mathbf{S}_{22\cdot 1}^{2-1}\mathbf{X}_{2\cdot 1}^{2\prime}\mathbf{y} = (y_{1\cdot} - y_{r\cdot} \cdots y_{r-1\cdot} - y_{r\cdot})\mathbf{C}^{-1}\begin{pmatrix} y_{1\cdot} - y_{r\cdot} \\ \cdot \\ \cdot \\ \cdot \\ y_{r-1\cdot} - y_{r\cdot} \end{pmatrix}$$

$$= \mathbf{y}_{\text{row}}'\mathbf{C}^{-1}\mathbf{y}_{\text{row}},$$

and $\hat{\boldsymbol{\beta}}_3^2$ is given by (10.40) and $\mathbf{S}_{33\cdot 12}^2$ by (10.37).

The original row and column effect parameters, under H_2, are estimated by

$$\hat{\boldsymbol{\rho}}^2 = \mathbf{G}_1\hat{\boldsymbol{\gamma}}_1^2, \qquad \boldsymbol{\tau}^2 = \mathbf{0}. \tag{10.44}$$

With the foregoing values, we now have

$$\hat{\sigma}^2\hat{J}(H_1, H_2) = \hat{\boldsymbol{\beta}}^{1\prime}\mathbf{S}^1\hat{\boldsymbol{\beta}}^1 - \hat{\boldsymbol{\beta}}^{2\prime}\mathbf{S}^2\hat{\boldsymbol{\beta}}^2 \tag{10.45}$$

where $(rc - 1 - r + 1 - c + 1 - p)\hat{\sigma}^2 = \mathbf{y}'\mathbf{y} - \hat{\boldsymbol{\beta}}^{1\prime}\mathbf{S}^1\hat{\boldsymbol{\beta}}^1$
$= ((r-1)(c-1) - p)\hat{\sigma}^2.$

We summarize the foregoing in the analysis of variance table 10.1. $\hat{J}(H_1, H_2) = (c-1)F$, where F has the analysis of variance distribution with $n_1 = c - 1$ and $n_2 = (r-1)(c-1) - p$ degrees of freedom under the null hypothesis H_2 in (10.4).

TABLE 10.1

Variation due to	Sum of Squares	D.F.
H_2:(10.4)	$\hat{\boldsymbol{\beta}}^{2\prime}\mathbf{S}^2\hat{\boldsymbol{\beta}}^2 = \frac{(y_{..})^2}{rc} + \mathbf{y}_{\text{row}}'\mathbf{C}^{-1}\mathbf{y}_{\text{row}} + \hat{\boldsymbol{\beta}}_3^{2\prime}\mathbf{S}_{33\cdot 12}^2\hat{\boldsymbol{\beta}}_3^2$	$r + p$
Difference	$\mathbf{y}_{\text{col}}'\mathbf{R}^{-1}\mathbf{y}_{\text{col}} + \hat{\boldsymbol{\beta}}_3^{1\prime}\mathbf{S}_{33\cdot 12}^1\hat{\boldsymbol{\beta}}_3^1 - \hat{\boldsymbol{\beta}}_3^{2\prime}\mathbf{S}_{33\cdot 12}^2\hat{\boldsymbol{\beta}}_3^2 = \hat{\sigma}^2\hat{J}(H_1, H_2)$	$c - 1$
H_1:(10.3)	$\hat{\boldsymbol{\beta}}^{1\prime}\mathbf{S}^1\hat{\boldsymbol{\beta}}^1 = \frac{(y_{..})^2}{rc} + \mathbf{y}_{\text{row}}'\mathbf{C}^{-1}\mathbf{y}_{\text{row}} + \mathbf{y}_{\text{col}}'\mathbf{R}^{-1}\mathbf{y}_{\text{col}} + \hat{\boldsymbol{\beta}}_3^{1\prime}\mathbf{S}_{33\cdot 12}^1\hat{\boldsymbol{\beta}}_3^1$	$r + c - 1 + p$
Difference	$\mathbf{y}'\mathbf{y} - \hat{\boldsymbol{\beta}}^{1\prime}\mathbf{S}^1\hat{\boldsymbol{\beta}}^1 = ((r-1)(c-1) - p)\hat{\sigma}^2$	$(r-1)(c-1) - p$
Total	$\mathbf{y}'\mathbf{y}$	rc

In particular, for the usual two-way table with no regression, that is, $\boldsymbol{\beta}_3^1 = \boldsymbol{\beta}_3^2 = \mathbf{0}$, table 10.1 yields the analysis summarized in the analysis of variance table 10.2.

TABLE 10.2

Variation due to	Sum of Squares	D.F.
Mean	$(y_{..})^2/rc = \hat{\sigma}^2\hat{J}(H_\mu, H_2)$	1
Rows	$\mathbf{y}'_{\text{row}}\mathbf{C}^{-1}\mathbf{y}_{\text{row}} = \hat{\sigma}^2\hat{J}(H_R, H_2)$	$r-1$
Columns	$\mathbf{y}'_{\text{col}}\mathbf{R}^{-1}\mathbf{y}_{\text{col}} = \hat{\sigma}^2\hat{J}(H_C, H_2)$	$c-1$
H_1	$\dfrac{(y_{..})^2}{rc} + \mathbf{y}'_{\text{row}}\mathbf{C}^{-1}\mathbf{y}_{\text{row}} + \mathbf{y}'_{\text{col}}\mathbf{R}^{-1}\mathbf{y}_{\text{col}} = \hat{\sigma}^2\hat{J}(H_1, H_2)$	$r+c-1$
Difference	$\mathbf{y}'\mathbf{y} - \dfrac{(y_{..})^2}{rc} - \mathbf{y}'_{\text{row}}\mathbf{C}^{-1}\mathbf{y}_{\text{row}} - \mathbf{y}'_{\text{col}}\mathbf{R}^{-1}\mathbf{y}_{\text{col}} = (r-1)(c-1)\hat{\sigma}^2$	$(r-1)(c-1)$
Total	$\mathbf{y}'\mathbf{y}$	rc

It may be shown that [see (10.23) and problem 11.5]

$$\mathbf{y}'_{\text{row}}\mathbf{C}^{-1}\mathbf{y}_{\text{row}} = \sum_{i=1}^{r}\frac{y_{i\cdot}^2}{c} - \frac{y_{\cdot\cdot}^2}{rc}, \tag{10.46}$$

$$\mathbf{y}'_{\text{col}}\mathbf{R}^{-1}\mathbf{y}_{\text{col}} = \sum_{j=1}^{c}\frac{y_{\cdot j}^2}{r} - \frac{y_{\cdot\cdot}^2}{rc}.$$

Note that here the alternative hypothesis H_1 may be expressed as the intersection of three independent hypotheses, $H_1 = H_\mu \cap H_R \cap H_C$, where H_μ is the hypothesis that $\mu \neq 0$, H_R is the hypothesis that $\boldsymbol{\rho} \neq \mathbf{0}$, and H_C is the hypothesis that $\boldsymbol{\tau} \neq \mathbf{0}$. Against a null hypothesis H_2, $\mu = 0$, $\boldsymbol{\rho} = \mathbf{0}$, $\boldsymbol{\tau} = \mathbf{0}$, we see that

$$\hat{J}(H_1, H_2) = \hat{J}(H_\mu, H_2) + \hat{J}(H_R, H_2) + \hat{J}(H_C, H_2),$$

where
$$\hat{J}(H_\mu, H_2) = F(n_1 = 1, n_2 = (r-1)(c-1)),$$
$$\hat{J}(H_R, H_2) = (r-1)F(n_1 = r-1, n_2 = (r-1)(c-1)),$$
$$\hat{J}(H_C, H_2) = (c-1)F(n_1 = c-1, n_2 = (r-1)(c-1)),$$
$$\hat{J}(H_1, H_2) = (r+c-1)F(n_1 = r+c-1, n_2 = (r-1)(c-1)),$$

where $F(n_1, n_2)$ has the analysis of variance distribution with n_1 and n_2 degrees of freedom under the null hypothesis.

For $p = 1$, we get from (10.23), (10.19), (10.24), (10.37), and (10.40) the following values for use in table 10.1 in addition to those in (10.46) (see problem 11.6):

(10.47) $$\mathbf{X}'_{\text{row}}\mathbf{C}^{-1}\mathbf{X}_{\text{row}} = \sum_{i=1}^{r}\frac{x_{i\cdot}^2}{c} - \frac{x_{\cdot\cdot}^2}{rc},$$

$$\mathbf{X}'_{\text{col}}\mathbf{R}^{-1}\mathbf{X}_{\text{col}} = \sum_{j=1}^{c}\frac{x_{\cdot j}^2}{r} - \frac{x_{\cdot\cdot}^2}{rc},$$

$$\mathbf{X}'_{\text{row}}\mathbf{C}^{-1}\mathbf{y}_{\text{row}} = \sum_{i=1}^{r}\frac{x_{i\cdot}y_{i\cdot}}{c} - \frac{x_{\cdot\cdot}y_{\cdot\cdot}}{rc},$$

$$\mathbf{X}'_{\text{col}}\mathbf{R}^{-1}\mathbf{y}_{\text{col}} = \sum_{j=1}^{c}\frac{x_{\cdot j}y_{\cdot j}}{r} - \frac{x_{\cdot\cdot}y_{\cdot\cdot}}{rc},$$

$$S_{33\cdot 12}^{1} = \sum_{i=1}^{r}\sum_{j=1}^{c}x_{ij}^2 - \frac{x_{\cdot\cdot}^2}{rc} - \sum_{i=1}^{r}\frac{x_{i\cdot}^2}{c} + \frac{x_{\cdot\cdot}^2}{rc} - \sum_{j=1}^{c}\frac{x_{\cdot j}^2}{r} + \frac{x_{\cdot\cdot}^2}{rc},$$

$$\hat{\boldsymbol{\beta}}_3^1 = \frac{\sum_{i=1}^{r}\sum_{j=1}^{c}x_{ij}y_{ij} - \sum_{i=1}^{r}\frac{x_{i\cdot}y_{i\cdot}}{c} - \sum_{j=1}^{c}\frac{x_{\cdot j}y_{\cdot j}}{r} + \frac{x_{\cdot\cdot}y_{\cdot\cdot}}{rc}}{\sum_{i=1}^{r}\sum_{j=1}^{c}x_{ij}^2 - \sum_{i=1}^{r}\frac{x_{i\cdot}^2}{c} - \sum_{j=1}^{c}\frac{x_{\cdot j}^2}{r} + \frac{x_{\cdot\cdot}^2}{rc}},$$

$$S_{33\cdot 12}^{2} = \sum_{i=1}^{r}\sum_{j=1}^{c}x_{ij}^2 - \sum_{i=1}^{r}\frac{x_{i\cdot}^2}{c},$$

$$\hat{\boldsymbol{\beta}}_3^2 = \frac{\sum_{i=1}^{r}\sum_{j=1}^{c}x_{ij}y_{ij} - \sum_{i=1}^{r}\frac{x_{i\cdot}y_{i\cdot}}{c}}{\sum_{i=1}^{r}\sum_{j=1}^{c}x_{ij}^2 - \sum_{i=1}^{r}\frac{x_{i\cdot}^2}{c}}.$$

11. PROBLEMS

11.1. What is the distribution of $\hat{J}(H_1, H_2)$ in (4.3) if the null hypothesis is not satisfied? [Cf. Anderson (1958, p. 107).]

11.2. Show that $S_{11}^{-1}S_{12}S_{22\cdot 1}^{-1} = S_{11\cdot 2}^{-1}S_{12}S_{22}^{-1}$, where the matrices are defined in (5.6), (5.7), (5.8).

11.3. Give the specification of the matrices $\mathbf{X}^{\alpha}$, $\boldsymbol{\beta}^{\alpha}$ for the hypotheses H_{α}, $\alpha = 2, 3, 4, 6, 7$, in section 8.

11.4. Confirm the results given in (10.23).

11.5. Confirm the results given in (10.46).

11.6. Confirm the results given in (10.47).

11.7. Verify that the asymptotic behavior of the minimum discrimination information statistics in chapter 10 is in accordance with the results of chapter 5.

11.8. Show that

(*a*) $\hat{J}(H_1, H_2)$ in table 5.2 is equal to $(n - p)r^2_{y\cdot 2\cdots p}/(1 - r^2_{y\cdot 2\cdots p})$, where $r_{y\cdot 2\cdots p}$ is the multiple correlation of y with $x_2, \cdots, x_p$.

(*b*) $\hat{J}(H_1, H_2)$ is an estimate of $J(1, 2)$ in (7.18) of chapter 9 for a sample of n observations.

11.9. Suppose that table 5.1 applies to a sample of $n + 1$ observations, that is, the y's and x's are centered about their respective sample averages, and that $q = p - 1$. Show that here $\hat{J}(H_1, H_2) = (n - p)r^2_{y1\cdot 23\cdots p}/(1 - r^2_{y1\cdot 23\cdots p})$, where $r_{y1\cdot 23\cdots p}$ is the partial correlation of y with x_1.

11.10. Show that $\hat{\boldsymbol{\beta}}_1$ in (7.7) may also be expressed as $\hat{\boldsymbol{\beta}}_1 = \mathbf{S}^{-1}_{11\cdot 2}\mathbf{X}'_{1\cdot 2}\mathbf{y}$, where $\mathbf{S}_{11\cdot 2} = \mathbf{S}_{11} - \mathbf{S}_{12}\mathbf{S}^{-1}_{22}\mathbf{S}_{21}$ and $\mathbf{X}'_{1\cdot 2} = \mathbf{X}'_1 - \mathbf{S}_{12}\mathbf{S}^{-1}_{22}\mathbf{X}'_2$.

CHAPTER 11

Multivariate Analysis; The Multivariate Linear Hypothesis

1. INTRODUCTION

In this chapter we examine tests of linear hypotheses for samples from multivariate normal populations, thus extending the analyses of the previous chapter. In the next chapter we apply the general ideas to the analysis of samples from multivariate normal populations under hypotheses on the covariance matrices and on the means other than those included in the linear hypothesis. The treatment in this chapter is not intended to be exhaustive, and has wider applicability than to the specific cases considered.

2. BACKGROUND

Suppose two simple statistical hypotheses, say H_1 and H_2, specify respectively the means of n k-variate normal populations with common covariance matrix $\mathbf{\Sigma} = (\sigma_{ij})$, $i, j = 1, 2, \cdots, k$. For n independent observations ($1 \times k$ matrices or vectors), one from each of the populations, (2.17) in chapter 9 becomes ($n_1 = n_2 = \cdots = n_r = 1$, $r = n$)

$$(2.1) \qquad 2I(1:2; O_n) = J(1, 2; O_n) = \operatorname{tr} \mathbf{\Sigma}^{-1}(\boldsymbol{\delta}_1\boldsymbol{\delta}_1' + \cdots + \boldsymbol{\delta}_n\boldsymbol{\delta}_n'),$$

where $\boldsymbol{\delta}_i = \boldsymbol{\mu}_i^1 - \boldsymbol{\mu}_i^2$, with $\boldsymbol{\mu}_i^\alpha$, $\alpha = 1, 2$, the one-column matrices (vectors) of means of the ith population under H_α, and $\boldsymbol{\mu}_i' = (\mu_{i1}, \mu_{i2}, \cdots, \mu_{ik})$, $i = 1, 2, \cdots, n$. (This was still further specialized in section 2 of chapter 10.)

3. THE MULTIVARIATE LINEAR HYPOTHESIS

3.1. Specification

For the ith observation, we have the regression model

$$(3.1) \qquad \mathbf{z}_i = \mathbf{y}_i - \mathbf{B}\mathbf{x}_i, \qquad i = 1, 2, \cdots, n,$$

where $\mathbf{z}_i' = (z_{i1}, z_{i2}, \cdots, z_{ik_2})$, $\mathbf{y}_i' = (y_{i1}, y_{i2}, \cdots, y_{ik_2})$, $\mathbf{x}_i' = (x_{i1}, x_{i2}, \cdots, x_{ik_1})$, $\mathbf{B} = (\beta_{rs})$, $r = 1, 2, \cdots, k_2$, $s = 1, 2, \cdots, k_1$, $k_1 < n$, $k_2 < n$, $\mathbf{B}$ of rank min (k_1, k_2). We may also express the n regressions in (3.1) as the one over-all regression model

$$\mathbf{Z} = \mathbf{Y} - \mathbf{X}\mathbf{B}', \tag{3.2}$$

where $\mathbf{Z}' = (\mathbf{z}_1, \mathbf{z}_2, \cdots, \mathbf{z}_n)$, $\mathbf{Y}' = (\mathbf{y}_1, \mathbf{y}_2, \cdots, \mathbf{y}_n)$, $\mathbf{X}' = (\mathbf{x}_1, \mathbf{x}_2, \cdots, \mathbf{x}_n)$, with $\mathbf{Z}'$ and $\mathbf{Y}'$ $k_2 \times n$ matrices and $\mathbf{X}'$ a $k_1 \times n$ matrix.

We assume that:

(*a*) the $\mathbf{z}_i$ are independent normal random $k_2 \times 1$ matrices (vectors) with zero means and common covariance matrix $\mathbf{\Sigma}$,

(*b*) the x_{ij}, $i = 1, 2, \cdots, n$, $j = 1, 2, \cdots, k_1$, are known,

(*c*) $\mathbf{X}$ is of rank k_1,

(*d*) $\mathbf{B} = \mathbf{B}^1$ and $\mathbf{B} = \mathbf{B}^2$ are parameter matrices specified respectively by the hypotheses H_1 and H_2,

(*e*) the $\mathbf{y}_i$ are stochastic $k_2 \times 1$ matrices, and $E_1(\mathbf{Y}) = \mathbf{X}\mathbf{B}^{1\prime}$, $E_2(\mathbf{Y}) = \mathbf{X}\mathbf{B}^{2\prime}$.

Under the foregoing assumptions (2.1) becomes

$$\begin{aligned} 2I(1:2; O_n) = J(1, 2; O_n) &= \operatorname{tr} \mathbf{\Sigma}^{-1}((\mathbf{B}^1\mathbf{x}_1 - \mathbf{B}^2\mathbf{x}_1)(\mathbf{B}^1\mathbf{x}_1 - \mathbf{B}^2\mathbf{x}_1)' \\ &\quad + \cdots + (\mathbf{B}^1\mathbf{x}_n - \mathbf{B}^2\mathbf{x}_n)(\mathbf{B}^1\mathbf{x}_n - \mathbf{B}^2\mathbf{x}_n)') \\ &= \operatorname{tr} \mathbf{\Sigma}^{-1}(\mathbf{B}^1 - \mathbf{B}^2)(\mathbf{x}_1\mathbf{x}_1' + \cdots + \mathbf{x}_n\mathbf{x}_n')(\mathbf{B}^1 - \mathbf{B}^2)' \\ &= \operatorname{tr} \mathbf{\Sigma}^{-1}(\mathbf{B}^1 - \mathbf{B}^2)\mathbf{X}'\mathbf{X}(\mathbf{B}^1 - \mathbf{B}^2)'. \end{aligned} \tag{3.3}$$

As in chapter 10, we shall see that suitable specification of the matrices $\mathbf{X}$ and $\mathbf{B}$ provides the appropriate model for many statistical problems of interest. [Cf. Anderson (1958, pp. 211–212; 215–216), Roy (1957, p. 82), Wilks (1943, pp. 245–252).]

3.2. Linear Discriminant Function

We generalize here the presentation in section 5 of chapter 9. Suppose we take $w_i = \boldsymbol{\alpha}'\mathbf{y}_i = \alpha_1 y_{i1} + \alpha_2 y_{i2} + \cdots + \alpha_{k_2} y_{ik_2}$, $i = 1, 2, \cdots, n$, the same linear compound of the y's for each observation. Since the w's are normally distributed with $\sigma_{w_i}^2 = \boldsymbol{\alpha}'\mathbf{\Sigma}\boldsymbol{\alpha}$, (3.3) yields for the w's [cf. (5.2) in chapter 9]

$$\begin{aligned} &J(1, 2; w) \\ &= \frac{\boldsymbol{\alpha}'(\mathbf{B}^1\mathbf{x}_1 - \mathbf{B}^2\mathbf{x}_1)(\mathbf{B}^1\mathbf{x}_1 - \mathbf{B}^2\mathbf{x}_1)'\boldsymbol{\alpha} + \cdots + \boldsymbol{\alpha}'(\mathbf{B}^1\mathbf{x}_n - \mathbf{B}^2\mathbf{x}_n)(\mathbf{B}^1\mathbf{x}_n - \mathbf{B}^2\mathbf{x}_n)'\boldsymbol{\alpha}}{\boldsymbol{\alpha}'\mathbf{\Sigma}\boldsymbol{\alpha}} \\ &= \frac{\boldsymbol{\alpha}'(\mathbf{B}^1 - \mathbf{B}^2)\mathbf{X}'\mathbf{X}(\mathbf{B}^1 - \mathbf{B}^2)'\boldsymbol{\alpha}}{\boldsymbol{\alpha}'\mathbf{\Sigma}\boldsymbol{\alpha}}. \end{aligned} \tag{3.4}$$

For the linear compound maximizing $J(1, 2; w)$, we find (by the usual calculus procedures) that $\boldsymbol{\alpha}$ satisfies $(\mathbf{B}^1 - \mathbf{B}^2)\mathbf{X}'\mathbf{X}(\mathbf{B}^1 - \mathbf{B}^2)'\boldsymbol{\alpha} = \lambda\boldsymbol{\Sigma}\boldsymbol{\alpha}$, where λ is the largest root of the determinantal equation

$$|(\mathbf{B}^1 - \mathbf{B}^2)\mathbf{X}'\mathbf{X}(\mathbf{B}^1 - \mathbf{B}^2)' - \lambda\boldsymbol{\Sigma}| = 0.$$

The rank of $(\mathbf{B}^1 - \mathbf{B}^2)\mathbf{X}'\mathbf{X}(\mathbf{B}^1 - \mathbf{B}^2)'$ is not greater than $\min(k_1, k_2)$; thus the determinantal equation has $p \leqq \min(k_1, k_2)$ nonzero roots designated in descending order as $\lambda_1, \lambda_2, \cdots, \lambda_p$. We thus have

$$(3.5)\quad \begin{aligned} J(1, 2; O_n) &= \operatorname{tr} \boldsymbol{\Sigma}^{-1}(\mathbf{B}^1 - \mathbf{B}^2)\mathbf{X}'\mathbf{X}(\mathbf{B}^1 - \mathbf{B}^2)' = \lambda_1 + \lambda_2 + \cdots + \lambda_p \\ &= J(1, 2; \lambda_1) + \cdots + J(1, 2; \lambda_p), \end{aligned}$$

where $\lambda_i = J(1, 2; \lambda_i)$ is the value of (3.4) for $\boldsymbol{\alpha}$ associated with $\lambda = \lambda_i$.

4. THE MINIMUM DISCRIMINATION INFORMATION STATISTIC

We first state some facts about estimates of the parameters $\mathbf{B}$ and $\boldsymbol{\Sigma}$ of section 3.1. [Cf. Anderson (1951, pp. 103–104; 1958, pp. 179–183), Lawley (1938, pp. 185–186), Wilks (1943, pp. 245–250).] The classical least squares procedure of minimizing $\operatorname{tr} \mathbf{Z}'\mathbf{Z} = \operatorname{tr}(\mathbf{Y}' - \mathbf{B}\mathbf{X}')(\mathbf{Y} - \mathbf{X}\mathbf{B}')$ with respect to the β_{rs} leads to the normal equations:

$$(4.1)\quad \mathbf{X}'\mathbf{X}\hat{\mathbf{B}}' = \mathbf{X}'\mathbf{Y}, \quad \text{or} \quad \hat{\mathbf{B}}\mathbf{X}'\mathbf{X} = \mathbf{Y}'\mathbf{X}, \quad \hat{\mathbf{B}} = (\hat{\beta}_{rs}) = \mathbf{Y}'\mathbf{X}(\mathbf{X}'\mathbf{X})^{-1}.$$

The $\hat{\beta}_{rs}$, $r = 1, 2, \cdots, k_2$, $s = 1, 2, \cdots, k_1$, (k_1k_2 linear functions of the y's), are normal, minimum variance, unbiased, sufficient estimates of the β_{rs}. These properties are derived in section 10, as is also the fact that the covariance matrix of the k_1k_2 values of $\hat{\beta}_{rs}$ ordered as $\hat{\beta}_{11}, \hat{\beta}_{12}, \cdots, \hat{\beta}_{1k_1}, \hat{\beta}_{21}, \cdots, \hat{\beta}_{2k_1}, \cdots, \hat{\beta}_{k_21}, \cdots, \hat{\beta}_{k_2k_1}$ is the $k_2k_1 \times k_2k_1$ matrix,

$$(4.2)\quad (\boldsymbol{\Sigma}) \times \cdot (\mathbf{X}'\mathbf{X})^{-1} = \begin{pmatrix} \sigma_{11}(\mathbf{X}'\mathbf{X})^{-1} & \sigma_{12}(\mathbf{X}'\mathbf{X})^{-1} \cdots & \sigma_{1k_2}(\mathbf{X}'\mathbf{X})^{-1} \\ \sigma_{21}(\mathbf{X}'\mathbf{X})^{-1} & \sigma_{22}(\mathbf{X}'\mathbf{X})^{-1} \cdots & \sigma_{2k_2}(\mathbf{X}'\mathbf{X})^{-1} \\ \cdot\ \cdot\ \cdot & \cdot\ \cdot\ \cdot & \cdot\ \cdot\ \cdot \\ \sigma_{k_21}(\mathbf{X}'\mathbf{X})^{-1} & \sigma_{k_22}(\mathbf{X}'\mathbf{X})^{-1} \cdots & \sigma_{k_2k_2}(\mathbf{X}'\mathbf{X})^{-1} \end{pmatrix},$$

where $(\boldsymbol{\Sigma}) \times \cdot (\mathbf{X}'\mathbf{X})^{-1}$ means the Kronecker or direct product of the matrices [MacDuffee (1946, pp. 81–88), see also Anderson (1958, pp. 347–348), Cornish (1957)]. An unbiased estimate of $\boldsymbol{\Sigma}$ with $(n - k_1)$ degrees of freedom is obtained from

$$\begin{aligned} (n - k_1)\hat{\boldsymbol{\Sigma}} &= \hat{\mathbf{Z}}'\hat{\mathbf{Z}} = (\mathbf{Y} - \mathbf{X}\hat{\mathbf{B}}')'(\mathbf{Y} - \mathbf{X}\hat{\mathbf{B}}') = \mathbf{Y}'\mathbf{Y} - \hat{\mathbf{B}}\mathbf{X}'\mathbf{X}\hat{\mathbf{B}}' \\ &= \mathbf{Y}'\mathbf{Y} - (\mathbf{Y}'\mathbf{X})(\mathbf{X}'\mathbf{X})^{-1}(\mathbf{X}'\mathbf{Y}). \end{aligned}$$

(See problems 12.15 and 12.16.)

The minimum discrimination information statistic is obtained by replacing the population parameters in $I(1:2)$ by best unbiased estimates under the hypotheses. (Details on the conjugate distribution of a multivariate normal distribution are in sections 2 and 3.1 of chapter 12.)

Suppose the hypothesis H_1 imposes no restriction on $\mathbf{B}$ and the null hypothesis H_2 specifies $\mathbf{B} = \mathbf{B}^2$. Writing $\hat{\mathbf{B}}^1$ to indicate the solution of (4.1) under H_1, we have

$$(4.3)\qquad 2\hat{I}(H_1:H_2;O_n) = \hat{J}(H_1,H_2;O_n) = \operatorname{tr}\hat{\mathbf{\Sigma}}^{-1}(\hat{\mathbf{B}}^1 - \mathbf{B}^2)\mathbf{X}'\mathbf{X}(\hat{\mathbf{B}}^1 - \mathbf{B}^2)',$$

where

$$\begin{aligned}(n-k_1)\hat{\mathbf{\Sigma}} = \hat{\mathbf{Z}}'\hat{\mathbf{Z}} &= (\mathbf{Y} - \mathbf{X}\hat{\mathbf{B}}^{1\prime})'(\mathbf{Y} - \mathbf{X}\hat{\mathbf{B}}^{1\prime}) = \mathbf{Y}'\mathbf{Y} - \hat{\mathbf{B}}^1\mathbf{X}'\mathbf{X}\hat{\mathbf{B}}^{1\prime}\\ &= \mathbf{Y}'\mathbf{Y} - (\mathbf{Y}'\mathbf{X})(\mathbf{X}'\mathbf{X})^{-1}(\mathbf{X}'\mathbf{Y}).\end{aligned}$$

Statistics of the form in (4.3) were introduced by Lawley (1938), Hotelling (1947).

In section 10, we also show that

$$(4.4)\qquad \operatorname{tr}\hat{\mathbf{\Sigma}}^{-1}(\hat{\mathbf{B}} - \mathbf{B})\mathbf{X}'\mathbf{X}(\hat{\mathbf{B}} - \mathbf{B})' = (\hat{\beta}_{11} - \beta_{11}, \cdots, \hat{\beta}_{1k_1} - \beta_{1k_1}, \cdots, \hat{\beta}_{k_21} - \beta_{k_21}, \cdots, \hat{\beta}_{k_2k_1} - \beta_{k_2k_1})((\hat{\mathbf{\Sigma}}^{-1}) \times \cdot (\mathbf{X}'\mathbf{X}))\begin{pmatrix}\hat{\beta}_{11} - \beta_{11}\\ \cdot\\ \cdot\\ \hat{\beta}_{1k_1} - \beta_{1k_1}\\ \cdot\\ \cdot\\ \cdot\\ \hat{\beta}_{k_21} - \beta_{k_21}\\ \cdot\\ \cdot\\ \cdot\\ \hat{\beta}_{k_2k_1} - \beta_{k_2k_1}\end{pmatrix}.$$

Since the inverse of the covariance matrix in (4.2) is the direct product of the inverses of the matrices, that is, $((\mathbf{\Sigma}) \times \cdot (\mathbf{X}'\mathbf{X})^{-1})^{-1} = (\mathbf{\Sigma}^{-1}) \times \cdot (\mathbf{X}'\mathbf{X})$, [MacDuffee (1946, p. 82)], we see from (4.4) that

(*a*) The divergence $[2I(1:2;O_n)]$ in (3.3) is equivalent to that between two k_1k_2-variate normal populations with respective means $(\beta_{11}^{\alpha}, \cdots, \beta_{1k_1}^{\alpha}, \cdots, \beta_{k_21}^{\alpha}, \cdots, \beta_{k_2k_1}^{\alpha})$, $\alpha = 1, 2$, and common covariance matrix $(\mathbf{\Sigma}) \times \cdot (\mathbf{X}'\mathbf{X})^{-1}$ (see the remark at the end of section 3 of chapter 10).

(*b*) The right-hand side of (4.4) is the quadratic form in the exponent of the k_1k_2-variate normal distribution of the $\hat{\beta}_{rs}$, $r = 1, 2, \cdots, k_2$, $s = 1, 2, \cdots, k_1$, with the covariance matrix replaced by an unbiased estimate with $(n - k_1)$ degrees of freedom. $\hat{J}(H_1, H_2; O_n)$ in (4.3) is therefore a form of Hotelling's generalized Student ratio (Hotelling's T^2).

Lawley (1938) has essentially shown that for $k_1 \neq 1$, $k_2 \neq 1$, and n large, approximately,

$$(4.5)\quad \hat{J}(H_1, H_2; O_n) = \operatorname{tr} \hat{\mathbf{\Sigma}}^{-1}(\hat{\mathbf{B}}^1 - \mathbf{B}^2)\mathbf{X}'\mathbf{X}(\hat{\mathbf{B}}^1 - \mathbf{B}^2)' = \frac{k_1k_2(n - k_1)}{n - k_1 - k_2 + 1} F,$$

where F has the analysis of variance distribution under the null hypothesis H_2 with degrees of freedom $n_1 = [(1 + c)k_1k_2]$ and $n_2 = [(1 + c)(n - k_1 - k_2 + 1)]$, where $c = (k_1 - 1)(k_2 - 1)/(n - k_1)$, and [ ] means to the nearest integer. When $k_1 = 1$, or $k_2 = 1$, (4.5) is exact. [In (4.4) of chapter 10, $k_2 = 1$, $k_1 = p$.] Pillai (1955) has shown that, approximately,

$$(4.6)\quad \hat{J}(H_1, H_2; O_n) = \operatorname{tr} \hat{\mathbf{\Sigma}}^{-1}(\hat{\mathbf{B}}^1 - \mathbf{B}^2)\mathbf{X}'\mathbf{X}(\hat{\mathbf{B}}^1 - \mathbf{B}^2)' = \frac{k_1k_2(n - k_1)}{n - k_1 - k_2 - 1 + 2/k_2} F,$$

where F has the analysis of variance distribution under the null hypothesis H_2 with $n_1 = k_1k_2$ and $n_2 = k_2(n - k_1 - k_2 - 1) + 2$ degrees of freedom. In accordance with the asymptotic theory, $\hat{J}(H_1, H_2; O_n)$ is asymptotically distributed as χ^2 with k_1k_2 degrees of freedom. [Cf. Anderson (1958, p. 224).]

On the other hand, under one of the alternatives, (4.5) still holds but F now has the noncentral analysis of variance distribution, with the same degrees of freedom as under the null hypothesis, and noncentrality parameter $J(H_1, H_2) = \operatorname{tr} \mathbf{\Sigma}^{-1}(\mathbf{B}^1 - \mathbf{B}^2)\mathbf{X}'\mathbf{X}(\mathbf{B}^1 - \mathbf{B}^2)'$. In accordance with the asymptotic theory, $\hat{J}(H_1, H_2; O_n)$ is asymptotically distributed as noncentral χ^2 with k_1k_2 degrees of freedom and noncentrality parameter $J(H_1, H_2)$ when the null hypothesis is not true. [For the noncentral distributions see, for example, Anderson (1958, pp. 112–115), Fisher (1928), Fix (1949), Hsu (1938), Kempthorne (1952, pp. 219–222), Patnaik (1949), Pearson and Hartley (1951), Rao (1952, p. 50), Simaika (1941), Tang (1938), Weibull (1953), Wijsman (1957), and section 6.1 of chapter 12.]

5. SUBHYPOTHESES

5.1. Two-Partition Subhypothesis

Suppose we partition the parameters into two sets, and instead of (3.1) we now consider

$$(5.1)\quad \mathbf{z}_i = \mathbf{y}_i - \mathbf{B}_1\mathbf{x}_{1i} - \mathbf{B}_2\mathbf{x}_{2i}, \qquad i = 1, 2, \cdots, n,$$

where $\mathbf{x}_i' = (\mathbf{x}_{1i}', \mathbf{x}_{2i}')$, $\mathbf{x}_{1i}' = (x_{i1}, x_{i2}, \cdots, x_{iq_1})$, $\mathbf{x}_{2i}' = (x_{iq_1+1}, \cdots, x_{iq_1+q_2})$, $q_1 + q_2 = k_1$, $\mathbf{B} = (\mathbf{B}_1, \mathbf{B}_2)$, with $\mathbf{B}_1$ and $\mathbf{B}_2$ respectively $k_2 \times q_1$ and $k_2 \times q_2$ matrices. We may also express the n regressions in (5.1) as the one over-all regression model

$$\mathbf{Z} = \mathbf{Y} - \mathbf{X}_1\mathbf{B}_1' - \mathbf{X}_2\mathbf{B}_2', \tag{5.2}$$

where $\mathbf{Y}$, $\mathbf{Z}$ are defined as in (3.2), and

$$\mathbf{X}' = (\mathbf{x}_1, \cdots, \mathbf{x}_n) = \begin{pmatrix} \mathbf{x}_{11}, \cdots, \mathbf{x}_{1n} \\ \mathbf{x}_{21}, \cdots, \mathbf{x}_{2n} \end{pmatrix} = \begin{pmatrix} \mathbf{X}_1' \\ \mathbf{X}_2' \end{pmatrix},$$

with $\mathbf{X}_1' = (\mathbf{x}_{11}, \mathbf{x}_{12}, \cdots, \mathbf{x}_{1n})$, $\mathbf{X}_2' = (\mathbf{x}_{21}, \mathbf{x}_{22}, \cdots, \mathbf{x}_{2n})$, and $\mathbf{X}_1$ and $\mathbf{X}_2$ respectively of ranks q_1 and $q_2 = k_1 - q_1$.

With the same assumptions about the $\mathbf{z}_i$ as in section 3.1, we now consider the hypotheses:

$$\begin{aligned} H_1&: E_1(\mathbf{Y}) = \mathbf{X}_1\mathbf{B}_1^{1'} + \mathbf{X}_2\mathbf{B}_2^{1'} \\ H_2&: E_2(\mathbf{Y}) = \mathbf{X}_1\mathbf{B}_1^{2'} + \mathbf{X}_2\mathbf{B}_2^{2'}. \end{aligned} \tag{5.3}$$

Now (3.3) yields

$$\begin{aligned} 2I(1:2; O_n) &= J(1, 2; O_n) \\ &= \operatorname{tr} \mathbf{\Sigma}^{-1}(\mathbf{B}_1^1 - \mathbf{B}_1^2, \mathbf{B}_2^1 - \mathbf{B}_2^2) \begin{pmatrix} \mathbf{S}_{11} & \mathbf{S}_{12} \\ \mathbf{S}_{21} & \mathbf{S}_{22} \end{pmatrix} \begin{pmatrix} (\mathbf{B}_1^1 - \mathbf{B}_1^2)' \\ (\mathbf{B}_2^1 - \mathbf{B}_2^2)' \end{pmatrix}, \end{aligned} \tag{5.4}$$

where

$$\mathbf{X}'\mathbf{X} = \begin{pmatrix} \mathbf{X}_1' \\ \mathbf{X}_2' \end{pmatrix} (\mathbf{X}_1 \quad \mathbf{X}_2) = \begin{pmatrix} \mathbf{X}_1'\mathbf{X}_1 & \mathbf{X}_1'\mathbf{X}_2 \\ \mathbf{X}_2'\mathbf{X}_1 & \mathbf{X}_2'\mathbf{X}_2 \end{pmatrix} = \begin{pmatrix} \mathbf{S}_{11} & \mathbf{S}_{12} \\ \mathbf{S}_{21} & \mathbf{S}_{22} \end{pmatrix} = \mathbf{S}.$$

The normal equations (4.1) under H_1 become

$$(\hat{\mathbf{B}}_1^1, \hat{\mathbf{B}}_2^1) \begin{pmatrix} \mathbf{S}_{11} & \mathbf{S}_{12} \\ \mathbf{S}_{21} & \mathbf{S}_{22} \end{pmatrix} = (\mathbf{Y}'\mathbf{X}_1, \mathbf{Y}'\mathbf{X}_2), \tag{5.5}$$

or

$$\begin{aligned} \hat{\mathbf{B}}_1^1\mathbf{S}_{11} + \hat{\mathbf{B}}_2^1\mathbf{S}_{21} &= \mathbf{Y}'\mathbf{X}_1 \\ \hat{\mathbf{B}}_1^1\mathbf{S}_{12} + \hat{\mathbf{B}}_2^1\mathbf{S}_{22} &= \mathbf{Y}'\mathbf{X}_2. \end{aligned} \tag{5.6}$$

From (5.6) we find [cf. (5.7) and (5.8) of chapter 10]

$$\hat{\mathbf{B}}_2^1 = \mathbf{Y}'\mathbf{X}_{2\cdot1}\mathbf{S}_{22\cdot1}^{-1}, \qquad \hat{\mathbf{B}}_1^1 = \mathbf{Y}'\mathbf{X}_1\mathbf{S}_{11}^{-1} - \hat{\mathbf{B}}_2^1\mathbf{S}_{21}\mathbf{S}_{11}^{-1}, \tag{5.7}$$

where $\mathbf{X}_{2\cdot1} = \mathbf{X}_2 - \mathbf{X}_1\mathbf{S}_{11}^{-1}\mathbf{S}_{12}$, $\mathbf{S}_{22\cdot1} = \mathbf{S}_{22} - \mathbf{S}_{21}\mathbf{S}_{11}^{-1}\mathbf{S}_{12}$.

For the estimate of $\mathbf{\Sigma}$ we have from (4.3)

$$(n - k_1)\hat{\mathbf{\Sigma}} = \mathbf{Y}'\mathbf{Y} - (\hat{\mathbf{B}}_1^1, \hat{\mathbf{B}}_2^1) \begin{pmatrix} \mathbf{S}_{11} & \mathbf{S}_{12} \\ \mathbf{S}_{21} & \mathbf{S}_{22} \end{pmatrix} \begin{pmatrix} \hat{\mathbf{B}}_1^{1'} \\ \hat{\mathbf{B}}_2^{1'} \end{pmatrix}. \tag{5.8}$$

Suppose, now, that in particular we want to test the null hypothesis

$$H_3: \mathbf{B} = \mathbf{B}^3 = (\mathbf{B}_1^3, \mathbf{0}), \tag{5.9}$$

that is, $\mathbf{B}_2^3 = \mathbf{0}$, with no restrictions on $\mathbf{B}_1^3$, against the alternative hypothesis

$$(5.10) \qquad H_1: \mathbf{B} = \mathbf{B}^1 = (\mathbf{B}_1^1, \mathbf{B}_2^1),$$

with no restrictions on the parameters.

Under H_1 we have (5.7) and (5.8). Under H_3 the normal equations (4.1) now yield

$$(5.11) \qquad \hat{\mathbf{B}}_1^3\mathbf{S}_{11} = \mathbf{Y}'\mathbf{X}_1, \qquad \hat{\mathbf{B}}_1^3 = \mathbf{Y}'\mathbf{X}_1\mathbf{S}_{11}^{-1}.$$

We estimate $J(1, 2; O_n)$ in (5.4) by replacing the parameters by best unbiased estimates under the hypotheses, so that

$$(5.12) \quad \hat{J}(H_1, H_3) = \mathrm{tr}\, \hat{\mathbf{\Sigma}}^{-1}(\hat{\mathbf{B}}_1^1 - \hat{\mathbf{B}}_1^3, \hat{\mathbf{B}}_2^1) \begin{pmatrix} \mathbf{S}_{11} & \mathbf{S}_{12} \\ \mathbf{S}_{21} & \mathbf{S}_{22} \end{pmatrix} \begin{pmatrix} (\hat{\mathbf{B}}_1^1 - \hat{\mathbf{B}}_1^3)' \\ \hat{\mathbf{B}}_2^{1'} \end{pmatrix}.$$

From (5.7) and (5.11), we find [cf. (5.10)–(5.13) in chapter 10]

$$(5.13) \quad (\hat{\mathbf{B}}_1^1, \hat{\mathbf{B}}_2^1) \begin{pmatrix} \mathbf{S}_{11} & \mathbf{S}_{12} \\ \mathbf{S}_{21} & \mathbf{S}_{22} \end{pmatrix} \begin{pmatrix} \hat{\mathbf{B}}_1^{1'} \\ \hat{\mathbf{B}}_2^{1'} \end{pmatrix} = \hat{\mathbf{B}}_1^{3'}\mathbf{S}_{11}\hat{\mathbf{B}}_1^{3'} + \hat{\mathbf{B}}_2^1\mathbf{S}_{22\cdot 1}\hat{\mathbf{B}}_2^{1'}$$
$$= \mathbf{Y}'\mathbf{X}_1\mathbf{S}_{11}^{-1}\mathbf{X}_1'\mathbf{Y} + \mathbf{Y}'\mathbf{X}_{2\cdot 1}\mathbf{S}_{22\cdot 1}^{-1}\mathbf{X}_{2\cdot 1}'\mathbf{Y},$$

$$(5.14) \quad (\hat{\mathbf{B}}_1^1 - \hat{\mathbf{B}}_1^3, \hat{\mathbf{B}}_2^1) \begin{pmatrix} \mathbf{S}_{11} & \mathbf{S}_{12} \\ \mathbf{S}_{21} & \mathbf{S}_{22} \end{pmatrix} \begin{pmatrix} (\hat{\mathbf{B}}_1^1 - \hat{\mathbf{B}}_1^3)' \\ \hat{\mathbf{B}}_2^{1'} \end{pmatrix} = \hat{\mathbf{B}}_2^1\mathbf{S}_{22\cdot 1}\hat{\mathbf{B}}_2^{1'},$$

$$(5.15) \qquad \hat{J}(H_1, H_3) = \mathrm{tr}\, \hat{\mathbf{\Sigma}}^{-1}\hat{\mathbf{B}}^1\mathbf{S}\hat{\mathbf{B}}^{1'} - \mathrm{tr}\, \hat{\mathbf{\Sigma}}^{-1}\hat{\mathbf{B}}_1^3\mathbf{S}_{11}\hat{\mathbf{B}}_1^{3'}.$$

It may be verified that

$$(5.16) \qquad \mathbf{X}_1\mathbf{S}_{11}^{-1}\mathbf{X}_1'\mathbf{X}_{2\cdot 1}\mathbf{S}_{22\cdot 1}^{-1}\mathbf{X}_{2\cdot 1}' = \mathbf{0},$$

and since $\mathbf{X}_{2\cdot 1}'\mathbf{X}_{2\cdot 1} = \mathbf{S}_{22\cdot 1}$,

$$(5.17) \qquad (\mathbf{I}_n - \mathbf{X}_1\mathbf{S}_{11}^{-1}\mathbf{X}_1' - \mathbf{X}_{2\cdot 1}\mathbf{S}_{22\cdot 1}^{-1}\mathbf{X}_{2\cdot 1}')\mathbf{X}_{2\cdot 1}\mathbf{S}_{22\cdot 1}^{-1}\mathbf{X}_{2\cdot 1}' = \mathbf{0},$$

where $\mathbf{I}_n$ is the $n \times n$ identity matrix, that is, the two factors in $\hat{J}(H_1, H_3)$ are independent.

The foregoing is summarized in tables 5.1 and 5.2.

TABLE 5.1

Variation due to	Generalized Sum of Squares	D.F.
$H_3: \mathbf{B}^3 = (\mathbf{B}_1^3, \mathbf{0})$	$\hat{\mathbf{B}}_1^3\mathbf{S}_{11}\hat{\mathbf{B}}_1^{3'} = \mathbf{Y}'\mathbf{X}_1\mathbf{S}_{11}^{-1}\mathbf{X}_1'\mathbf{Y}$	q_1
Difference	$\hat{\mathbf{B}}_2^1\mathbf{S}_{22\cdot 1}\hat{\mathbf{B}}_2^{1'} = \mathbf{Y}'\mathbf{X}_{2\cdot 1}\mathbf{S}_{22.1}^{-1}\mathbf{X}_{2\cdot 1}'\mathbf{Y}$	q_2
$H_1: \mathbf{B}^1 = (\mathbf{B}_1^1, \mathbf{B}_2^1)$	$\hat{\mathbf{B}}^1\mathbf{X}'\mathbf{X}\hat{\mathbf{B}}^{1'} = \mathbf{Y}'\mathbf{X}_1\mathbf{S}_{11}^{-1}\mathbf{X}_1'\mathbf{Y} + \mathbf{Y}'\mathbf{X}_{2\cdot 1}\mathbf{S}_{22\cdot 1}^{-1}\mathbf{X}_{2\cdot 1}'\mathbf{Y}$	k_1
Difference	$\mathbf{Y}'\mathbf{Y} - \hat{\mathbf{B}}^1\mathbf{X}'\mathbf{X}\hat{\mathbf{B}}^{1'} = (n - k_1)\hat{\mathbf{\Sigma}}$	$n - k_1$
Total	$\mathbf{Y}'\mathbf{Y}$	n

TABLE 5.2

Test	Distribution on the	Null Hypothesis
$\hat{J}(H_1, H_2) = \mathrm{tr}\, \hat{\mathbf{\Sigma}}^{-1}\hat{\mathbf{B}}^1\mathbf{X}'\mathbf{X}\hat{\mathbf{B}}^{1'}$	$\hat{J}(H_1, H_2) = \dfrac{k_1k_2(n-k_1)}{n-k_1-k_2+1}F,$ $n_1 = [k_1k_2(1+c_1)]$ $n_2 = [(n-k_1-k_2+1)(1+c_1)]$ $c_1 = (k_1-1)(k_2-1)/(n-k_1)$	$\mathbf{B} = \mathbf{B}^2 = \mathbf{0}$, i.e., $\mathbf{B}_1^2 = \mathbf{0}, \mathbf{B}_2^2 = \mathbf{0}$
$\hat{J}(H_1, H_3) = \mathrm{tr}\, \hat{\mathbf{\Sigma}}^{-1}\hat{\mathbf{B}}_2^1\mathbf{S}_{22\cdot 1}\hat{\mathbf{B}}_2^{1'}$	$\hat{J}(H_1, H_3) = \dfrac{q_2k_2(n-k_1)}{n-k_1-k_2+1}F,$ $n_1 = [q_2k_2(1+c_2)]$ $n_2 = [(n-k_1-k_2+1)(1+c_2)]$ $c_2 = (q_2-1)(k_2-1)/(n-k_1)$	$\mathbf{B} = \mathbf{B}^3 = (\mathbf{B}_1^3, \mathbf{0})$, i.e., $\mathbf{B}_2^3 = \mathbf{0}$

5.2. Three-Partition Subhypothesis

(Cf. section 5.2 of chapter 10.) When the subhypothesis requires partitioning the matrices $\mathbf{X}$ and $\mathbf{B}$ into three submatrices, $\mathbf{X} = (\mathbf{X}_1, \mathbf{X}_2, \mathbf{X}_3)$, $\mathbf{B} = (\mathbf{B}_1, \mathbf{B}_2, \mathbf{B}_3)$, we obtain from (4.1) the solutions

$$(5.18)\qquad \begin{aligned}\hat{\mathbf{B}}_3 &= \mathbf{Y}'\mathbf{X}_{3\cdot 12}\mathbf{S}_{33\cdot 12}^{-1}\\ \hat{\mathbf{B}}_2 &= (\mathbf{Y}'\mathbf{X}_{2\cdot 1} - \hat{\mathbf{B}}_3\mathbf{S}_{32\cdot 1})\mathbf{S}_{22\cdot 1}^{-1}\\ \hat{\mathbf{B}}_1 &= (\mathbf{Y}'\mathbf{X}_1 - \hat{\mathbf{B}}_2\mathbf{S}_{21} - \hat{\mathbf{B}}_3\mathbf{S}_{31})\mathbf{S}_{11}^{-1},\end{aligned}$$

where

$$\mathbf{S} = \begin{pmatrix}\mathbf{S}_{11} & \mathbf{S}_{12} & \mathbf{S}_{13}\\ \mathbf{S}_{21} & \mathbf{S}_{22} & \mathbf{S}_{23}\\ \mathbf{S}_{31} & \mathbf{S}_{32} & \mathbf{S}_{33}\end{pmatrix}, \qquad \mathbf{S}_{tu} = \mathbf{X}_t'\mathbf{X}_u, \qquad t, u = 1, 2, 3,$$

and

$$\mathbf{S}_{33\cdot 12} = \mathbf{S}_{33\cdot 1} - \mathbf{S}_{32\cdot 1}\mathbf{S}_{22\cdot 1}^{-1}\mathbf{S}_{23\cdot 1},$$

$$\mathbf{S}_{33\cdot 1} = \mathbf{S}_{33} - \mathbf{S}_{31}\mathbf{S}_{11}^{-1}\mathbf{S}_{13}, \qquad \mathbf{S}_{32\cdot 1} = \mathbf{S}_{32} - \mathbf{S}_{31}\mathbf{S}_{11}^{-1}\mathbf{S}_{12} = \mathbf{S}_{23\cdot 1}',$$

$$\mathbf{S}_{22\cdot 1} = \mathbf{S}_{22} - \mathbf{S}_{21}\mathbf{S}_{11}^{-1}\mathbf{S}_{12}, \qquad \mathbf{Y}'\mathbf{X}_{2\cdot 1} = \mathbf{Y}'(\mathbf{X}_2 - \mathbf{X}_1\mathbf{S}_{11}^{-1}\mathbf{S}_{12}),$$

$$\mathbf{Y}'\mathbf{X}_{3\cdot 12} = \mathbf{Y}'(\mathbf{X}_{3\cdot 1} - \mathbf{X}_{2\cdot 1}\mathbf{S}_{22\cdot 1}^{-1}\mathbf{S}_{23\cdot 1}), \qquad \mathbf{Y}'\mathbf{X}_{3\cdot 1} = \mathbf{Y}'(\mathbf{X}_3 - \mathbf{X}_1\mathbf{S}_{11}^{-1}\mathbf{S}_{13}).$$

We also have [cf. (5.21) in chapter 10]

$$(5.19)\qquad \begin{aligned}\hat{\mathbf{B}}\mathbf{S}\hat{\mathbf{B}}' &= \mathbf{Y}'\mathbf{X}_1\mathbf{S}_{11}^{-1}\mathbf{X}_1'\mathbf{Y} + \mathbf{Y}'\mathbf{X}_{2\cdot 1}\mathbf{S}_{22\cdot 1}^{-1}\mathbf{X}_{2\cdot 1}'\mathbf{Y} + \hat{\mathbf{B}}_3\mathbf{S}_{33\cdot 12}\hat{\mathbf{B}}_3'\\ &= \mathbf{Y}'\mathbf{X}_1\mathbf{S}_{11}^{-1}\mathbf{X}_1'\mathbf{Y} + \mathbf{Y}'\mathbf{X}_{2\cdot 1}\hat{\mathbf{B}}_2' + \mathbf{Y}'\mathbf{X}_{3\cdot 1}\hat{\mathbf{B}}_3',\end{aligned}$$

where the last version is convenient when the data are raw observations and $x_{i1} = 1$ for all i. (See problems 12.11, 12.12, 12.13.)

6. SPECIAL CASES

To illustrate sections 3, 4, and 5 we examine several interesting special cases.

6.1. Hotelling's Generalized Student Ratio (Hotelling's T^2)

Suppose we have a random sample of n independent observations from a multivariate normal population, and we want to test a null hypothesis H_2, specifying the population means, against an alternative hypothesis H_1, that the population means are not as specified. (See section 3.1 of chapter 12.)

The matrices in the regression models (3.1) and (3.2) are specified by

$$(6.1) \qquad H_1\colon \mathbf{B} = \mathbf{B}^1 = \begin{pmatrix} \beta_{11} \\ \beta_{21} \\ \cdot \\ \cdot \\ \cdot \\ \beta_{k_21} \end{pmatrix}, \qquad H_2\colon \mathbf{B} = \mathbf{B}^2 = \begin{pmatrix} \beta_{11}^2 \\ \beta_{21}^2 \\ \cdot \\ \cdot \\ \cdot \\ \beta_{k_21}^2 \end{pmatrix},$$

$$\mathbf{x}_i = (1), \qquad \mathbf{X}' = (1, 1, \cdots, 1),$$

with $\mathbf{X}'$ a $1 \times n$ matrix. We find that $\mathbf{X}'\mathbf{X} = n$,

$$\mathbf{Y}'\mathbf{X} = \mathbf{y}_1 + \mathbf{y}_2 + \cdots + \mathbf{y}_n = n\bar{\mathbf{y}} = \begin{pmatrix} n\bar{y}_1 \\ n\bar{y}_2 \\ \cdot \\ \cdot \\ \cdot \\ n\bar{y}_{k_2} \end{pmatrix},$$

$$n\bar{y}_j = y_{1j} + y_{2j} + \cdots + y_{nj}, \qquad i = 1, 2, \cdots, k_2.$$

The normal equations (4.1) thus yield

$$(6.2) \qquad \hat{\mathbf{B}}^1 = \begin{pmatrix} \hat{\beta}_{11} \\ \hat{\beta}_{21} \\ \cdot \\ \cdot \\ \cdot \\ \hat{\beta}_{k_21} \end{pmatrix} = \begin{pmatrix} \bar{y}_1 \\ \bar{y}_2 \\ \cdot \\ \cdot \\ \cdot \\ \bar{y}_{k_2} \end{pmatrix} = \bar{\mathbf{y}},$$

and [see (4.3)]

$$(6.3) \qquad (n-1)\hat{\mathbf{\Sigma}} = \mathbf{Y}'\mathbf{Y} - n\bar{\mathbf{y}}\bar{\mathbf{y}}' = \left(\sum_{i=1}^{n}(y_{ij} - \bar{y}_j)(y_{il} - \bar{y}_l)\right)$$
$$= N\mathbf{S}_{yy}, \qquad N = n - 1, \qquad j, l = 1, 2, \cdots, k_2,$$

where $\mathbf{S}_{yy}$ is the $k_2 \times k_2$ unbiased covariance matrix of the y's.

Since $\mathbf{B}^2$ is specified, (4.3) yields

$$\hat{J}(H_1, H_2) = \operatorname{tr} \hat{\boldsymbol{\Sigma}}^{-1}(\bar{\mathbf{y}} - \mathbf{B}^2)n(\bar{\mathbf{y}} - \mathbf{B}^2)' = n(\bar{\mathbf{y}} - \mathbf{B}^2)'\hat{\boldsymbol{\Sigma}}^{-1}(\bar{\mathbf{y}} - \mathbf{B}^2). \tag{6.4}$$

Note that $\hat{J}(H_1, H_2)$ is Hotelling's generalized Student ratio (Hotelling's T^2) (see section 4 in this chapter and section 4 in chapter 10), or from (4.5) with $k_1 = 1$,

$$\hat{J}(H_1, H_2) = \frac{(n-1)k_2}{n-k_2} F, \tag{6.5}$$

where F has the analysis of variance distribution with $n_1 = k_2$ and $n_2 = n - k_2$ degrees of freedom under the null hypothesis H_2 in (6.1). [Cf. Anderson (1958, p. 107), Rao (1952, p. 243).]

6.2. Centering

[Cf. (5.14)–(5.19) in chapter 10.] We may explicitly introduce the mean value of $\mathbf{y}_i$ in (3.1) by taking $x_{i1} = 1$, $i = 1, 2, \cdots, n$, so that the matrix $\mathbf{X}'$ of (5.2) is partitioned as

$$\mathbf{X}' = \begin{pmatrix} 1 & 1 & \cdots & 1 \\ \hdashline x_{12} & x_{22} & \cdots & x_{n2} \\ \cdot & \cdot & & \cdot \\ \cdot & \cdot & & \cdot \\ \cdot & \cdot & & \cdot \\ x_{1k_1} & x_{2k_1} & \cdots & x_{nk_1} \end{pmatrix} = \begin{pmatrix} \mathbf{X}_1' \\ \mathbf{X}_2' \end{pmatrix}, \quad \mathbf{X}_1' = (1, 1, \cdots, 1), \quad 1 \times n. \tag{6.6}$$

As in (5.14)–(5.17) in chapter 10, we have then that

$$\mathbf{X}'\mathbf{X} = \begin{pmatrix} \mathbf{X}_1' \\ \mathbf{X}_2' \end{pmatrix}(\mathbf{X}_1 \quad \mathbf{X}_2) = \left(\begin{array}{c:c} n & n\bar{x}_2 \cdots n\bar{x}_{k_1} \\ \hdashline n\bar{x}_2 & \\ \cdot & \\ \cdot & \mathbf{S}_{22} \\ \cdot & \\ n\bar{x}_{k_1} & \end{array}\right), \tag{6.7}$$

$$\mathbf{S}_{11} = n, \qquad \mathbf{S}_{12} = (n\bar{x}_2, \cdots, n\bar{x}_{k_1}),$$

$$\mathbf{S}_{22\cdot 1} = \left(\sum_{i=1}^{n} (x_{ij} - \bar{x}_j)(x_{il} - \bar{x}_l)\right) = N\mathbf{S}_{xx},$$

$$j, l = 2, \cdots, k_1, \qquad N = n - 1,$$

($\mathbf{S}_{xx}$ is the $(k_1 - 1) \times (k_1 - 1)$ unbiased covariance matrix of the x's)

$$\mathbf{X}_{2\cdot 1} = \mathbf{X}_2 - \mathbf{X}_1\mathbf{S}_{11}^{-1}\mathbf{S}_{12} = ((x_{ij} - \bar{x}_j)), \quad i = 1, 2, \cdots, n, \quad j = 2, 3, \cdots, k_1.$$

We also find, as in (6.1), and from (5.11), that [cf. (5.18) and (5.19) in chapter 10]

$$(6.8) \qquad \mathbf{Y}'\mathbf{X}_1 = (\mathbf{y}_1, \mathbf{y}_2, \cdots, \mathbf{y}_n)\begin{pmatrix}1\\1\\ \cdot\\ \cdot\\ \cdot\\ 1\end{pmatrix} = n\bar{\mathbf{y}} = n\begin{pmatrix}\bar{y}_1\\ \bar{y}_2\\ \cdot\\ \cdot\\ \cdot\\ \bar{y}_{k_2}\end{pmatrix},$$

$$\hat{\mathbf{B}}_1^2 = \bar{\mathbf{y}},$$

$$\mathbf{Y}'\mathbf{X}_{2\cdot 1} = \mathbf{Y}'\mathbf{X}_2 - n\begin{pmatrix}\bar{y}_1\\ \bar{y}_2\\ \cdot\\ \cdot\\ \cdot\\ \bar{y}_{k_2}\end{pmatrix}(\bar{x}_2, \bar{x}_3, \cdots, \bar{x}_{k_1})$$

$$= \left(\sum_{i=1}^{n}(y_{ij} - \bar{y}_j)(x_{il} - \bar{x}_l)\right) = N\mathbf{S}_{yx},$$

$$j = 1, 2, \cdots, k_2, \qquad l = 2, 3, \cdots, k_1,$$

($\mathbf{S}_{yx}$ is the $k_2 \times (k_1 - 1)$ unbiased covariance matrix of the y's with the x's, with $\mathbf{S}'_{yx} = \mathbf{S}_{xy}$).

For the partitioning given by (6.6) the analysis summarized in table 5.1 becomes table 6.1.

TABLE 6.1

Variation due to	Generalized Sum of Squares	D.F.
Means $H_2{:}\mathbf{B}^2 = (\mathbf{B}_1^2, \mathbf{0})$	$\hat{\mathbf{B}}_1^2\mathbf{S}_{11}\hat{\mathbf{B}}_1^{2'} = n\bar{\mathbf{y}}\bar{\mathbf{y}}'$	1
Difference	$\hat{\mathbf{B}}_2^1\mathbf{S}_{22\cdot 1}\hat{\mathbf{B}}_2^{1'} = \mathbf{Y}'\mathbf{X}_{2\cdot 1}\mathbf{S}_{22\cdot 1}^{-1}\mathbf{X}'_{2\cdot 1}\mathbf{Y}$	$k_1 - 1$
$H_1{:}\mathbf{B}^1 = (\mathbf{B}_1^1, \mathbf{B}_2^1)$	$\hat{\mathbf{B}}^1\mathbf{X}'\mathbf{X}\hat{\mathbf{B}}^{1'} = n\bar{\mathbf{y}}\bar{\mathbf{y}}' + \mathbf{Y}'\mathbf{X}_{2\cdot 1}\mathbf{S}_{22\cdot 1}^{-1}\mathbf{X}'_{2\cdot 1}\mathbf{Y}$	k_1
Difference	$\mathbf{Y}'\mathbf{Y} - \hat{\mathbf{B}}^1\mathbf{X}'\mathbf{X}\hat{\mathbf{B}}^{1'} = (n - k_1)\hat{\mathbf{\Sigma}}$	$n - k_1$
Total	$\mathbf{Y}'\mathbf{Y}$	n

If we center the y's and x's about their respective sample averages, the analysis in table 6.1 may be condensed into table 6.2. $\hat{J}(H_1, H_2) = N \operatorname{tr} \hat{\mathbf{\Sigma}}^{-1}\mathbf{S}_{yx}\mathbf{S}_{xx}^{-1}\mathbf{S}_{xy} = \dfrac{(k_1 - 1)k_2(n - k_1)}{n - k_1 - k_2 + 1}F$, where F has the analysis of variance distribution with $[(k_1-1)k_2(1+c)]$ and $[(n-k_1-k_2+1)(1+c)]$ degrees of freedom, $c = (k_1 - 2)(k_2 - 1)/(n - k_1)$, under the null hypothesis $\mathbf{B} = \mathbf{B}^2 = \mathbf{0}$. [$\hat{J}(H_1, H_2)$ is asymptotically distributed as χ^2 with $(k_1 - 1)k_2$ degrees of freedom.]

TABLE 6.2

Variation due to	Generalized Sum of Squares	D.F.
Multivariate regression	$N\mathbf{S}_{yx}\mathbf{S}_{xx}^{-1}\mathbf{S}_{xy}$	$k_1 - 1$
Difference	$(n - k_1)\hat{\mathbf{\Sigma}}$	$n - k_1$
Total	$N\mathbf{S}_{yy}$	$n - 1$

More generally, if we center the y's and x's about their respective sample averages, the analysis in table 5.1, for what is essentially a three-partition subhypothesis, would be similar except that n would be replaced by $n - 1$ and k_1 by $k_1 - 1$ and of course $q_1 + q_2 = k_1 - 1$.

6.3. Homogeneity of r Samples

Suppose we have r independent samples respectively of n_i, $i = 1, 2, \cdots, r$, independent observations, from multivariate normal populations with a common covariance matrix. We want to test a null hypothesis H_2, the r population mean matrices (vectors) are equal, against an alternative hypothesis H_1, the population mean matrices are not all equal. [Cf. (6.15)–(6.24) in chapter 10.]

For the ith sample the regression model is [cf. (3.2)]

$$\mathbf{Z}_i = \mathbf{Y}_i - \mathbf{X}_i\mathbf{B}_i', \tag{6.9}$$

where

$$\begin{aligned} \mathbf{Z}_i' &= (\mathbf{z}_{i1}, \mathbf{z}_{i2}, \cdots, \mathbf{z}_{in_i}), & \mathbf{z}_{ij}' &= (z_{ij1}, z_{ij2}, \cdots, z_{ijk_2}), \\ \mathbf{Y}_i' &= (\mathbf{y}_{i1}, \mathbf{y}_{i2}, \cdots, \mathbf{y}_{in_i}), & \mathbf{y}_{ij}' &= (y_{ij1}, y_{ij2}, \cdots, y_{ijk_2}), \\ \mathbf{X}_i' &= (1, 1, \cdots, 1), \quad 1 \times n_i, & \mathbf{B}_i' &= (\beta_{i1}, \beta_{i2}, \cdots, \beta_{ik_2}), \\ i &= 1, 2, \cdots, r \text{ samples}, & j &= 1, 2, \cdots, n_i \text{ observations}. \end{aligned}$$

The alternative hypothesis is

$$H_1\!:\!\mathbf{B}_i^{1'} = (\beta_{i1}, \beta_{i2}, \cdots, \beta_{ik_2}), \qquad i = 1, 2, \cdots, r, \tag{6.10}$$

and the null hypothesis of homogeneity is

$$(6.11)\qquad H_2: \mathbf{B}_i^{2'} = \mathbf{B}'_{\cdot} = (\beta_{\cdot 1}, \beta_{\cdot 2}, \cdots, \beta_{\cdot k_2}), \qquad i = 1, 2, \cdots, r.$$

We may write the regression model for the r samples combined, under H_1, as

$$(6.12)\qquad \mathbf{Z} = \mathbf{Y} - \mathbf{X}^1\mathbf{B}^{1'},$$

where

$$\mathbf{Z}' = (\mathbf{Z}'_1, \mathbf{Z}'_2, \cdots, \mathbf{Z}'_r), \qquad \mathbf{Y}' = (\mathbf{Y}'_1, \mathbf{Y}'_2, \cdots, \mathbf{Y}'_r),$$

$$\mathbf{X}^{1'} = \begin{pmatrix} \mathbf{X}'_1 & \mathbf{0} & \cdots & \mathbf{0} \\ \mathbf{0} & \mathbf{X}'_2 & \cdots & \mathbf{0} \\ \cdot & \cdot & & \cdot \\ \cdot & \cdot & & \cdot \\ \cdot & \cdot & & \cdot \\ \mathbf{0} & \mathbf{0} & \cdots & \mathbf{X}'_r \end{pmatrix}, \qquad \mathbf{B}^1 = (\mathbf{B}_1, \mathbf{B}_2, \cdots, \mathbf{B}_r).$$

Under H_2, the regression model for the r samples combined is

$$(6.13)\qquad \mathbf{Z} = \mathbf{Y} - \mathbf{X}^2\mathbf{B}^{2'},$$

where $\mathbf{Z}$ and $\mathbf{Y}$ are defined as in (6.12) and $\mathbf{X}^{2'} = (\mathbf{X}'_1, \mathbf{X}'_2, \cdots, \mathbf{X}'_r)$, $\mathbf{B}^2 = \mathbf{B}_{\cdot\cdot}$

We thus have under H_1

$$(6.14)\qquad \mathbf{X}^{1'}\mathbf{X}^1 = \begin{pmatrix} \mathbf{X}'_1\mathbf{X}_1 & \cdots & & \mathbf{0} \\ \cdot & \mathbf{X}'_2\mathbf{X}_2 & & \cdot \\ \cdot & \cdot & & \cdot \\ \cdot & & \cdot & \cdot \\ \mathbf{0} & \cdots & & \mathbf{X}'_r\mathbf{X}_r \end{pmatrix} = \begin{pmatrix} n_1 & 0 & \cdots & 0 \\ \cdot & \cdot & & \cdot \\ \cdot & & \cdot & \cdot \\ \cdot & & & \cdot \\ 0 & 0 & \cdots & n_r \end{pmatrix},$$

$$(6.15)\qquad \mathbf{Y}'\mathbf{X}^1 = (\mathbf{Y}'_1\mathbf{X}_1, \mathbf{Y}'_2\mathbf{X}_2, \cdots, \mathbf{Y}'_r\mathbf{X}_r) = (n_1\bar{\mathbf{y}}_1, n_2\bar{\mathbf{y}}_2, \cdots, n_r\bar{\mathbf{y}}_r),$$

where $\bar{\mathbf{y}}'_i = (\bar{y}_{i1}, \bar{y}_{i2}, \cdots, \bar{y}_{ik_2})$, $n_i\bar{y}_{il} = y_{i1l} + y_{i2l} + \cdots + y_{in_il}$.

The normal equations (4.1) are

$$(6.16)\qquad \hat{\mathbf{B}}^1 \begin{pmatrix} n_1 & 0 & \cdots & 0 \\ 0 & n_2 & \cdots & 0 \\ \cdot & \cdot & \cdot & \cdot \\ 0 & 0 & \cdots & n_r \end{pmatrix} = (n_1\bar{\mathbf{y}}_1, n_2\bar{\mathbf{y}}_2, \cdots, n_r\bar{\mathbf{y}}_r),$$

or

$$(n_1\hat{\mathbf{B}}_1, n_2\hat{\mathbf{B}}_2, \cdots, n_r\hat{\mathbf{B}}_r) = (n_1\bar{\mathbf{y}}_1, n_2\bar{\mathbf{y}}_2, \cdots, n_r\bar{\mathbf{y}}_r).$$

that is, $\hat{\mathbf{B}}_i = \bar{\mathbf{y}}_{i\cdot}$. From (4.3) the estimate of $\mathbf{\Sigma}$ is

$$(6.17)\quad \begin{aligned}(n-r)\hat{\mathbf{\Sigma}} &= \mathbf{Y}'\mathbf{Y} - \hat{\mathbf{B}}^1\mathbf{X}^{1\prime}\mathbf{X}^1\hat{\mathbf{B}}^{1\prime} \\ &= \mathbf{Y}_1'\mathbf{Y}_1 + \cdots + \mathbf{Y}_r'\mathbf{Y}_r \\ &\qquad - (\bar{\mathbf{y}}_1, \bar{\mathbf{y}}_2, \cdots, \bar{\mathbf{y}}_r)\begin{pmatrix} n_1 & \cdots & 0 \\ \cdot & \cdot & \cdot \\ 0 & \cdots & n_r\end{pmatrix}\begin{pmatrix}\bar{\mathbf{y}}_1' \\ \cdot \\ \cdot \\ \cdot \\ \bar{\mathbf{y}}_r'\end{pmatrix} \\ &= \mathbf{Y}_1'\mathbf{Y}_1 - n_1\bar{\mathbf{y}}_1\bar{\mathbf{y}}_1' + \cdots + \mathbf{Y}_r'\mathbf{Y}_r - n_r\bar{\mathbf{y}}_r\bar{\mathbf{y}}_r' \\ &= N_1\mathbf{S}_1 + \cdots + N_r\mathbf{S}_r = N\mathbf{S},\end{aligned}$$

where $N_i = n_i - 1$, $n = n_1 + n_2 + \cdots + n_r$, $N = N_1 + N_2 + \cdots + N_r = n - r$, and $\mathbf{S}_i$ is the unbiased covariance matrix of the y's within the ith sample.

Under H_2 we have

$$(6.18)\quad \mathbf{X}^{2\prime}\mathbf{X}^2 = \mathbf{X}_1'\mathbf{X}_1 + \cdots + \mathbf{X}_r'\mathbf{X}_r = n_1 + n_2 + \cdots + n_r = n,$$

$$(6.19)\quad \begin{aligned}\mathbf{Y}'\mathbf{X}^2 &= \mathbf{Y}_1'\mathbf{X}_1 + \mathbf{Y}_2'\mathbf{X}_2 + \cdots + \mathbf{Y}_r'\mathbf{X}_r = n_1\bar{\mathbf{y}}_1 + n_2\bar{\mathbf{y}}_2 + \cdots + n_r\bar{\mathbf{y}}_r \\ &= n\bar{\mathbf{y}}, \\ \bar{\mathbf{y}}' &= (\bar{y}_{\cdot 1}, \bar{y}_{\cdot 2}, \cdots, \bar{y}_{\cdot k_2}), \qquad n\bar{y}_{\cdot l} = n_1\bar{y}_{1l} + n_2\bar{y}_{2l} + \cdots + n_r\bar{y}_{rl}, \\ &\qquad l = 1, 2, \cdots, k_2.\end{aligned}$$

The normal equations (4.1) now yield

$$(6.20)\quad n\hat{\mathbf{B}}_{\cdot} = n\bar{\mathbf{y}}.$$

We therefore have [cf. (2.17) in chapter 9]

$$(6.21)\quad \begin{aligned}\hat{J}(H_1, H_2) &= \operatorname{tr} \hat{\mathbf{\Sigma}}^{-1}(\hat{\mathbf{B}}_1 - \hat{\mathbf{B}}_{\cdot}, \hat{\mathbf{B}}_2 - \hat{\mathbf{B}}_{\cdot}, \cdots, \hat{\mathbf{B}}_r - \hat{\mathbf{B}}_{\cdot})\begin{pmatrix} n_1 & 0 & \cdots & 0 \\ 0 & n_2 & \cdots & 0 \\ \cdot & \cdot & \cdot & \cdot \\ 0 & 0 & \cdots & n_r\end{pmatrix}\begin{pmatrix}(\hat{\mathbf{B}}_1 - \hat{\mathbf{B}}_{\cdot})' \\ (\hat{\mathbf{B}}_2 - \hat{\mathbf{B}}_{\cdot})' \\ \cdot \\ \cdot \\ \cdot \\ (\hat{\mathbf{B}}_r - \hat{\mathbf{B}}_{\cdot})'\end{pmatrix} \\ &= \operatorname{tr} \hat{\mathbf{\Sigma}}^{-1}(n_1\mathbf{d}_1\mathbf{d}_1' + \cdots + n_r\mathbf{d}_r\mathbf{d}_r') \\ &= \operatorname{tr} \mathbf{S}^{-1}\mathbf{S}^*,\end{aligned}$$

where $\mathbf{d}_i = \bar{\mathbf{y}}_i - \bar{\mathbf{y}}$, $\mathbf{S}$ is defined in (6.17), and $\mathbf{S}^* = n_1\mathbf{d}_1\mathbf{d}_1' + \cdots + n_r\mathbf{d}_r\mathbf{d}_r'$ is $(r-1)$ times the unbiased covariance matrix of the $\bar{y}$'s between samples.

Note that

$$(6.22)\qquad (\hat{\mathbf{B}}_1 - \hat{\mathbf{B}}_{.}, \cdots, \hat{\mathbf{B}}_r - \hat{\mathbf{B}}_{.})\begin{pmatrix} n_1 & 0 \cdots 0 \\ \cdot & \cdot \quad \cdot \quad \cdot \\ 0 & 0 \cdots n_r \end{pmatrix}\begin{pmatrix} (\hat{\mathbf{B}}_1 - \hat{\mathbf{B}}_{.})' \\ \cdot \\ \cdot \\ \cdot \\ (\hat{\mathbf{B}}_r - \hat{\mathbf{B}}_{.})' \end{pmatrix}$$

$$= (\hat{\mathbf{B}}_1, \cdots, \hat{\mathbf{B}}_r)\begin{pmatrix} n_1 & 0 \cdots 0 \\ \cdot & \cdot \quad \cdot \quad \cdot \\ 0 & 0 \cdots n_r \end{pmatrix}\begin{pmatrix} \hat{\mathbf{B}}_1' \\ \cdot \\ \cdot \\ \cdot \\ \hat{\mathbf{B}}_r' \end{pmatrix} - n\hat{\mathbf{B}}_{.}\hat{\mathbf{B}}_{.}'$$

$$= \hat{\mathbf{B}}^1\mathbf{X}^{1\prime}\mathbf{X}^1\hat{\mathbf{B}}^{1\prime} - \hat{\mathbf{B}}^2\mathbf{X}^{2\prime}\mathbf{X}^2\hat{\mathbf{B}}^{2\prime}.$$

We may write [cf. (6.6) in chapter 10]

$$(6.23)\qquad \hat{J}(H_1, H_2) = \operatorname{tr} \hat{\mathbf{\Sigma}}^{-1}\hat{\mathbf{B}}^1\mathbf{X}^{1\prime}\mathbf{X}^1\hat{\mathbf{B}}^{1\prime} - \operatorname{tr} \hat{\mathbf{\Sigma}}^{-1}\hat{\mathbf{B}}^2\mathbf{X}^{2\prime}\mathbf{X}^2\hat{\mathbf{B}}^{2\prime}.$$

The foregoing is summarized in table 6.3 (cf. table 6.2 in chapter 10).

TABLE 6.3

Variation due to	Generalized Sum of Squares	D.F.
H_2:$\mathbf{B}_{.}$	$n\bar{\mathbf{y}}\bar{\mathbf{y}}'$	1
Difference, between	$n_1\mathbf{d}_1\mathbf{d}_1' + \cdots + n_r\mathbf{d}_r\mathbf{d}_r' = \mathbf{S}^*$	$r - 1$
H_1:$\mathbf{B}_i$	$n_1\bar{\mathbf{y}}_1\bar{\mathbf{y}}_1' + \cdots + n_r\bar{\mathbf{y}}_r\bar{\mathbf{y}}_r'$	r
Difference, within	$\mathbf{Y}'\mathbf{Y} - n_1\bar{\mathbf{y}}_1\bar{\mathbf{y}}_1' - \cdots - n_r\bar{\mathbf{y}}_r\bar{\mathbf{y}}_r' = N_1\mathbf{S}_1 + \cdots + N_r\mathbf{S}_r = N\mathbf{S}$	$n - r$
Total	$\mathbf{Y}'\mathbf{Y}$	n

Writing table 6.3 in the usual analysis of variance form, we have table 6.4. $\hat{J}(H_1, H_2) = \operatorname{tr} \mathbf{S}^{-1}\mathbf{S}^* = \dfrac{(r-1)k_2(n-r)}{n-r-k_2+1}F$, where F has the analysis of variance distribution with $[(r-1)k_2(1+c)]$ and $[(n-r-k_2+1)(1+c)]$ degrees of freedom, $c = (r-2)(k_2-1)/(n-r)$, under the null hypothesis H_2 of (6.11). Asymptotically, $\hat{J}(H_1, H_2)$ is distributed as χ^2 with $k_2(r-1)$

degrees of freedom. [Cf. the direct derivation in Kullback (1956, section 5).] For $r = 2$ see problem 12.14 and Anderson (1958, pp. 108–109), Rao (1952, pp. 73–74).

TABLE 6.4

Variation due to	Generalized Sum of Squares	D.F.
Between	$n_1\mathbf{d}_1\mathbf{d}_1' + \cdots + n_r\mathbf{d}_r\mathbf{d}_r' = \mathbf{S}^*$	$r - 1$
Within	$N_1\mathbf{S}_1 + \cdots + N_r\mathbf{S}_r = N\mathbf{S}$	$n - r$
Total	$\mathbf{Y}'\mathbf{Y} - n\bar{\mathbf{y}}\bar{\mathbf{y}}'$	$n - 1$

Statistics of the form tr $\mathbf{S}^{-1}\mathbf{S}^*$ were first introduced by Lawley (1938) and Hotelling (1947, 1951). The asymptotic behavior of the distribution of this statistic was investigated by Ito (1956), who gives the percentage points of the distribution as an asymptotic expression in terms of the corresponding percentage points of the χ^2-distribution with $(r - 1)k_2$ degrees of freedom.

6.4. *r* Samples with Covariance

Suppose we have r independent samples, respectively, of n_i, $i = 1, 2, \cdots, r$, independent observations, from multivariate normal populations with a common covariance matrix. We shall examine some hypotheses more general than those of section 6.3.

6.4.1. Test of Regression. Suppose we want to test a null hypothesis H_2, there is no linear regression, against an alternative hypothesis H_1, there is a common linear regression in the r samples.

For the ith sample the regression model is [cf. (3.2), (6.9)],

$$\mathbf{Z}_i = \mathbf{Y}_i - \mathbf{X}_{i1}\mathbf{B}_{i1}' - \mathbf{X}_{i2}\mathbf{B}_{\cdot 2}', \tag{6.24}$$

where $\mathbf{Z}_i$, $\mathbf{Y}_i$ are defined in (6.9),

$$\mathbf{X}_{i1}' = (1, 1, \cdots, 1), \quad 1 \times n_i, \qquad \mathbf{B}_{i1}' = (\beta_{i11}, \beta_{i21}, \cdots, \beta_{ik_21}),$$
$$\mathbf{X}_{i2}' = (\mathbf{x}_{i1}, \mathbf{x}_{i2}, \cdots, \mathbf{x}_{in_i}), \qquad \mathbf{x}_{ij}' = (x_{ij2}, \cdots, x_{ijk_1}),$$
$$i = 1, 2, \cdots, r \text{ samples}, \qquad j = 1, 2, \cdots, n_i \text{ observations},$$
$$\mathbf{B}_{\cdot 2} = (\beta_{pq}), \qquad p = 1, 2, \cdots, k_2, \qquad q = 2, 3, \cdots, k_1.$$

The alternative hypothesis of a common linear regression is

$$H_1\colon \mathbf{B}_{i1}^{1'} = (\beta_{i11}^1, \beta_{i21}^1, \cdots, \beta_{ik_21}^1), \qquad \mathbf{B}_{\cdot 2}^1 = (\beta_{pq}), \tag{6.25}$$
$$p = 1, 2, \cdots, k_2, \qquad q = 2, 3, \cdots, k_1,$$

and the null hypothesis of no regression is

(6.26) $$H_2:\mathbf{B}_{i1}^{2'} = (\beta_{i11}^2, \beta_{i21}^2, \cdots, \beta_{ik_21}^2), \qquad \mathbf{B}_{\cdot 2}^2 = \mathbf{0}.$$

We may write the regression model for the r samples combined, under H_1, as

(6.27) $$\mathbf{Z} = \mathbf{Y} - \mathbf{X}_1^1\mathbf{B}_1^{1'} - \mathbf{X}_2^1\mathbf{B}_2^{1'},$$

where $\mathbf{Z}$ and $\mathbf{Y}$ are defined in (6.12),

$$\mathbf{X}_1^{1'} = \begin{pmatrix} \mathbf{X}_{11}' & \mathbf{0} & \cdots & \mathbf{0} \\ \mathbf{0} & \mathbf{X}_{21}' & & \mathbf{0} \\ \cdot & & \cdot & \cdot \\ \cdot & & \cdot & \cdot \\ \cdot & & \cdot & \cdot \\ \mathbf{0} & \mathbf{0} & \cdots & \mathbf{X}_{r1}' \end{pmatrix}, \qquad \mathbf{X}_2^{1'} = (\mathbf{X}_{12}', \mathbf{X}_{22}', \cdots, \mathbf{X}_{r2}'),$$

$$\mathbf{B}_1^1 = (\mathbf{B}_{11}^1, \mathbf{B}_{21}^1, \cdots, \mathbf{B}_{r1}^1), \qquad \mathbf{B}_2^1 = \mathbf{B}_{\cdot 2}.$$

Under H_2, the regression model for the r samples combined is

(6.28) $$\mathbf{Z} = \mathbf{Y} - \mathbf{X}_1^2\mathbf{B}_1^{2'},$$

where $\mathbf{Z}$ and $\mathbf{Y}$ are defined in (6.27), $\mathbf{X}_1^{2'} = \mathbf{X}_1^{1'}$, and $\mathbf{B}_1^2 = (\mathbf{B}_{11}^2, \mathbf{B}_{21}^2, \cdots, \mathbf{B}_{r1}^2)$.

We thus have [cf. (6.14)]

(6.29) $$\mathbf{X}_1^{1'}\mathbf{X}_1^1 = \mathbf{X}_1^{2'}\mathbf{X}_1^2 = \begin{pmatrix} n_1 & 0 & \cdots & 0 \\ 0 & n_2 & \cdots & 0 \\ \cdot & \cdot & \cdot & \cdot \\ 0 & 0 & \cdots & n_r \end{pmatrix} = \mathbf{S}_{11}^1 = \mathbf{S}_{11}^2,$$

(6.30) $$\mathbf{X}_1^{1'}\mathbf{X}_2^1 = \begin{pmatrix} \mathbf{X}_{11}'\mathbf{X}_{12} \\ \mathbf{X}_{21}'\mathbf{X}_{22} \\ \cdot \\ \cdot \\ \cdot \\ \mathbf{X}_{r1}'\mathbf{X}_{r2} \end{pmatrix} = \begin{pmatrix} n_1\bar{\mathbf{x}}_1' \\ n_2\bar{\mathbf{x}}_2' \\ \cdot \\ \cdot \\ \cdot \\ n_r\bar{\mathbf{x}}_r' \end{pmatrix} = \mathbf{S}_{12}^1,$$

where $n_i\bar{\mathbf{x}}_i' = \mathbf{x}_{i1}' + \mathbf{x}_{i2}' + \cdots + \mathbf{x}_{in_i}' = (x_{i\cdot 2}, x_{i\cdot 3}, \cdots, x_{i\cdot k_1})$, $x_{i\cdot p} = x_{i1p} + x_{i2p} + \cdots + x_{in_ip}$,

(6.31) $$\mathbf{Y}'\mathbf{X}_1^1 = (\mathbf{Y}_1'\mathbf{X}_{11}, \mathbf{Y}_2'\mathbf{X}_{21}, \cdots, \mathbf{Y}_r'\mathbf{X}_{r1}) = (n_1\bar{\mathbf{y}}_1, n_2\bar{\mathbf{y}}_2, \cdots, n_r\bar{\mathbf{y}}_r) = \mathbf{Y}'\mathbf{X}_1^2,$$

where $\bar{\mathbf{y}}_i$ is defined in (6.15),

(6.32) $$\mathbf{Y}'\mathbf{X}_2^1 = (\mathbf{Y}_1'\mathbf{X}_{12} + \mathbf{Y}_2'\mathbf{X}_{22} + \cdots + \mathbf{Y}_r'\mathbf{X}_{r2}) = \sum_{i=1}^{r}\sum_{j=1}^{n_i} \mathbf{y}_{ij}\mathbf{x}_{ij}'$$
$$= \left(\sum_{i=1}^{r}\sum_{j=1}^{n_i} y_{ijt}x_{iju}\right), \quad t = 1, 2, \cdots, k_2, \quad u = 2, 3, \cdots, k_1,$$

(6.33)

$$\mathbf{X}_{2\cdot 1}^1 = \mathbf{X}_2^1 - \mathbf{X}_1^1\mathbf{S}_{11}^{1^{-1}}\mathbf{S}_{12}^1$$

$$= \begin{pmatrix} \mathbf{X}_{12} \\ \mathbf{X}_{22} \\ \cdot \\ \cdot \\ \cdot \\ \mathbf{X}_{r2} \end{pmatrix} - \begin{pmatrix} \mathbf{X}_{11} & \mathbf{0} & \cdots & \mathbf{0} \\ \mathbf{0} & \mathbf{X}_{21} & \cdots & \mathbf{0} \\ \cdot & \cdot & & \cdot \\ \cdot & \cdot & & \cdot \\ \cdot & \cdot & & \cdot \\ \mathbf{0} & \mathbf{0} & \cdots & \mathbf{X}_{r1} \end{pmatrix} \begin{pmatrix} \bar{\mathbf{x}}_1' \\ \bar{\mathbf{x}}_2' \\ \cdot \\ \cdot \\ \cdot \\ \bar{\mathbf{x}}_r' \end{pmatrix}$$

$$= \begin{pmatrix} \mathbf{X}_{12} - \mathbf{X}_{11}\bar{\mathbf{x}}_1' \\ \mathbf{X}_{22} - \mathbf{X}_{21}\bar{\mathbf{x}}_2' \\ \cdot \\ \cdot \\ \cdot \\ \mathbf{X}_{r2} - \mathbf{X}_{r1}\bar{\mathbf{x}}_r' \end{pmatrix}$$

$$= \begin{pmatrix} x_{112} - \bar{x}_{12} & x_{113} - \bar{x}_{13} & \cdots & x_{11k_1} - \bar{x}_{1k_1} \\ \cdot & \cdot & & \cdot \\ \cdot & \cdot & & \cdot \\ \cdot & \cdot & & \cdot \\ x_{1n_12} - \bar{x}_{12} & x_{1n_13} - \bar{x}_{13} & \cdots & x_{1n_1k_1} - \bar{x}_{1k_1} \\ x_{212} - \bar{x}_{22} & x_{213} - \bar{x}_{23} & \cdots & x_{21k_1} - \bar{x}_{2k_1} \\ \cdot & \cdot & & \cdot \\ \cdot & \cdot & & \cdot \\ \cdot & \cdot & & \cdot \\ x_{2n_22} - \bar{x}_{22} & x_{2n_23} - \bar{x}_{23} & \cdots & x_{2n_2k_1} - \bar{x}_{2k_1} \\ \cdot & \cdot & & \cdot \\ \cdot & \cdot & & \cdot \\ \cdot & \cdot & & \cdot \\ x_{r12} - \bar{x}_{r2} & x_{r13} - \bar{x}_{r3} & \cdots & x_{r1k_1} - \bar{x}_{rk_1} \\ \cdot & \cdot & & \cdot \\ \cdot & \cdot & & \cdot \\ \cdot & \cdot & & \cdot \\ x_{rn_r2} - \bar{x}_{r2} & x_{rn_r3} - \bar{x}_{r3} & \cdots & x_{rn_rk_1} - \bar{x}_{rk_1} \end{pmatrix}, \; n \times (k_1 - 1)$$

that is, $\mathbf{X}_{2\cdot 1}^1$ is an $n \times (k_1 - 1)$ matrix of the x's centered about their respective sample averages. From (6.33) and (6.31) we have

$$\text{(6.34)} \quad \mathbf{Y}'\mathbf{X}_{2\cdot 1}^1 = \mathbf{Y}_1'\mathbf{X}_{12} - \mathbf{Y}_1'\mathbf{X}_{11}\bar{\mathbf{x}}_1' + \cdots + \mathbf{Y}_r'\mathbf{X}_{r2} - \mathbf{Y}_r'\mathbf{X}_{r1}\bar{\mathbf{x}}_r'$$
$$= \mathbf{Y}_1'\mathbf{X}_{12} - n_1\bar{\mathbf{y}}_1\bar{\mathbf{x}}_1' + \cdots + \mathbf{Y}_r'\mathbf{X}_{r2} - n_r\bar{\mathbf{y}}_r\bar{\mathbf{x}}_r'$$
$$= N_1\mathbf{S}_{1yx} + \cdots + N_r\mathbf{S}_{ryx} = N\mathbf{S}_{yx},$$

where $N_i = n_i - 1$, $N = N_1 + N_2 + \cdots + N_r$, $\mathbf{S}_{iyx}$ is the $k_2 \times (k_1 - 1)$ unbiased covariance matrix of the y's and x's within the ith sample, and $\mathbf{S}'_{yx} = \mathbf{S}_{xy}$,

$$(6.35) \qquad \mathbf{S}^1_{22} = \mathbf{X}^{1\prime}_2\mathbf{X}^1_2 = \mathbf{X}'_{12}\mathbf{X}_{12} + \mathbf{X}'_{22}\mathbf{X}_{22} + \cdots + \mathbf{X}'_{r2}\mathbf{X}_{r2} = \mathbf{S}_{122} + \mathbf{S}_{222} + \cdots + \mathbf{S}_{r22},$$

where $\mathbf{S}_{i22} = \mathbf{X}'_{i2}\mathbf{X}_{i2}$, $i = 1, 2, \cdots, r$,

$$(6.36) \qquad \mathbf{S}^1_{22\cdot 1} = \mathbf{S}^1_{22} - \mathbf{S}^1_{21}\mathbf{S}^{1^{-1}}_{11}\mathbf{S}^1_{12} = \mathbf{S}^1_{22} - (n_1\bar{\mathbf{x}}_1, n_2\bar{\mathbf{x}}_2, \cdots, n_r\bar{\mathbf{x}}_r)\begin{pmatrix}\bar{\mathbf{x}}'_1\\ \bar{\mathbf{x}}'_2\\ \cdot\\ \cdot\\ \cdot\\ \bar{\mathbf{x}}'_r\end{pmatrix}$$

$$= (\mathbf{S}_{122} - n_1\bar{\mathbf{x}}_1\bar{\mathbf{x}}'_1) + \cdots + (\mathbf{S}_{r22} - n_r\bar{\mathbf{x}}_r\bar{\mathbf{x}}'_r),$$
$$= N_1\mathbf{S}_{1xx} + \cdots + N_r\mathbf{S}_{rxx} = N\mathbf{S}_{xx},$$

where $\mathbf{S}_{ixx}$ is the unbiased covariance matrix of the x's within the ith sample.

From (5.7) and (5.11) we have

$$(6.37) \qquad \hat{\mathbf{B}}^1_2 = \mathbf{Y}'\mathbf{X}^1_{2\cdot 1}\mathbf{S}^{1^{-1}}_{22\cdot 1}, \qquad \hat{\mathbf{B}}^1_1 = (\bar{\mathbf{y}}_1, \bar{\mathbf{y}}_2, \cdots, \bar{\mathbf{y}}_r) - \hat{\mathbf{B}}^1_2(\bar{\mathbf{x}}_1, \bar{\mathbf{x}}_2, \cdots, \bar{\mathbf{x}}_r),$$
$$\hat{\mathbf{B}}^2_1 = (\bar{\mathbf{y}}_1, \bar{\mathbf{y}}_2, \cdots, \bar{\mathbf{y}}_r).$$

From (5.8) and (5.13) we have

$$(6.38) \qquad (n - k_1 + 1 - r)\hat{\mathbf{\Sigma}} = \mathbf{Y}'\mathbf{Y} - n_1\bar{\mathbf{y}}_1\bar{\mathbf{y}}'_1 - \cdots - n_r\bar{\mathbf{y}}_r\bar{\mathbf{y}}'_r - \hat{\mathbf{B}}^1_2\mathbf{S}^1_{22\cdot 1}\hat{\mathbf{B}}^{1\prime}_2 = N\mathbf{S}_{yy} - \hat{\mathbf{B}}^1_2\mathbf{S}^1_{22\cdot 1}\hat{\mathbf{B}}^{1\prime}_2,$$

where $N\mathbf{S}_{yy} = N_1\mathbf{S}_{1yy} + \cdots + N_r\mathbf{S}_{ryy}$, and S_{iyy} is the unbiased covariance matrix of the y's within the ith sample [cf. (6.17)].

TABLE 6.5

Variation due to	Generalized Sum of Squares	D.F.
H_2:(6.26)	$n_1\bar{\mathbf{y}}_1\bar{\mathbf{y}}'_1 + \cdots + n_r\bar{\mathbf{y}}_r\bar{\mathbf{y}}'_r$	r
Difference	$\hat{\mathbf{B}}^1_2\mathbf{S}^1_{22\cdot 1}\hat{\mathbf{B}}^{1\prime}_2 = N\mathbf{S}_{yx}\mathbf{S}^{-1}_{xx}\mathbf{S}_{xy}$	$k_1 - 1$
H_1:(6.25)	$n_1\bar{\mathbf{y}}_1\bar{\mathbf{y}}'_1 + \cdots + n_r\bar{\mathbf{y}}_r\bar{\mathbf{y}}'_r + \hat{\mathbf{B}}^1_2\mathbf{S}^1_{22\cdot 1}\hat{\mathbf{B}}^{1\prime}_2$	$k_1 - 1 + r$
Difference	$N\mathbf{S}_{yy} - \hat{\mathbf{B}}^1_2\mathbf{S}^1_{22\cdot 1}\hat{\mathbf{B}}^{1\prime}_2 = (n - k_1 + 1 - r)\hat{\mathbf{\Sigma}}$	$n - k_1 + 1 - r$
Total	$\mathbf{Y}'\mathbf{Y}$	n

We summarize the preceding analysis in table 6.5 (cf. table 5.1). $\hat{J}(H_1, H_2) = \text{tr}\, \hat{\mathbf{\Sigma}}^{-1}\hat{\mathbf{B}}_2^1\mathbf{S}_{22\cdot1}^1\hat{\mathbf{B}}_2^{1'} = \frac{(k_1 - 1)k_2(n - k_1 + 1 - r)}{n - k_1 + 1 - r - k_2 + 1} F$, where F has the analysis of variance distribution with $[(k_1 - 1)k_2(1 + c)]$ and $[(n - k_1 - k_2 - r + 2)(1 + c)]$ degrees of freedom, $c = (k_1 - 2)(k_2 - 1)/(n - k_1 + 1 - r)$, under the null hypothesis H_2 of (6.26). Asymptotically, $\hat{J}(H_1, H_2)$ is distributed as χ^2 with $k_2(k_1 - 1)$ degrees of freedom.

6.4.2. Test of Homogeneity of Means and Regression. If instead of the null hypothesis H_2, there is no regression [see (6.26)], we want to test a null hypothesis that there is no regression and the means are homogeneous, against the alternative hypothesis H_1 in (6.25), then we must examine the null hypothesis H_3

$$\text{(6.39)} \qquad H_3: \mathbf{B}_{i1}^{3'} = \mathbf{B}'_{\cdot1} = (\beta_{\cdot11}, \beta_{\cdot21}, \cdots, \beta_{\cdot k_2 1}), \qquad \mathbf{B}_{\cdot2}^3 = \mathbf{0}.$$

The results under H_1 are those derived in section 6.4.1. The results under H_3 are similar to those in section 6.3 under H_2, that is,

$$\text{(6.40)} \qquad \begin{aligned} \mathbf{X}_1^{3'} &= (\mathbf{X}'_{11}, \mathbf{X}'_{21}, \cdots, \mathbf{X}'_{r1}), \\ \mathbf{S}_{11}^3 &= \mathbf{X}_1^{3'}\mathbf{X}_1^3 = \mathbf{X}'_{11}\mathbf{X}_{11} + \cdots + \mathbf{X}'_{r1}\mathbf{X}_{r1} = n_1 + \cdots + n_r = n, \\ \mathbf{Y}'\mathbf{X}_1^3 &= \mathbf{Y}'_1\mathbf{X}_{11} + \cdots + \mathbf{Y}'_r\mathbf{X}_{r1} = n_1\bar{\mathbf{y}}_1 + n_2\bar{\mathbf{y}}_2 + \cdots + n_r\bar{\mathbf{y}}_r = n\bar{\mathbf{y}}, \\ n\hat{\mathbf{B}}_{\cdot1} &= n\bar{\mathbf{y}}. \end{aligned}$$

We summarize the analysis covering H_1, H_2, H_3 in table 6.6, where $\mathbf{S}_{yy}^*$ is the matrix $\mathbf{S}^*$ in table 6.3 (to show its relation to the y's). $\hat{J}(H_1, H_3) =$

TABLE 6.6

Variation due to	Generalized Sum of Squares	D.F.
H_3:(6.39)	$n\bar{\mathbf{y}}\bar{\mathbf{y}}'$	1
Difference	$n_1\mathbf{d}_1\mathbf{d}'_1 + \cdots + n_r\mathbf{d}_r\mathbf{d}'_r = \mathbf{S}_{yy}^*$	$r - 1$
H_2:(6.26)	$n_1\bar{\mathbf{y}}_1\bar{\mathbf{y}}'_1 + \cdots + n_r\bar{\mathbf{y}}_r\bar{\mathbf{y}}'_r$	r
Difference	$\hat{\mathbf{B}}_2^1\mathbf{S}_{22\cdot1}^1\hat{\mathbf{B}}_2^{1'} = N\mathbf{S}_{yx}\mathbf{S}_{xx}^{-1}\mathbf{S}_{xy}$	$k_1 - 1$
H_1:(6.25)	$n_1\bar{\mathbf{y}}_1\bar{\mathbf{y}}'_1 + \cdots + n_r\bar{\mathbf{y}}_r\bar{\mathbf{y}}'_r + \hat{\mathbf{B}}_2^1\mathbf{S}_{22\cdot1}^1\hat{\mathbf{B}}_2^{1'}$	$k_1 - 1 + r$
Difference	$N\mathbf{S}_{yy} - \hat{\mathbf{B}}_2^1\mathbf{S}_{22\cdot1}^1\hat{\mathbf{B}}_2^{1'} = (N - k_1 + 1)\hat{\mathbf{\Sigma}}$	$N - k_1 + 1$
Total	$\mathbf{Y}'\mathbf{Y}$	n

$\operatorname{tr} \hat{\mathbf{\Sigma}}^{-1}(\mathbf{S}^*_{yy} + \hat{\mathbf{B}}^1_2\mathbf{S}^1_{22\cdot1}\hat{\mathbf{B}}^{1'}_2) = \dfrac{(k_1 + r - 2)k_2(N - k_1 + 1)}{(N - k_1 + 1 - k_2 + 1)} F$, where F has the analysis of variance distribution with $[(k_1 + r - 2)k_2(1 + c)]$ and $[(N - k_1 - k_2 + 2)(1 + c)]$ degrees of freedom, $c = (k_1 + r - 3)(k_2 - 1)/(N - k_1 + 1)$, under the null hypothesis H_3 of (6.39). Asymptotically, $\hat{J}(H_1, H_3)$ is distributed as χ^2 with $(k_1 + r - 2)k_2$ degrees of freedom, $\operatorname{tr} \hat{\mathbf{\Sigma}}^{-1}\mathbf{S}^*_{yy}$, with $(r - 1)k_2$ degrees of freedom, is a test of the homogeneity, and $\operatorname{tr} \hat{\mathbf{\Sigma}}^{-1}\hat{\mathbf{B}}^1_2\mathbf{S}^1_{22\cdot1}\hat{\mathbf{B}}^{1'}_2$, with $(k_1 - 1)k_2$ degrees of freedom, is a test of the regression.

6.4.3. Test of Homogeneity, Assuming Regression. Suppose we assume that there is a common linear regression in the r samples. We want to test a null hypothesis of homogeneity of the sample means. The alternative hypothesis is H_1 in (6.25), and the null hypothesis is

$$(6.41)\qquad H_4 : \mathbf{B}^{4'}_{i1} = \mathbf{B}'_{\cdot1} = (\beta_{\cdot11}, \beta_{\cdot21}, \cdots, \beta_{\cdot k_21}), \qquad \mathbf{B}^4_{\cdot2} = (\beta^4_{pq}),$$
$$p = 1, \cdots, k_2, \qquad q = 2, \cdots, k_1.$$

The results under H_1 are those derived in section 6.4.1.

Under H_4 we see that [cf. (6.27) and (6.40)]

$$(6.42)\qquad \mathbf{X}^{4'}_1 = (\mathbf{X}'_{11}, \mathbf{X}'_{21}, \cdots, \mathbf{X}'_{r1}), \qquad \mathbf{X}^{4'}_2 = (\mathbf{X}'_{12}, \mathbf{X}'_{22}, \cdots, \mathbf{X}'_{r2}),$$

so that [cf. (6.40)]

$$(6.43)\qquad \mathbf{S}^4_{11} = n, \qquad \mathbf{Y}'\mathbf{X}^4_1 = n\bar{\mathbf{y}}$$

and [cf. (6.32), (6.35)]

$$(6.44)\quad \mathbf{S}^4_{22} = \mathbf{S}_{122} + \mathbf{S}_{222} + \cdots + \mathbf{S}_{r22}, \qquad \mathbf{Y}'\mathbf{X}^4_2 = \left(\sum_{i=1}^{r}\sum_{j=1}^{n_i} y_{ijt}x_{iju}\right),$$
$$t = 1, 2, \cdots, k_2, \qquad u = 2, 3, \cdots, k_1.$$

We also find that [cf. (6.30)]

$$(6.45)\qquad \mathbf{X}^{4'}_1\mathbf{X}^4_2 = \mathbf{S}^4_{12} = \mathbf{X}'_{11}\mathbf{X}_{12} + \mathbf{X}'_{21}\mathbf{X}_{22} + \cdots + \mathbf{X}'_{r1}\mathbf{X}_{r2}$$
$$= n_1\bar{\mathbf{x}}'_1 + n_2\bar{\mathbf{x}}'_2 + \cdots + n_r\bar{\mathbf{x}}'_r = n\bar{\mathbf{x}}'.$$

We thus have [cf. (6.33)]

$$(6.46)\quad \mathbf{X}^4_{2\cdot1} = \mathbf{X}^4_2 - \mathbf{X}^4_1\mathbf{S}^{4^{-1}}_{11}\mathbf{S}^4_{12} = \begin{pmatrix}\mathbf{X}_{12}\\ \mathbf{X}_{22}\\ \cdot\\ \cdot\\ \cdot\\ \mathbf{X}_{r2}\end{pmatrix} - \begin{pmatrix}\mathbf{X}_{11}\\ \mathbf{X}_{21}\\ \cdot\\ \cdot\\ \cdot\\ \mathbf{X}_{r1}\end{pmatrix}\bar{\mathbf{x}}' = \begin{pmatrix}\mathbf{X}_{12} - \mathbf{X}_{11}\bar{\mathbf{x}}'\\ \mathbf{X}_{22} - \mathbf{X}_{21}\bar{\mathbf{x}}'\\ \cdot\\ \cdot\\ \cdot\\ \mathbf{X}_{r2} - \mathbf{X}_{r1}\bar{\mathbf{x}}'\end{pmatrix},$$

that is, $\mathbf{X}_{2\cdot1}^4$ is an $n \times (k_1 - 1)$ matrix of the x's centered about their respective combined sample averages, and [cf. (6.34)]

$$(6.47)\quad \begin{aligned}\mathbf{Y}'\mathbf{X}_{2\cdot1}^4 &= \mathbf{Y}_1'\mathbf{X}_{12} - \mathbf{Y}_1'\mathbf{X}_{11}\bar{\mathbf{x}}' + \cdots + \mathbf{Y}_r'\mathbf{X}_{r2} - \mathbf{Y}_r'\mathbf{X}_{r1}\bar{\mathbf{x}}' \\ &= \mathbf{Y}_1'\mathbf{X}_{12} - n_1\bar{\mathbf{y}}_1\bar{\mathbf{x}}' + \cdots + \mathbf{Y}_r'\mathbf{X}_{r2} - n_r\bar{\mathbf{y}}_r\bar{\mathbf{x}}' \\ &= \mathbf{Y}_1'\mathbf{X}_{12} + \cdots + \mathbf{Y}_r'\mathbf{X}_{r2} - n\bar{\mathbf{y}}\bar{\mathbf{x}}' \\ &= \mathbf{Y}_1'\mathbf{X}_{12} - n_1\bar{\mathbf{y}}_1\bar{\mathbf{x}}_1' + \cdots + \mathbf{Y}_r'\mathbf{X}_{r2} - n_r\bar{\mathbf{y}}_r\bar{\mathbf{x}}_r' \\ &\qquad + n_1\bar{\mathbf{y}}_1\bar{\mathbf{x}}_1' + \cdots + n_r\bar{\mathbf{y}}_r\bar{\mathbf{x}}_r' - n\bar{\mathbf{y}}\bar{\mathbf{x}}' \\ &= N\mathbf{S}_{yx} + \mathbf{S}_{yx}^*,\end{aligned}$$

where $\mathbf{S}_{yx}$ is defined in (6.34), $\mathbf{S}_{yx}^* = \sum_{i=1}^{r} n_i(\bar{\mathbf{y}}_i - \bar{\mathbf{y}})(\bar{\mathbf{x}}_i - \bar{\mathbf{x}})'$ with $\mathbf{S}_{yx}^{*\prime} = \mathbf{S}_{xy}^*$, $\mathbf{S}_{yx}^*$ is a $k_2 \times (k_1 - 1)$ matrix proportional to the between unbiased covariance matrix of the y's with the x's, and [cf. (6.36)]

$$(6.48)\quad \begin{aligned}\mathbf{S}_{22\cdot1}^4 &= \mathbf{S}_{22}^4 - \mathbf{S}_{21}^4\mathbf{S}_{11}^{4^{-1}}\mathbf{S}_{12}^4 = \mathbf{S}_{22}^4 - n\bar{\mathbf{x}}\bar{\mathbf{x}}' = \mathbf{S}_{122} + \cdots + \mathbf{S}_{r22} - n\bar{\mathbf{x}}\bar{\mathbf{x}}' \\ &= \mathbf{S}_{122} - n_1\bar{\mathbf{x}}_1\bar{\mathbf{x}}_1' + \cdots + \mathbf{S}_{r22} - n_r\bar{\mathbf{x}}_r\bar{\mathbf{x}}_r' + n_1\bar{\mathbf{x}}_1\bar{\mathbf{x}}_1' + \cdots \\ &\qquad + n_r\bar{\mathbf{x}}_r\bar{\mathbf{x}}_r' - n\bar{\mathbf{x}}\bar{\mathbf{x}}' \\ &= N\mathbf{S}_{xx} + \mathbf{S}_{xx}^* = \mathbf{S}_{22\cdot1}^1 + \mathbf{S}_{xx}^*,\end{aligned}$$

where $\mathbf{S}_{xx}$ is defined in (6.36) and $\mathbf{S}_{xx}^* = \sum_{i=1}^{r} n_i(\bar{\mathbf{x}}_i - \bar{\mathbf{x}})(\bar{\mathbf{x}}_i - \bar{\mathbf{x}})'$.

From (5.7) we then have

$$(6.49)\qquad \hat{\mathbf{B}}_2^4 = \mathbf{Y}'\mathbf{X}_{2\cdot1}^4\mathbf{S}_{22\cdot1}^{4^{-1}}, \qquad \hat{\mathbf{B}}_1^4 = \bar{\mathbf{y}} - \hat{\mathbf{B}}_2^4\bar{\mathbf{x}},$$

where $\mathbf{Y}'\mathbf{X}_{2\cdot1}^4$ and $\mathbf{S}_{22\cdot1}^4$ are given respectively in (6.47) and (6.48).

TABLE 6.7

Variation due to	Generalized Sum of Squares	D.F.
H_4:(6.41)	$n\bar{\mathbf{y}}\bar{\mathbf{y}}' + \hat{\mathbf{B}}_2^4\mathbf{S}_{22\cdot1}^4\hat{\mathbf{B}}_2^{4'}$	k_1
Difference	$\mathbf{S}_{yy}^* + \hat{\mathbf{B}}_2^1\mathbf{S}_{22\cdot1}^1\hat{\mathbf{B}}_2^{1'} - \hat{\mathbf{B}}_2^4\mathbf{S}_{22\cdot1}^4\hat{\mathbf{B}}_2^{4'}$	$r - 1$
H_1:(6.25)	$n_1\bar{\mathbf{y}}_1\bar{\mathbf{y}}_1' + \cdots + n_r\bar{\mathbf{y}}_r\bar{\mathbf{y}}_r' + \hat{\mathbf{B}}_2^1\mathbf{S}_{22\cdot1}^1\hat{\mathbf{B}}_2^{1'}$	$k_1 - 1 + r$
Difference	$N\mathbf{S}_{yy} - \hat{\mathbf{B}}_2^1\mathbf{S}_{22\cdot1}^1\hat{\mathbf{B}}_2^{1'} = (N - k_1 + 1)\hat{\mathbf{\Sigma}}$	$N - k_1 + 1$
Total	$\mathbf{Y}'\mathbf{Y}$	n

We summarize the analysis covering H_1 and H_4 in table 6.7. $\hat{J}(H_1, H_4) = \operatorname{tr}\hat{\mathbf{\Sigma}}^{-1}(\mathbf{S}_{yy}^* + \hat{\mathbf{B}}_2^1\mathbf{S}_{22\cdot1}^1\hat{\mathbf{B}}_2^{1'} - \hat{\mathbf{B}}_2^4\mathbf{S}_{22\cdot1}^4\hat{\mathbf{B}}_2^{4'}) = \dfrac{(r-1)k_2(N-k_1+1)}{(N-k_1+1-k_2+1)}F$, where F

has the analysis of variance distribution with $[(r-1)k_2(1+c)]$ and $[(N-k_1-k_2+2)(1+c)]$ degrees of freedom, $c=(r-2)(k_2-1)/(N-k_1+1)$, under the null hypothesis H_4 of (6.41). Asymptotically, $\hat{J}(H_1, H_4)$ is distributed as χ^2 with $k_2(r-1)$ degrees of freedom.

Note that in the usual analysis of variance relation for sums of squares, total = within + between, we may write

$$\text{(6.50)}\qquad \begin{aligned} \mathbf{Y}'\mathbf{Y} - n\bar{\mathbf{y}}\bar{\mathbf{y}}' = \mathbf{S}^{**}_{yy} &= N\mathbf{S}_{yy} + \mathbf{S}^{*}_{yy}, \\ \mathbf{S}^{**}_{xx} &= N\mathbf{S}_{xx} + \mathbf{S}^{*}_{xx}, \\ \mathbf{S}^{**}_{yx} &= N\mathbf{S}_{yx} + \mathbf{S}^{*}_{yx}. \end{aligned}$$

$(N-k_1+1)\hat{\mathbf{\Sigma}} = N\mathbf{S}_{yy} - N\mathbf{S}_{yx}\mathbf{S}_{xx}^{-1}\mathbf{S}_{xy}$ is computed in terms of within values, and

$$\mathbf{S}^{*}_{yy} + \hat{\mathbf{B}}_2^1\mathbf{S}^1_{22\cdot1}\hat{\mathbf{B}}_2^{1'} - \hat{\mathbf{B}}_2^4\mathbf{S}^4_{22\cdot1}\hat{\mathbf{B}}_2^{4'} = \mathbf{S}^{*}_{yy} - (\mathbf{S}^{**}_{yx}\mathbf{S}^{**^{-1}}_{xx}\mathbf{S}^{**}_{xy} - N\mathbf{S}_{yx}\mathbf{S}_{xx}^{-1}\mathbf{S}_{xy})$$

is computed in terms of between values and the difference between an expression in total values and within values.

7. CANONICAL CORRELATION

We shall now examine tests of hypotheses associated with the canonical correlations defined in section 7 of chapter 9. We shall need the analysis summarized in table 6.2.

For the y's and x's centered about their respective sample averages, we have, according to the analysis in table 6.2,

$$\text{(7.1)}\qquad \hat{J}(H_1, H_2) = (n-k_1)\,\mathrm{tr}\,(\mathbf{S}_{yy} - \mathbf{S}_{yx}\mathbf{S}_{xx}^{-1}\mathbf{S}_{xy})^{-1}\mathbf{S}_{yx}\mathbf{S}_{xx}^{-1}\mathbf{S}_{xy}.$$

Suppose that, as in section 7 of chapter 9, we take the y's as the second set of k_2 variates and the x's as the first set of (k_1-1) variates into which a population of $(k_1-1)+k_2$ variates has been partitioned. If we write, in accordance with the notation in section 7 of chapter 9, $\mathbf{S}_{yy} = \mathbf{S}_{22}$, $\mathbf{S}_{yx} = \mathbf{S}_{21}$, $\mathbf{S}_{xx} = \mathbf{S}_{11}$, $\mathbf{S}_{yy} - \mathbf{S}_{yx}\mathbf{S}_{xx}^{-1}\mathbf{S}_{xy} = \mathbf{S}_{22} - \mathbf{S}_{21}\mathbf{S}_{11}^{-1}\mathbf{S}_{12} = \mathbf{S}_{22\cdot1}$, then (7.1) becomes

$$\text{(7.2)}\qquad \hat{J}(H_1, H_2) = (n-k_1)\,\mathrm{tr}\,\mathbf{S}_{22\cdot1}^{-1}\mathbf{S}_{21}\mathbf{S}_{11}^{-1}\mathbf{S}_{12},$$

an estimate for the parametric value in (7.5) of chapter 9.

We may also express $\hat{J}(H_1, H_2)$ as $(n-k_1)$ times the sum of the k_2 roots (almost everywhere positive) of the determinantal equation

$$\text{(7.3)}\qquad |\mathbf{S}_{21}\mathbf{S}_{11}^{-1}\mathbf{S}_{12} - l\mathbf{S}_{22\cdot1}| = 0,$$

where we have assumed that $k_2 \le k_1 - 1$ so that the rank of the $k_2 \times k_2$ matrix $\mathbf{S}_{21}\mathbf{S}_{11}^{-1}\mathbf{S}_{12}$ is k_2.

Replacing $\mathbf{S}_{22\cdot 1}$ in (7.3) by $\mathbf{S}_{22} - \mathbf{S}_{21}\mathbf{S}_{11}^{-1}\mathbf{S}_{12}$, we find

$$(7.4) \qquad |\mathbf{S}_{21}\mathbf{S}_{11}^{-1}\mathbf{S}_{12} - l\mathbf{S}_{22\cdot 1}| = 0 = |\mathbf{S}_{21}\mathbf{S}_{11}^{-1}\mathbf{S}_{12} - r^2\mathbf{S}_{22}|,$$

where $l = r^2/(1 - r^2)$, $r^2 = l/(1 + l)$. The r's thus defined are the observed values of Hotelling's canonical correlation coefficients [Hotelling (1936); cf. (7.11) in chapter 9].

Accordingly, we may also write (7.2) as [cf. (7.16) in chapter 9]

$$(7.5) \quad \hat{J}(H_1, H_2) = (n - k_1) \operatorname{tr} \mathbf{S}_{22\cdot 1}^{-1}\mathbf{S}_{21}\mathbf{S}_{11}^{-1}\mathbf{S}_{12}$$
$$= (n - k_1)(l_1 + l_2 + \cdots + l_{k_2})$$
$$= (n - k_1)\left(\frac{r_1^2}{1 - r_1^2} + \frac{r_2^2}{1 - r_2^2} + \cdots + \frac{r_{k_2}^2}{1 - r_{k_2}^2}\right).$$

Under the null hypothesis H_2: $\mathbf{B}^2 = \mathbf{0}$, the results are equivalent to those under the null hypothesis that in a $((k_1 - 1) + k_2)$-variate normal population the set of the first $(k_1 - 1)$ variates is independent of the set of the last k_2 variates, the hypothesis considered in section 7 of chapter 9. [Cf. Anderson (1958, p. 242), Hsu (1949, pp. 391–392).] (See section 3.6 in chapter 12.)

Note that the terms in (7.5) depend only on the sample correlation coefficients, for if the elements of the matrices $\mathbf{S}_{11}$, $\mathbf{S}_{12}$, $\mathbf{S}_{22}$ are expressed in terms of the standard deviations and correlation coefficients, it may be shown (this is left to the reader) that the standard deviations divide out and

$$(7.6) \qquad \hat{J}(H_1, H_2) = (n - k_1) \operatorname{tr} \mathbf{R}_{22\cdot 1}^{-1}\mathbf{R}_{21}\mathbf{R}_{11}^{-1}\mathbf{R}_{12},$$

in terms of the related correlation matrices.

8. LINEAR DISCRIMINANT FUNCTIONS

8.1. Homogeneity of r Samples

The samples and hypotheses are those specified in section 6.3. We want to examine the analysis of the linear discriminant function described in section 5 of chapter 9 with population parameters. We seek the linear discriminant function

$$(8.1) \qquad w_{ij} = \boldsymbol{\alpha}'\mathbf{y}_{ij} = \alpha_1 y_{ij1} + \alpha_2 y_{ij2} + \cdots + \alpha_{k_2} y_{ijk_2},$$

$i = 1, 2, \cdots, r$, $j = 1, 2, \cdots, n_i$, where $\mathbf{y}_{ij}$ is defined in (6.9), that is, the same linear compound of the y's for each sample.

[Cf. Binet and Watson (1956), Roy (1957, pp. 95–104).]

We thus get for the w's, as the estimate of the parameter in (5.5) of chapter 9, and corresponding to (6.21),

$$\hat{J}(H_1, H_2; w) = \frac{\boldsymbol{\alpha}'\mathbf{S}^*\boldsymbol{\alpha}}{\boldsymbol{\alpha}'\mathbf{S}\boldsymbol{\alpha}}. \tag{8.2}$$

The value of $\boldsymbol{\alpha}$ for which $\hat{J}(H_1, H_2; w)$ is a maximum satisfies (by the usual calculus procedures)

$$\mathbf{S}^*\boldsymbol{\alpha} = l\mathbf{S}\boldsymbol{\alpha}, \tag{8.3}$$

where l is the largest root of the determinantal equation

$$|\mathbf{S}^* - l\mathbf{S}| = 0, \tag{8.4}$$

which has (almost everywhere) p positive and $(k_2 - p)$ zero roots, with $p \leqq \min(k_2, r - 1)$. Denoting the positive roots in descending order as $l_1, l_2, \cdots, l_p$,

$$\begin{aligned} \hat{J}(H_1, H_2) &= \operatorname{tr} \mathbf{S}^{-1}\mathbf{S}^* = l_1 + l_2 + \cdots + l_p \\ &= \hat{J}(H_1, H_2; l_1) + \cdots + \hat{J}(H_1, H_2; l_p), \end{aligned} \tag{8.5}$$

where $\hat{J}(H_1, H_2; l_i) = l_i$ is (8.2) for $\boldsymbol{\alpha}$ satisfying (8.3) with $l = l_i$.

The discrimination efficiency of the linear compound associated with l_i may be defined as (see section 6 of chapter 3 and section 5 of chapter 9)

$$\text{Eff.}(l_i) = \frac{\hat{J}(H_1, H_2; l_i)}{\hat{J}(H_1, H_2)} = \frac{l_i}{l_1 + l_2 + \cdots + l_p}. \tag{8.6}$$

Asymptotically, under the null hypothesis of homogeneity H_2 in (6.11), we have the χ^2 decomposition [cf. Rao (1952, p. 373)]

(8.7)

$\hat{J}(H_1, H_2; l_p)$	$= l_p$	$\lvert k_2 - (r-1)\rvert + 1$	d.f.
$\hat{J}(H_1, H_2; l_{p-1})$	$= l_{p-1}$	$\lvert k_2 - (r-1)\rvert + 3$	d.f.
. . .	. . .	. . .	.
$\hat{J}(H_1, H_2)$	$= l_1 + l_2 + \cdots + l_p = \operatorname{tr} \mathbf{S}^{-1}\mathbf{S}^*$	$k_2(r-1)$	d.f.

This is to be taken in the sense that $l_{m+1} + \cdots + l_p$ is distributed asymptotically as χ^2 with $(k_2 - m)(r - 1 - m)$ degrees of freedom, not that $l_{m+1}, \cdots, l_p$ have asymptotic independent χ^2-distributions. (See section 6.4 of chapter 12.)

8.2. Canonical Correlation

[Cf. Marriott (1952).] The sample and hypotheses are those specified in section 7. We want to examine the analysis of the linear discriminant function described in section 3.2 with population parameters. We seek the linear discriminant function

$$(8.8)\quad w_i = \boldsymbol{\alpha}'\mathbf{y}_i = \alpha_1 y_{i1} + \alpha_2 y_{i2} + \cdots + \alpha_{k_2} y_{ik_2}, \qquad i = 1, 2, \cdots, n,$$

that is, the same linear compound of the y's for each observation.

We thus get for the w's, as the estimate of the parameter in (3.4), corresponding to the hypotheses and notation of (7.2),

$$(8.9)\qquad \hat{J}(H_1, H_2; w) = \frac{\boldsymbol{\alpha}'\hat{\mathbf{B}}^1\mathbf{X}'\mathbf{X}\hat{\mathbf{B}}^{1'}\boldsymbol{\alpha}}{\boldsymbol{\alpha}'\hat{\boldsymbol{\Sigma}}\boldsymbol{\alpha}} = (n - k_1)\frac{\boldsymbol{\alpha}'\mathbf{S}_{21}\mathbf{S}_{11}^{-1}\mathbf{S}_{12}\boldsymbol{\alpha}}{\boldsymbol{\alpha}'\mathbf{S}_{22\cdot 1}\boldsymbol{\alpha}}.$$

The value of $\boldsymbol{\alpha}$ for which $\hat{J}(H_1, H_2; w)$ in (8.9) is a maximum satisfies (by the usual calculus procedures) [cf. (7.10) in chapter 9]

$$(8.10)\qquad \mathbf{S}_{21}\mathbf{S}_{11}^{-1}\mathbf{S}_{12}\boldsymbol{\alpha} = l\mathbf{S}_{22\cdot 1}\boldsymbol{\alpha},$$

where l is the largest root of the determinantal equation

$$(8.11)\qquad |\mathbf{S}_{21}\mathbf{S}_{11}^{-1}\mathbf{S}_{12} - l\mathbf{S}_{22\cdot 1}| = 0.$$

Note that (8.11) is the same as (7.4), and (8.10) is the same as $\mathbf{S}_{21}\mathbf{S}_{11}^{-1}\mathbf{S}_{12}\boldsymbol{\alpha} = r^2\mathbf{S}_{22}\boldsymbol{\alpha}$. Denoting the k_2 (almost everywhere) positive roots in descending order as $l_1, l_2, \cdots, l_{k_2}$, we may also write the decomposition in (7.5) as

$$(8.12)\qquad \hat{J}(H_1, H_2) = \hat{J}(H_1, H_2; l_1) + \cdots + \hat{J}(H_1, H_2; l_{k_2}),$$

where $\hat{J}(H_1, H_2; l_i) = (n - k_1)l_i = (n - k_1)r_i^2/(1 - r_i^2)$ is (8.9) for $\boldsymbol{\alpha}$ satisfying (8.10) with $l = l_i$.

The discrimination efficiency of the linear compound associated with l_i may be defined as in (8.6).

Asymptotically, under the null hypothesis H_2: $\mathbf{B} = \mathbf{B}^2 = \mathbf{0}$, we have the χ^2 decomposition

$$(8.13)\quad \begin{array}{llll}
\hat{J}(H_1, H_2; l_{k_2}) = (n - k_1)l_{k_2} & = (n - k_1)r_{k_2}^2/(1 - r_{k_2}^2) & k_1 - k_2 & \text{d.f.} \\
\hat{J}(H_1, H_2; l_{k_2-1}) = (n - k_1)l_{k_2-1} & = (n - k_1)r_{k_2-1}^2/(1 - r_{k_2-1}^2) & k_1 - k_2 + 2 & \text{d.f.} \\
\cdots & \cdots & \cdots & \cdots \\
\hat{J}(H_1, H_2; l_1) = (n - k_1)l_1 & = (n - k_1)r_1^2/(1 - r_1^2) & k_1 + k_2 - 2 & \text{d.f.} \\
\hline
\hat{J}(H_1, H_2) = (n - k_1)\sum_{i=1}^{k_2} l_i & = (n - k_1)\sum_{i=1}^{k_2} r_i^2/(1 - r_i^2) & (k_1 - 1)k_2 & \text{d.f.}
\end{array}$$

As in (8.7), this is to be taken in the sense that $(n - k_1)(l_{m+1} + \cdots + l_{k_2})$ is asymptotically distributed as χ^2 with $(k_1 - 1 - m)(k_2 - m)$ degrees of freedom, not that $(n - k_1)l_{m+1}, \cdots, (n - k_1)l_{k_2}$ have asymptotic independent χ^2-distributions. (See section 6.4 of chapter 12.)

8.3. Hotelling's Generalized Student Ratio (Hotelling's T^2)

The sample and hypotheses are those specified in section 6.1. We want to examine the analysis of the linear discriminant function described in (5.2) of chapter 9 with population parameters. We may treat this as a special case of that in section 8.2 by specifying H_2 in (6.1) with $\mathbf{B}^2 = \mathbf{0}$ and denoting the values in (6.1), (6.2), (6.3) as

$$\mathbf{X}'\mathbf{X} = n\mathbf{S}_{11} = n, \qquad \mathbf{Y}'\mathbf{X} = n\mathbf{S}_{21} = n\bar{\mathbf{y}}, \qquad \mathbf{Y}'\mathbf{Y} = n\mathbf{S}_{22},$$

$$\mathbf{S}_{22\cdot 1} = \frac{1}{n}\mathbf{Y}'\mathbf{Y} - \bar{\mathbf{y}}\bar{\mathbf{y}}' = \frac{n-1}{n}\hat{\mathbf{\Sigma}} = \frac{N}{n}\mathbf{S}_{vv},$$

so that the coefficients of the linear discriminant function (8.8) must satisfy, as in (8.10) and (8.11),

$$\bar{\mathbf{y}}\bar{\mathbf{y}}'\boldsymbol{\alpha} = l\frac{N}{n}\mathbf{S}_{vv}\boldsymbol{\alpha}, \tag{8.14}$$

where l is the largest root of

$$\left|\bar{\mathbf{y}}\bar{\mathbf{y}}' - l\frac{N}{n}\mathbf{S}_{vv}\right| = 0 = \left|\bar{\mathbf{y}}\bar{\mathbf{y}}' - r^2\frac{1}{n}\mathbf{Y}'\mathbf{Y}\right|. \tag{8.15}$$

Here there is just a single root [cf. Anderson (1958, p. 108)]

$$l = \frac{n}{N}\bar{\mathbf{y}}'\mathbf{S}_{vv}^{-1}\bar{\mathbf{y}}, \qquad \hat{J}(H_1, H_2) = (n-1)l = n\bar{\mathbf{y}}'\mathbf{S}_{vv}^{-1}\bar{\mathbf{y}} = n \operatorname{tr} \mathbf{S}_{vv}^{-1}\bar{\mathbf{y}}\bar{\mathbf{y}}',$$

the canonical correlation squared is

$$r^2 = n\bar{\mathbf{y}}'(\mathbf{Y}'\mathbf{Y})^{-1}\bar{\mathbf{y}} = n \operatorname{tr} (\mathbf{Y}'\mathbf{Y})^{-1}\bar{\mathbf{y}}\bar{\mathbf{y}}' = \frac{\frac{n}{N}\bar{\mathbf{y}}'\mathbf{S}_{vv}^{-1}\bar{\mathbf{y}}}{1 + \frac{n}{N}\bar{\mathbf{y}}'\mathbf{S}_{vv}^{-1}\bar{\mathbf{y}}},$$

and the coefficients of the linear discriminant function are $\boldsymbol{\alpha} = \mathbf{S}_{vv}^{-1}\bar{\mathbf{y}}$. [Cf. the discussion following (5.2) in chapter 9.] The linear discriminant function is thus $w = \boldsymbol{\alpha}'\mathbf{y} = \bar{\mathbf{y}}'\mathbf{S}_{vv}^{-1}\mathbf{y}$, and the coefficient vector of the linear function of the x's, whose correlation with $w = \boldsymbol{\alpha}'\mathbf{y}$ yields the canonical correlation r above, is proportional to $\boldsymbol{\alpha}'\hat{\mathbf{B}} = \bar{\mathbf{y}}'\mathbf{S}_{vv}^{-1}\bar{\mathbf{y}}$. [Cf. Fisher (1938).]

9. EXAMPLES

We shall illustrate some of the particulars in the preceding sections with numerical examples. The aim is the illustration of the computational procedure, not the complete analysis per se of the problem from which the data may have arisen.

9.1. Homogeneity of Sample Means

Pearson and Wilks (1933) give some data from Shewhart (1931) for five samples of 12 observations each on tensile strength, y_1, and hardness, y_2, in aluminum die-castings. It is desired to test whether the sample averages are homogeneous. (A test for the homogeneity of the covariance matrices, to be discussed in chapter 12, leads us to accept a null hypothesis that the covariance matrices are the same.) This corresponds to the analysis in table 6.3 with $k_2 = 2$, $r = 5$, $n_1 = n_2 = \cdots = n_5 = 12$, $n = 60$.

The five sample averages are [Pearson and Wilks (1933, p. 356)]:

Strength	Hardness
$\bar{y}_{11} = 33.399$	$\bar{y}_{12} = 68.49$
$\bar{y}_{21} = 28.216$	$\bar{y}_{22} = 68.02$
$\bar{y}_{31} = 30.313$	$\bar{y}_{32} = 66.57$
$\bar{y}_{41} = 33.150$	$\bar{y}_{42} = 76.12$
$\bar{y}_{51} = 34.269$	$\bar{y}_{52} = 69.92$

The elements of the matrices corresponding to the generalized sums of squares are:

	D.F.	y_1^2	y_2^2	y_1y_2
Between	$r - 1 = 4$	306.089	662.77	214.86
Within	$n - r = 55$	636.165	7653.42	1697.52
Total	$n - 1 = 59$	942.254	8316.19	1912.38

that is,

$$\mathbf{S} = \frac{1}{55}\begin{pmatrix} 636.165 & 1697.52 \\ 1697.52 & 7653.42 \end{pmatrix} = \begin{pmatrix} 11.5666 & 30.9004 \\ 30.9004 & 139.153 \end{pmatrix},$$

$$\mathbf{S}^* = \begin{pmatrix} 306.089 & 214.86 \\ 214.86 & 662.77 \end{pmatrix}.$$

$\hat{J}(H_1, H_2) = \text{tr}\, \mathbf{S}^{-1}\mathbf{S}^* = 56.3 = \dfrac{4 \times 2 \times 55}{55 - 2 + 1} F$ or $F = 6.91$, exceeding the 0.001 point of the F-distribution for $n_1 = 8$ and $n_2 = 57$ degrees of freedom. For $4 \times 2 = 8$ degrees of freedom, we find from tables of the χ^2-distribution that Prob $(\chi^2 \geqq 56.3) < 0.00001$. We therefore reject the null hypothesis of homogeneity. (Pearson and Wilks use a different statistic, denoted by L_2, with observed value 0.6896 and for which Prob

$(L_2 < 0.6896) = 0.0000019$.) To find the linear discriminant functions for this example, the determinantal equation (8.4) is

$$\begin{vmatrix} 306.089 - 11.5666l & 214.86 - 30.9004l \\ 214.86 - 30.9004l & 662.77 - 139.153l \end{vmatrix} = 0,$$

and the quadratic equation yields the roots $l_1 = 51.702$, $l_2 = 4.614$. The decomposition corresponding to (8.7) is therefore

$$\begin{array}{lll} \hat{J}(H_1, H_2; l_2) = & 4.6 & 3 \text{ d.f.} \\ \hat{J}(H_1, H_2; l_1) = & 51.7 & 5 \text{ d.f.} \\ \hline \hat{J}(H_1, H_2) \quad = & 56.3 & 8 \text{ d.f.} \end{array}$$

The root l_2 is not significant and we proceed to find the coefficients of the linear discriminant function associated with l_1. With $l = 51.7 = l_1$, the equations (8.3) become

$$\begin{pmatrix} -291.906 & -1380.809 \\ -1380.809 & -6531.445 \end{pmatrix} \begin{pmatrix} \alpha_1 \\ \alpha_2 \end{pmatrix} = \mathbf{0},$$

that is,

$$\begin{aligned} 291.906\alpha_1 + 1380.809\alpha_2 &= 0 \\ 1380.809\alpha_1 + 6531.445\alpha_2 &= 0 \end{aligned}$$

yielding $\alpha_2/\alpha_1 = -0.211$. Thus, the only significant linear discriminant function, that associated with the root $l_1 = 51.7$, is $w = y_1 - 0.211y_2$.

9.2. Canonical Correlation

Hotelling (1936) considered the following data, given by Kelley (1928, p. 100) for a sample of 140 seventh-grade school children, in which x_1 and x_2 refer to reading speed and reading power respectively, and y_1 and y_2 to arithmetic speed and arithmetic power respectively. The data have been normalized and the correlation matrix of the 140 observations is*

$$\mathbf{R} = \left(\begin{array}{cc:cc} 1.0000 & 0.6328 & 0.2412 & 0.0586 \\ 0.6328 & 1.0000 & -0.0553 & 0.0655 \\ \hdashline 0.2412 & -0.0553 & 1.0000 & 0.4248 \\ 0.0586 & 0.0655 & 0.4248 & 1.0000 \end{array}\right) = \begin{pmatrix} \mathbf{R}_{11} & \mathbf{R}_{12} \\ \mathbf{R}_{21} & \mathbf{R}_{22} \end{pmatrix}.$$

We find that

$$\mathbf{R}_{21}\mathbf{R}_{11}^{-1}\mathbf{R}_{12} = \begin{pmatrix} 0.1303 & 0.0043 \\ 0.0043 & 0.0048 \end{pmatrix},$$

* Reprinted from *Crossroads in the Mind of Man* by Truman L. Kelley with the permission of the publishers, Stanford University Press. Copyright 1928 by the Board of Trustees of Leland Stanford Junior University.

and the determinantal equation corresponding to (7.4),

$$\begin{vmatrix} 0.1303 - r^2 & 0.0043 - 0.4248r^2 \\ 0.0043 - 0.4248r^2 & 0.0048 - r^2 \end{vmatrix} = 0,$$

yields the roots $r_1^2 = 0.1556$, $r_2^2 = 0.0047$.

The decomposition corresponding to (8.13) is therefore

$$\begin{array}{lrl} \hat{J}(H_1, H_2; r_2^2) = 137\dfrac{r_2^2}{1 - r_2^2} & = 0.6439 & 1 \text{ d.f.} \\ \hat{J}(H_1, H_2; r_1^2) = 137\dfrac{r_1^2}{1 - r_1^2} & = 25.2491 & 3 \text{ d.f.} \\ \hline \hat{J}(H_1, H_2) & = 25.8930 & 4 \text{ d.f.} \end{array}$$

$\hat{J}(H_1, H_2)$ and $\hat{J}(H_1, H_2; r_1^2)$ are significant at the 0.005 level. There is thus only one significant canonical correlation and the coefficients of the associated linear discriminant function must satisfy (8.10), or the equivalent

$$\begin{pmatrix} 0.1303 - 0.1556 & 0.0043 - 0.1556\,(0.4248) \\ 0.0043 - 0.1556\,(0.4248) & 0.0048 - 0.1556 \end{pmatrix} \begin{pmatrix} \alpha_1 \\ \alpha_2 \end{pmatrix} = 0,$$

that is,

$$-0.0253\alpha_1 - 0.0618\alpha_2 = 0$$
$$-0.0618\alpha_1 - 0.1508\alpha_2 = 0,$$

or $\alpha_1/\alpha_2 = -2.44$. The linear discriminant function is $w = -2.44y_1 + y_2$. [This corresponds to the second of the pair of linear discriminant functions in (7.19) of chapter 9.] We reject the null hypothesis that the y's (arithmetic speed and arithmetic power) are independent of the x's (reading speed and reading power). We now test the subhypothesis that reading power is not relevant, that is, the coefficient of x_2 in the regressions of y_1 and y_2 on x_1 and x_2 is zero. We therefore compute the values needed for the analysis of table 5.1, keeping in mind the remark at the end of section 6.2.

In the notation of section 5.1 we have

$$\mathbf{Y}'\mathbf{Y} = \begin{pmatrix} 1.0000 & 0.4248 \\ 0.4248 & 1.0000 \end{pmatrix}, \quad \begin{pmatrix} \mathbf{S}_{11} & \mathbf{S}_{12} \\ \mathbf{S}_{21} & \mathbf{S}_{22} \end{pmatrix} = \begin{pmatrix} 1.0000 & 0.6328 \\ 0.6328 & 1.0000 \end{pmatrix},$$

$$\mathbf{Y}'\mathbf{X}_1 = \begin{pmatrix} 0.2412 \\ 0.0586 \end{pmatrix}, \quad \mathbf{Y}'\mathbf{X}_2 = \begin{pmatrix} -0.0553 \\ 0.0655 \end{pmatrix},$$

$$\mathbf{S}_{22\cdot 1} = 1.0000 - (0.6328)^2 = 0.599564,$$

$$\hat{\mathbf{B}}^1 = (\mathbf{Y}'\mathbf{X})(\mathbf{X}'\mathbf{X})^{-1} = \begin{pmatrix} 0.2412 & -0.0553 \\ 0.0586 & 0.0655 \end{pmatrix} \begin{pmatrix} 1.0000 & 0.6328 \\ 0.6328 & 1.0000 \end{pmatrix}^{-1}$$

$$= \begin{pmatrix} 0.2412 & -0.0553 \\ 0.0586 & 0.0655 \end{pmatrix} \begin{pmatrix} 1.667878 & -1.055433 \\ -1.055433 & 1.667878 \end{pmatrix}$$

$$= \begin{pmatrix} 0.460658 & -0.346804 \\ 0.028607 & 0.047398 \end{pmatrix},$$

$$\hat{\mathbf{B}}^1(\mathbf{X}'\mathbf{X})\hat{\mathbf{B}}^{1'} = (\mathbf{Y}'\mathbf{X})(\mathbf{X}'\mathbf{X})^{-1}(\mathbf{X}\mathbf{Y}')$$

$$= \begin{pmatrix} 0.2412 & -0.0553 \\ 0.0586 & 0.0655 \end{pmatrix} \begin{pmatrix} 1.0000 & 0.6328 \\ 0.6328 & 1.0000 \end{pmatrix}^{-1} \begin{pmatrix} 0.2412 & 0.0586 \\ -0.0553 & 0.0655 \end{pmatrix}$$

$$= \begin{pmatrix} 0.1303 & 0.0043 \\ 0.0043 & 0.0048 \end{pmatrix},$$

$$\hat{\mathbf{B}}_1^2 = \mathbf{Y}'\mathbf{X}_1\mathbf{S}_{11}^{-1} = \begin{pmatrix} 0.2412 \\ 0.0586 \end{pmatrix}, \qquad \hat{\mathbf{B}}_1^2\mathbf{S}_{11}\hat{\mathbf{B}}_1^{2'} = \begin{pmatrix} 0.2412 \\ 0.0586 \end{pmatrix}(0.2412, \quad 0.0586)$$

$$= \begin{pmatrix} 0.0582 & 0.0141 \\ 0.0141 & 0.0034 \end{pmatrix}.$$

Table 9.1 corresponds to table 5.1 and provides the appropriate analysis. We find that

$$\operatorname{tr} 137 \begin{pmatrix} 0.8697 & 0.4205 \\ 0.4205 & 0.9952 \end{pmatrix}^{-1} \begin{pmatrix} 0.1303 & 0.0043 \\ 0.0043 & 0.0048 \end{pmatrix}$$

$$= \operatorname{tr} 137 \begin{pmatrix} 1.4450 & -0.6106 \\ -0.6106 & 1.2628 \end{pmatrix} \begin{pmatrix} 0.1303 & 0.0043 \\ 0.0043 & 0.0048 \end{pmatrix} = 25.8930,$$

TABLE 9.1

Variation due to	Generalized Sum of Squares	D.F.
$H_2:\mathbf{B}^2 = (\mathbf{B}_1^2, \mathbf{0})$	$\hat{\mathbf{B}}_1^2\mathbf{S}_{11}\hat{\mathbf{B}}_1^{2'} = \begin{pmatrix} 0.0582 & 0.0141 \\ 0.0141 & 0.0034 \end{pmatrix}$	1
Difference	$\hat{\mathbf{B}}_2^1\mathbf{S}_{22\cdot 1}\hat{\mathbf{B}}_2^{1'} = \begin{pmatrix} 0.0721 & -0.0098 \\ -0.0098 & 0.0014 \end{pmatrix}$	1
$H_1:\mathbf{B}^1 = (\mathbf{B}_1^1, \mathbf{B}_2^1)$	$\hat{\mathbf{B}}^1\mathbf{X}'\mathbf{X}\hat{\mathbf{B}}^{1'} = \begin{pmatrix} 0.1303 & 0.0043 \\ 0.0043 & 0.0048 \end{pmatrix}$	2
Difference	$137\hat{\mathbf{\Sigma}} = \begin{pmatrix} 0.8697 & 0.4205 \\ 0.4205 & 0.9952 \end{pmatrix}$	137
Total	$\mathbf{Y}'\mathbf{Y} = \begin{pmatrix} 1.0000 & 0.4248 \\ 0.4248 & 1.0000 \end{pmatrix}$	139

which is of course the value already found in terms of the canonical correlations, and

$$\operatorname{tr} 137 \begin{pmatrix} 0.8697 & 0.4205 \\ 0.4205 & 0.9952 \end{pmatrix}^{-1} \begin{pmatrix} 0.0721 & -0.0098 \\ -0.0098 & 0.0014 \end{pmatrix}$$

$$= 16.16 = \frac{1 \times 2 \times 137}{137 - 2 + 1} F \quad \text{or} \quad F = 8.02,$$

exceeding the 0.001 point of the F-distribution for $n_1 = 2$ and $n_2 = 136$ degrees of freedom. We therefore reject the null subhypothesis that x_2 is not relevant. A similar test can be made of a subhypothesis with respect to x_1, but we leave this to the reader.

The coefficient vector of the linear function of the x's whose correlation with the linear function of the y's, $w = -2.44y_1 + y_2$, yields the canonical correlation r_1, is proportional to $\boldsymbol{\alpha}'\hat{\mathbf{B}}$, that is [cf. (7.10) in chapter 9],

$$(-2.44, \quad 1) \begin{pmatrix} 0.460658 & -0.346804 \\ 0.028607 & 0.047398 \end{pmatrix} = (-1.095, \quad 0.894),$$

or $v = -1.095x_1 + 0.894x_2$.

9.3. Subhypothesis

Consider the following correlation matrix used by Thomson (1947, p. 30) to illustrate the computation of canonical correlations and by Bartlett (1948) to illustrate the relevant significance tests, assuming $n = 20$:

$$\mathbf{R} = \left(\begin{array}{ccc|cc} 1.0 & 0.1 & 0.6 & 0.7 & 0.2 \\ 0.1 & 1.0 & 0.4 & 0.3 & 0.8 \\ 0.6 & 0.4 & 1.0 & 0.5 & 0.3 \\ \hline 0.7 & 0.3 & 0.5 & 1.0 & 0.4 \\ 0.2 & 0.8 & 0.3 & 0.4 & 1.0 \end{array}\right) = \begin{pmatrix} \mathbf{R}_{11} & \mathbf{R}_{12} \\ \mathbf{R}_{21} & \mathbf{R}_{22} \end{pmatrix}.$$

We associate the first three rows with x_1, x_2, x_3, and the last two rows with y_1, y_2. Because of the relatively large values of the correlation of x_3 with x_1 (0.6) and with x_2 (0.4), we want to test a null subhypothesis that x_3 does not contribute significantly, in addition to x_1 and x_2, in the regression of the y's on the x's.

The determinantal equation corresponding to (7.4) is

$$\begin{vmatrix} 0.5434 - r^2 & 0.3210 - 0.4r^2 \\ 0.3210 - 0.4r^2 & 0.6693 - r^2 \end{vmatrix} = 0,$$

and yields the roots $r_1^2 = 0.6850$, $r_2^2 = 0.4530$. The decomposition corresponding to (8.13) is

$$\hat{J}(H_1, H_2; r_2^2) = 16\frac{r_2^2}{1 - r_2^2} = 13.28 \quad 2 \text{ d.f.}$$

$$\hat{J}(H_1, H_2; r_1^2) = 16\frac{r_1^2}{1 - r_1^2} = 34.72 \quad 4 \text{ d.f.}$$

$$\hat{J}(H_1, H_2) = 48.00 \quad 6 \text{ d.f.}$$

Here all values are significant at the 0.005 level, both canonical correlations are significant, and there are two significant linear discriminant functions.

In the notation of section 5.1 we have

$$\mathbf{Y}'\mathbf{Y} = \begin{pmatrix} 1.0 & 0.4 \\ 0.4 & 1.0 \end{pmatrix}, \qquad \begin{pmatrix} \mathbf{S}_{11} & \mathbf{S}_{12} \\ \mathbf{S}_{21} & \mathbf{S}_{22} \end{pmatrix} = \left(\begin{array}{cc|c} 1.0 & 0.1 & 0.6 \\ 0.1 & 1.0 & 0.4 \\ \hline 0.6 & 0.4 & 1.0 \end{array}\right),$$

$$\mathbf{Y}'\mathbf{X}_1 = \begin{pmatrix} 0.7 & 0.3 \\ 0.2 & 0.8 \end{pmatrix}, \qquad \mathbf{Y}'\mathbf{X}_2 = \begin{pmatrix} 0.5 \\ 0.3 \end{pmatrix},$$

$$\mathbf{S}_{22\cdot 1} = 1.0 - (0.6, \ 0.4)\begin{pmatrix} 1.0 & 0.1 \\ 0.1 & 1.0 \end{pmatrix}^{-1}\begin{pmatrix} 0.6 \\ 0.4 \end{pmatrix} = 0.523232,$$

$$\hat{\mathbf{B}}^1(\mathbf{X}'\mathbf{X})\hat{\mathbf{B}}^{1'} = (\mathbf{Y}'\mathbf{X})(\mathbf{X}'\mathbf{X})^{-1}(\mathbf{X}'\mathbf{Y})$$

$$= \begin{pmatrix} 0.7 & 0.3 & 0.5 \\ 0.2 & 0.8 & 0.3 \end{pmatrix}\begin{pmatrix} 1.0 & 0.1 & 0.6 \\ 0.1 & 1.0 & 0.4 \\ 0.6 & 0.4 & 1.0 \end{pmatrix}^{-1}\begin{pmatrix} 0.7 & 0.2 \\ 0.3 & 0.8 \\ 0.5 & 0.3 \end{pmatrix}$$

$$= \begin{pmatrix} 0.5434 & 0.3210 \\ 0.3210 & 0.6693 \end{pmatrix},$$

$$\hat{\mathbf{B}}_1^2\mathbf{S}_{11}\hat{\mathbf{B}}_1^{2'} = \mathbf{Y}'\mathbf{X}_1\mathbf{S}_{11}^{-1}\mathbf{X}_1'\mathbf{Y} = \begin{pmatrix} 0.7 & 0.3 \\ 0.2 & 0.8 \end{pmatrix}\begin{pmatrix} 1.0 & 0.1 \\ 0.1 & 1.0 \end{pmatrix}^{-1}\begin{pmatrix} 0.7 & 0.2 \\ 0.3 & 0.8 \end{pmatrix}$$

$$= \begin{pmatrix} 0.5434 & 0.3212 \\ 0.3212 & 0.6545 \end{pmatrix}.$$

Table 9.2 corresponds to table 5.1 and provides the appropriate analysis.

$$\operatorname{tr} 16\begin{pmatrix} 0.4566 & 0.0790 \\ 0.0790 & 0.3307 \end{pmatrix}^{-1}\begin{pmatrix} 0.5434 & 0.3210 \\ 0.3210 & 0.6693 \end{pmatrix}$$

$$= \operatorname{tr} 16\begin{pmatrix} 2.2845 & -0.5457 \\ -0.5457 & 3.1543 \end{pmatrix}\begin{pmatrix} 0.5434 & 0.3210 \\ 0.3210 & 0.6693 \end{pmatrix}$$

$$= 48.00 = \frac{3 \times 2 \times 16}{16 - 2 + 1}F \quad \text{or} \quad F = 7.50,$$

exceeding the 0.001 point of the F-distribution for $n_1 = 7$ and $n_2 = 17$ degrees of freedom. $\hat{J}(H_1, H_2) = 48.00$ is also the value obtained by the use of the canonical correlations.

$$\operatorname{tr} 16 \begin{pmatrix} 0.4566 & 0.0790 \\ 0.0790 & 0.3307 \end{pmatrix}^{-1} \begin{pmatrix} 0.0000 & -0.0002 \\ -0.0002 & 0.0148 \end{pmatrix}$$

$$= 0.75 = \frac{1 \times 2 \times 16}{16 - 2 + 1} F \quad \text{or} \quad F = 0.35,$$

not exceeding 3.683, the 0.05 point of the F-distribution for $n_1 = 2$ and $n_2 = 15$ degrees of freedom. We therefore accept the null subhypothesis that the x_3 variate contributes no significant information.

TABLE 9.2

Variation due to	Generalized Sum of Squares	D.F.
$H_2: \mathbf{B}^2 = (\mathbf{B}_1^2, \mathbf{0})$	$\hat{\mathbf{B}}_1^2 \mathbf{S}_{11} \hat{\mathbf{B}}_1^{2'} = \begin{pmatrix} 0.5434 & 0.3212 \\ 0.3212 & 0.6545 \end{pmatrix}$	2
Difference	$\hat{\mathbf{B}}_2^1 \mathbf{S}_{22\cdot1} \hat{\mathbf{B}}_2^{1'} = \begin{pmatrix} 0.0000 & -0.0002 \\ -0.0002 & 0.0148 \end{pmatrix}$	1
$H_1: \mathbf{B}^1 = (\mathbf{B}_1^1, \mathbf{B}_2^1)$	$\hat{\mathbf{B}}^1 \mathbf{X}'\mathbf{X} \hat{\mathbf{B}}^{1'} = \begin{pmatrix} 0.5434 & 0.3210 \\ 0.3210 & 0.6693 \end{pmatrix}$	3
Difference	$16\hat{\mathbf{\Sigma}} = \begin{pmatrix} 0.4566 & 0.0790 \\ 0.0790 & 0.3307 \end{pmatrix}$	16
Total	$\mathbf{Y}'\mathbf{Y} = \begin{pmatrix} 1.0 & 0.4 \\ 0.4 & 1.0 \end{pmatrix}$	19

To carry out the test of a similar subhypothesis on the pair of variables x_2, x_3, we have:

$$\begin{pmatrix} \mathbf{S}_{11} & \mathbf{S}_{12} \\ \mathbf{S}_{21} & \mathbf{S}_{22} \end{pmatrix} = \left(\begin{array}{c|cc} 1.0 & 0.1 & 0.6 \\ \hline 0.1 & 1.0 & 0.4 \\ 0.6 & 0.4 & 1.0 \end{array}\right), \quad \mathbf{Y}'\mathbf{X}_1 = \begin{pmatrix} 0.7 \\ 0.2 \end{pmatrix}, \quad \mathbf{Y}'\mathbf{X}_2 = \begin{pmatrix} 0.3 & 0.5 \\ 0.8 & 0.3 \end{pmatrix},$$

$$\mathbf{S}_{22\cdot1} = \begin{pmatrix} 1.0 & 0.4 \\ 0.4 & 1.0 \end{pmatrix} - \begin{pmatrix} 0.1 \\ 0.6 \end{pmatrix} (0.1, \ 0.6) = \begin{pmatrix} 0.99 & 0.34 \\ 0.34 & 0.64 \end{pmatrix},$$

$$\hat{\mathbf{B}}_1^2 \mathbf{S}_{11} \hat{\mathbf{B}}_1^{2'} = \mathbf{Y}'\mathbf{X}_1 \mathbf{S}_{11}^{-1} \mathbf{X}_1'\mathbf{Y} = \begin{pmatrix} 0.7 \\ 0.2 \end{pmatrix} (0.7, \ 0.2) = \begin{pmatrix} 0.49 & 0.14 \\ 0.14 & 0.04 \end{pmatrix}.$$

Table 9.3 corresponds to table 5.1 and provides the appropriate analysis.

TABLE 9.3

Variation due to	Generalized Sum of Squares	D.F.
$H_2: \mathbf{B}^2 = (\mathbf{B}_1^2, \mathbf{0})$	$\hat{\mathbf{B}}_1^2 \mathbf{S}_{11} \hat{\mathbf{B}}_1^{2'} = \begin{pmatrix} 0.49 & 0.14 \\ 0.14 & 0.04 \end{pmatrix}$	1
Difference	$\hat{\mathbf{B}}_2^1 \mathbf{S}_{22\cdot 1} \hat{\mathbf{B}}_2^{1'} = \begin{pmatrix} 0.0534 & 0.1810 \\ 0.1810 & 0.6293 \end{pmatrix}$	2
$H_1: \mathbf{B}^1 = (\mathbf{B}_1^1, \mathbf{B}_2^1)$	$\hat{\mathbf{B}}^1 \mathbf{X}'\mathbf{X} \hat{\mathbf{B}}^{1'} = \begin{pmatrix} 0.5434 & 0.3210 \\ 0.3210 & 0.6693 \end{pmatrix}$	3
Difference	$16\hat{\mathbf{\Sigma}} = \begin{pmatrix} 0.4566 & 0.0790 \\ 0.0790 & 0.3307 \end{pmatrix}$	16
Total	$\mathbf{Y}'\mathbf{Y} = \begin{pmatrix} 1.0 & 0.4 \\ 0.4 & 1.0 \end{pmatrix}$	19

$$\operatorname{tr} 16 \begin{pmatrix} 0.4566 & 0.0790 \\ 0.0790 & 0.3307 \end{pmatrix}^{-1} \begin{pmatrix} 0.0534 & 0.1810 \\ 0.1810 & 0.6293 \end{pmatrix}$$
$$= 30.54 = \frac{2 \times 2 \times 16}{16 - 2 + 1} F \quad \text{or} \quad F = 7.15,$$

between 4.772 and 7.944, the 0.01 and 0.001 points of the F-distribution for $n_1 = 4$ and $n_2 = 16$ degrees of freedom. We therefore reject the null subhypothesis that both x_2 and x_3 are not relevant.

Finally, if we consider a three-partition subhypothesis on the x's, then in the notation of section 5.2 we have:

$$\begin{pmatrix} \mathbf{S}_{11} & \mathbf{S}_{12} & \mathbf{S}_{13} \\ \mathbf{S}_{21} & \mathbf{S}_{22} & \mathbf{S}_{23} \\ \mathbf{S}_{31} & \mathbf{S}_{32} & \mathbf{S}_{33} \end{pmatrix} = \begin{pmatrix} 1.0 & 0.1 & 0.6 \\ 0.1 & 1.0 & 0.4 \\ 0.6 & 0.4 & 1.0 \end{pmatrix}, \qquad \mathbf{S}_{33\cdot 1} = 1.0 - (0.6)^2 = 0.64,$$

$$\mathbf{S}_{32\cdot 1} = 0.4 - (0.6)(0.1) = 0.34, \qquad \mathbf{S}_{22\cdot 1} = 1.0 - (0.1)^2 = 0.99,$$

$$\mathbf{S}_{33\cdot 12} = 0.64 - \frac{(0.34)(0.34)}{0.99} = 0.523232,$$

$$\mathbf{Y}'\mathbf{X}_{2\cdot 1} = \begin{pmatrix} 0.3 \\ 0.8 \end{pmatrix} - \begin{pmatrix} 0.7 \\ 0.2 \end{pmatrix}(0.1) = \begin{pmatrix} 0.23 \\ 0.78 \end{pmatrix},$$

$$\mathbf{Y}'\mathbf{X}_{3\cdot 1} = \begin{pmatrix} 0.5 \\ 0.3 \end{pmatrix} - \begin{pmatrix} 0.7 \\ 0.2 \end{pmatrix}(0.6) = \begin{pmatrix} 0.08 \\ 0.18 \end{pmatrix},$$

$$\mathbf{Y}'\mathbf{X}_{3\cdot 12} = \begin{pmatrix}0.08\\0.18\end{pmatrix} - \begin{pmatrix}0.23\\0.78\end{pmatrix}\frac{1}{0.99}(0.34) = \begin{pmatrix}0.0010\\-0.0879\end{pmatrix},$$

$$\mathbf{Y}'\mathbf{X}_1\mathbf{S}_{11}^{-1}\mathbf{X}_1'\mathbf{Y} = \begin{pmatrix}0.7\\0.2\end{pmatrix}(0.7, 0.2) = \begin{pmatrix}0.49 & 0.14\\0.14 & 0.04\end{pmatrix},$$

$$\mathbf{Y}'\mathbf{X}_{2\cdot 1}\mathbf{S}_{22\cdot 1}^{-1}\mathbf{X}_{2\cdot 1}'\mathbf{Y} = \begin{pmatrix}0.23\\0.78\end{pmatrix}\frac{1}{0.99}(0.23, 0.78) = \begin{pmatrix}0.0534 & 0.1812\\0.1812 & 0.6145\end{pmatrix},$$

$$\mathbf{Y}'\mathbf{X}_{3\cdot 12}\mathbf{S}_{33\cdot 12}^{-1}\mathbf{X}_{3\cdot 12}'\mathbf{Y} = \begin{pmatrix}0.0010\\-0.0879\end{pmatrix}\frac{1}{0.5232}(0.0010, -0.0879)$$
$$= \begin{pmatrix}0.0000 & -0.0002\\-0.0002 & 0.0148\end{pmatrix}.$$

Because of (5.19) we summarize these results and tables 9.2 and 9.3 in table 9.4.

TABLE 9.4

Variation due to	Generalized Sum of Squares	D.F.
x_1	$\mathbf{Y}'\mathbf{X}_1\mathbf{S}_{11}^{-1}\mathbf{X}_1'\mathbf{Y} = \begin{pmatrix}0.49 & 0.14\\0.14 & 0.04\end{pmatrix}$	1
$x_{2\cdot 1}$	$\mathbf{Y}'\mathbf{X}_{2\cdot 1}\mathbf{S}_{22\cdot 1}^{-1}\mathbf{X}_{2\cdot 1}'\mathbf{Y} = \begin{pmatrix}0.0534 & 0.1812\\0.1812 & 0.6145\end{pmatrix}$	1
$x_{3\cdot 12}$	$\mathbf{Y}'\mathbf{X}_{3\cdot 12}\mathbf{S}_{33\cdot 12}^{-1}\mathbf{X}_{3\cdot 12}'\mathbf{Y} = \begin{pmatrix}0.0000 & -0.0002\\-0.0002 & 0.0148\end{pmatrix}$	1
$\mathbf{B}^1$	$\mathbf{Y}'\mathbf{X}(\mathbf{X}'\mathbf{X})^{-1}\mathbf{X}'\mathbf{Y} = \begin{pmatrix}0.5434 & 0.3210\\0.3210 & 0.6693\end{pmatrix}$	3
Difference	$16\hat{\mathbf{\Sigma}} = \begin{pmatrix}0.4566 & 0.0790\\0.0790 & 0.3307\end{pmatrix}$	16
Total	$\mathbf{Y}'\mathbf{Y} = \begin{pmatrix}1.0 & 0.4\\0.4 & 1.0\end{pmatrix}$	19

$$\operatorname{tr} 16\begin{pmatrix}0.4566 & 0.0790\\0.0790 & 0.3307\end{pmatrix}^{-1}\begin{pmatrix}0.49 & 0.14\\0.14 & 0.04\end{pmatrix} = 17.46 \quad 2 \text{ d.f.}$$

$$\operatorname{tr} 16\begin{pmatrix}0.4566 & 0.0790\\0.0790 & 0.3307\end{pmatrix}^{-1}\begin{pmatrix}0.0534 & 0.1812\\0.1812 & 0.6145\end{pmatrix} = 29.79 \quad 2 \text{ d.f.}$$

$$\operatorname{tr} 16\begin{pmatrix}0.4566 & 0.0790\\0.0790 & 0.3307\end{pmatrix}^{-1}\begin{pmatrix}0.0000 & -0.0002\\-0.0002 & 0.0148\end{pmatrix} = 0.75 \quad 2 \text{ d.f.}$$

10. REPARAMETRIZATION

10.1. Hypotheses Not of Full Rank

(Cf. section 9 of chapter 10.) Suppose that the components of the rows of **B** in (3.1) are not linearly independent, but, for each row are linear functions of the same $p < k_1$ parameters, that is,

$$\mathbf{B} = \mathbf{\Gamma G}', \tag{10.1}$$

where $\mathbf{\Gamma} = (\gamma_{ij})$, $\mathbf{G}' = (g_{jk})$, $i = 1, 2, \cdots, k_2$, $j = 1, 2, \cdots, p$, $k = 1, 2, \cdots, k_1$, $\mathbf{G}'$ is of rank $p < k_1$, and $\mathbf{\Gamma}$ of rank min (p, k_2). This implies that the matrix **X** in (3.2) is of rank $p < k_1$, and conversely, so that $\mathbf{X'X}$ is now a positive (not positive definite) matrix of rank $p < k_1$, is therefore singular and has no inverse, so that we must re-examine the solution for $\hat{\mathbf{B}}$ in (4.1). We may write (3.2) as

$$\mathbf{Z} = \mathbf{Y} - \mathbf{XG\Gamma}' = \mathbf{Y} - \mathbf{A\Gamma}', \tag{10.2}$$

where $\mathbf{A} = \mathbf{XG}$ is an $n \times p$ matrix of rank p. The least squares estimate of $\mathbf{\Gamma}$ is derived from the normal equations [cf. (4.1)]

$$\hat{\mathbf{\Gamma}}\mathbf{A'A} = \mathbf{Y'A} \qquad \text{or} \qquad \hat{\mathbf{\Gamma}}\mathbf{G'X'XG} = \mathbf{Y'XG}. \tag{10.3}$$

The estimate of **B** is obtained from $\hat{\mathbf{B}} = \hat{\mathbf{\Gamma}}\mathbf{G}'$, or

$$\hat{\mathbf{B}} = \mathbf{Y'XG(G'X'XG)^{-1}G'}. \tag{10.4}$$

From (10.2) and (10.3), we see that

$$\hat{\mathbf{\Gamma}} = \mathbf{Y'A(A'A)^{-1}} = (\mathbf{Z}' + \mathbf{\Gamma A}')\mathbf{A(A'A)^{-1}},$$

so that $E(\hat{\mathbf{\Gamma}}) = \mathbf{\Gamma}$ and $E(\hat{\mathbf{B}}) = E(\hat{\mathbf{\Gamma}})\mathbf{G}' = \mathbf{\Gamma G}' = \mathbf{B}$, that is, $\hat{\mathbf{\Gamma}}$ and $\hat{\mathbf{B}}$ are unbiased estimates of $\mathbf{\Gamma}$ and **B** respectively.

Corresponding to (4.3), we have

$$\begin{aligned} \hat{J}(H_1, H_2; O_n) &= \operatorname{tr} \hat{\mathbf{\Sigma}}^{-1}(\hat{\mathbf{\Gamma}}^1 - \mathbf{\Gamma}^2)\mathbf{A'A}(\hat{\mathbf{\Gamma}}^1 - \mathbf{\Gamma}^2)' \\ &= \operatorname{tr} \hat{\mathbf{\Sigma}}^{-1}(\hat{\mathbf{\Gamma}}^1 - \mathbf{\Gamma}^2)\mathbf{G'X'XG}(\hat{\mathbf{\Gamma}}^1 - \mathbf{\Gamma}^2)' \\ &= \operatorname{tr} \hat{\mathbf{\Sigma}}^{-1}(\hat{\mathbf{B}}^1 - \mathbf{B}^2)\mathbf{X'X}(\hat{\mathbf{B}}^1 - \mathbf{B}^2)' \end{aligned} \tag{10.5}$$

where $(n - p)\hat{\mathbf{\Sigma}} = \mathbf{Y'Y} - \hat{\mathbf{\Gamma}}^1\mathbf{A'A}\hat{\mathbf{\Gamma}}^{1'} = \mathbf{Y'Y} - \hat{\mathbf{B}}^1\mathbf{X'X}\hat{\mathbf{B}}^{1'}$.

Note from (10.3) that $\hat{\mathbf{B}}\mathbf{X'XG} = \mathbf{Y'XG}$ represents k_2p linear functions of the y's that are normally distributed and that are also linear functions of the $\hat{\beta}$'s. These are unbiased estimates of the same linear functions of the β's. Since $\hat{\mathbf{B}}\mathbf{X'XG} = \mathbf{Y'XG} = \hat{\mathbf{\Gamma}}\mathbf{G'X'XG}$, we may make similar statements about the γ's and their estimates. Consider now any other set of k_2p linear functions of the y's, say $\mathbf{Y}'\mathbf{L}$, where $\mathbf{L}$ is an $n \times p$ matrix of rank p. Since

$$E(\mathbf{Y}'\mathbf{L}) = E(\mathbf{Z}' + \mathbf{B}\mathbf{X}')\mathbf{L} = \mathbf{B}\mathbf{X}'\mathbf{L} = \mathbf{\Gamma}\mathbf{G}'\mathbf{X}'\mathbf{L}, \tag{10.6}$$

$\mathbf{Y}'\mathbf{L}$ is an unbiased estimate of $\mathbf{\Gamma}$ if $\mathbf{G}'\mathbf{X}'\mathbf{L} = \mathbf{I}_p$, the $p \times p$ identity matrix. To obtain the covariance matrix of the linear functions of the y's we proceed as follows. Instead of the partitioning of the matrix $\mathbf{Y}'$ given for (3.2), consider the partitioning

$$\mathbf{Y}' = \begin{pmatrix} \boldsymbol{\zeta}_1' \\ \boldsymbol{\zeta}_2' \\ \cdot \\ \cdot \\ \cdot \\ \boldsymbol{\zeta}_{k_2}' \end{pmatrix}, \qquad \boldsymbol{\zeta}_j' = (y_{1j}, y_{2j}, \cdots, y_{nj}), \tag{10.7}$$

so that

$$\mathbf{Y}'\mathbf{L} = \begin{pmatrix} \boldsymbol{\zeta}_1'\mathbf{L} \\ \boldsymbol{\zeta}_2'\mathbf{L} \\ \cdot \\ \cdot \\ \cdot \\ \boldsymbol{\zeta}_{k_2}'\mathbf{L} \end{pmatrix} \tag{10.8}$$

with $\boldsymbol{\zeta}_j'\mathbf{L}$ a $1 \times p$ matrix representing p linear functions of the n observed values of the jth y variable. Considering the $pk_2 \times 1$ matrix

$$\begin{pmatrix} \mathbf{L}'\boldsymbol{\zeta}_1 \\ \mathbf{L}'\boldsymbol{\zeta}_2 \\ \cdot \\ \cdot \\ \cdot \\ \mathbf{L}'\boldsymbol{\zeta}_{k_2} \end{pmatrix}, \tag{10.9}$$

the covariance matrix of the pk_2 linear functions in (10.8) is

$$\begin{pmatrix} \mathbf{L}'\operatorname{cov}(\boldsymbol{\zeta}_1\boldsymbol{\zeta}_1')\mathbf{L} & \mathbf{L}'\operatorname{cov}(\boldsymbol{\zeta}_1\boldsymbol{\zeta}_2')\mathbf{L} & \cdots & \mathbf{L}'\operatorname{cov}(\boldsymbol{\zeta}_1\boldsymbol{\zeta}_{k_2}')\mathbf{L} \\ \mathbf{L}'\operatorname{cov}(\boldsymbol{\zeta}_2\boldsymbol{\zeta}_1')\mathbf{L} & \mathbf{L}'\operatorname{cov}(\boldsymbol{\zeta}_2\boldsymbol{\zeta}_2')\mathbf{L} & \cdots & \mathbf{L}'\operatorname{cov}(\boldsymbol{\zeta}_2\boldsymbol{\zeta}_{k_2}')\mathbf{L} \\ \cdot & \cdot & \cdots & \cdot \\ \mathbf{L}'\operatorname{cov}(\boldsymbol{\zeta}_{k_2}\boldsymbol{\zeta}_1')\mathbf{L} & \mathbf{L}'\operatorname{cov}(\boldsymbol{\zeta}_{k_2}\boldsymbol{\zeta}_2')\mathbf{L} & \cdots & \mathbf{L}'\operatorname{cov}(\boldsymbol{\zeta}_{k_2}\boldsymbol{\zeta}_{k_2}')\mathbf{L} \end{pmatrix} \tag{10.10}$$

$$= \begin{pmatrix} \mathbf{L}'\sigma_{11}\mathbf{I}_n\mathbf{L} & \mathbf{L}'\sigma_{12}\mathbf{I}_n\mathbf{L} & \cdots & \mathbf{L}'\sigma_{1k_2}\mathbf{I}_n\mathbf{L} \\ \mathbf{L}'\sigma_{21}\mathbf{I}_n\mathbf{L} & \mathbf{L}'\sigma_{22}\mathbf{I}_n\mathbf{L} & \cdots & \mathbf{L}'\sigma_{2k_2}\mathbf{I}_n\mathbf{L} \\ \cdot & \cdot & \cdots & \cdot \\ \mathbf{L}'\sigma_{k_21}\mathbf{I}_n\mathbf{L} & \mathbf{L}'\sigma_{k_22}\mathbf{I}_n\mathbf{L} & \cdots & \mathbf{L}'\sigma_{k_2k_2}\mathbf{I}_n\mathbf{L} \end{pmatrix}$$

$$= \begin{pmatrix} \sigma_{11}\mathbf{L}'\mathbf{L} & \sigma_{12}\mathbf{L}'\mathbf{L} & \cdots & \sigma_{1k_2}\mathbf{L}'\mathbf{L} \\ \sigma_{21}\mathbf{L}'\mathbf{L} & \sigma_{22}\mathbf{L}'\mathbf{L} & \cdots & \sigma_{2k_2}\mathbf{L}'\mathbf{L} \\ \cdot & \cdot & \cdots & \cdot \\ \sigma_{k_21}\mathbf{L}'\mathbf{L} & \sigma_{k_22}\mathbf{L}'\mathbf{L} & \cdots & \sigma_{k_2k_2}\mathbf{L}'\mathbf{L} \end{pmatrix}$$

$$= (\mathbf{\Sigma}) \times \cdot (\mathbf{L}'\mathbf{L}),$$

with $\mathbf{I}_n$ the $n \times n$ identity matrix and $\mathbf{\Sigma}$ the covariance matrix of the y's. The notation $(\mathbf{\Sigma}) \times \cdot (\mathbf{L}'\mathbf{L})$ means the Kronecker or direct product of matrices [defined by the last two members of the equality, MacDuffee (1946, pp. 81–88), see also Anderson (1958, p. 347), Cornish (1957)], and $(\mathbf{\Sigma}) \times \cdot (\mathbf{L}'\mathbf{L})$ is a $pk_2 \times pk_2$ matrix. Similarly, writing

$$(10.11) \qquad \hat{\mathbf{\Gamma}} = \begin{pmatrix} \hat{\mathbf{\gamma}}_1' \\ \hat{\mathbf{\gamma}}_2' \\ \cdot \\ \cdot \\ \cdot \\ \hat{\mathbf{\gamma}}_{k_2}' \end{pmatrix} = \begin{pmatrix} \mathbf{\zeta}_1'\mathbf{A}(\mathbf{A}'\mathbf{A})^{-1} \\ \mathbf{\zeta}_2'\mathbf{A}(\mathbf{A}'\mathbf{A})^{-1} \\ \cdot \\ \cdot \\ \cdot \\ \mathbf{\zeta}_{k_2}'\mathbf{A}(\mathbf{A}'\mathbf{A})^{-1} \end{pmatrix} = \mathbf{Y}'\mathbf{A}(\mathbf{A}'\mathbf{A})^{-1}, \quad \hat{\mathbf{\gamma}}_j' = (\hat{\gamma}_{j1}, \hat{\gamma}_{j2}, \cdots, \hat{\gamma}_{jp}),$$

and considering the pk_2 elements of $\hat{\mathbf{\Gamma}}$ in their order in the $1 \times pk_2$ matrix $(\hat{\mathbf{\gamma}}_1', \hat{\mathbf{\gamma}}_2', \cdots, \hat{\mathbf{\gamma}}_{k_2}')$, we have for the estimates in $\hat{\mathbf{\Gamma}}$ the covariance matrix

$$(10.12) \quad \begin{pmatrix} (\mathbf{A}'\mathbf{A})^{-1}\mathbf{A}'\sigma_{11}\mathbf{I}_n\mathbf{A}(\mathbf{A}'\mathbf{A})^{-1} & \cdots & (\mathbf{A}'\mathbf{A})^{-1}\mathbf{A}'\sigma_{1k_2}\mathbf{I}_n\mathbf{A}(\mathbf{A}'\mathbf{A})^{-1} \\ (\mathbf{A}'\mathbf{A})^{-1}\mathbf{A}'\sigma_{21}\mathbf{I}_n\mathbf{A}(\mathbf{A}'\mathbf{A})^{-1} & \cdots & (\mathbf{A}'\mathbf{A})^{-1}\mathbf{A}'\sigma_{2k_2}\mathbf{I}_n\mathbf{A}(\mathbf{A}'\mathbf{A})^{-1} \\ \cdot & \cdots & \cdot \\ (\mathbf{A}'\mathbf{A})^{-1}\mathbf{A}'\sigma_{k_21}\mathbf{I}_n\mathbf{A}(\mathbf{A}'\mathbf{A})^{-1} & \cdots & (\mathbf{A}'\mathbf{A})^{-1}\mathbf{A}'\sigma_{k_2k_2}\mathbf{I}_n\mathbf{A}(\mathbf{A}'\mathbf{A})^{-1} \end{pmatrix}$$

$$= \begin{pmatrix} \sigma_{11}(\mathbf{A}'\mathbf{A})^{-1} & \sigma_{12}(\mathbf{A}'\mathbf{A})^{-1} & \cdots & \sigma_{1k_2}(\mathbf{A}'\mathbf{A})^{-1} \\ \sigma_{21}(\mathbf{A}'\mathbf{A})^{-1} & \sigma_{22}(\mathbf{A}'\mathbf{A})^{-1} & \cdots & \sigma_{2k_2}(\mathbf{A}'\mathbf{A})^{-1} \\ \cdot & \cdot & \cdots & \cdot \\ \sigma_{k_21}(\mathbf{A}'\mathbf{A})^{-1} & \sigma_{k_22}(\mathbf{A}'\mathbf{A})^{-1} & \cdots & \sigma_{k_2k_2}(\mathbf{A}'\mathbf{A})^{-1} \end{pmatrix} = (\mathbf{\Sigma}) \times \cdot (\mathbf{A}'\mathbf{A})^{-1}$$

$$= (\mathbf{\Sigma}) \times \cdot (\mathbf{G}'\mathbf{X}'\mathbf{X}\mathbf{G})^{-1},$$

a $pk_2 \times pk_2$ matrix. Similarly, writing

$$(10.13) \qquad \hat{\mathbf{B}} = \begin{pmatrix} \hat{\mathbf{\beta}}_1' \\ \hat{\mathbf{\beta}}_2' \\ \cdot \\ \cdot \\ \cdot \\ \hat{\mathbf{\beta}}_{k_2}' \end{pmatrix}, \qquad \hat{\mathbf{\beta}}_j' = (\hat{\beta}_{j1}, \hat{\beta}_{j2}, \cdots, \hat{\beta}_{jk_1}),$$

we get for the k_1k_2 elements of $\hat{\mathbf{B}}$ the covariance matrix

$$(10.14) \qquad (\mathbf{\Sigma}) \times \cdot (\mathbf{G}(\mathbf{G}'\mathbf{X}'\mathbf{X}\mathbf{G})^{-1}\mathbf{G}').$$

From lemma 5.4 of chapter 3 with $k = n$, $r = p$, $\mathbf{B} = \sigma_{ii}\mathbf{I}_n$, $i = 1, 2, \cdots, k_2$, $\mathbf{I}_n$ the $n \times n$ identity matrix, $\mathbf{C} = \mathbf{L}$, $\mathbf{U} = \mathbf{G}'\mathbf{X}'$, $\mathbf{UC} = \mathbf{G}'\mathbf{X}'\mathbf{L} = \mathbf{I}_p$,

$$\sigma_{ii}\mathbf{L}'\mathbf{L} \geqq \sigma_{ii}(\mathbf{G}'\mathbf{X}'\mathbf{X}\mathbf{G})^{-1}, \tag{10.15}$$

where (10.15) means that any quadratic form with matrix $\sigma_{ii}\mathbf{L}'\mathbf{L}$ is greater than or equal to the quadratic form with matrix $\sigma_{ii}(\mathbf{G}'\mathbf{X}'\mathbf{X}\mathbf{G})^{-1}$. From (10.15), (10.12), (10.10), and lemma 5.1 of chapter 3 we conclude that the variances of the components of $\hat{\mathbf{\Gamma}}$ are the smallest among all linear functions of the y's that are unbiased estimates of $\mathbf{\Gamma}$. Similarly, $\mathbf{Y}'\mathbf{L}\mathbf{G}'$ is an unbiased estimate of $\mathbf{B}$ if $\mathbf{G}'\mathbf{X}'\mathbf{L} = \mathbf{I}_p$, and as above, we may conclude that

$$\sigma_{ii}\mathbf{G}\mathbf{L}'\mathbf{L}\mathbf{G}' \geqq \sigma_{ii}\mathbf{G}(\mathbf{G}'\mathbf{X}'\mathbf{X}\mathbf{G})^{-1}\mathbf{G}', \tag{10.16}$$

from which we infer that the variances of the components of $\hat{\mathbf{B}}$ are the smallest among all linear functions of the y's that are unbiased estimates of $\mathbf{B}$.

The value of $J(1, 2; O_n)$ and its estimate is the same for any reparametrization, as is indicated in (10.5). Since there are only p linearly independent linear functions of the elements of a row of $\mathbf{B}$, any one such set of p linearly independent functions may be derived from any other such set by a nonsingular linear transformation. The information functions are invariant under nonsingular transformations (see section 4 of chapter 2, and also section 3 of chapter 9), hence our conclusion.

We show that the elements of $\hat{\mathbf{\Gamma}}$ are sufficient estimates as follows. For the model in (10.2), take $\mathbf{\Gamma}^2 = \mathbf{0}$ for convenience; then

$$J(1, 2; O_n) = \operatorname{tr} \mathbf{\Sigma}^{-1}\mathbf{\Gamma}\mathbf{A}'\mathbf{A}\mathbf{\Gamma}'. \tag{10.17}$$

We have seen that $(\hat{\boldsymbol{\gamma}}_1', \hat{\boldsymbol{\gamma}}_2', \cdots, \hat{\boldsymbol{\gamma}}_{k_2}')$ are normally distributed with mean $(\boldsymbol{\gamma}_1', \boldsymbol{\gamma}_2', \cdots, \boldsymbol{\gamma}_{k_2}')$ and covariance matrix $(\mathbf{\Sigma}) \times \cdot (\mathbf{A}'\mathbf{A})^{-1}$. Since the inverse of a direct product of matrices is the direct product of the inverses of the matrices [MacDuffee (1946, p. 82)], we have

$$J(1, 2; \hat{\mathbf{\Gamma}}) = \operatorname{tr}\left((\mathbf{\Sigma}^{-1}) \times \cdot (\mathbf{A}'\mathbf{A})\right)\begin{pmatrix} \boldsymbol{\gamma}_1 \\ \boldsymbol{\gamma}_2 \\ \cdot \\ \cdot \\ \cdot \\ \boldsymbol{\gamma}_{k_2} \end{pmatrix}(\boldsymbol{\gamma}_1', \boldsymbol{\gamma}_2', \cdots, \boldsymbol{\gamma}_{k_2}') \tag{10.18}$$

$$= \sum_{i=1}^{k_2}\sum_{j=1}^{k_2} \sigma^{ij} \operatorname{tr} \mathbf{A}'\mathbf{A}\boldsymbol{\gamma}_i\boldsymbol{\gamma}_j' = \sum_{i=1}^{k_2}\sum_{j=1}^{k_2} \sigma^{ij}\boldsymbol{\gamma}_j'\mathbf{A}'\mathbf{A}\boldsymbol{\gamma}_i.$$

But

$$\boldsymbol{\Gamma}\mathbf{A}'\mathbf{A}\boldsymbol{\Gamma}' = \begin{pmatrix} \boldsymbol{\gamma}_1' \\ \boldsymbol{\gamma}_2' \\ \cdot \\ \cdot \\ \cdot \\ \boldsymbol{\gamma}_{k_2}' \end{pmatrix} \mathbf{A}'\mathbf{A}(\boldsymbol{\gamma}_1, \boldsymbol{\gamma}_2, \cdots, \boldsymbol{\gamma}_{k_2})$$

$$= \begin{pmatrix} \boldsymbol{\gamma}_1'\mathbf{A}'\mathbf{A}\boldsymbol{\gamma}_1 & \boldsymbol{\gamma}_1'\mathbf{A}'\mathbf{A}\boldsymbol{\gamma}_2 & \cdots & \boldsymbol{\gamma}_1'\mathbf{A}'\mathbf{A}\boldsymbol{\gamma}_{k_2} \\ \cdot & \cdot & \cdot & \cdot \\ \boldsymbol{\gamma}_{k_2}'\mathbf{A}'\mathbf{A}\boldsymbol{\gamma}_1 & \boldsymbol{\gamma}_{k_2}'\mathbf{A}'\mathbf{A}\boldsymbol{\gamma}_2 & \cdots & \boldsymbol{\gamma}_{k_2}'\mathbf{A}'\mathbf{A}\boldsymbol{\gamma}_{k_2} \end{pmatrix},$$

so that

$$\text{(10.19)} \qquad \operatorname{tr} \boldsymbol{\Sigma}^{-1}\boldsymbol{\Gamma}\mathbf{A}'\mathbf{A}\boldsymbol{\Gamma}' = \sum_{i=1}^{k_2}\sum_{j=1}^{k_2} \sigma^{ij}\boldsymbol{\gamma}_i'\mathbf{A}'\mathbf{A}\boldsymbol{\gamma}_j,$$

and since $\sigma^{ij} = \sigma^{ji}$, we have from (10.19) and (10.18),

$$\text{(10.20)} \qquad J(1, 2; O_n) = J(1, 2; \hat{\boldsymbol{\Gamma}}).$$

From theorem 4.2 of chapter 2 we conclude that $\hat{\boldsymbol{\Gamma}}$ is a sufficient estimate.

Example 10.1. Using the data in section 9.2, that is,

$$\hat{\boldsymbol{\Sigma}}^{-1} = 137 \begin{pmatrix} 1.4450 & -0.6106 \\ -0.6106 & 1.2628 \end{pmatrix}, \qquad \mathbf{X}'\mathbf{X} = \begin{pmatrix} 1.0000 & 0.6328 \\ 0.6328 & 1.0000 \end{pmatrix},$$

we have

$(\hat{\boldsymbol{\Sigma}}^{-1}) \times \cdot (\mathbf{X}'\mathbf{X})$

$$= 137 \begin{pmatrix} 1.4450 \begin{pmatrix} 1.0000 & 0.6328 \\ 0.6328 & 1.0000 \end{pmatrix} & -0.6106 \begin{pmatrix} 1.0000 & 0.6328 \\ 0.6328 & 1.0000 \end{pmatrix} \\ -0.6106 \begin{pmatrix} 1.0000 & 0.6328 \\ 0.6328 & 1.0000 \end{pmatrix} & 1.2628 \begin{pmatrix} 1.0000 & 0.6328 \\ 0.6328 & 1.0000 \end{pmatrix} \end{pmatrix}$$

$$= 137 \begin{pmatrix} 1.44500000 & 0.91439600 & -0.61060000 & -0.38638768 \\ 0.91439600 & 1.44500000 & -0.38638768 & -0.61060000 \\ -0.61060000 & -0.38638768 & 1.26280000 & 0.79909984 \\ -0.38638768 & -0.61060000 & 0.79909984 & 1.26280000 \end{pmatrix}.$$

We find that

$$137(0.460658, -0.346804, 0.028607, 0.047398)$$
$$\begin{pmatrix} 1.44500000 & 0.91439600 & -0.61060000 & -0.38638768 \\ 0.91439600 & 1.44500000 & -0.38638768 & -0.61060000 \\ -0.61060000 & -0.38638768 & 1.26280000 & 0.79909984 \\ -0.38638768 & -0.61060000 & 0.79909984 & 1.26280000 \end{pmatrix} \begin{pmatrix} 0.460658 \\ -0.346804 \\ 0.028607 \\ 0.047398 \end{pmatrix}$$

$$= 137 \operatorname{tr} \begin{pmatrix} 1.4450 & -0.6106 \\ -0.6106 & 1.2628 \end{pmatrix} \begin{pmatrix} 0.1303 & 0.0043 \\ 0.0043 & 0.0048 \end{pmatrix} = 25.8930,$$

verifying (10.19).

10.2. Partition

When the hypotheses call for a partitioning of the parameters into two sets, for example as in (5.2), it is possible that linear relations exist among the rows of the parameter matrix in one of the partitioned sets only. Here it is necessary to apply the procedures of section 10.1 only to the partitioned set not of full rank. Thus, suppose that in (5.2) the $n \times q_1$ matrix $\mathbf{X}_1$ is of rank $m < q_1$. This implies [cf. (10.1)] that

$$\mathbf{B}_1 = \mathbf{\Gamma}_1\mathbf{G}_1', \tag{10.21}$$

where $\mathbf{\Gamma}_1 = (\gamma_{ij})$, $\mathbf{G}_1' = (g_{jk})$, $i = 1, 2, \cdots, k_2$, $j = 1, 2, \cdots, m$, $k = 1, 2, \cdots, q_1$, $\mathbf{G}_1'$ is of rank $m < q_1$, and $\mathbf{\Gamma}_1$ of rank min (m, k_2). The results of section 5.1 are applicable if $\mathbf{B}_1$ and $\hat{\mathbf{B}}_1$ are replaced in the various formulas by $\mathbf{\Gamma}_1$ and $\hat{\mathbf{\Gamma}}_1$ respectively, $\mathbf{X}_1$ by $\mathbf{X}_1\mathbf{G}_1$, and $(n - q_1)$ degrees of freedom by $(n - m)$ degrees of freedom. The estimate $\hat{\mathbf{B}}_1$ is obtained from $\hat{\mathbf{B}}_1 = \hat{\mathbf{\Gamma}}_1\mathbf{G}_1'$. Thus, for example, $\mathbf{S}_{11}$ in (5.6) is to be replaced by $\mathbf{G}_1'\mathbf{S}_{11}\mathbf{G}_1$, where $\mathbf{S}_{11} = \mathbf{X}_1'\mathbf{X}_1$, and $\mathbf{S}_{12}$ by $\mathbf{G}_1'\mathbf{S}_{12}$, where $\mathbf{S}_{12} = \mathbf{X}_1'\mathbf{X}_2$.

Similar remarks also apply for a partitioning into three sets as in section 5.2, when one of the sets may not be of full rank.

11. REMARK

The reader doubtlessly has noted the similarities between the argument and results in chapters 10 and 11. As a matter of fact, we shall now indicate how the multivariate analogue of an analysis of variance table may be derived from that corresponding to appropriate specification of the linear regression model in (3.1) of chapter 10.

Consider the multivariate regression model (3.2), $\mathbf{Z} = \mathbf{Y} - \mathbf{X}\mathbf{B}'$. With $\boldsymbol{\alpha}' = (\alpha_1, \alpha_2, \cdots, \alpha_{k_2})$ any real $1 \times k_2$ matrix such that at least one of the α's is not zero,

$$\mathbf{Z}\boldsymbol{\alpha} = \mathbf{Y}\boldsymbol{\alpha} - \mathbf{X}\mathbf{B}'\boldsymbol{\alpha}, \tag{11.1}$$

derived from (3.2), is equivalent to the regression model in (3.1) of chapter 10, by setting

$$\mathbf{z} = \mathbf{Z}\boldsymbol{\alpha}, \qquad \mathbf{y} = \mathbf{Y}\boldsymbol{\alpha}, \qquad \boldsymbol{\beta} = \mathbf{B}'\boldsymbol{\alpha}. \tag{11.2}$$

Replace $\mathbf{y}$ by $\mathbf{Y}\boldsymbol{\alpha}$, and any specification of $\hat{\boldsymbol{\beta}}$ by the corresponding $\hat{\mathbf{B}}'\boldsymbol{\alpha}$, in any of the sum of squares columns in the analyses tabulated in chapter 10, or derived by the methods of chapter 10. The results are quadratic forms in the α's. Since the relations among these quadratic forms are identically true in the α's, we have the corresponding generalized sums of squares columns for the multivariate analogue with the matrices of the quadratic forms of the α's. This is evident if we compare table 5.1 in chapter 10 and table 5.1 in chapter 11, recalling that k_1 in chapter 11 is p in chapter 10 and q_1 in chapter 11 is q in chapter 10.

Similar remarks apply to the reparametrization, since from (10.2) we have

$$\mathbf{Z}\boldsymbol{\alpha} = \mathbf{Y}\boldsymbol{\alpha} - \mathbf{X}\mathbf{G}\boldsymbol{\Gamma}'\boldsymbol{\alpha}, \tag{11.3}$$

which is equivalent to (9.2) of chapter 10 by setting,

$$\mathbf{z} = \mathbf{Z}\boldsymbol{\alpha}, \qquad \mathbf{y} = \mathbf{Y}\boldsymbol{\alpha}, \qquad \boldsymbol{\gamma} = \boldsymbol{\Gamma}'\boldsymbol{\alpha}, \qquad \mathbf{A} = \mathbf{X}\mathbf{G}. \tag{11.4}$$

12. PROBLEMS

12.1. Derive the normal equations (4.1).

12.2. Verify (5.16) and (5.17).

12.3. Verify (5.18) and (5.19).

12.4. Verify (7.6).

12.5. In section 9.2 test the null subhypothesis that the coefficient of x_1 in the regressions of y_1 and y_2 on x_1 and x_2 is zero.

12.6. Consider the following data from a problem discussed by Bartlett (1947, p. 177); here $r = 8$, $k_2 = 2$, $n = n_1 + \cdots + n_8 = 57$,

$$49\mathbf{S} = \begin{pmatrix} 136{,}972.6 & 58{,}549.0 \\ 58{,}549.0 & 71{,}496.1 \end{pmatrix}, \qquad \mathbf{S}^* = \begin{pmatrix} 12{,}496.8 & -6{,}786.6 \\ -6{,}786.6 & 32{,}985.0 \end{pmatrix}.$$

(*a*) Are the eight samples homogeneous?

(*b*) Compute the value(s) for the significant linear discriminant function(s), if any.

12.7. Consider the following correlation matrix, assuming $n = 20$:

$$\mathbf{R} = \left(\begin{array}{ccc|cc} 1.0 & 0.5 & 0.3 & 0.8 & 0.8 \\ 0.5 & 1.0 & 0.4 & 0.7 & 0.3 \\ 0.3 & 0.4 & 1.0 & 0.2 & 0.1 \\ \hline 0.8 & 0.7 & 0.2 & 1.0 & 0.5 \\ 0.8 & 0.3 & 0.1 & 0.5 & 1.0 \end{array}\right) = \begin{pmatrix} \mathbf{R}_{11} & \mathbf{R}_{12} \\ \mathbf{R}_{21} & \mathbf{R}_{22} \end{pmatrix}.$$

Carry out an analysis similar to that of section 9.3.

12.8. Foster and Rees (1957, p. 241) give the following sample unbiased covariance matrix based on 82 degrees of freedom:

$$\mathbf{S} = 10^{-4}\left(\begin{array}{ccc|cc} 13.03 & 5.77 & 4.90 & 3.83 & -1.95 \\ 5.77 & 12.36 & 8.33 & 39.14 & -44.75 \\ 4.90 & 8.33 & 11.88 & 28.38 & -30.95 \\ \hline 3.83 & 39.14 & 28.38 & 229.36 & -261.52 \\ -1.95 & -44.75 & -30.95 & -261.52 & 388.31 \end{array}\right) = \begin{pmatrix} \mathbf{S}_{11} & \mathbf{S}_{12} \\ \mathbf{S}_{21} & \mathbf{S}_{22} \end{pmatrix}.$$

If the first three rows are associated with x_1, x_2, x_3 and the last two rows with y_1, y_2, are the regressions of y_1 and y_2 on x_1, x_2, x_3 significant?

12.9. Verify (4.4) with the data in section 9.3 assuming $\mathbf{B} = \mathbf{0}$.

12.10. Cornish (1957, p. 25) gives the following matrices [I have redesignated them according to the notation in (4.4); this does not imply the same interpretation for Cornish's problem]:

$$\mathbf{B} = \mathbf{0}, \qquad \hat{\mathbf{B}} = \begin{pmatrix} 0.072948 & -0.000524 \\ 0.022898 & 0.000619 \\ -0.089651 & -0.001473 \end{pmatrix},$$

$$\mathbf{X}'\mathbf{X} = \begin{pmatrix} 175.2654 & -722.3850 \\ -722.3850 & 19855.5000 \end{pmatrix},$$

$$\hat{\mathbf{\Sigma}}^{-1} = \begin{pmatrix} 1138.265050 & -161.151320 & 215.304630 \\ -161.151320 & 534.296632 & -125.495288 \\ 215.304630 & -125.495288 & 199.183242 \end{pmatrix}.$$

Cornish (1957) computed the value of the right-hand side of (4.4) as 950.06. Verify by computing the value of the left-hand side of (4.4).

12.11. In the notation of section 5, show that:

(*a*) $\mathbf{X}_1'\mathbf{X}_{2\cdot1} = \mathbf{0}$.
(*b*) $\mathbf{X}_2'\mathbf{X}_{2\cdot1} = \mathbf{S}_{22\cdot1} = \mathbf{X}_{2\cdot1}'\mathbf{X}_{2\cdot1}$.
(*c*) $\hat{\mathbf{B}}_2^1 = \mathbf{Y}'\mathbf{X}_{2\cdot1}\mathbf{S}_{22\cdot1}^{-1} = \mathbf{Z}'\mathbf{X}_{2\cdot1}\mathbf{S}_{22\cdot1}^{-1} + \mathbf{B}_2^1$.
(*d*) The covariance matrix of the k_2q_2 elements of $\hat{\mathbf{B}}_2^1$ is $(\mathbf{\Sigma}) \times \cdot (\mathbf{S}_{22\cdot1}^{-1})$.
(*e*) $\mathbf{X}_{3\cdot12}'\mathbf{X}_{3\cdot12} = \mathbf{S}_{33\cdot12}$.
(*f*) $\hat{\mathbf{B}}_3 = \mathbf{Y}'\mathbf{X}_{3\cdot12}\mathbf{S}_{33\cdot12}^{-1} = \mathbf{Z}'\mathbf{X}_{3\cdot12}\mathbf{S}_{33\cdot12}^{-1} + \mathbf{B}_3$.
(*g*) The covariance matrix of the k_2q_3 elements of $\hat{\mathbf{B}}_3$ is $(\mathbf{\Sigma}) \times \cdot (\mathbf{S}_{33\cdot12}^{-1})$.
(*h*) $|\mathbf{S}| = |\mathbf{S}_{11}| \cdot |\mathbf{S}_{22\cdot1}| \cdot |\mathbf{S}_{33\cdot12}|$.
(*i*) $\mathbf{X}_{2\cdot1}'\mathbf{X}_{3\cdot12} = \mathbf{0}$.

12.12. Summarize section 5.2 in a table similar to table 5.1, with $H_1:\mathbf{B}^1 = (\mathbf{B}_1^1, \mathbf{B}_2^1, \mathbf{B}_3^1)$, $H_2:\mathbf{B}^2 = (\mathbf{B}_1^2, \mathbf{B}_2^2, \mathbf{0})$, $H_3:\mathbf{B}^3 = (\mathbf{B}_1^3, \mathbf{0}, \mathbf{0})$.

12.13. Develop the results corresponding to section 5.2 for a four-partition subhypothesis.

12.14. In section 6.3, for two samples $r = 2$, show that:

(*a*) $\mathbf{S}^* = \dfrac{n_1n_2}{n_1 + n_2}(\bar{\mathbf{y}}_1 - \bar{\mathbf{y}}_2)(\bar{\mathbf{y}}_1 - \bar{\mathbf{y}}_2)'$.

(*b*) $\hat{J}(H_1, H_2) = \text{tr}\, \mathbf{S}^{-1}\mathbf{S}^* = \dfrac{n_1n_2}{n_1 + n_2}(\bar{\mathbf{y}}_1 - \bar{\mathbf{y}}_2)'\mathbf{S}^{-1}(\bar{\mathbf{y}}_1 - \bar{\mathbf{y}}_2)$.

(*c*) $\dfrac{(n_1 + n_2 - k_2 - 1)n_1n_2}{k_2(n_1 + n_2 - 2)(n_1 + n_2)}(\bar{\mathbf{y}}_1 - \bar{\mathbf{y}}_2)'\mathbf{S}^{-1}(\bar{\mathbf{y}}_1 - \bar{\mathbf{y}}_2) = F$, where F has the analysis of variance distribution with k_2 and $n_1 + n_2 - k_2 - 1$ degrees of freedom. [Cf. Anderson (1958, pp. 108–109), Rao (1952, pp. 73–74, 246–248).]

12.15. Use lemma 5.4 of chapter 3 to show that $\mathbf{Y}'\mathbf{Y} \geq (\mathbf{Y}'\mathbf{X})(\mathbf{X}'\mathbf{X})^{-1}(\mathbf{X}'\mathbf{Y})$, where $\mathbf{X}$, $\mathbf{Y}$ are defined in section 3. (Note the remark following lemma 5.1 of chapter 3.)

12.16. Show that (see section 4) $|(n - k_1)\hat{\mathbf{\Sigma}}| = \dfrac{\begin{vmatrix} \mathbf{Y}'\mathbf{Y} & \mathbf{Y}'\mathbf{X} \\ \mathbf{X}'\mathbf{Y} & \mathbf{X}'\mathbf{X} \end{vmatrix}}{|\mathbf{X}'\mathbf{X}|}$. (Cf. problem 4.6 in chapter 10.)

CHAPTER 12

Multivariate Analysis: Other Hypotheses

1. INTRODUCTION

In the preceding chapter, we studied tests of linear hypotheses for samples from multivariate normal populations, with the underlying assumption that all populations had a common covariance matrix. We shall now drop the assumption about common covariance matrices, and also consider certain hypotheses on the covariance matrices themselves.

2. BACKGROUND

In sections 1 and 2 of chapter 9 we saw that for two k-variate normal populations $N(\boldsymbol{\mu}_i, \boldsymbol{\Sigma}_i)$, $i = 1, 2$,

$$I(1:2; O_n) = nI(1:2) = I(1:2; \bar{\mathbf{x}}) + I(1:2; \mathbf{S}), \tag{2.1}$$

where $I(1:2)$, $I(1:2; \bar{\mathbf{x}})$, and $I(1:2; \mathbf{S})$ are given respectively in (1.2), (2.1), and (2.4) in chapter 9.

Consider a sample O_n of n independent observations from a k-variate normal population $N(\boldsymbol{\mu}, \boldsymbol{\Sigma})$, with mean $\boldsymbol{\mu}' = (\mu_1, \mu_2, \cdots, \mu_k)$ and covariance matrix $\boldsymbol{\Sigma} = (\sigma_{ij})$, $i, j = 1, 2, \cdots, k$. The moment generating function of the sample averages $\bar{\mathbf{x}}' = (\bar{x}_1, \bar{x}_2, \cdots, \bar{x}_k)$ and the elements of the sample unbiased covariance matrix $\mathbf{S} = (s_{ij})$, $i, j = 1, 2, \cdots, k$, with N degrees of freedom, is known to be [Anderson (1958, pp. 36, 53, 160), Wilks (1943, p. 121)]

$$M(\boldsymbol{\tau}, \mathbf{T}) = \left|\mathbf{I}_k - 2\frac{1}{N}\boldsymbol{\Sigma}\mathbf{T}\right|^{-N/2} \exp\left(\boldsymbol{\tau}'\boldsymbol{\mu} + \frac{1}{2}\boldsymbol{\tau}'\frac{1}{n}\boldsymbol{\Sigma}\boldsymbol{\tau}\right), \tag{2.2}$$

where $\boldsymbol{\tau}' = (\tau_1, \tau_2, \cdots, \tau_k)$, $\mathbf{T} = (\tau_{ij})$, $i, j = 1, 2, \cdots, k$.

For the conjugate distribution of $N(\boldsymbol{\mu}_2, \boldsymbol{\Sigma}_2)$, with mean $\boldsymbol{\mu}^*$ (see section 4 of chapter 3),

$$I(*:2; \bar{\mathbf{x}}) = \boldsymbol{\tau}'\boldsymbol{\mu}^* - \boldsymbol{\tau}'\boldsymbol{\mu}_2 - \frac{1}{2}\boldsymbol{\tau}'\frac{1}{n}\boldsymbol{\Sigma}_2\boldsymbol{\tau}, \tag{2.3}$$

where (cf. example 4.2 in chapter 3),

$$\boldsymbol{\mu}^* = \boldsymbol{\mu}_2 + \frac{1}{n}\boldsymbol{\Sigma}_2\boldsymbol{\tau}. \tag{2.4}$$

[For the matrix differentiation needed for (2.4) and (2.7) see problems 10.2 and 10.3 in chapter 9, Deemer and Olkin (1951, p. 364).] From (2.4), $\boldsymbol{\tau} = n\boldsymbol{\Sigma}_2^{-1}(\boldsymbol{\mu}^* - \boldsymbol{\mu}_2)$, and (2.3) yields

$$I(*:2; \bar{\mathbf{x}}) = \frac{n}{2}(\boldsymbol{\mu}^* - \boldsymbol{\mu}_2)'\boldsymbol{\Sigma}_2^{-1}(\boldsymbol{\mu}^* - \boldsymbol{\mu}_2). \tag{2.5}$$

Note that $I(1:2; \bar{\mathbf{x}}) > I(*:2; \bar{\mathbf{x}})$ for $\boldsymbol{\mu}^* = \boldsymbol{\mu}_1$ and $\boldsymbol{\Sigma}_1 \neq \boldsymbol{\Sigma}_2$, and that the conjugate distribution is a k-variate normal distribution $N(\boldsymbol{\mu}^*, \boldsymbol{\Sigma}_2)$.

For the conjugate distribution of $N(\boldsymbol{\mu}_2, \boldsymbol{\Sigma}_2)$, with covariance matrix $\boldsymbol{\Sigma}^*$,

$$I(*:2; \mathbf{S}) = \operatorname{tr} \mathbf{T}\boldsymbol{\Sigma}^* + \frac{N}{2}\log\left|\mathbf{I}_k - 2\frac{1}{N}\boldsymbol{\Sigma}_2\mathbf{T}\right|, \tag{2.6}$$

where (cf. example 4.4 in chapter 3, see problem 10.3 in chapter 9)

$$\boldsymbol{\Sigma}^* = \left(\mathbf{I}_k - 2\frac{1}{N}\boldsymbol{\Sigma}_2\mathbf{T}\right)^{-1}\boldsymbol{\Sigma}_2. \tag{2.7}$$

From (2.7), $\mathbf{T} = \frac{N}{2}(\boldsymbol{\Sigma}_2^{-1} - \boldsymbol{\Sigma}^{*-1})$, and (2.6) yields

$$I(*:2; \mathbf{S}) = \frac{N}{2}\left(\log\frac{|\boldsymbol{\Sigma}_2|}{|\boldsymbol{\Sigma}^*|} - k + \operatorname{tr} \boldsymbol{\Sigma}^*\boldsymbol{\Sigma}_2^{-1}\right). \tag{2.8}$$

Note that $I(1:2; \mathbf{S}) = I(*:2; \mathbf{S})$ for $\boldsymbol{\Sigma}^* = \boldsymbol{\Sigma}_1$.

Because of the independence of $\bar{\mathbf{x}}$ and $\mathbf{S}$ in a sample from a multivariate normal population, we have (cf. example 4.3 in chapter 3)

$$I(*:2; \bar{\mathbf{x}}, \mathbf{S}) = \boldsymbol{\tau}'\boldsymbol{\mu}^* - \boldsymbol{\tau}'\boldsymbol{\mu}_2 - \frac{1}{2}\boldsymbol{\tau}'\frac{1}{n}\boldsymbol{\Sigma}_2\boldsymbol{\tau} + \operatorname{tr} \mathbf{T}\boldsymbol{\Sigma}^* + \frac{N}{2}\log\left|\mathbf{I}_k - 2\frac{1}{N}\boldsymbol{\Sigma}_2\mathbf{T}\right|, \tag{2.9}$$

where $\boldsymbol{\tau}$ and $\mathbf{T}$ are given in (2.4) and (2.7) respectively, or

$$\begin{aligned} I(*:2; \bar{\mathbf{x}}, \mathbf{S}) &= I(*:2; \bar{\mathbf{x}}) + I(*:2; \mathbf{S}) \\ &= \frac{n}{2}(\boldsymbol{\mu}^* - \boldsymbol{\mu}_2)'\boldsymbol{\Sigma}_2^{-1}(\boldsymbol{\mu}^* - \boldsymbol{\mu}_2) \\ &\quad + \frac{N}{2}\left(\log\frac{|\boldsymbol{\Sigma}_2|}{|\boldsymbol{\Sigma}^*|} - k + \operatorname{tr} \boldsymbol{\Sigma}^*\boldsymbol{\Sigma}_2^{-1}\right). \end{aligned} \tag{2.10}$$

3. SINGLE SAMPLE

Suppose we have a random sample of n independent observations from k-variate normal populations. Let $\bar{\mathbf{x}}' = (\bar{x}_1, \bar{x}_2, \cdots, \bar{x}_k)$ and $\mathbf{S} = (s_{ij})$, $i, j = 1, 2, \cdots, k$, respectively, be the sample averages and sample unbiased variances and covariances with N degrees of freedom. We now examine tests of certain hypotheses on the normal populations from which the sample was drawn.

3.1. Homogeneity of the Sample

Suppose we want to test a null hypothesis of homogeneity, the observations in the sample are from the same k-variate normal population with specified covariance matrix $\boldsymbol{\Sigma}$, against an alternative hypothesis, the observations are from k-variate normal populations with different means but the same specified covariance matrix $\boldsymbol{\Sigma}$ (cf. example 4.1 in chapter 5). We denote the null hypothesis by

$$(3.1) \qquad H_2(\boldsymbol{\mu}|\boldsymbol{\Sigma}), \qquad \text{or} \qquad H_2(\cdot|\boldsymbol{\Sigma}),$$

according as the common mean is, or is not, specified, and the alternative hypothesis by

$$(3.2) \qquad H_1(\boldsymbol{\mu}_i|\boldsymbol{\Sigma}), \qquad \text{or} \qquad H_1(\cdot|\boldsymbol{\Sigma}),$$

according as the different means are, or are not, specified. With the sample values as the statistic $T(x)$ and

$$f_2(x) = \prod_{i=1}^{n} \frac{1}{|2\pi\boldsymbol{\Sigma}|^{1/2}} \exp\left(-\tfrac{1}{2}(\mathbf{x}_i - \boldsymbol{\mu})'\boldsymbol{\Sigma}^{-1}(\mathbf{x}_i - \boldsymbol{\mu})\right),$$

we have [cf. (4.8) in chapter 5 and (2.3) and (2.4) in this chapter]

$$(3.3) \qquad \hat{I}(*:2; O_n) = \sum_{i=1}^{n} (\hat{\boldsymbol{\tau}}_i'\mathbf{x}_i - \hat{\boldsymbol{\tau}}_i'\boldsymbol{\mu} - \tfrac{1}{2}\hat{\boldsymbol{\tau}}_i'\boldsymbol{\Sigma}\hat{\boldsymbol{\tau}}_i),$$

where $\hat{\boldsymbol{\tau}}_i$ satisfies $\mathbf{x}_i = \boldsymbol{\mu} + \boldsymbol{\Sigma}\hat{\boldsymbol{\tau}}_i$. We thus have

$$(3.4) \qquad \hat{I}(*:H_2(\boldsymbol{\mu}|\boldsymbol{\Sigma})) = \sum_{i=1}^{n} \tfrac{1}{2}(\mathbf{x}_i - \boldsymbol{\mu})'\boldsymbol{\Sigma}^{-1}(\mathbf{x}_i - \boldsymbol{\mu}).$$

If $\boldsymbol{\mu}$ is not specified, $\hat{I}(*:H_2(\cdot|\boldsymbol{\Sigma})) = \min_{\boldsymbol{\mu}} \hat{I}(*:H_2(\boldsymbol{\mu}|\boldsymbol{\Sigma}))$ is

$$(3.5) \qquad \hat{I}(*:H_2(\cdot|\boldsymbol{\Sigma})) = \sum_{i=1}^{n} \tfrac{1}{2}(\mathbf{x}_i - \bar{\mathbf{x}})'\boldsymbol{\Sigma}^{-1}(\mathbf{x}_i - \bar{\mathbf{x}}),$$

where $\bar{\mathbf{x}}' = (\bar{x}_1, \bar{x}_2, \cdots, \bar{x}_k)$.

On the other hand, with the same statistic $T(x)$ but with

$$f_2(x) = \prod_{i=1}^{n} \frac{1}{|2\pi\boldsymbol{\Sigma}|^{1/2}} \exp\left(-\tfrac{1}{2}(\mathbf{x}_i - \boldsymbol{\mu}_i)'\boldsymbol{\Sigma}^{-1}(\mathbf{x}_i - \boldsymbol{\mu}_i)\right),$$

we have [cf. (4.11) in chapter 5 and (2.3) and (2.4) in this chapter]

$$\hat{I}(*:2; O_n) = \sum_{i=1}^{n} (\hat{\boldsymbol{\tau}}_i'\mathbf{x}_i - \hat{\boldsymbol{\tau}}_i'\boldsymbol{\mu}_i - \tfrac{1}{2}\hat{\boldsymbol{\tau}}_i'\boldsymbol{\Sigma}\hat{\boldsymbol{\tau}}_i), \tag{3.6}$$

where $\hat{\boldsymbol{\tau}}_i$ satisfies $\mathbf{x}_i = \boldsymbol{\mu}_i + \boldsymbol{\Sigma}\hat{\boldsymbol{\tau}}_i$. We thus have

$$\hat{I}(*: H_1(\boldsymbol{\mu}_i|\boldsymbol{\Sigma})) = \sum_{i=1}^{n} \tfrac{1}{2}(\mathbf{x}_i - \boldsymbol{\mu}_i)'\boldsymbol{\Sigma}^{-1}(\mathbf{x}_i - \boldsymbol{\mu}_i). \tag{3.7}$$

If the $\boldsymbol{\mu}_i$ are not specified, $\hat{I}(*: H_1(\cdot|\boldsymbol{\Sigma})) = \min_{\boldsymbol{\mu}_i} \hat{I}(*: H_1(\boldsymbol{\mu}_i|\boldsymbol{\Sigma}))$ is

$$\hat{I}(*: H_1(\cdot|\boldsymbol{\Sigma})) = 0. \tag{3.8}$$

If the conjugate distribution in (3.3) is to range over k-variate normal populations with a common mean, then $\boldsymbol{\mu}_1^* = \cdots = \boldsymbol{\mu}_n^*$ implies that $\boldsymbol{\mu} + \boldsymbol{\Sigma}\boldsymbol{\tau}_1 = \cdots = \boldsymbol{\mu} + \boldsymbol{\Sigma}\boldsymbol{\tau}_n$, or only values $\boldsymbol{\tau}_1 = \cdots = \boldsymbol{\tau}_n = \boldsymbol{\tau}$ are admissible. With this restriction, (3.3) yields

$$\hat{I}(H_2(\cdot|\boldsymbol{\Sigma}):2; O_n) = n\hat{\boldsymbol{\tau}}'\bar{\mathbf{x}} - n\hat{\boldsymbol{\tau}}'\boldsymbol{\mu} - \frac{n}{2}\hat{\boldsymbol{\tau}}'\boldsymbol{\Sigma}\hat{\boldsymbol{\tau}}, \tag{3.9}$$

where $\hat{\boldsymbol{\tau}}$ satisfies $\bar{\mathbf{x}} = \boldsymbol{\mu} + \boldsymbol{\Sigma}\hat{\boldsymbol{\tau}}$, and (3.9) becomes

$$\hat{I}(H_2(\cdot|\boldsymbol{\Sigma}):2; O_n) = \frac{n}{2}(\bar{\mathbf{x}} - \boldsymbol{\mu})'\boldsymbol{\Sigma}^{-1}(\bar{\mathbf{x}} - \boldsymbol{\mu}). \tag{3.10}$$

Note that [cf. (4.17) in chapter 5]

$$\sum_{i=1}^{n} (\mathbf{x}_i - \boldsymbol{\mu})'\boldsymbol{\Sigma}^{-1}(\mathbf{x}_i - \boldsymbol{\mu}) = \sum_{i=1}^{n} (\mathbf{x}_i - \bar{\mathbf{x}})'\boldsymbol{\Sigma}^{-1}(\mathbf{x}_i - \bar{\mathbf{x}}) + n(\bar{\mathbf{x}} - \boldsymbol{\mu})'\boldsymbol{\Sigma}^{-1}(\bar{\mathbf{x}} - \boldsymbol{\mu}) \tag{3.11}$$

that is,

$$\hat{I}(*: H_2(\boldsymbol{\mu}|\boldsymbol{\Sigma})) = \hat{I}(*: H_2(\cdot|\boldsymbol{\Sigma})) + \hat{I}(H_2(\cdot|\boldsymbol{\Sigma}):2; O_n). \tag{3.12}$$

The hypothesis $H_2(\boldsymbol{\mu}|\boldsymbol{\Sigma})$ is the intersection of two hypotheses: (i) the sample is homogeneous; and (ii) the mean for the homogeneous sample is $\boldsymbol{\mu}$. $2\hat{I}(*: H_2(\cdot|\boldsymbol{\Sigma})) = \sum_{i=1}^{n} (\mathbf{x}_i - \bar{\mathbf{x}})'\boldsymbol{\Sigma}^{-1}(\mathbf{x}_i - \bar{\mathbf{x}})$, which is distributed as χ^2 with $(n-1)k$ degrees of freedom under the null hypothesis, tests the homogeneity. $2\hat{I}(H_2(\cdot|\boldsymbol{\Sigma}):2; O_n) = n(\bar{\mathbf{x}} - \boldsymbol{\mu})'\boldsymbol{\Sigma}^{-1}(\bar{\mathbf{x}} - \boldsymbol{\mu})$, which is distributed as χ^2 with k degrees of freedom under the null hypothesis, tests the specified mean given a homogeneous sample.

Suppose we assume now that the sample is homogeneous, namely, all the observations are from the same k-variate normal population, and we want to test a hypothesis about the mean, with no specification of the covariance matrix (cf. example 4.2 in chapter 5). Let the hypothesis $H_2(\boldsymbol{\mu}, \boldsymbol{\Sigma})$ imply that the sample is from a specified k-variate normal population $N(\boldsymbol{\mu}, \boldsymbol{\Sigma})$, and the hypothesis $H_2(\boldsymbol{\mu})$ imply that the sample is from a k-variate normal population with specified mean $\boldsymbol{\mu}$ but unspecified covariance matrix. Suppose the alternative hypothesis H_1 implies that the sample is from an unspecified k-variate normal population. With $T(x) = (\bar{\mathbf{x}}, \mathbf{S})$, where $\bar{\mathbf{x}}$ and $\mathbf{S}$ are defined in section 2, and

$$f_2(x) = \prod_{i=1}^{n} \frac{1}{|2\pi\boldsymbol{\Sigma}|^{1/2}} \exp\left(-\tfrac{1}{2}(\mathbf{x}_i - \boldsymbol{\mu})'\boldsymbol{\Sigma}^{-1}(\mathbf{x}_i - \boldsymbol{\mu})\right)$$

we have [cf. (2.9)]

$$\hat{I}(*: H_2(\boldsymbol{\mu}, \boldsymbol{\Sigma})) = \hat{\boldsymbol{\tau}}'\bar{\mathbf{x}} - \hat{\boldsymbol{\tau}}'\boldsymbol{\mu} - \frac{1}{2}\hat{\boldsymbol{\tau}}'\frac{1}{n}\boldsymbol{\Sigma}\hat{\boldsymbol{\tau}} + \operatorname{tr}\hat{\mathbf{T}}\mathbf{S} + \frac{N}{2}\log\left|\mathbf{I}_k - 2\frac{1}{N}\boldsymbol{\Sigma}\hat{\mathbf{T}}\right|,$$

with $\bar{\mathbf{x}} = \boldsymbol{\mu} + \frac{1}{n}\boldsymbol{\Sigma}\hat{\boldsymbol{\tau}}$, $\mathbf{S} = \left(\mathbf{I}_k - 2\frac{1}{N}\boldsymbol{\Sigma}\hat{\mathbf{T}}\right)^{-1}\boldsymbol{\Sigma}$, or

$$\hat{I}(*: H_2(\boldsymbol{\mu}, \boldsymbol{\Sigma})) = \frac{n}{2}(\bar{\mathbf{x}} - \boldsymbol{\mu})'\boldsymbol{\Sigma}^{-1}(\bar{\mathbf{x}} - \boldsymbol{\mu}) + \frac{N}{2}\left(\log\frac{|\boldsymbol{\Sigma}|}{|\mathbf{S}|} - k + \operatorname{tr}\mathbf{S}\boldsymbol{\Sigma}^{-1}\right).$$

In accordance with the general asymptotic theory, under the null hypothesis $H_2(\boldsymbol{\mu}, \boldsymbol{\Sigma})$, $2\hat{I}(*: H_2(\boldsymbol{\mu}, \boldsymbol{\Sigma}))$ is asymptotically distributed as χ^2 with $k + k(k + 1)/2$ degrees of freedom [cf. Anderson (1958, p. 268), Hoyt (1953)].

If the k-variate normal populations have the same covariance matrix under H_1 and H_2, we see from (2.7) that $\boldsymbol{\Sigma}^* = \boldsymbol{\Sigma}_2$ implies that $\mathbf{T} = \mathbf{0}$ is the only admissible value. This is equivalent to requiring that for samples from the conjugate distribution the covariance matrix parameters in the distribution of $\bar{\mathbf{x}}$ and $\mathbf{S}$ are the same. Accordingly, for $\hat{I}(*: H_2(\boldsymbol{\mu}))$, $\bar{\mathbf{x}} = \boldsymbol{\mu} + \frac{1}{n}\boldsymbol{\Sigma}\hat{\boldsymbol{\tau}}$ and $\hat{\mathbf{T}} = \mathbf{0}$ or $\mathbf{S} = \boldsymbol{\Sigma}$, and we have instead of $\hat{I}(*: H_2(\boldsymbol{\mu}, \boldsymbol{\Sigma}))$:

$$\hat{I}(*: H_2(\boldsymbol{\mu})) = \frac{n}{2}(\bar{\mathbf{x}} - \boldsymbol{\mu})'\mathbf{S}^{-1}(\bar{\mathbf{x}} - \boldsymbol{\mu}).$$

Note that this is (2.10) for $\boldsymbol{\mu}_2 = \boldsymbol{\mu}$ and $\boldsymbol{\Sigma}_2 = \boldsymbol{\Sigma}^* = \mathbf{S}$. We see that $\hat{I}(*: H_1) = 0$, and the test of the hypothesis $H_2(\boldsymbol{\mu})$ depends only on the value of $2\hat{I}(*: H_2(\boldsymbol{\mu}))$, Hotelling's generalization of Student's t-test. (See section 6.1 of chapter 11.)

3.2. The Hypothesis that a k-Variate Normal Population Has a Specified Covariance Matrix

We now examine the test for a null hypothesis H_2 that specifies the population covariance matrix, with no specification of the mean, against an alternative hypothesis H_1 that does not specify the covariance matrix or the mean, that is,

$$H_1:\boldsymbol{\Sigma}, \boldsymbol{\mu};\ H_2:\boldsymbol{\Sigma} = \boldsymbol{\Sigma}_2, \boldsymbol{\mu}. \tag{3.13}$$

We take the conjugate distribution with parameters the same as the observed best unbiased sample estimates, that is, $\boldsymbol{\mu}^* = \bar{\mathbf{x}}$, $\boldsymbol{\Sigma}^* = \mathbf{S}$, and (2.10) becomes

$$\hat{I}(*:2) = \frac{n}{2}(\bar{\mathbf{x}} - \boldsymbol{\mu})'\boldsymbol{\Sigma}_2^{-1}(\bar{\mathbf{x}} - \boldsymbol{\mu}) + \frac{N}{2}\left(\log\frac{|\boldsymbol{\Sigma}_2|}{|\mathbf{S}|} - k + \operatorname{tr}\mathbf{S}\boldsymbol{\Sigma}_2^{-1}\right). \tag{3.14}$$

Since the null hypothesis does not specify the mean, writing $\hat{I}(H_1:H_2) = \min_{\boldsymbol{\mu}} \hat{I}(*:2)$, we find that the minimum discrimination information statistic is

$$2\hat{I}(H_1:H_2) = N\left(\log\frac{|\boldsymbol{\Sigma}_2|}{|\mathbf{S}|} - k + \operatorname{tr}\mathbf{S}\boldsymbol{\Sigma}_2^{-1}\right). \tag{3.15}$$

(See problems 8.32 and 8.33.)

In accordance with the general asymptotic theory, under the null hypothesis H_2 in (3.13), $2\hat{I}(H_1:H_2)$ in (3.15) is asymptotically distributed as χ^2 with $k(k+1)/2$ degrees of freedom. Using the characteristic function of the distribution of $2\hat{I}(H_1:H_2)$, it may be shown (see section 6.2) that a better approximation to the distribution is R. A. Fisher's B-distribution [Fisher (1928, p. 665)], the noncentral χ^2-distribution, where for Fisher's distribution $\beta^2 = (2k^3 + 3k^2 - k)/12N$, $B^2 = 2\hat{I}(H_1:H_2)$, with $k(k+1)/2$ degrees of freedom [cf. Hoyt (1953)]. The table computed by Fisher in terms of β and B has been recalculated for convenience, in terms of β^2 and B^2 and is Table III, on page 380. For degrees of freedom greater than 7, the largest tabulated, instead of the noncentral χ^2-distribution, $2\hat{I}(H_1:H_2)(1 - (2k^3 + 3k^2 - k)/6Nk(k+1))$ may be treated as a χ^2 with $k(k+1)/2$ degrees of freedom. (See section 6.2.)

For tests of significance in factor analysis, Bartlett (1950, 1954), using a "homogeneous" likelihood function, and Rippe (1951), using the likelihood-ratio procedure for the test of significance of components in matrix factorization, arrived at the statistic $2\hat{I}(H_1:H_2)$ and the same conclusion as to its asymptotic χ^2-distribution. [Cf. Anderson (1958, pp. 264–267).]

3.3. The Hypothesis of Independence

When the null hypothesis H_2 implies that the variates are independent, that is,

$$(3.16) \qquad H_2: \mathbf{\Sigma}_2 = (\sigma_{ij}), \qquad \sigma_{ij} = 0, \qquad i \neq j, \qquad i, j = 1, 2, \cdots, k,$$

so that $\mathbf{P} = (\rho_{ij}) = \mathbf{I}_k$, where $\mathbf{P}$ is the matrix of population correlation coefficients, we may write (3.15) as

$$(3.17) \qquad 2\hat{I}(H_1:H_2) = -N \log |\mathbf{R}| + N \sum_{i=1}^{k} \left(\frac{s_{ii}}{\sigma_{ii}} + \log \frac{\sigma_{ii}}{s_{ii}} - 1 \right),$$

with $\mathbf{R}$ the matrix of sample correlation coefficients. The hypothesis H_2 in (3.16) is the intersection of two hypotheses, $H_2 = H_2' \cap H_2''$, with H_2' the hypothesis of independence that $\mathbf{P} = \mathbf{I}_k$, and H_2'' the hypothesis specifying the variances. We may thus write (3.17) as

$$(3.18) \qquad 2\hat{I}(H_1:H_2) = 2\hat{I}(H_1:H_2') + 2\hat{I}(H_1:H_2''),$$

with $2\hat{I}(H_1:H_2') = -N \log |\mathbf{R}|$ the minimum discrimination information statistic for the test of independence [see (6.12) in chapter 9], and $2\hat{I}(H_1:H_2'') = N \sum_{i=1}^{k} \left(\frac{s_{ii}}{\sigma_{ii}} + \log \frac{\sigma_{ii}}{s_{ii}} - 1 \right)$ the minimum discrimination information statistic for the test of specified variances. [Note that $2\hat{I}(H_1:H_2'')$ is the sum of k single-variate statistics.] It is known that, under (3.16), the s_{ii} and r_{ij} are independent [Wilks (1932)], so that $2\hat{I}(H_1:H_2')$ and $2\hat{I}(H_1:H_2'')$ are independent. In accordance with the general asymptotic theory, under the null hypothesis H_2 of (3.16), $2\hat{I}(H_1:H_2')$ is asymptotically distributed as χ^2 with $k(k-1)/2$ degrees of freedom and $2\hat{I}(H_1:H_2'')$ is asymptotically distributed as χ^2 with k degrees of freedom. It may be shown (see section 6.3) that a better approximation to the distribution of $2\hat{I}(H_1:H_2')$ is Fisher's B-distribution [Fisher (1928, p. 665)] with $\beta^2 = k(k-1)(2k+5)/12N$, $B^2 = 2\hat{I}(H_1:H_2')$, with $k(k-1)/2$ degrees of freedom [cf. Bartlett (1950, 1951b, 1954), Lawley (1940)] and a better approximation to the distribution of $2\hat{I}(H_1:H_2'')$ is Fisher's B-distribution with $\beta^2 = k/3N$, $B^2 = 2\hat{I}(H_1:H_2'')$, with k degrees of freedom. Note that the degrees of freedom and the values of β^2 for the distributions of the three terms in (3.18) are additive, that is, $k(k+1)/2 = k(k-1)/2 + k$ and $(2k^3 + 3k^2 - k)/12N = k(k-1)(2k+5)/12N + k/3N$, a property of the noncentral χ^2 [cf. Bateman (1949), Laha (1954)]. (See problems 8.21 and 8.22.)

Example 3.1. In section 9.2 of chapter 11, we had the correlation matrix

$$\mathbf{R} = \begin{pmatrix} 1.0000 & 0.6328 & 0.2412 & 0.0586 \\ 0.6328 & 1.0000 & -0.0553 & 0.0655 \\ 0.2412 & -0.0553 & 1.0000 & 0.4248 \\ 0.0586 & 0.0655 & 0.4248 & 1.0000 \end{pmatrix},$$

from a sample of 140 observations. To test a null hypothesis that the four variates are independent we compute $2\hat{I}(H_1:H_2') = -N \log |\mathbf{R}| = -139 \log 0.4129 = 139(0.88431) = 122.92$, $k(k-1)(2k+5)/12N = 4(3)(13)/12(139) = 0.0935$. For 6 degrees of freedom the 5% points for B^2 corresponding to $\beta^2 = 0.04$ and 0.16 are respectively 12.6750 and 12.9247, and the observed value of $2\hat{I}(H_1:H_2')$ is clearly significant. We reject the null hypothesis of independence, as we should, in view of the conclusions in section 9.2 of chapter 11.

3.4. Hypothesis on the Correlation Matrix

When the null hypothesis $H_2''':\mathbf{\Sigma}_2 = (\sigma_{ij}) = \mathbf{D}_\sigma\mathbf{P}_2\mathbf{D}_\sigma$ specifies the matrix of correlation coefficients $\mathbf{P}_2$, but not the diagonal matrix of standard deviations

$$\mathbf{D}_\sigma = \begin{pmatrix} \sigma_1 & 0 & \cdots & 0 \\ 0 & \sigma_2 & \cdots & 0 \\ \cdot & \cdot & \cdot & \cdot \\ 0 & 0 & \cdots & \sigma_k \end{pmatrix}, \text{ using } \mathbf{D}_{\hat{\sigma}^2} = \mathbf{D}_{s^2} = \begin{pmatrix} s_1^2 & 0 & \cdots & 0 \\ 0 & s_2^2 & \cdots & 0 \\ \cdot & \cdot & \cdot & \cdot \\ 0 & 0 & \cdots & s_k^2 \end{pmatrix},$$

we have from (3.15)

$$2\hat{I}(H_1:H_2''') = N\left(\log \frac{|\mathbf{P}_2|}{|\mathbf{R}|} - k + \operatorname{tr} \mathbf{R}\mathbf{P}_2^{-1}\right). \tag{3.19}$$

$2\hat{I}(H_1:H_2''')$ in (3.19), asymptotically, is distributed as χ^2 with $k(k-1)/2$ degrees of freedom under the null hypothesis H_2'''. Note that (3.19) is (2.8) of chapter 9, with $\mathbf{P}_1 = \mathbf{R}$, and yields $2\hat{I}(H_1:H_2')$ when $\mathbf{P}_2 = \mathbf{I}_k$.

For bivariate populations, $k = 2$, (3.19) yields

$$(3.20)\quad 2\hat{I}(H_1:H_2''')$$

$$= N\left[\log \frac{1-\rho_2^2}{1-r^2} - 2 + \operatorname{tr}\begin{pmatrix} 1 & r \\ r & 1 \end{pmatrix}\begin{pmatrix} \dfrac{1}{1-\rho_2^2} & -\dfrac{\rho_2}{1-\rho_2^2} \\ -\dfrac{\rho_2}{1-\rho_2^2} & \dfrac{1}{1-\rho_2^2} \end{pmatrix}\right]$$

$$= N\left(\log \frac{1-\rho_2^2}{1-r^2} + \frac{2\rho_2(\rho_2 - r)}{1-\rho_2^2}\right),$$

which is asymptotically distributed as χ^2 with 1 degree of freedom. Note that (3.20) is (4.33) in example 4.6 of chapter 3, with N for n, and r for ρ_1. See the remark in example 5.7 of chapter 5 about a confidence interval for ρ.

3.5. Linear Discriminant Function

The estimates of the linear discriminant functions in section 6 of chapter 9 may be derived by the same procedure as for the information statistics. There is some tutorial value, however, in paralleling the discussion with the appropriate sample values.

We first examine the null hypothesis that specifies $\mathbf{\Sigma}_2$. [See (3.15).] We want the linear discriminant function

$$y = \alpha_1 x_1 + \alpha_2 x_2 + \cdots + \alpha_k x_k = \boldsymbol{\alpha}'\mathbf{x}, \tag{3.21}$$

the same linear compound for each observation. We seek the α's so as to maximize

$$2\hat{I}(H_1:H_2; y) = N\left(\log \frac{\boldsymbol{\alpha}'\mathbf{\Sigma}_2\boldsymbol{\alpha}}{\boldsymbol{\alpha}'\mathbf{S}\boldsymbol{\alpha}} - 1 + \frac{\boldsymbol{\alpha}'\mathbf{S}\boldsymbol{\alpha}}{\boldsymbol{\alpha}'\mathbf{\Sigma}_2\boldsymbol{\alpha}}\right), \tag{3.22}$$

the equivalent of (3.15) for y. We are thereby led to conclusions similar to (6.4) and (6.5) in chapter 9, namely, that $\boldsymbol{\alpha}$ must satisfy

$$\mathbf{S}\boldsymbol{\alpha} = F\mathbf{\Sigma}_2\boldsymbol{\alpha}, \tag{3.23}$$

where F is a root of the determinantal equation

$$|\mathbf{S} - F\mathbf{\Sigma}_2| = 0 = |N\mathbf{S} - l\mathbf{\Sigma}_2|, \qquad F = l/N, \tag{3.24}$$

with roots almost everywhere real and positive. (See section 6.4 for the distribution of these roots.) Designating these roots as $F_1, F_2, \cdots, F_k$ in descending order, the discussion in section 6 of chapter 9 is applicable (taking suitable account of the ordering). In particular, we have the decomposition of (3.15)

$$\begin{aligned} 2\hat{I}(H_1:H_2) &= N\left(\log \frac{|\mathbf{\Sigma}_2|}{|\mathbf{S}|} - k + \operatorname{tr} \mathbf{S}\mathbf{\Sigma}_2^{-1}\right) \\ &= 2\hat{I}(H_1:H_2; y_1) + \cdots + 2\hat{I}(H_1:H_2; y_k), \end{aligned} \tag{3.25}$$

where y_i is the linear discriminant function associated with F_i. From (3.22) we see that

$$\begin{aligned} 2\hat{I}(H_1:H_2; y_i) &= N(-\log F_i - 1 + F_i) \\ &= N \log N - N - N \log l_i + l_i. \end{aligned} \tag{3.26}$$

When the values of $2\hat{I}(H_1:H_2; y_i)$ are arranged in descending order of magnitude, under the null hypothesis that the sample is from a normal population with covariance matrix $\mathbf{\Sigma}_2$, the sum of the last $(k - m)$ of the $2\hat{I}(H_1:H_2; y_i)$ asymptotically is distributed as χ^2 with $(k - m)(k - m + 1)/2$ degrees of freedom. (See section 6.4.) A better approximation to the distribution is R. A. Fisher's B-distribution [Fisher (1928, p. 665)], the noncentral χ^2-distribution, where for Fisher's distribution $\beta^2 = ((2k^3 + 3k^2 - k) - (2m^3 + 3m^2 - m))/12N$, B^2 is the sum of the last $(k - m)$ of the $2\hat{I}(H_1:H_2; y_i)$, with $(k - m)(k - m + 1)/2$ degrees of freedom.

3.6. Independence of Sets of Variates †

[Cf. Anderson (1958, pp. 230–245), Hsu (1949, pp. 373–376), Wald and Brookner (1941), Wilks (1935b, 1943, pp. 242–245).] Suppose we partition the variates of a k-variate normal population into m sets of $k_1, k_2, \cdots, k_m$ variates, $k_1 + k_2 + \cdots + k_m = k$. We now want to test a null hypothesis H_2, the sets of variates are mutually independent, against an alternative hypothesis H_1, the sets are not independent, with no specification of the means, that is,

$$(3.27)\qquad H_1:\mathbf{\Sigma} = (\sigma_{ij}), \qquad i, j = 1, 2, \cdots, k,$$

$$(3.28)\qquad H_2:\mathbf{\Sigma} = \begin{pmatrix} \mathbf{\Sigma}_{11} & \mathbf{0} & \cdots & \mathbf{0} \\ \mathbf{0} & \mathbf{\Sigma}_{22} & \cdots & \mathbf{0} \\ \cdot & \cdot & \cdots & \cdot \\ \mathbf{0} & \mathbf{0} & \cdots & \mathbf{\Sigma}_{mm} \end{pmatrix}, \qquad \mathbf{\Sigma}_{ii} = (\sigma_{\alpha\beta}),$$

$$\alpha, \beta = k_1 + k_2 + \cdots + k_{i-1} + 1, \cdots, k_1 + k_2 + \cdots + k_i.$$

The discussion in section 7 of chapter 9 is for two sets, $m = 2$. (See problems 10.13–10.19 in chapter 9.)

Denoting the hypothesis of (3.28) by $H_2(\mathbf{\Sigma}_{ii})$ when $\mathbf{\Sigma}_{11}, \cdots, \mathbf{\Sigma}_{mm}$ are specified, we get from (3.15),

$$(3.29)\quad 2\hat{I}(H_1:H_2(\mathbf{\Sigma}_{ii})) =$$

$$N\left(\log \frac{|\mathbf{\Sigma}_{11}||\mathbf{\Sigma}_{22}| \cdots |\mathbf{\Sigma}_{mm}|}{|\mathbf{S}|} - k + \operatorname{tr}(\mathbf{S}_{11}\mathbf{\Sigma}_{11}^{-1} + \cdots + \mathbf{S}_{mm}\mathbf{\Sigma}_{mm}^{-1})\right),$$

with $\mathbf{S}_{ii}$ the best unbiased sample covariance matrix of the variates in the ith set. Denoting the hypothesis of (3.28) with no specification of the matrices $\mathbf{\Sigma}_{ii}$, $i = 1, 2, \cdots, m$, by $H_2(\cdot)$, we find that (3.29) is a minimum for $\hat{\mathbf{\Sigma}}_{ii} = \mathbf{S}_{ii}$, and

$$(3.30)\quad 2\hat{I}(H_1:H_2(\cdot)) = N \log \frac{|\mathbf{S}_{11}| \cdots |\mathbf{S}_{mm}|}{|\mathbf{S}|} = N \log \frac{|\mathbf{R}_{11}| \cdots |\mathbf{R}_{mm}|}{|\mathbf{R}|},$$

with $\mathbf{R}_{ii}$ and $\mathbf{R}$ respectively the sample correlation matrices of the variates in the ith set and the entire set. The last member in (3.30) is obtained by factoring out the standard deviations in the numerator and denominator terms. In accordance with the general asymptotic theory, under the null hypothesis, $2\hat{I}(H_1:H_2(\cdot))$ is asymptotically distributed as χ^2 with

† see Appendix page 390

$k(k + 1)/2 - \sum_{i=1}^{m} k_i(k_i + 1)/2 = \sum_{i<j} k_i k_j$ degrees of freedom. It may be shown (see section 6.3) that a better approximation to the distribution of $2\hat{I}(H_1:H_2(\cdot))$ is R. A. Fisher's B-distribution [Fisher (1928, p. 665)], the noncentral χ^2-distribution, where for Fisher's distribution $\beta^2 = \left((2k^3 + 3k^2 - k) - \sum_{i=1}^{m}(2k_i^3 + 3k_i^2 - k_i)\right)/12N$, $B^2 = 2\hat{I}(H_1:H_2(\cdot))$, with $\sum_{i<j} k_i k_j$ degrees of freedom. We summarize the analysis of the minimum discrimination information statistic of (3.29) in table 3.1. Note that the degrees of freedom and the values of the noncentrality parameter β^2 in table 3.1 are additive, properties of the χ^2-distribution, central and noncentral.

We remark that when $k_1 = \cdots = k_m = 1$, $2\hat{I}(H_1:H_2(\cdot)) = 2\hat{I}(H_1:H_2')$ of section 3.3, the "between" component in table 3.1 is $2\hat{I}(H_1:H_2'')$ of section 3.3, and the degrees of freedom and the values of β^2 are those given in section 3.3. (See problems 8.19, 8.25–8.29, 8.34.)

Example 3.2. Consider the correlation matrix in example 3.1, with the partitioning of the four variates into two sets as in section 9.2 of chapter 11. To test a null hypothesis that the sets are independent, we compute, $|\mathbf{R}_{11}| = 0.5996$, $|\mathbf{R}_{22}| = 0.8195$, $2\hat{I}(H_1:H_2(\cdot)) = 139 \log((0.5996)(0.8195)/0.4129) = 24.16$, $\sum_{i<j} k_i k_j = 4$, $\beta^2 = (172 - 26 - 26)/12(139) = 0.0719$. For 4 degrees of freedom, the 5% points for B^2 corresponding to $\beta^2 = 0.04$ and $\beta^2 = 0.16$ are respectively 9.5821 and 9.8627. The observed value of $2\hat{I}(H_1:H_2(\cdot)) = 24.16$ is clearly significant, and we reject the null hypothesis, as we should, in view of the conclusions in section 9.2 of chapter 11. [Cf. Kullback (1952, pp. 98–99).]

3.7. Independence and Equality of Variances

[Cf. Anderson (1958, pp. 259–261), Hsu (1949, pp. 376–378).] We want to test the null hypothesis in (3.16), with the specification that $\sigma_{11} = \sigma_{22} = \cdots = \sigma_{kk} = \sigma^2$. Denote by $H_2''(\sigma^2)$ the hypothesis H_2'' in (3.18) with the common variance σ^2 specified, and denote by $H_2''(\cdot)$ the hypothesis of equality of the variances. From (3.17) and (3.18) (with the more common notation $s_{ii} = s_i^2$ for the variance) we see that

$$(3.31)\qquad 2\hat{I}(H_1:H_2''(\sigma^2)) = N\sum_{i=1}^{k}\left(\frac{s_i^2}{\sigma^2} + \log\frac{\sigma^2}{s_i^2} - 1\right).$$

Since the minimum of (3.31) is given for $\hat{\sigma}^2 = (s_1^2 + \cdots + s_k^2)/k = s^2$, we have that $\hat{I}(H_1:H_2''(\cdot)) = \min_{\sigma^2} \hat{I}(H_1:H_2''(\sigma^2))$ is

$$(3.32)\qquad 2\hat{I}(H_1:H_2''(\cdot)) = N\sum_{i=1}^{k}\log\frac{s^2}{s_i^2}.$$

TABLE 3.1

Component due to	Information	D.F.	β^2
Between $\mathbf{S}_{ii}$ against $\mathbf{\Sigma}_{ii}$	$N\sum_{i=1}^{m}\left(\log\frac{\lvert\mathbf{\Sigma}_{ii}\rvert}{\lvert\mathbf{S}_{ii}\rvert} - k_i + \mathrm{tr}\,\mathbf{S}_{ii}\mathbf{\Sigma}_{ii}^{-1}\right)$	$\sum_{i=1}^{m}\frac{k_i(k_i+1)}{2}$	$\sum_{i=1}^{m}\frac{2k_i^3+3k_i^2-k_i}{12N}$
Within $2\hat{I}(H_1:H_2(\cdot))$	$N\log\frac{\lvert\mathbf{S}_{11}\rvert\cdots\lvert\mathbf{S}_{mm}\rvert}{\lvert\mathbf{S}\rvert} = N\log\frac{\lvert\mathbf{R}_{11}\rvert\cdots\lvert\mathbf{R}_{mm}\rvert}{\lvert\mathbf{R}\rvert}$	$\sum_{i<j}k_ik_j$	$\frac{2k^3+3k^2-k-\sum_{i=1}^{m}(2k_i^3+3k_i^2-k_i)}{12N}$
Total $2\hat{I}(H_1:H_2(\mathbf{\Sigma}_{ii}))$	$N\left(\log\frac{\lvert\mathbf{\Sigma}_{11}\rvert\cdots\lvert\mathbf{\Sigma}_{mm}\rvert}{\lvert\mathbf{S}\rvert} - k + \sum_{i=1}^{m}\mathrm{tr}\,\mathbf{S}_{ii}\mathbf{\Sigma}_{ii}^{-1}\right)$	$\frac{k(k+1)}{2}$	$\frac{2k^3+3k^2-k}{12N}$

A summary of the analysis of $2\hat{I}(H_1:H_2''(\sigma^2))$, with the appropriate degrees of freedom and noncentrality parameters, is given in table 3.2.

TABLE 3.2

Component due to	Information	D.F.	β^2
Between, s^2 against σ^2	$Nk\left(\frac{s^2}{\sigma^2} + \log\frac{\sigma^2}{s^2} - 1\right)$	1	$1/3Nk$
Within, $2\hat{I}(H_1:H_2''(\cdot))$	$N\sum_{i=1}^{k}\log\frac{s^2}{s_i^2}$	$k - 1$	$(k^2 - 1)/3Nk$
Total, $2\hat{I}(H_1:H_2''(\sigma^2))$	$N\sum_{i=1}^{k}\left(\frac{s_i^2}{\sigma^2} + \log\frac{\sigma^2}{s_i^2} - 1\right)$	k	$k/3N$

Under the null hypothesis, $2\hat{I}(H_1:H_2''(\cdot))$ is asymptotically distributed as χ^2 with $k - 1$ degrees of freedom. A better approximation to the distribution is R. A. Fisher's B-distribution [Fisher (1928, p. 665)], the noncentral χ^2-distribution, with $\beta^2 = (k^2 - 1)/3Nk$, $B^2 = 2\hat{I}(H_1:H_2''(\cdot))$, and $k - 1$ degrees of freedom. We remark that $2\hat{I}(H_1:H_2''(\cdot))$ above is a special case of the more general result to be derived in section 5.3, and is (5.16) with $r = k$, $N_1 = \cdots = N_r = N$. [Note that in (5.16) $N = N_1 + N_2 + \cdots + N_r$ is Nk here.]

(See problem 8.35.)

4. HOMOGENEITY OF MEANS

We now want to consider the problem of testing a hypothesis about the equality of r means for each of k variates for r k-variate normal samples, but with no assumption that the population covariance matrices are equal. We first deal with two samples, $r = 2$, for its intrinsic interest and expository value.

4.1. Two Samples

Suppose we have two independent samples of n_1 and n_2 independent observations from k-variate normal populations with covariance matrices $\mathbf{\Sigma}_1$ and $\mathbf{\Sigma}_2$. We want to test the null hypothesis H_2, the population mean vectors (matrices) are equal, with no specification about $\mathbf{\Sigma}_1$ and $\mathbf{\Sigma}_2$, against the alternative hypothesis H_1, the means are not equal, that is,

$$H_2: \boldsymbol{\mu}_1 = \boldsymbol{\mu}_2 = \boldsymbol{\mu}, \mathbf{\Sigma}_1, \mathbf{\Sigma}_2, \qquad H_1: \boldsymbol{\mu}_1, \boldsymbol{\mu}_2, \mathbf{\Sigma}_1, \mathbf{\Sigma}_2. \tag{4.1}$$

For the conjugate distribution with $\theta^* = (\bar{\mathbf{x}}_1, \bar{\mathbf{x}}_2, \mathbf{S}_1, \mathbf{S}_2)$, and with the notation in section 2, we have

$$(4.2)\quad \hat{I}(*:2) = \hat{\boldsymbol{\tau}}_1'\bar{\mathbf{x}}_1 - \hat{\boldsymbol{\tau}}_1'\boldsymbol{\mu} - \frac{1}{2}\hat{\boldsymbol{\tau}}_1'\frac{1}{n_1}\boldsymbol{\Sigma}_1\hat{\boldsymbol{\tau}}_1 + \operatorname{tr}\hat{\mathbf{T}}_1\mathbf{S}_1$$
$$+ \frac{N_1}{2}\log\left|\mathbf{I}_k - \frac{2}{N_1}\boldsymbol{\Sigma}_1\hat{\mathbf{T}}_1\right| + \hat{\boldsymbol{\tau}}_2'\bar{\mathbf{x}}_2 - \hat{\boldsymbol{\tau}}_2'\boldsymbol{\mu} - \frac{1}{2}\hat{\boldsymbol{\tau}}_2'\frac{1}{n_2}\boldsymbol{\Sigma}_2\hat{\boldsymbol{\tau}}_2$$
$$+ \operatorname{tr}\hat{\mathbf{T}}_2\mathbf{S}_2 + \frac{N_2}{2}\log\left|\mathbf{I}_k - \frac{2}{N_2}\boldsymbol{\Sigma}_2\hat{\mathbf{T}}_2\right|.$$

Following the procedure in section 2, we find that [cf. (2.4) and (2.7)]

$$(4.3)\qquad \hat{\boldsymbol{\tau}}_1 = n_1\boldsymbol{\Sigma}_1^{-1}(\bar{\mathbf{x}}_1 - \boldsymbol{\mu}), \qquad \hat{\boldsymbol{\tau}}_2 = n_2\boldsymbol{\Sigma}_2^{-1}(\bar{\mathbf{x}}_2 - \boldsymbol{\mu}),$$
$$\hat{\mathbf{T}}_1 = \frac{N_1}{2}(\boldsymbol{\Sigma}_1^{-1} - \mathbf{S}_1^{-1}), \qquad \hat{\mathbf{T}}_2 = \frac{N_2}{2}(\boldsymbol{\Sigma}_2^{-1} - \mathbf{S}_2^{-1}),$$

and (4.2) becomes

$$(4.4)\quad \hat{I}(*:2) = \frac{n_1}{2}(\bar{\mathbf{x}}_1 - \boldsymbol{\mu})'\boldsymbol{\Sigma}_1^{-1}(\bar{\mathbf{x}}_1 - \boldsymbol{\mu}) + \frac{n_2}{2}(\bar{\mathbf{x}}_2 - \boldsymbol{\mu})'\boldsymbol{\Sigma}_2^{-1}(\bar{\mathbf{x}}_2 - \boldsymbol{\mu})$$
$$+ \frac{N_1}{2}\left(\log\frac{|\boldsymbol{\Sigma}_1|}{|\mathbf{S}_1|} - k + \operatorname{tr}\mathbf{S}_1\boldsymbol{\Sigma}_1^{-1}\right) + \frac{N_2}{2}\left(\log\frac{|\boldsymbol{\Sigma}_2|}{|\mathbf{S}_2|} - k + \operatorname{tr}\mathbf{S}_2\boldsymbol{\Sigma}_2^{-1}\right).$$

The null hypothesis H_2 specifies equality of the means with no specification on the covariance matrices. For variations of $\boldsymbol{\Sigma}_1$ and $\boldsymbol{\Sigma}_2$, $\hat{I}(*:2)$ is a minimum for $\hat{\boldsymbol{\Sigma}}_1 = \mathbf{S}_1$, $\hat{\boldsymbol{\Sigma}}_2 = \mathbf{S}_2$, and for $\hat{\boldsymbol{\mu}}$ satisfying

$$(4.5)\qquad \mathbf{0} = n_1\mathbf{S}_1^{-1}(\bar{\mathbf{x}}_1 - \hat{\boldsymbol{\mu}}) + n_2\mathbf{S}_2^{-1}(\bar{\mathbf{x}}_2 - \hat{\boldsymbol{\mu}}),$$

or

$$(4.6)\qquad \hat{\boldsymbol{\mu}} = (n_1\mathbf{S}_1^{-1} + n_2\mathbf{S}_2^{-1})^{-1}(n_1\mathbf{S}_1^{-1}\bar{\mathbf{x}}_1 + n_2\mathbf{S}_2^{-1}\bar{\mathbf{x}}_2).$$

For convenience let $\mathbf{d} = \bar{\mathbf{x}}_1 - \bar{\mathbf{x}}_2$, $\mathbf{A} = n_1\mathbf{S}_1^{-1}$, $\mathbf{B} = n_2\mathbf{S}_2^{-1}$, and substituting in (4.4) we get

$$(4.7)\quad 2\hat{I}(H_1:H_2) = \operatorname{tr}[(\mathbf{B}(\mathbf{A}+\mathbf{B})^{-1}\mathbf{A}(\mathbf{A}+\mathbf{B})^{-1}\mathbf{B}$$
$$+ \mathbf{A}(\mathbf{A}+\mathbf{B})^{-1}\mathbf{B}(\mathbf{A}+\mathbf{B})^{-1}\mathbf{A})\mathbf{d}\mathbf{d}'].$$

But

$$\mathbf{B}(\mathbf{A}+\mathbf{B})^{-1}\mathbf{A} = (\mathbf{A}^{-1}(\mathbf{A}+\mathbf{B})\mathbf{B}^{-1})^{-1} = (\mathbf{B}^{-1} + \mathbf{A}^{-1})^{-1}$$

and

$$\mathbf{A}(\mathbf{A}+\mathbf{B})^{-1}\mathbf{B} = (\mathbf{B}^{-1}(\mathbf{A}+\mathbf{B})\mathbf{A}^{-1})^{-1} = (\mathbf{B}^{-1} + \mathbf{A}^{-1})^{-1},$$

so that finally

$$(4.8) \qquad 2\hat{I}(H_1:H_2) = \mathrm{tr}\,[(\mathbf{B}^{-1} + \mathbf{A}^{-1})^{-1}\mathbf{dd}'] = \mathbf{d}'(\mathbf{B}^{-1} + \mathbf{A}^{-1})^{-1}\mathbf{d} = (\bar{\mathbf{x}}_1 - \bar{\mathbf{x}}_2)'\left(\frac{1}{n_1}\mathbf{S}_1 + \frac{1}{n_2}\mathbf{S}_2\right)^{-1}(\bar{\mathbf{x}}_1 - \bar{\mathbf{x}}_2).$$

We find that here $\hat{J}(H_1, H_2) = 2\hat{I}(H_1:H_2)$.

[For single-variate populations cf. Fisher (1939a), Gronow (1951), Welch (1938). For the multivariate Behrens–Fisher problem, cf. Anderson (1958, pp. 118–122), James (1954, pp. 37–38).]

The distribution of $2\hat{I}(H_1:H_2)$ is given for r samples in section 4.3.

4.2. Linear Discriminant Function

Consider $y = \boldsymbol{\alpha}'\mathbf{x} = \alpha_1 x_1 + \alpha_2 x_2 + \cdots + \alpha_k x_k$, the same linear compound for each sample. Since y is normally distributed, we seek $\boldsymbol{\alpha}$ maximizing

$$(4.9) \qquad 2\hat{I}(H_1:H_2; y) = \frac{\boldsymbol{\alpha}'\mathbf{dd}'\boldsymbol{\alpha}}{\boldsymbol{\alpha}'\left(\frac{1}{n_1}\mathbf{S}_1 + \frac{1}{n_2}\mathbf{S}_2\right)\boldsymbol{\alpha}}.$$

As may be determined (cf. section 5 of chapter 9), the maximum occurs for $\boldsymbol{\alpha} = \left(\frac{1}{n_1}\mathbf{S}_1 + \frac{1}{n_2}\mathbf{S}_2\right)^{-1}\mathbf{d}$ and $2\hat{I}(H_1:H_2; y) = 2\hat{I}(H_1:H_2)$.

4.3. r Samples

Suppose we have r independent samples of n_i, $i = 1, 2, \cdots, r$, independent observations from k-variate normal populations with covariance matrices $\boldsymbol{\Sigma}_i$, $i = 1, 2, \cdots, r$. We want to test the null hypothesis H_2, the population mean vectors (matrices) are equal, with no specification about the $\boldsymbol{\Sigma}_i$, against the alternative hypothesis H_1, the means are not equal, that is,

$$(4.10) \qquad H_2: \boldsymbol{\mu}_1 = \boldsymbol{\mu}_2 = \cdots = \boldsymbol{\mu}_r = \boldsymbol{\mu}, \boldsymbol{\Sigma}_1, \boldsymbol{\Sigma}_2, \cdots, \boldsymbol{\Sigma}_r,$$
$$H_1: \boldsymbol{\mu}_1, \cdots, \boldsymbol{\mu}_r, \boldsymbol{\Sigma}_1, \cdots, \boldsymbol{\Sigma}_r.$$

Without repeating the details, we find here that

$$(4.11) \qquad 2\hat{I}(*:H_2) = \sum_{i=1}^{r} n_i(\bar{\mathbf{x}}_i - \boldsymbol{\mu})'\mathbf{S}_i^{-1}(\bar{\mathbf{x}}_i - \boldsymbol{\mu}).$$

As in other tests of homogeneity for several samples, here too the null hypothesis can be expressed as the intersection of two hypotheses, one specifying the homogeneity and the other specifying the common parameters of the populations. Let $H_2(\cdot)$ be the null hypothesis specifying homogeneity, and $H_2(\boldsymbol{\mu})$ the null hypothesis specifying the population means of the homogeneous samples, in each case with no specification of the covariance matrices, so that $H_2 = H_2(\cdot) \cap H_2(\boldsymbol{\mu})$.

Since the minimum of $2\hat{I}(*:H_2)$ in (4.11) is given for

$$\hat{\boldsymbol{\mu}} = \left(\sum_{i=1}^{r} n_i \mathbf{S}_i^{-1}\right)^{-1}\left(\sum_{i=1}^{r} n_i \mathbf{S}_i^{-1}\bar{\mathbf{x}}_i\right) = \hat{\mathbf{x}},$$

we have that $\hat{I}(H_1:H_2(\cdot)) = \min_{\boldsymbol{\mu}} \hat{I}(*:H_2)$ is

$$(4.12) \qquad 2\hat{I}(H_1:H_2(\cdot)) = \sum_{i=1}^{r} n_i(\bar{\mathbf{x}}_i - \hat{\mathbf{x}})'\mathbf{S}_i^{-1}(\bar{\mathbf{x}}_i - \hat{\mathbf{x}})$$
$$= \sum_{i=1}^{r} n_i \bar{\mathbf{x}}_i'\mathbf{S}_i^{-1}\bar{\mathbf{x}}_i - \hat{\mathbf{x}}'\left(\sum_{i=1}^{r} n_i \mathbf{S}_i^{-1}\right)\hat{\mathbf{x}}.$$

From (4.11) and (4.12), we have

$$(4.13)\ 2\hat{I}(*:H_2) = \sum_{i=1}^{r} n_i(\bar{\mathbf{x}}_i - \boldsymbol{\mu})'\mathbf{S}_i^{-1}(\bar{\mathbf{x}}_i - \boldsymbol{\mu})$$
$$= \sum_{i=1}^{r} n_i(\bar{\mathbf{x}}_i - \hat{\mathbf{x}})'\mathbf{S}_i^{-1}(\bar{\mathbf{x}}_i - \hat{\mathbf{x}}) + (\hat{\mathbf{x}} - \boldsymbol{\mu})'\left(\sum_{i=1}^{r} n_i \mathbf{S}_i^{-1}\right)(\hat{\mathbf{x}} - \boldsymbol{\mu})$$
$$= 2\hat{I}(H_1:H_2(\cdot)) + 2\hat{I}(H_2(\cdot):H_2(\boldsymbol{\mu})),$$

with $2\hat{I}(H_1:H_2(\cdot))$ a test for the homogeneity and $2\hat{I}(H_2(\cdot):H_2(\boldsymbol{\mu}))$ a test for the means of the homogeneous samples. This analysis is summarized in table 4.1.

TABLE 4.1

Component due to	Information	D.F.
Between, $\hat{\mathbf{x}}$ against $\boldsymbol{\mu}$	$(\hat{\mathbf{x}} - \boldsymbol{\mu})'\left(\sum_{i=1}^{r} n_i \mathbf{S}_i^{-1}\right)(\hat{\mathbf{x}} - \boldsymbol{\mu})$	k
Within, $2\hat{I}(H_1:H_2(\cdot))$	$\sum_{i=1}^{r} n_i(\bar{\mathbf{x}}_i - \hat{\mathbf{x}})'\mathbf{S}_i^{-1}(\bar{\mathbf{x}}_i - \hat{\mathbf{x}})$	$(r-1)k$
Total, $2\hat{I}(*:H_2)$	$\sum_{i=1}^{r} n_i(\bar{\mathbf{x}}_i - \boldsymbol{\mu})'\mathbf{S}_i^{-1}(\bar{\mathbf{x}}_i - \boldsymbol{\mu})$	rk

The degrees of freedom in table 4.1 are those of the asymptotic χ^2-distributions under the null hypothesis. [Cf. Hsu (1949, pp. 394–396), James (1954, pp. 39–40).]

James (1954) has shown that a better approximation to the distribution is obtained by comparing $2\hat{I}(H_1:H_2(\cdot))$, for a $100\alpha\%$ significance level, with $\chi_\alpha^2(A + B\chi_\alpha^2)$, rather than with χ_α^2, where

$$(4.14)\quad A = 1 + \frac{1}{2k(r-1)}\sum_{i=1}^{r}\frac{1}{(n_i-1)}\left[\operatorname{tr}\left(\mathbf{I}_k - \left(\sum_{i=1}^{r} n_i\mathbf{S}_i^{-1}\right)^{-1} n_i\mathbf{S}_i^{-1}\right)\right]^2,$$

$$B = \frac{1}{k(r-1)(k(r-1)+2)}\left[\sum_{i=1}^{r}\frac{1}{(n_i-1)}\operatorname{tr}\left(\mathbf{I}_k - \left(\sum_{i=1}^{r} n_i\mathbf{S}_i^{-1}\right)^{-1} n_i\mathbf{S}_i^{-1}\right)^2 + (A-1)k(r-1)\right].$$

Example 4.1. Kossack (1945) discussed the problem of classifying an A.S.T.P. (Army Specialized Training Program) pre-engineering trainee as to whether he would do unsatisfactory or satisfactory work in his first-term mathematics course. The three variables are x_1, a mathematics placement test score; x_2, a high-school mathematics score; x_3, the Army General Classification Test score. There were 96 trainees who did unsatisfactory work and 209 who performed satisfactory work. We shall find the linear discriminant function as in section 4.2. Here $k = 3$, $n_1 = 96$, $n_2 = 209$. Kossack (1945, p. 96) gives the following data:

$$\mathbf{d}' = (-17.5972,\quad -1.7997,\quad -5.3308),$$

$$\mathbf{S}_1 = \begin{pmatrix} 133.8592 & 7.0572 & 2.0717 \\ 7.0572 & 4.1288 & -2.0109 \\ 2.0717 & -2.0109 & 27.7016 \end{pmatrix}, \qquad \mathbf{S}_2 = \begin{pmatrix} 217.1505 & 14.0692 & 35.7085 \\ 14.0692 & 3.9820 & 0.4031 \\ 35.7085 & 0.4031 & 72.7206 \end{pmatrix}.$$

We now calculate (the computations were carried out by J. H. Kullback)

$$\frac{1}{n_1}\mathbf{S}_1 = \begin{pmatrix} 1.39436676 & 0.073512939 & 0.021580263 \\ 0.073512939 & 0.043008772 & -0.020946382 \\ 0.021580263 & -0.020946382 & 0.288558772 \end{pmatrix},$$

$$\frac{1}{n_2}\mathbf{S}_2 = \begin{pmatrix} 1.038997768 & 0.067316709 & 0.170853859 \\ 0.067316709 & 0.019052470 & 0.001928528 \\ 0.170853859 & 0.001928528 & 0.347945253 \end{pmatrix},$$

$$\left(\frac{1}{n_1}\mathbf{S}_1 + \frac{1}{n_2}\mathbf{S}_2\right)^{-1} = \begin{pmatrix} 0.493634948 & -1.176664865 & -0.184397647 \\ -1.176664865 & 19.066796354 & 0.925430152 \\ -0.184397647 & 0.925430152 & 1.654481546 \end{pmatrix},$$

$$\begin{pmatrix} \alpha_1 \\ \alpha_2 \\ \alpha_3 \end{pmatrix} = \begin{pmatrix} 0.493634948 & -1.176664865 & -0.184397647 \\ -1.176664865 & 19.066796354 & 0.925430152 \\ -0.184397647 & 0.925430152 & 1.654481546 \end{pmatrix}\begin{pmatrix} -17.5972 \\ -1.7997 \\ -5.3308 \end{pmatrix}$$

$$= \begin{pmatrix} -5.58596 \\ -18.54179 \\ -7.24032 \end{pmatrix},$$

$$2\hat{I}(H_1:H_2;y) = 2\hat{I}(H_1:H_2) = (-17.5972,\ -1.7997,\ -5.3308)\begin{pmatrix} -5.58596 \\ -18.54179 \\ -7.24032 \end{pmatrix}$$

$$= 170.2637.$$

The linear discriminant function may be expressed as $y = x_1 + 3.32x_2 + 1.29x_3$ with the ratios of the α's to α_1 as coefficients.

Kossack (1945) obtained the coefficients of a linear discriminant function from $\boldsymbol{\alpha} = \mathbf{S}^{-1}\mathbf{d}$, where $N\mathbf{S} = N_1\mathbf{S}_1 + N_2\mathbf{S}_2$, $N = N_1 + N_2$. [Cf. Fisher (1936).] This is the procedure, for $r = 2$, discussed in section 8.1 of chapter 11, when the population covariance matrices are assumed to be equal. The linear discriminant function obtained by Kossack (1945) may be written as $y = x_1 + 3.69x_2 + 0.93x_3$. Using Kossack's pooling procedure and his result that $\mathbf{d}'\mathbf{S}^{-1}\mathbf{d} = 1.9890$, we compute

$$2\hat{I}(H_1:H_2) = \frac{n_1 n_2}{n_1 + n_2}\mathbf{d}'\mathbf{S}^{-1}\mathbf{d} = \frac{96 \times 209}{305}1.9890 = 130.8637,$$

a smaller value than that computed above when the covariance matrices were not pooled. (We shall see in example 5.2 that the null hypothesis that the population covariance matrices are equal should be rejected.)

Example 4.2. To illustrate the test for the null hypothesis of homogeneity of means, we use the following data and computations from James (1954, pp. 42–43). (I have expressed the results in the notation of section 4.3.) There are three bivariate samples, with $n_1 = 16$, $n_2 = 11$, $n_3 = 11$:

$$\bar{\mathbf{x}}_1 = \begin{pmatrix} 9.82 \\ 15.06 \end{pmatrix}, \quad \bar{\mathbf{x}}_2 = \begin{pmatrix} 13.05 \\ 22.57 \end{pmatrix}, \quad \bar{\mathbf{x}}_3 = \begin{pmatrix} 14.67 \\ 25.17 \end{pmatrix},$$

$$\mathbf{S}_1 = \begin{pmatrix} 120.0 & -16.3 \\ -16.3 & 17.8 \end{pmatrix}, \quad \mathbf{S}_2 = \begin{pmatrix} 81.8 & 32.1 \\ 32.1 & 53.8 \end{pmatrix}, \quad \mathbf{S}_3 = \begin{pmatrix} 100.3 & 23.2 \\ 23.2 & 97.1 \end{pmatrix},$$

$$n_1\mathbf{S}_1^{-1} = \begin{pmatrix} 0.1523 & 0.1396 \\ 0.1396 & 1.0272 \end{pmatrix}, \quad n_2\mathbf{S}_2^{-1} = \begin{pmatrix} 0.1756 & -0.1048 \\ -0.1048 & 0.2670 \end{pmatrix},$$

$$n_3\mathbf{S}_3^{-1} = \begin{pmatrix} 0.1161 & -0.0277 \\ -0.0277 & 0.1199 \end{pmatrix},$$

$$\sum_{i=1}^{3} n_i\mathbf{S}_i^{-1} = \begin{pmatrix} 0.4440 & 0.0071 \\ 0.0071 & 1.4141 \end{pmatrix}, \quad \left(\sum_{i=1}^{3} n_i\mathbf{S}_i^{-1}\right)^{-1} = \begin{pmatrix} 2.2524 & -0.0113 \\ -0.0113 & 0.7072 \end{pmatrix},$$

$$n_1\mathbf{S}_1^{-1}\bar{\mathbf{x}}_1 = \begin{pmatrix} 3.5980 \\ 16.8405 \end{pmatrix}, \quad n_2\mathbf{S}_2^{-1}\bar{\mathbf{x}}_2 = \begin{pmatrix} -0.0738 \\ 4.6586 \end{pmatrix}, \quad n_3\mathbf{S}_3^{-1}\bar{\mathbf{x}}_3 = \begin{pmatrix} 1.0060 \\ 2.6115 \end{pmatrix},$$

$$\sum_{i=1}^{3} n_i\mathbf{S}_i^{-1}\bar{\mathbf{x}}_i = \begin{pmatrix} 4.5302 \\ 24.1106 \end{pmatrix}, \quad \hat{\mathbf{x}} = \begin{pmatrix} 2.2524 & -0.0113 \\ -0.0113 & 0.7072 \end{pmatrix}\begin{pmatrix} 4.5302 \\ 24.1106 \end{pmatrix} = \begin{pmatrix} 9.9314 \\ 16.9998 \end{pmatrix},$$

$$2\hat{I}(H_1:H_2(\cdot)) = (9.82, 15.06)\begin{pmatrix} 3.5980 \\ 16.8405 \end{pmatrix} + (13.05, 22.57)\begin{pmatrix} -0.0738 \\ 4.6586 \end{pmatrix}$$

$$+ (14.67, 25.17)\begin{pmatrix} 1.0060 \\ 2.6115 \end{pmatrix} - (9.9314, 16.9998)\begin{pmatrix} 0.4440 & 0.0071 \\ 0.0071 & 1.4141 \end{pmatrix}\begin{pmatrix} 9.9314 \\ 16.9998 \end{pmatrix}$$

$$= 18.75.$$

Asymptotically, $2\hat{I}(H_1:H_2(\cdot)) = 18.75$ is a χ^2 with $(r-1)k = 4$ degrees of freedom. For a better approximation to the significance levels, we find

$$\begin{pmatrix} 1 & 0 \\ 0 & 1 \end{pmatrix} - \begin{pmatrix} 2.2524 & -0.0113 \\ -0.0113 & 0.7072 \end{pmatrix}\begin{pmatrix} 0.1523 & 0.1396 \\ 0.1396 & 1.0272 \end{pmatrix} = \begin{pmatrix} 0.6585 & -0.3028 \\ -0.0970 & 0.2751 \end{pmatrix},$$

$$\begin{pmatrix} 1 & 0 \\ 0 & 1 \end{pmatrix} - \begin{pmatrix} 2.2524 & -0.0113 \\ -0.0113 & 0.7072 \end{pmatrix}\begin{pmatrix} 0.1756 & -0.1048 \\ -0.1048 & 0.2670 \end{pmatrix} = \begin{pmatrix} 0.6033 & 0.2391 \\ 0.0761 & 0.8100 \end{pmatrix},$$

$$\begin{pmatrix} 1 & 0 \\ 0 & 1 \end{pmatrix} - \begin{pmatrix} 2.2524 & -0.0113 \\ -0.0113 & 0.7072 \end{pmatrix}\begin{pmatrix} 0.1161 & -0.0277 \\ -0.0277 & 0.1199 \end{pmatrix} = \begin{pmatrix} 0.7382 & 0.0637 \\ 0.0209 & 0.9149 \end{pmatrix},$$

$$\operatorname{tr}\begin{pmatrix} 0.6585 & -0.3028 \\ -0.0970 & 0.2751 \end{pmatrix}^2 = 0.5680, \qquad \operatorname{tr}\begin{pmatrix} 0.6033 & 0.2391 \\ 0.0761 & 0.8100 \end{pmatrix}^2 = 1.0565,$$

$$\operatorname{tr}\begin{pmatrix} 0.7382 & 0.0637 \\ 0.0209 & 0.9149 \end{pmatrix}^2 = 1.3846, \qquad \left[\operatorname{tr}\begin{pmatrix} 0.6585 & -0.3028 \\ -0.0970 & 0.2751 \end{pmatrix}\right]^2 = 0.8716,$$

$$\left[\operatorname{tr}\begin{pmatrix} 0.6033 & 0.2391 \\ 0.0761 & 0.8100 \end{pmatrix}\right]^2 = 1.9974, \qquad \left[\operatorname{tr}\begin{pmatrix} 0.7382 & 0.0637 \\ 0.0209 & 0.9149 \end{pmatrix}\right]^2 = 2.7327,$$

$$\frac{0.5680}{15} + \frac{1.0565}{10} + \frac{1.3846}{10} = 0.2820, \qquad \frac{0.8716}{15} + \frac{1.9974}{10} + \frac{2.7327}{10} = 0.5311,$$

$$A = 1 + \tfrac{1}{8}(0.5311) = 1.0664, \qquad B = \tfrac{1}{24}(0.2820 + \tfrac{1}{2}(0.5311)) = 0.02281.$$

The appropriate comparison value for the 5%, 1%, 0.1% significance level is then obtained from:

Significance Level	χ^2, 4 d.f.	$A + B\chi^2$	$\chi^2(A + B\chi^2)$
5%	9.488	1.283	12.17
1%	13.277	1.369	18.18
0.1%	18.467	1.488	27.48

that is, the corrected comparison value for the 5% level is 12.17, for the 1% level is 18.18, and for the 0.1% level is 27.48. The null hypothesis of homogeneity would be rejected at the 1% level.

5. HOMOGENEITY OF COVARIANCE MATRICES

We shall now examine the test for the null hypothesis of equality of the covariance matrices of r k-variate normal populations. For its own interest, and as an introduction, we consider two samples first and then r samples.

5.1. Two Samples

Suppose we have two independent samples with n_1 and n_2 independent observations from k-variate normal populations with no specification about the means. For the population covariance matrices we have the two hypotheses $H_1: \mathbf{\Sigma}_1 \neq \mathbf{\Sigma}_2$ and $H_2: \mathbf{\Sigma}_1 = \mathbf{\Sigma}_2 = \mathbf{\Sigma}$.

For the conjugate distribution with $\theta^* = (\bar{\mathbf{x}}_1, \bar{\mathbf{x}}_2, \mathbf{S}_1, \mathbf{S}_2)$, and with the notation in section 2, we have [cf. (4.2)]

$$(5.1)\quad \hat{I}(*:2) = \hat{\boldsymbol{\tau}}_1'\bar{\mathbf{x}}_1 - \hat{\boldsymbol{\tau}}_1'\boldsymbol{\mu}_1 - \frac{1}{2}\hat{\boldsymbol{\tau}}_1'\frac{1}{n_1}\mathbf{\Sigma}\hat{\boldsymbol{\tau}}_1 + \operatorname{tr}\hat{\mathbf{T}}_1\mathbf{S}_1 + \frac{N_1}{2}\log\left|\mathbf{I}_k - \frac{2}{N_1}\mathbf{\Sigma}\hat{\mathbf{T}}_1\right| + \hat{\boldsymbol{\tau}}_2'\bar{\mathbf{x}}_2 - \hat{\boldsymbol{\tau}}_2'\boldsymbol{\mu}_2 - \frac{1}{2}\hat{\boldsymbol{\tau}}_2'\frac{1}{n_2}\mathbf{\Sigma}\hat{\boldsymbol{\tau}}_2 + \operatorname{tr}\hat{\mathbf{T}}_2\mathbf{S}_2 + \frac{N_2}{2}\log\left|\mathbf{I}_k - \frac{2}{N_2}\mathbf{\Sigma}\hat{\mathbf{T}}_2\right|.$$

Using the same procedure as for (4.2), we find that [cf. (4.3)]

$$(5.2)\qquad \hat{\boldsymbol{\tau}}_1 = n_1\boldsymbol{\Sigma}^{-1}(\bar{\mathbf{x}}_1 - \boldsymbol{\mu}_1),\qquad \hat{\boldsymbol{\tau}}_2 = n_2\boldsymbol{\Sigma}^{-1}(\bar{\mathbf{x}}_2 - \boldsymbol{\mu}_2),$$

$$\hat{\mathbf{T}}_1 = \frac{N_1}{2}(\boldsymbol{\Sigma}^{-1} - \mathbf{S}_1^{-1}),\qquad \hat{\mathbf{T}}_2 = \frac{N_2}{2}(\boldsymbol{\Sigma}^{-1} - \mathbf{S}_2^{-1}),$$

and (5.1) becomes [cf. (4.4)]

$$(5.3)\quad \hat{I}(*:2) = \frac{n_1}{2}(\bar{\mathbf{x}}_1 - \boldsymbol{\mu}_1)'\boldsymbol{\Sigma}^{-1}(\bar{\mathbf{x}}_1 - \boldsymbol{\mu}_1) + \frac{n_2}{2}(\bar{\mathbf{x}}_2 - \boldsymbol{\mu}_2)'\boldsymbol{\Sigma}^{-1}(\bar{\mathbf{x}}_2 - \boldsymbol{\mu}_2)$$
$$+ \frac{N_1}{2}\left(\log\frac{|\boldsymbol{\Sigma}|}{|\mathbf{S}_1|} - k + \operatorname{tr}\mathbf{S}_1\boldsymbol{\Sigma}^{-1}\right) + \frac{N_2}{2}\left(\log\frac{|\boldsymbol{\Sigma}|}{|\mathbf{S}_2|} - k + \operatorname{tr}\mathbf{S}_2\boldsymbol{\Sigma}^{-1}\right).$$

For variations of $\boldsymbol{\mu}_1$, $\boldsymbol{\mu}_2$, and $\boldsymbol{\Sigma}$, $\hat{I}(*:2)$ will be a minimum for $\hat{\boldsymbol{\mu}}_1$, $\hat{\boldsymbol{\mu}}_2$, and $\hat{\boldsymbol{\Sigma}}$ satisfying [see problems 10.2, 10.3 in chapter 9, Deemer and Olkin (1951), for the matrix differentiation]

$$(5.4)\quad n_1\hat{\boldsymbol{\Sigma}}^{-1}(\bar{\mathbf{x}}_1 - \hat{\boldsymbol{\mu}}_1) = \mathbf{0},\qquad n_2\hat{\boldsymbol{\Sigma}}^{-1}(\bar{\mathbf{x}}_2 - \hat{\boldsymbol{\mu}}_2) = \mathbf{0},$$

$$\mathbf{0} = -\frac{n_1}{2}(\bar{\mathbf{x}}_1 - \hat{\boldsymbol{\mu}}_1)'\hat{\boldsymbol{\Sigma}}^{-1}(d\boldsymbol{\Sigma})\hat{\boldsymbol{\Sigma}}^{-1}(\bar{\mathbf{x}}_1 - \hat{\boldsymbol{\mu}}_1) - \frac{n_2}{2}(\bar{\mathbf{x}}_2 - \hat{\boldsymbol{\mu}}_2)'\hat{\boldsymbol{\Sigma}}^{-1}(d\boldsymbol{\Sigma})\hat{\boldsymbol{\Sigma}}^{-1}(\bar{\mathbf{x}}_2 - \hat{\boldsymbol{\mu}}_2)$$
$$+ \frac{N_1}{2}\operatorname{tr}\hat{\boldsymbol{\Sigma}}^{-1}(d\boldsymbol{\Sigma}) - \frac{N_1}{2}\operatorname{tr}\mathbf{S}_1\hat{\boldsymbol{\Sigma}}^{-1}(d\boldsymbol{\Sigma})\hat{\boldsymbol{\Sigma}}^{-1}$$
$$+ \frac{N_2}{2}\operatorname{tr}\hat{\boldsymbol{\Sigma}}^{-1}(d\boldsymbol{\Sigma}) - \frac{N_2}{2}\operatorname{tr}\mathbf{S}_2\hat{\boldsymbol{\Sigma}}^{-1}(d\boldsymbol{\Sigma})\hat{\boldsymbol{\Sigma}}^{-1}.$$

We find that

$$(5.5)\quad \hat{\boldsymbol{\mu}}_1 = \bar{\mathbf{x}}_1,\qquad \hat{\boldsymbol{\mu}}_2 = \bar{\mathbf{x}}_2,\qquad (N_1 + N_2)\hat{\boldsymbol{\Sigma}} = N_1\mathbf{S}_1 + N_2\mathbf{S}_2 = N\mathbf{S},$$

where $N = N_1 + N_2$, and consequently [cf. Wilks (1932, p. 489)]

$$(5.6)\qquad 2\hat{I}(H_1:H_2) = N_1\log\frac{|\mathbf{S}|}{|\mathbf{S}_1|} + N_2\log\frac{|\mathbf{S}|}{|\mathbf{S}_2|}.$$

It is found that the estimate $\hat{J}(H_1, H_2)$ is [cf. Kullback (1952, p. 91), and equation (1.7) in chapter 9]

$$(5.7)\qquad \hat{J}(H_1, H_2) = \frac{N_1N_2}{2(N_1 + N_2)}(\operatorname{tr}\mathbf{S}_1\mathbf{S}_2^{-1} + \operatorname{tr}\mathbf{S}_2\mathbf{S}_1^{-1} - 2k).$$

In accordance with the general asymptotic theory, under the null hypothesis H_2, $2\hat{I}(H_1:H_2)$ in (5.6) asymptotically is distributed as χ^2 with $k(k + 1)/2$ degrees of freedom. Using the characteristic function of the distribution of $2\hat{I}(H_1:H_2)$, it may be shown (see section 6.1) that a better approximation to the distribution is R. A. Fisher's B-distribution [Fisher (1928, p. 665)], the noncentral χ^2-distribution, where for Fisher's distribution $\beta^2 = \frac{(2k^3 + 3k^2 - k)}{12}\left(\frac{1}{N_1} + \frac{1}{N_2} - \frac{1}{N}\right)$, $B^2 = 2\hat{I}(H_1:H_2)$, with $k(k+1)/2$ degrees of freedom.

5.2. Linear Discriminant Function

(Cf. section 3.5.) We seek a linear compound, the same for both samples, $y = \boldsymbol{\alpha}'\mathbf{x} = \alpha_1 x_1 + \alpha_2 x_2 + \cdots + \alpha_k x_k$, that maximizes [see (5.7)]

$$\hat{J}(H_1, H_2; y) = \frac{N_1 N_2}{2(N_1 + N_2)}\left(\frac{\boldsymbol{\alpha}'\mathbf{S}_1\boldsymbol{\alpha}}{\boldsymbol{\alpha}'\mathbf{S}_2\boldsymbol{\alpha}} + \frac{\boldsymbol{\alpha}'\mathbf{S}_2\boldsymbol{\alpha}}{\boldsymbol{\alpha}'\mathbf{S}_1\boldsymbol{\alpha}} - 2\right). \tag{5.8}$$

We find (by the usual calculus procedures) that $\boldsymbol{\alpha}$ satisfies $\mathbf{S}_1\boldsymbol{\alpha} = F\mathbf{S}_2\boldsymbol{\alpha}$, where F is a root of the determinantal equation $|\mathbf{S}_1 - F\mathbf{S}_2| = |N_1\mathbf{S}_1 - lN_2\mathbf{S}_2| = 0$, and $F = N_2 l/N_1$ (cf. section 6 of chapter 9). The same linear function results from maximizing [see (5.6)]

$$\hat{I}(H_1:H_2; y) = \frac{N_1}{2}\log\frac{\boldsymbol{\alpha}'\mathbf{S}\boldsymbol{\alpha}}{\boldsymbol{\alpha}'\mathbf{S}_1\boldsymbol{\alpha}} + \frac{N_2}{2}\log\frac{\boldsymbol{\alpha}'\mathbf{S}\boldsymbol{\alpha}}{\boldsymbol{\alpha}'\mathbf{S}_2\boldsymbol{\alpha}}. \tag{5.9}$$

If the roots of the determinantal equation, which are almost everywhere positive, are $F_1, F_2, \cdots, F_k$ arranged in ascending order, then, as was shown in section 6 of chapter 9, the maximum of $\hat{J}(H_1, H_2; y)$ occurs for the linear compound associated with F_1 or F_k according as $F_1F_k < 1$ or $F_1F_k > 1$.

It may also be shown that

$$\hat{I}(H_1:H_2) = \hat{I}(H_1:H_2; l_1) + \hat{I}(H_1:H_2; l_2) + \cdots + \hat{I}(H_1:H_2; l_k), \tag{5.10}$$
$$\hat{J}(H_1, H_2) = \hat{J}(H_1, H_2; F_1) + \hat{J}(H_1, H_2; F_2) + \cdots + \hat{J}(H_1, H_2; F_k),$$

where

$$\hat{I}(H_1:H_2; l_i) = \frac{N_1}{2}\log\frac{N_1}{N_1 + N_2}\frac{1 + l_i}{l_i} + \frac{N_2}{2}\log\frac{N_2}{N_1 + N_2}(1 + l_i) \tag{5.11}$$
$$= \frac{N_1}{2}\log\frac{N_1}{N_1 + N_2} + \frac{N_2}{2}\log\frac{N_2}{N_1 + N_2}$$
$$+ \frac{N_1 + N_2}{2}\log(1 + l_i) - \frac{N_1}{2}\log l_i,$$

$$\hat{J}(H_1, H_2; F_i) = \frac{N_1 N_2}{2(N_1 + N_2)}\frac{(F_i - 1)^2}{F_i}.$$

Asymptotically, when the population parameters have the null hypothesis values, $2\hat{I}(H_1:H_2; l_{m+1}) + \cdot \cdot \cdot + 2\hat{I}(H_1:H_2; l_k)$ (the summands arranged in descending order of magnitude) is distributed as χ^2 with $(k - m)(k - m + 1)/2$ degrees of freedom. A better approximation is R. A. Fisher's B-distribution [Fisher (1928, p. 665)], the noncentral χ^2-distribution, where for Fisher's distribution

$$\beta^2 = \frac{(2k^3 + 3k^2 - k) - (2m^3 + 3m^2 - m)}{12}\left(\frac{1}{N_1} + \frac{1}{N_2} - \frac{1}{N}\right),$$

$N = N_1 + N_2$, $B^2 = \sum_{i=m+1}^{k} 2\hat{I}(H_1:H_2; l_i)$, with $(k - m)(k - m + 1)/2$ degrees of freedom. (See section 6.4.) [Cf. Anderson (1958, p. 259).]

5.3. r Samples

Suppose we have r independent samples of $n_1, n_2, \cdot \cdot \cdot, n_r$ independent observations from k-variate normal populations with no specification about the means. For the population covariance matrices we have the two hypotheses $H_1:\mathbf{\Sigma}_1, \mathbf{\Sigma}_2, \cdot \cdot \cdot, \mathbf{\Sigma}_r$ and $H_2:\mathbf{\Sigma}_1 = \mathbf{\Sigma}_2 = \cdot \cdot \cdot = \mathbf{\Sigma}_r = \mathbf{\Sigma}$.

Without repeating the details, as in section 5.1, we find that for the conjugate distribution with $\theta^* = (\bar{\mathbf{x}}_1, \cdot \cdot \cdot, \bar{\mathbf{x}}_r, \mathbf{S}_1, \cdot \cdot \cdot, \mathbf{S}_r)$

$$(5.12)\quad \hat{I}(*:2) = \sum_{i=1}^{r} \frac{n_i}{2}(\bar{\mathbf{x}}_i - \boldsymbol{\mu}_i)'\mathbf{\Sigma}^{-1}(\bar{\mathbf{x}}_i - \boldsymbol{\mu}_i) + \sum_{i=1}^{r} \frac{N_i}{2}\left(\log \frac{|\mathbf{\Sigma}|}{|\mathbf{S}_i|} - k + \operatorname{tr} \mathbf{S}_i\mathbf{\Sigma}^{-1}\right).$$

When the null hypothesis $H_2(\mathbf{\Sigma})$ specifies $\mathbf{\Sigma}$, the minimum of $\hat{I}(*:2)$ in (5.12) for variations of the $\boldsymbol{\mu}_i$, $i = 1, 2, \cdot \cdot \cdot, r$, is

$$(5.13)\quad \hat{I}(*:H_2(\mathbf{\Sigma})) = \sum_{i=1}^{r} \frac{N_i}{2}\left(\log \frac{|\mathbf{\Sigma}|}{|\mathbf{S}_i|} - k + \operatorname{tr} \mathbf{S}_i\mathbf{\Sigma}^{-1}\right).$$

This is (2.18) in chapter 9 with $\mathbf{\Sigma}$ for $\mathbf{\Sigma}_{2j}$ and $\mathbf{S}_j$ for $\mathbf{\Sigma}_{1j}$, $j = 1, 2, \cdot \cdot \cdot, r$. When the null hypothesis $H_2(\cdot)$ does not specify $\mathbf{\Sigma}$ but only the homogeneity, the minimum of $\hat{I}(*:H_2(\mathbf{\Sigma}))$ in (5.13) for variations of $\mathbf{\Sigma}$ is given for $N\hat{\mathbf{\Sigma}} = N_1\mathbf{S}_1 + \cdot \cdot \cdot + N_r\mathbf{S}_r = N\mathbf{S}$, $N = N_1 + N_2 + \cdot \cdot \cdot + N_r$, and $\hat{I}(H_1:H_2(\cdot)) = \min_{\mathbf{\Sigma}} \hat{I}(*:H_2(\mathbf{\Sigma}))$ is

$$(5.14)\quad \hat{I}(H_1:H_2(\cdot)) = \sum_{i=1}^{r} \frac{N_i}{2}\left(\log \frac{|\mathbf{S}|}{|\mathbf{S}_i|} - k + \operatorname{tr} \mathbf{S}_i\mathbf{S}^{-1}\right) = \sum_{i=1}^{r} \frac{N_i}{2} \log \frac{|\mathbf{S}|}{|\mathbf{S}_i|}.$$

[Cf. Anderson (1958, p. 249), Box (1949), Wilks (1932, p. 489).]

Note that the estimate of $J(H_1, H_2(\cdot))$ may be obtained from (2.19) of chapter 9 by replacing $\mathbf{\Sigma}_{1j}$ by $\mathbf{S}_j$ and $\mathbf{\Sigma}_{2j}$ by $\mathbf{S}$, $j = 1, 2, \cdots, r$, yielding

$$(5.15) \qquad \hat{J}(H_1, H_2(\cdot)) = \sum_{i=1}^{r} \frac{N_i}{2}(\text{tr}\, \mathbf{S}_i\mathbf{S}^{-1} + \text{tr}\, \mathbf{S}\mathbf{S}_i^{-1}) - kN$$

$$= \sum_{i=1}^{r} \frac{N_i}{2} \text{tr}\, \mathbf{S}\mathbf{S}_i^{-1} - \frac{kN}{2}$$

$$= \sum_{i<j} \frac{N_iN_j}{2N}(\text{tr}\, \mathbf{S}_i\mathbf{S}_j^{-1} + \text{tr}\, \mathbf{S}_j\mathbf{S}_i^{-1} - 2k).$$

In accordance with the general asymptotic theory, under the null hypothesis H_2, $2\hat{I}(H_1\!:H_2(\cdot))$ in (5.14) asymptotically is distributed as χ^2 with $(r - 1)k(k + 1)/2$ degrees of freedom. Using the characteristic function of the distribution of $2\hat{I}(H_1\!:H_2(\cdot))$, it may be shown (see section 6.1) that a better approximation to the distribution is R. A. Fisher's B-distribution [Fisher (1928, p. 665)], the noncentral χ^2-distribution, where for Fisher's distribution $\beta^2 = \dfrac{2k^3 + 3k^2 - k}{12}\left(\sum_{i=1}^{r} 1/N_i - 1/N\right)$, $B^2 = 2\hat{I}(H_1\!:H_2(\cdot))$, with $(r - 1)k(k + 1)/2$ degrees of freedom. For degrees of freedom greater than 7 (the largest tabulated by Fisher),

$$2\hat{I}(H_1\!:H_2(\cdot))(1 - 2\beta^2/(r - 1)k(k + 1))$$

may be treated as a χ^2 with $(r - 1)k(k + 1)/2$ degrees of freedom.

For the single-variate case, $k = 1$, we have

$$(5.16) \qquad 2\hat{I}(H_1\!:H_2(\cdot)) = \sum_{i=1}^{r} N_i \log \frac{s^2}{s_i^2},$$

where $Ns^2 = N_1s_1^2 + \cdots + N_rs_r^2$, $N = N_1 + N_2 + \cdots + N_r$, $\beta^2 = \dfrac{1}{3}\left(\sum_{i=1}^{r}\dfrac{1}{N_i} - \dfrac{1}{N}\right)$, and $\dfrac{2\beta^2}{(r - 1)k(k + 1)} = \dfrac{1}{3(r - 1)}\left(\sum_{i=1}^{r}\dfrac{1}{N_i} - \dfrac{1}{N}\right)$. These are the results for Bartlett's test for the homogeneity of variances [Bartlett (1937, 1954), Box (1949), Kempthorne (1952, p. 21), Lawley (1956)]. See the remark at the end of section 3.7.

We summarize the analysis of the discrimination information statistic in (5.13) in table 5.1.

Note that the between component in table 5.1 is the discrimination information statistic for the test of the null hypothesis $H_2(\cdot|\mathbf{\Sigma})$, the covariance matrix of homogeneous samples is $\mathbf{\Sigma}$. The analysis in table 5.1 is a reflection of the fact that $H_2(\mathbf{\Sigma}) = H_2(\cdot) \cap H_2(\cdot|\mathbf{\Sigma})$, and may be written as $2\hat{I}(*:H_2(\mathbf{\Sigma})) = 2\hat{I}(H_1:H_2(\cdot)) + 2\hat{I}(H_2(\cdot):H_2(\cdot|\mathbf{\Sigma}))$.

The degrees of freedom are those of the asymptotic χ^2-distribution or those of the better approximation given by Fisher's B-distribution, the noncentral χ^2-distribution with noncentrality parameter β^2.

TABLE 5.1

Component due to	Information	D.F.	β^2
Between S against $\mathbf{\Sigma}$	$N\left(\log \frac{\|\mathbf{\Sigma}\|}{\|\mathbf{S}\|} - k + \operatorname{tr} \mathbf{S}\mathbf{\Sigma}^{-1}\right)$	$\frac{k(k+1)}{2}$	$\frac{2k^3+3k^2-k}{12N}$
Within $2\hat{I}(H_1:H_2(\cdot))$	$\sum_{i=1}^{r} N_i \log \frac{\|\mathbf{S}\|}{\|\mathbf{S}_i\|}$	$\frac{(r-1)k(k+1)}{2}$	$\frac{2k^3+3k^2-k}{12}\left(\sum_{i=1}^{r}\frac{1}{N_i} - \frac{1}{N}\right)$
Total $2\hat{I}(*:H_2(\mathbf{\Sigma}))$	$\sum_{i=1}^{r} N_i\left(\log \frac{\|\mathbf{\Sigma}\|}{\|\mathbf{S}_i\|} - k + \operatorname{tr} \mathbf{S}_i\mathbf{\Sigma}^{-1}\right)$	$\frac{rk(k+1)}{2}$	$\frac{2k^3+3k^2-k}{12}\sum_{i=1}^{r}\frac{1}{N_i}$

5.4. Correlation Matrices

By using the minimum discrimination information statistic in (3.19) and the convexity property we may derive a test for the null hypothesis that the correlation matrices of m populations are equal. Suppose there are m independent samples of $n_1 = N_1 + 1$, $n_2 = N_2 + 1, \cdots, n_m = N_m + 1$ independent observations each from k-variate normal populations. Denote the sample correlation matrices by $\mathbf{R}_1, \mathbf{R}_2, \cdots, \mathbf{R}_m$ and the corresponding population correlation matrices by $\mathbf{P}_1, \mathbf{P}_2, \cdots, \mathbf{P}_m$. Let H_1 denote the alternative hypothesis that the population correlation matrices are not all equal, that is,

$$H_1: \mathbf{P}_1, \mathbf{P}_2, \cdots, \mathbf{P}_m; \tag{5.17}$$

let $H_2(\mathbf{P})$ denote the null hypothesis that the population correlation matrices are equal to $\mathbf{P}$, that is,

$$H_2(\mathbf{P}): \mathbf{P}_1 = \mathbf{P}_2 = \cdots = \mathbf{P}_m = \mathbf{P}; \tag{5.18}$$

and let $H_2(\cdot)$ denote the null hypothesis of homogeneity that the population covariance matrices are equal but unspecified. Since $H_2(\mathbf{P})$ is equivalent to the intersection of two hypotheses, (i) the observed correlation matrices are homogeneous and (ii) the common value of the population correlation matrix is $\mathbf{P}$, we may set up the analysis in table 5.2.

TABLE 5.2

Component due to	Information	D.F.
$\mathbf{P}$	$N(\log \frac{\lvert\mathbf{P}\rvert}{\lvert\mathbf{R}\rvert} - k + \mathrm{tr}\mathbf{R}\mathbf{P}^{-1})$	$\frac{k(k-1)}{2}$
$H_2(\cdot)$	$\sum_{i=1}^{m} N_i \log \frac{\lvert\mathbf{R}\rvert}{\lvert\mathbf{R}_i\rvert}$	$\frac{(m-1)k(k-1)}{2}$
$H_2(\mathbf{P})$	$\sum_{i=1}^{m} N_i\left(\log \frac{\lvert\mathbf{P}\rvert}{\lvert\mathbf{R}_i\rvert} - k + \mathrm{tr}\mathbf{R}_i\mathbf{P}^{-1}\right)$	$\frac{mk(k-1)}{2}$

In table 5.2,

$$(5.19)\quad N = N_1 + N_2 + \cdots + N_m,\ N\mathbf{R} = N_1\mathbf{R}_1 + N_2\mathbf{R}_2 + \cdots + N_m\mathbf{R}_m,$$

and the degrees of freedom are those of the asymptotic χ^2-distributions under the null hypothesis. The convexity property insures that

$$(5.20)\quad N\sum_{i=1}^{m}\frac{N_i}{N}\left(\log \frac{\lvert\mathbf{P}\rvert}{\lvert\mathbf{R}_i\rvert} - k + \mathrm{tr}\mathbf{R}_i\mathbf{P}^{-1}\right) \geqq N\left(\log \frac{\lvert\mathbf{P}\rvert}{\lvert\mathbf{R}\rvert} - k + \mathrm{tr}\mathbf{R}\mathbf{P}^{-1}\right).$$

For bivariate populations, $k = 2$, we have

$$(5.21)\quad 2\hat{I}(H_1 . H_2(\cdot)) = \sum_{i=1}^{m} N_i \log \frac{1 - r_{12}^2}{1 - r_{i12}^2}$$

where $Nr_{12} = \sum_{i=1}^{m} N_i r_{i12}$ and r_{i12} is the correlation coefficient in the i-th sample. The degrees of freedom for $2\hat{I}(H_1 : H_2(\cdot))$ in (5.21) are $m - 1$.

Example 5.1. We illustrate the test of a null hypothesis of homogeneity of covariance matrices with data given by Smith (1947, Table 2, p. 277) to calculate a linear discriminant function for a group of 25 normal persons and 25 psychotics. Here $k = 2$, $r = 2$, $N_1 = N_2 = 24$, $N = 48$,

$$\mathbf{S}_1 = \begin{pmatrix} 6.92 & -5.27 \\ -5.27 & 40.89 \end{pmatrix}, \qquad \mathbf{S}_2 = \begin{pmatrix} 36.75 & 13.92 \\ 13.92 & 287.92 \end{pmatrix}, \qquad \mathbf{S} = \begin{pmatrix} 21.83 & 4.33 \\ 4.33 & 164.40 \end{pmatrix},$$

$$|\mathbf{S}_1| = 255.1859, \qquad |\mathbf{S}_2| = 10387.2936, \qquad |\mathbf{S}| = 3570.1031,$$

$$2\hat{I}(H_1:H_2(\cdot)) = 24 \log (3570.1031/255.1859) + 24 \log (3570.1031/10387.2936) = 37.7019 = B^2,$$

$$\beta^2 = \frac{16 + 12 - 2}{12}\left(\frac{2}{24} - \frac{1}{48}\right) = 0.135416,$$

$$\frac{(2-1)2 \times 3}{2} = 3 \text{ degrees of freedom.}$$

In Fisher's B^2 table, Table III on page 380, the 5% values for $n = 3$ and $\beta^2 = 0.04$ and 0.16 are respectively 7.9186 and 8.2254. We therefore reject the null hypothesis of equality of the population covariance matrices. Smith (1947) does remark that the correlations are not significant, but that the variances of the psychotics are significantly greater than those of the normals.

Example 5.2. We now justify the comment at the end of example 4.1. In addition to $\mathbf{S}_1$ and $\mathbf{S}_2$ in example 4.1, we also have

$$\mathbf{S} = \begin{pmatrix} 191.04 & 11.871 & 25.162 \\ 11.871 & 4.0280 & -0.35378 \\ 25.162 & -0.35378 & 58.606 \end{pmatrix},$$

$$|\mathbf{S}_1| = 13313, \qquad |\mathbf{S}_2| = 43779, \qquad |\mathbf{S}| = 34053,$$

$$2\hat{I}(H_1:H_2(\cdot)) = 95 \log \frac{34053}{13313} + 208 \log \frac{34053}{43779} = 36.96 = B^2,$$

$$\beta^2 = \frac{54 + 27 - 3}{12}\left(\frac{1}{95} + \frac{1}{208} - \frac{1}{303}\right) = 0.0782,$$

$$\frac{(2-1)3 \times 4}{2} = 6 \text{ degrees of freedom.}$$

In Fisher's B^2 table, Table III on page 380, the 5% values for $n = 6$ and $\beta^2 = 0.04$ and 0.16 are respectively 12.6750 and 12.9247. We therefore reject the null hypothesis of equality of the population covariance matrices.

Example 5.3. We use data given by Pearson and Wilks (1933) for five samples of 12 observations each on the strength and hardness in aluminum die-castings. (See section 9.1 of chapter 11.) Based on their data (note that they did not use the unbiased estimates), details not being repeated here, $k = 2$, $r = 5$, $N_1 = \cdots = N_5 = 11$, $N = 55$,

$$\log |\mathbf{S}_1| = 5.82588, \qquad \log |\mathbf{S}_2| = 6.63942, \qquad \log |\mathbf{S}_3| = 5.31904,$$

$$\log |\mathbf{S}_4| = 6.66973, \qquad \log |\mathbf{S}_5| = 5.35937, \qquad \log |\mathbf{S}| = 6.13953,$$

$$2\hat{I}(H_1:H_2(\cdot)) = 55(6.13953) - 11(29.81344) = 9.726 = B^2,$$

$$\beta^2 = \frac{16 + 12 - 2}{12}\left(\frac{5}{11} - \frac{1}{55}\right) = 0.945454,$$

$$n = \frac{(5 - 1)2 \times 3}{2} = 12 \text{ degrees of freedom.}$$

In Fisher's B^2 table, Table III on page 380, the 5% values for $n = 7$ (the largest there tabulated) and $\beta^2 = 0.64$ and 1.0 are respectively 15.3225 and 16.0040. Since the tabulated values increase with increasing n for a fixed β^2, here we do not reject the null hypothesis of equality of population covariance matrices. We could also test $9.726\left(1 - \frac{0.945454}{12}\right) = 8.96$ as a χ^2 with 12 degrees of freedom, with the same conclusion, accept the null hypothesis of equality of the population covariance matrices. This agrees with Pearson and Wilks (1933). [Cf. Anderson (1958, p. 256).]

Example 5.4. To illustrate section 5.4, we shall compute $2\hat{I}(H_1:H_2(\cdot))$ in (5.21) for the five samples of example 5.3, so that $k = 2$, $r = 5$, $N_1 = \cdots = N_5 = 11$, $N = 55$. From the data given by Pearson and Wilks (1933, p. 370) we make the computations shown in table 5.3.

TABLE 5.3

i	r_{i12}	$1 - r_{i12}^2$
1	0.68257	0.534106
2	0.87601	0.232617
3	0.71372	0.490595
4	0.71496	0.488835
5	0.80505	0.351891

$$1 - r_{12}^2 = 0.424735, \qquad n = 4 \text{ degrees of freedom,}$$

$$2\hat{I}(H_1:H_2(\cdot)) = 11 \log \frac{0.424735}{0.534106} + \cdots + 11 \log \frac{0.424735}{0.351891} = 3.0498.$$

The 5% value for chi-square for 4 degrees of freedom is 9.4877 so that, consistent with example 5.3, we accept the null hypothesis of homogeneity of the correlation coefficients.

Example 5.5. As another illustration of section 5.4, let us consider the data given by Pearson and Wilks (1933, pp. 372–375) consisting of standard measurements of length and breadth of skull in millimeters obtained for 20 adult males from each of 30 different races or groups, so that $k = 2$, $r = 30$, $N_1 = \cdots = N_{30} = 19$, $N = 570$. From the data given by Pearson and Wilks (1933, p. 373) we make the computations shown in table 5.4.

TABLE 5.4

i	r_{i12}	$1 - r_{i12}^2$	i	r_{i12}	$1 - r_{i12}^2$	i	r_{i12}	$1 - r_{i12}^2$
1	0.097	0.990591	11	0.219	0.952039	21	0.178	0.968316
2	0.198	0.960796	12	−0.152	0.976896	22	0.763	0.417831
3	0.576	0.668224	13	0.319	0.898239	23	0.101	0.989799
4	−0.015	0.999775	14	0.310	0.903900	24	0.449	0.798399
5	0.173	0.970071	15	0.019	0.999639	25	0.245	0.939975
6	0.764	0.416304	16	0.445	0.801975	26	0.360	0.870400
7	−0.037	0.998631	17	0.410	0.831900	27	0.592	0.649536
8	0.667	0.555111	18	0.946	0.105084	28	−0.515	0.734775
9	0.014	0.999804	19	0.018	0.999676	29	0.023	0.999471
10	−0.112	0.987456	20	0.160	0.974400	30	0.254	0.935484

$$1 - r_{12}^2 = 0.937999, \qquad n = 29 \text{ degrees of freedom},$$

$$2\hat{I}(H_1:H_2(\cdot)) = 19 \log \frac{0.937999}{0.990591} + \cdots + 19 \log \frac{0.937999}{0.935484} = 98.$$

Since 98 as a chi-square with 29 degrees of freedom is significant, we reject the null hypothesis of homogeneity of the correlation coefficients, a conclusion consistent with that reached by Pearson and Wilks using an *ad hoc* approach not generalizable to the k-variate case. For this data Pearson and Wilks (1933, p. 374), using Fisher's z-test [Fisher (1921)], computed $\chi^2 = \sum_{i=1}^{30} (n_i - 3)(z_i - \bar{z})^2$, where $z_i = \frac{1}{2}[\log_e(1 + r_{i12}) - \log_e (1 - r_{i12})]$ and $\bar{z} = \sum_{i=1}^{30} z_i/30$, obtaining $\chi^2 = 96.01$ with 29 degrees of freedom.

6. ASYMPTOTIC DISTRIBUTIONS

In this section we shall justify the statements made about the asymptotic behavior of the statistics in the previous sections of this chapter.

6.1. Homogeneity of Covariance Matrices

Under the hypothesis H_2 of section 5.3, we let

$$(6.1) \quad N_i\mathbf{S}_i = \mathbf{\Sigma}^{1/2}\mathbf{V}_i\mathbf{\Sigma}^{1/2}, \qquad N\mathbf{S} = \mathbf{\Sigma}^{1/2}\mathbf{V}\mathbf{\Sigma}^{1/2}, \qquad i = 1, 2, \cdots, r,$$

which define transformations linear in the elements of the matrices $\mathbf{S}_i$, $\mathbf{S}$ respectively by $\mathbf{V}_i$, $\mathbf{V}$. The Jacobians of these transformations are [cf.

Anderson (1958, p. 162), Deemer and Olkin (1951)]

$$\left|\frac{1}{N_i}\mathbf{\Sigma}\right|^{\frac{k+1}{2}} \quad \text{and} \quad \left|\frac{1}{N}\mathbf{\Sigma}\right|^{\frac{k+1}{2}}.$$

The Wishart distributions of the elements of $\mathbf{S}_i$, $\mathbf{S}$ are thereby transformed into the respective probability densities of the elements of $\mathbf{V}_i$, $\mathbf{V}$

$$(6.2) \qquad \frac{(\frac{1}{2})^{\frac{kN_i}{2}} e^{-\frac{1}{2}\mathrm{tr}\mathbf{V}_i} |\mathbf{V}_i|^{\frac{N_i-k-1}{2}}}{\pi^{\frac{k(k-1)}{4}} \prod_{\alpha=1}^{k} \Gamma(N_i+1-\alpha)/2} \quad \text{and} \quad \frac{(\frac{1}{2})^{\frac{kN}{2}} e^{-\frac{1}{2}\mathrm{tr}\mathbf{V}} |\mathbf{V}|^{\frac{N-k-1}{2}}}{\pi^{\frac{k(k-1)}{4}} \prod_{\alpha=1}^{k} \Gamma(N+1-\alpha)/2}.$$

Applying the transformations in (6.1) to $\hat{I}(H_1:H_2(\cdot))$ in (5.14), we get

$$(6.3) \qquad \hat{I}(H_1:H_2(\cdot)) = \sum_{\beta=1}^{r} \frac{N_\beta}{2}\left(\log\frac{|\mathbf{V}|}{|\mathbf{V}_\beta|} + k\log\frac{N_\beta}{N}\right).$$

Since the r samples are independent, the characteristic function of the distribution of

$$\sum_{\beta=1}^{r} N_\beta \log\frac{|\mathbf{V}|}{|\mathbf{V}_\beta|} = N\log|\mathbf{V}| - \sum_{\beta=1}^{r} N_\beta \log|\mathbf{V}_\beta|$$

is [cf. Box (1949, p. 321)]

$$(6.4) \quad \phi(t) = \int\left(\prod_{\beta=1}^{r}\frac{(\frac{1}{2})^{\frac{kN_\beta}{2}} e^{-\frac{1}{2}\mathrm{tr}\mathbf{V}_\beta}|\mathbf{V}_\beta|^{\frac{N_\beta(1-2it)-k-1}{2}}}{\pi^{\frac{k(k-1)}{4}}\prod_{\alpha=1}^{k}\Gamma(N_\beta+1-\alpha)/2}\right)|\mathbf{V}|^{Nit}\prod_{\beta=1}^{r}\prod_{\gamma,\delta=1}^{k} dv_{\beta\gamma\delta}$$

$$= \left(\prod_{\beta=1}^{r}\prod_{\alpha=1}^{k}\frac{\Gamma(N_\beta(1-2it)+1-\alpha)/2}{\Gamma(N_\beta+1-\alpha)/2}\right)\times$$

$$\int\frac{(\frac{1}{2})^{\frac{kN}{2}} e^{-\frac{1}{2}\mathrm{tr}\mathbf{V}}|\mathbf{V}|^{\frac{N(1-2it)-k-1}{2}+Nit}\prod_{\gamma,\delta=1}^{k} dv_{\gamma\delta}}{\pi^{\frac{k(k-1)}{4}}\prod_{\alpha=1}^{k}\Gamma(N(1-2it)+1-\alpha)/2}$$

$$= \prod_{\alpha=1}^{k}\left(\frac{\Gamma(N+1-\alpha)/2}{\Gamma(N(1-2it)+1-\alpha)/2}\prod_{\beta=1}^{r}\frac{\Gamma(N_\beta(1-2it)+1-\alpha)/2}{\Gamma(N_\beta+1-\alpha)/2}\right),$$

where the middle result follows from the reproductive property of the

Wishart distribution [Anderson (1958, p. 162), Wilks (1943, p. 232)]. We use Stirling's approximation,

$$\log \Gamma(p) = \tfrac{1}{2}\log 2\pi + (p - \tfrac{1}{2})\log p - p + \frac{1}{12p} - \frac{1}{360p^3} + O(1/p^5),$$

to get an approximate value for large N_β in (6.4). We have

$$(6.5)\quad \log \frac{\Gamma(N_\beta(1-2it) + 1 - \alpha)/2}{\Gamma(N_\beta + 1 - \alpha)/2}$$

$$= \frac{N_\beta(1-2it) - \alpha}{2}\log\frac{N_\beta(1-2it)+1-\alpha}{2} - \frac{N_\beta(1-2it)+1-\alpha}{2}$$

$$+ \frac{1}{6(N_\beta(1-2it)+1-\alpha)} - \frac{1}{45(N_\beta(1-2it)+1-\alpha)^3}$$

$$- \frac{N_\beta - \alpha}{2}\log\frac{N_\beta + 1 - \alpha}{2} + \frac{N_\beta + 1 - \alpha}{2}$$

$$- \frac{1}{6(N_\beta + 1 - \alpha)} + \frac{1}{45(N_\beta + 1 - \alpha)^3} + O(1/N_\beta{}^5),$$

and after some algebraic manipulation the right-hand member of (6.5) may be written as

$$-itN_\beta \log\frac{N_\beta}{2} + \frac{N_\beta(1-2it)-\alpha}{2}\log(1-2it) + N_\beta it$$

$$+ \frac{(3\alpha^2 - 1)2it}{12N_\beta(1-2it)} + O(1/N_\beta{}^2).$$

We therefore have

$$(6.6)\quad \log\phi(t) = \sum_{\alpha=1}^{k}\Bigg(itN\log\frac{N}{2} - \frac{N(1-2it)-\alpha}{2}\log(1-2it) - Nit$$

$$- \frac{(3\alpha^2-1)it}{6N(1-2it)} - O(1/N^2)\Bigg) + \sum_{\alpha=1}^{k}\sum_{\beta=1}^{r}\Bigg(-it\,N_\beta\log\frac{N_\beta}{2}$$

$$+ \frac{N_\beta(1-2it)-\alpha}{2}\log(1-2it) + N_\beta it$$

$$+ \frac{(3\alpha^2-1)it}{6N_\beta(1-2it)} + O(1/N_\beta{}^2)\Bigg)$$

$$= -it\sum_{\beta=1}^{r} kN_\beta \log\frac{N_\beta}{N} - \frac{(r-1)k(k+1)}{4}\log(1-2it)$$

$$+ \frac{it(2k^3 + 3k^2 - k)}{12(1-2it)}\left(\sum_{\beta=1}^{r}\frac{1}{N_\beta} - \frac{1}{N}\right) + \sum_{\beta=1}^{r} O(1/N_\beta{}^2) - O(1/N^2).$$

Neglecting the last term in (6.6), we have

$$(6.7)\quad \phi(t) = (1 - 2it)^{-(r-1)k(k+1)/4} \exp\left(-it\sum_{\beta=1}^{r} kN_\beta \log\frac{N_\beta}{N} + \frac{cit}{1 - 2it}\right),$$

where $c = (2k^3 + 3k^2 - k)\left(\sum_{\beta=1}^{r} 1/N_\beta - 1/N\right)/12$.

Because of (6.3) and (6.4), writing $\zeta = 2\hat{I}(H_1:H_2(\cdot))$, the probability density of ζ is

$$(6.8)\quad D(\zeta) = \frac{1}{2\pi}\int_{-\infty}^{\infty} \frac{\exp(-it\zeta + cit/(1 - 2it))\,dt}{(1 - 2it)^{(r-1)k(k+1)/4}}.$$

If we neglect the term with c, it follows that $D(\zeta)$ is the probability density of the χ^2-distribution with $(r - 1)k(k + 1)/2$ degrees of freedom; otherwise, by integrating (6.8) [see Laha (1954), McLachlan (1939, p. 86)] we get, since ζ is real and positive and $(r - 1)k(k + 1)/4 > 0$,

$$(6.9)\quad D(\zeta) = \tfrac{1}{2}e^{-c/2-\zeta/2}\left(\frac{\zeta}{c}\right)^{(n-1)/2} I_{n-1}(\sqrt{c\zeta}),$$

where $n = (r - 1)k(k + 1)/4$ and $I_{n-1}(\sqrt{c\zeta})$ is the Bessel function of purely imaginary argument [Watson (1944)]

$$I_{n-1}(\sqrt{c\zeta}) = \sum_{j=0}^{\infty} \frac{\left(\frac{1}{2}\right)^{\frac{n-1}{2}+j}\left(\frac{c\zeta}{2}\right)^{\frac{n-1}{2}+j}}{j!\,\Gamma(n + j)}.$$

The probability density (6.9) is that of the noncentral χ^2-distribution with $2n$ degrees of freedom and noncentrality parameter c, and is Fisher's B-distribution [Fisher (1928, p. 665)] with $c = \beta^2$, $\zeta = B^2$, $2n = n_1$.

The approximation to the logarithm of the characteristic function of ζ, that is, $-n\log(1 - 2it) + \dfrac{cit}{1 - 2it}$, corresponds to that of Box (1949, formula 29, p. 323), retaining only the first term in his sum; that is, his $\dfrac{\alpha_1}{\mu}\left(\dfrac{1}{1 - 2it} - 1\right)$ (there is a misprint in the formula) is $\dfrac{cit}{1 - 2it}$ here, as may be verified by using the appropriate formulas with $\beta = 0$ given by Box (1949, pp. 324–325).

For large n we may approximate $I_{n-1}(\sqrt{c\zeta})$ in (6.9) by writing

$$\begin{aligned} I_{n-1}(\sqrt{c\zeta}) &= \frac{(c\zeta/4)^{(n-1)/2}}{\Gamma(n)}\sum_{j=0}^{\infty}\frac{(c\zeta/4)^j\Gamma(n)}{j!\,\Gamma(n + j)} \\ &\approx \frac{(c\zeta/4)^{(n-1)/2}}{\Gamma(n)}\sum_{j=0}^{\infty}\frac{1}{j!}\left(\frac{c\zeta}{4n}\right)^j \\ &= \frac{(c\zeta/4)^{(n-1)/2}}{\Gamma(n)}\,e^{c\zeta/4n}, \end{aligned}$$

thereby getting

$$(6.10)\qquad D(\zeta) \approx \frac{1}{2}\frac{\exp\left(-c/2 - \zeta\left(1 - \frac{c}{2n}\right)/2\right)}{\Gamma(n)}\left(\frac{\zeta}{2}\right)^{n-1}$$

Setting $\zeta\left(1 - \frac{c}{2n}\right) = \chi^2$, (6.10) yields

$$(6.11)\qquad D(\chi^2)d\chi^2 = \frac{e^{-c/2}}{\left(1 - \frac{c}{2n}\right)^n}\cdot\frac{e^{-\chi^2/2}}{\Gamma(n)}\left(\frac{\chi^2}{2}\right)^{n-1} d\frac{\chi^2}{2}$$

$$\approx \frac{e^{-\chi^2/2}(\chi^2/2)^{n-1}\, d\chi^2/2}{\Gamma(n)},$$

or $\zeta\left(1 - \frac{c}{2n}\right)$ asymptotically is distributed as χ^2 with $2n = (r-1)k(k+1)/2$ degrees of freedom. It may be verified that $1 - \frac{c}{2n} = \rho$, the scale factor in the χ^2 approximation by Box (1949, p. 329). [Cf. Anderson (1958, p. 255).]

For other approximations to the noncentral χ^2-distribution see Abdel-Aty (1954), Tukey (1957).

6.2. Single Sample

For a single sample, we derived the value $2\hat{I}(H_1:H_2)$ in (3.15). With the same transformation as in (6.1), that is, $N\mathbf{S} = \mathbf{\Sigma}_2^{1/2}\mathbf{V}\mathbf{\Sigma}_2^{1/2}$, with Jacobian $\left|\frac{1}{N}\mathbf{\Sigma}_2\right|^{(k+1)/2}$, the probability density of the Wishart distribution of the elements of $\mathbf{S}$ is transformed into that in the right-hand member of the pair in (6.2) and

$$(6.12)\qquad 2\hat{I}(H_1:H_2) = N\left(\log\frac{|\mathbf{\Sigma}_2|}{|\mathbf{S}|} - k + \operatorname{tr}\mathbf{S}\mathbf{\Sigma}_2^{-1}\right)$$

$$= Nk\log N - N\log|\mathbf{V}| - Nk + \operatorname{tr}\mathbf{V}.$$

The characteristic function of the distribution of $2\hat{I}(H_1:H_2)$ is therefore

(6.13) $\phi(t)$

$$= \int\frac{(\frac{1}{2})^{\frac{kN}{2}}|\mathbf{V}|^{\frac{N(1-2it)-k-1}{2}}\exp\left(-\frac{1}{2}\operatorname{tr}(1-2it)\mathbf{V} + itNk\log N - itNk\right)\prod_{\gamma,\delta=1}^{k} dv_{\gamma\delta}}{\pi^{k(k-1)/4}\prod_{\alpha=1}^{k}\Gamma(N+1-\alpha)/2}$$

$$= \frac{(\frac{1}{2})^{itNk}\exp\left(itNk\log N - itNk\right)}{(1-2it)^{Nk(1-2it)/2}}\prod_{\alpha=1}^{k}\frac{\Gamma(N(1-2it)+1-\alpha)/2}{\Gamma(N+1-\alpha)/2}.$$

Using (6.5), we derive

$$\text{(6.14)}\qquad \log \phi(t) = -\frac{k(k+1)}{4}\log(1-2it) + \frac{it(2k^3+3k^2-k)}{12(1-2it)N} + O(1/N^2)$$

from which the conclusions stated in the preceding sections follow as in section 6.1.

6.3. The Hypothesis of Independence

It is known that the logarithm of the characteristic function of the distribution of $2\hat{I}(H_1:H_2') = -N\log|\mathbf{R}|$ [see (3.18)] is [see Bartlett (1950), Wilks (1932, p. 492)]:

$$\text{(6.15)}\qquad \log \phi(t) = (k-1)\log\frac{\Gamma(N/2)}{\Gamma(N(1-2it)/2)} + \sum_{\alpha=1}^{k-1}\log\frac{\Gamma(N(1-2it)-\alpha)/2}{\Gamma(N-\alpha)/2}.$$

Employing Stirling's approximation as in (6.5), and retaining comparable terms as in (6.7), we have

$$\text{(6.16)}\qquad \log \phi(t) = -\frac{k(k-1)}{4}\log(1-2it) + \frac{cit}{1-2it},$$

where $c = k(k-1)(2k+5)/12N$.

The statement at the end of section 3.3 then follows from (6.16), (6.8), and (6.9). From (6.11) we may also deduce that

$$2\hat{I}(H_1:H_2')\left(1-\frac{k(k-1)(2k+5)}{6Nk(k-1)}\right) = -(N-\tfrac{1}{6}(2k+5))\log|\mathbf{R}|$$

asymptotically is distributed as χ^2 with $k(k-1)/2$ degrees of freedom. The last result is given by Bartlett (1950).

The logarithm of the characteristic function of the distribution of $2\hat{I}(H_1:H_2(\cdot)) = N\log\dfrac{|\mathbf{R}_{11}|\cdots|\mathbf{R}_{mm}|}{|\mathbf{R}|}$ [see (3.30)] is [Wald and Brookner (1941), Wilks (1932, p. 493, 1943, p. 244)]:

$$\text{(6.17)}\qquad \log \phi(t) = \sum_{\beta=1}^{m}\sum_{\alpha=1}^{k_\beta}\log\frac{\Gamma(N+1-\alpha)/2}{\Gamma(N(1-2it)+1-\alpha)/2} + \sum_{\gamma=1}^{k}\log\frac{\Gamma(N(1-2it)+1-\gamma)/2}{\Gamma(N+1-\gamma)/2}.$$

Employing Stirling's approximation as in (6.5), and retaining comparable terms as in (6.7), we have

$$(6.18)\quad \log \phi(t) = -\frac{k(k+1) - \sum_{\beta=1}^{m} k_\beta(k_\beta+1)}{4} \log(1-2it) + \frac{cit}{1-2it},$$

where $c = \left((2k^3+3k^2-1) - \sum_{\beta=1}^{m}(2k_\beta^3+3k_\beta^2-k_\beta)\right)\Big/ 12N$, from which the results in table 3.1 follow.

Note that for $k_\beta = 1$, $\beta = 1, \cdots, m$, so that $m = k$, (6.17) becomes (6.15), and (6.18) becomes (6.16).

6.4. Roots of Determinantal Equations

From results derived by Fisher (1939b), Girshick (1939), Hsu (1939, 1941a, 1941b, 1941–42), Roy (1939, 1957) [see Anderson (1951, 1958, pp. 307–329), Mood (1951), Wilks (1943, pp. 260–270)], it is known that the probability density of the distribution of the roots of $|\mathbf{S}^* - l\mathbf{S}| = 0$ [see (8.4) in chapter 11], for $(n - r)$ large, is

$$(6.19)\quad \frac{(\frac{1}{2})^{(r-1)p/2}\pi^{p/2}}{\prod_{\alpha=1}^{p} \Gamma(r-\alpha)/2\,\Gamma(p+1-\alpha)/2} (l_1 \cdots l_p)^{(r-p-2)/2} \times e^{-\frac{1}{2}(l_1+\cdots+l_p)} \prod_{i>j} (l_j - l_i),$$

and that of the roots of $|\mathbf{S}_{21}\mathbf{S}_{11}^{-1}\mathbf{S}_{12} - l\mathbf{S}_{22\cdot 1}| = 0$ [see (7.4) in chapter 11], for $(n - k_1)$ large, is

$$(6.20)\quad \frac{(\frac{1}{2})^{(k_1-1)k_2/2}\pi^{k_2/2}}{\prod_{\alpha=1}^{k_2} \Gamma(k_1-\alpha)/2\,\Gamma(k_2+1-\alpha)/2} (v_1 \cdots v_{k_2})^{(k_1-k_2-2)/2} \times e^{-\frac{1}{2}(v_1+\cdots+v_{k_2})} \prod_{i>j} (v_j - v_i),$$

where $v_i = (n - k_1)l_i$.

The characteristic functions of the asymptotic distributions of $\hat{J}(H_1, H_2)$ in (8.5) of chapter 11 and (7.5) of chapter 11 may be derived from (6.19) and (6.20) as, respectively, $(1-2it)^{-(r-1)p/2}$ and $(1-2it)^{-(k_1-1)k_2/2}$, hence the conclusion as to their χ^2 distributions. The χ^2 decompositions in sections 8.1 and 8.2 in chapter 11 follow from the fact that, asymptotically, the distributions of $l_{m+1}, \cdots, l_p$ of (6.19) and $v_{m+1}, \cdots, v_{k_2}$ of (6.20), assuming the corresponding population parameters have the null hypothesis values, are independent of the distribution of the remaining roots and with probability densities given respectively by

$$(6.21)\quad \frac{(\tfrac{1}{2})^{(r-1-m)(p-m)/2}\pi^{(p-m)/2}}{\prod_{\alpha=1}^{p-m}\Gamma(r-m-\alpha)/2\Gamma(p-m+1-\alpha)/2}(l_{m+1}\cdots l_p)^{(r-p-2)/2}\times$$
$$e^{-\frac{1}{2}(l_{m+1}+\cdots+l_p)}\prod_{i>j}(l_j-l_i),$$

$$(6.22)\quad \frac{(\tfrac{1}{2})^{(k_1-1-m)(k_2-m)/2}\pi^{(k_2-m)/2}}{\prod_{\alpha=1}^{k_2-m}\Gamma(k_1-m-\alpha)/2\Gamma(k_2-m+1-\alpha)/2}(v_{m+1}\cdots v_{k_2})^{(k_1-k_2-2)/2}$$
$$\times\; e^{-\frac{1}{2}(v_{m+1}+\cdots+v_{k_2})}\prod_{i>j}(v_j-v_i).$$

When $\mathbf{S}_1$ and $\mathbf{S}_2$ are independent, unbiased estimates of the same covariance matrix with N_1 and N_2 degrees of freedom respectively, the probability density of the distribution of the roots of $|N_1\mathbf{S}_1 - lN_2\mathbf{S}_2| = 0$ is

$$(6.23)\quad \pi^{k/2}\left(\prod_{\alpha=1}^{k}\frac{\Gamma(N_1+N_2+1-\alpha)/2}{\Gamma(N_1+1-\alpha)/2\Gamma(N_2+1-\alpha)/2\Gamma(k+1-\alpha)/2}\right)\times$$
$$\frac{(l_1\cdots l_k)^{(N_1-k-1)/2}\prod_{i>j}(l_j-l_i)}{((1+l_1)\cdots(1+l_k))^{(N_1+N_2)/2}}.$$

When $\mathbf{S}$ is an unbiased estimate of $\mathbf{\Sigma}$ with N degrees of freedom, the probability density of the distribution of the roots of $|N\mathbf{S} - l\mathbf{\Sigma}| = 0$ is

$$(6.24)\quad \frac{\pi^{k/2}(\tfrac{1}{2})^{Nk/2}}{\prod_{\alpha=1}^{k}\Gamma(N+1-\alpha)/2\Gamma(k+1-\alpha)/2}(l_1\cdots l_k)^{(N-k-1)/2}\times$$
$$e^{-\frac{1}{2}(l_1+\cdots+l_k)}\prod_{i>j}(l_j-l_i).$$

The distribution of $l_{m+1},\cdots,l_k$ in (6.24), assuming the corresponding population parameters have the null hypothesis values, is independent of the distribution of the remaining roots, with probability density

$$(6.25)\quad \frac{\pi^{(k-m)/2}(\tfrac{1}{2})^{(N-m)(k-m)/2}}{\prod_{\alpha=1}^{k-m}\Gamma(N-m+1-\alpha)/2\Gamma(k-m+1-\alpha)/2}\times$$
$$(l_{m+1}\cdots l_k)^{(N-k-1)/2}e^{-\frac{1}{2}(l_{m+1}+\cdots+l_k)}\prod_{i>j}(l_j-l_i).$$

In section 3.5 we were concerned with the distribution of

$$N\sum_{i=m+1}^{k}(-\log F_i - 1 + F_i) = (k-m)N\log N - (k-m)N$$
$$+\sum_{i=m+1}^{k}(-N\log l_i + l_i),$$

where the l's are roots of $|N\mathbf{S} - l\mathbf{\Sigma}_2| = 0$. We find that the characteristic function of the desired distribution is [using (6.25)]

$$(6.26)\quad \phi(t) = \frac{(\frac{1}{2})^{N(k-m)it}\exp(it(k-m)N\log N - (k-m)Nit)}{(1-2it)^{(k-m)(N(1-2it)-m)/2}} \times \prod_{\alpha=1}^{k-m} \frac{\Gamma(N(1-2it)-m+1-\alpha)/2}{\Gamma(N-m+1-\alpha)/2}.$$

Note that when $m = 0$ the sum in question is $2\hat{I}(H_1:H_2)$ in (6.12), and the characteristic function derived in (6.13) is (6.26) for $m = 0$.

By using Stirling's approximation as in (6.5) and retaining comparable terms as in (6.7), we find that the logarithm of the characteristic function in (6.26) is

$$(6.27)\quad \log\phi(t) = -\frac{(k-m)(k-m+1)}{2}\log(1-2it) + \frac{cit}{1-2it},$$

where $c = (2k^3 + 3k^2 - k - (2m^3 + 3m^2 - m))/12N$, from which the statement about the distribution made in section 3.5 follows.

The distribution of $l_{m+1}, \cdots, l_k$ in (6.23), assuming the corresponding population parameters have the null hypothesis values, is independent of the distribution of the remaining roots, with probability density

$$(6.28)\quad \pi^{(k-m)/2} \times \left(\prod_{\alpha=1}^{k-m} \frac{\Gamma(N-m+1-\alpha)/2}{\Gamma(N_1-m+1-\alpha)/2\,\Gamma(N_2-m+1-\alpha)/2\,\Gamma(k-m+1-\alpha)/2}\right) \times \frac{(l_{m+1}\cdots l_k)^{(N_1-k-1)/2}\prod_{i>j}(l_j - l_i)}{((1+l_{m+1})\cdots(1+l_k))^{(N_1+N_2)/2}},$$

where $N = N_1 + N_2$. In section 5.2 we were concerned with the distribution of

$$(k-m)N_1\log\frac{N_1}{N_1+N_2} + (k-m)N_2\log\frac{N_2}{N_1+N_2} + \sum_{i=m+1}^{k}((N_1+N_2)\log(1+l_i) - N_1\log l_i),$$

where the l_i are the roots of $|N_1\mathbf{S}_1 - lN_2\mathbf{S}_2| = 0$. We find that the characteristic function of the desired distribution is [using (6.28)]

$$(6.29)\quad \phi(t) = \exp\left(it(k-m)\sum_{j=1}^{2} N_j\log\frac{N_j}{N}\right)\prod_{\alpha=1}^{k-m} \frac{\Gamma(N-m+1-\alpha)/2}{\Gamma(N(1-2it)-m+1-\alpha)/2} \times \frac{\Gamma(N_1(1-2it)-m+1-\alpha)/2}{\Gamma(N_1-m+1-\alpha)/2} \times \frac{\Gamma(N_2(1-2it)-m+1-\alpha)/2}{\Gamma(N_2-m+1-\alpha)/2}.$$

Similarly, as in (6.5), (6.6), and (6.7), we find

$$(6.30)\qquad \log \phi(t) = -\frac{(k-m)(k-m+1)}{2}(1-2it) + \frac{cit}{1-2it},$$

where $c = \dfrac{(2k^3+3k^2-k)-(2m^3+3m^2-m)}{12}\left(\dfrac{1}{N_1}+\dfrac{1}{N_2}-\dfrac{1}{N}\right)$, from which the statement about the distribution made in section 5.2 follows.

7. STUART'S TEST FOR HOMOGENEITY OF THE MARGINAL DISTRIBUTIONS IN A TWO-WAY CLASSIFICATION

We return to the test of the null hypothesis of equality of marginal distributions mentioned at the end of section 11 of chapter 8 and indicate Stuart's (1955a) procedure.

7.1. A Multivariate Normal Hypothesis

Consider the following alternative hypothesis H_1 and null hypothesis H_2 for the means and covariance matrices of multivariate normal populations:

$$(7.1)\qquad \begin{aligned} H_1: \boldsymbol{\mu}_1 &= n\boldsymbol{\Delta}, & \boldsymbol{\Sigma}_1 &= n\boldsymbol{\Sigma} - n\boldsymbol{\Delta}\boldsymbol{\Delta}', \\ H_2: \boldsymbol{\mu}_2 &= \mathbf{0}, & \boldsymbol{\Sigma}_2 &= n\boldsymbol{\Sigma}. \end{aligned}$$

From (1.2) of chapter 9 we then have

$$(7.2)\qquad \begin{aligned} I(1:2) &= \tfrac{1}{2}n\boldsymbol{\Delta}'\boldsymbol{\Sigma}^{-1}\boldsymbol{\Delta} + \tfrac{1}{2}\log\frac{|n\boldsymbol{\Sigma}|}{|n\boldsymbol{\Sigma} - n\boldsymbol{\Delta}\boldsymbol{\Delta}'|} - \frac{k}{2} \\ &\qquad + \tfrac{1}{2}\operatorname{tr}(n\boldsymbol{\Sigma} - n\boldsymbol{\Delta}\boldsymbol{\Delta}')\frac{1}{n}\boldsymbol{\Sigma}^{-1} \\ &= \tfrac{1}{2}n\boldsymbol{\Delta}'\boldsymbol{\Sigma}^{-1}\boldsymbol{\Delta} - \tfrac{1}{2}\log(1 - \boldsymbol{\Delta}'\boldsymbol{\Sigma}^{-1}\boldsymbol{\Delta}) - \tfrac{1}{2}\boldsymbol{\Delta}'\boldsymbol{\Sigma}^{-1}\boldsymbol{\Delta}, \end{aligned}$$

using the fact that [cf. Wilks (1943, pp. 237–238), problems 10.4 and 10.6 in chapter 9]

$$(7.3)\qquad \frac{1}{n}|n\boldsymbol{\Sigma} - n\boldsymbol{\Delta}\boldsymbol{\Delta}'| = \begin{vmatrix} \frac{1}{n} & \boldsymbol{\Delta}' \\ \boldsymbol{\Delta} & n\boldsymbol{\Sigma} \end{vmatrix} = \frac{1}{n}|n\boldsymbol{\Sigma}|(1 - \boldsymbol{\Delta}'\boldsymbol{\Sigma}^{-1}\boldsymbol{\Delta}).$$

Accordingly, for large n, we may use

$$(7.4)\qquad 2I(1:2) = n\boldsymbol{\Delta}'\boldsymbol{\Sigma}^{-1}\boldsymbol{\Delta} = (n\boldsymbol{\Delta}')(n\boldsymbol{\Sigma})^{-1}(n\boldsymbol{\Delta}),$$

equivalent to that under hypotheses specifying a common covariance matrix $n\boldsymbol{\Sigma}$ and differences of means $n\boldsymbol{\Delta}$.

7.2. The Contingency Table Problem

With the notation for a two-way contingency table in section 2 of chapter 8, since $x_{1\cdot} + \cdots + x_{c\cdot} = x_{\cdot 1} + \cdots + x_{\cdot c} = n$, this is a $(c-1)$-variate problem, and Stuart (1955a) defines the statistics of interest as

$$(7.5) \qquad d_i = x_{i\cdot} - x_{\cdot i}, \qquad i = 1, 2, \cdots, c-1.$$

It is known that the multinomial distribution tends to the multivariate normal distribution [Cramér (1946a, pp. 318, 418), Kendall (1943, pp. 290–291)], and Stuart (1955a, pp. 413–414) shows that

$$(7.6) \quad E(d_i) = n(p_{i\cdot} - p_{\cdot i}), \qquad \text{var}\,(d_i) = n[(p_{i\cdot} + p_{\cdot i} - 2p_{ii}) - (p_{i\cdot} - p_{\cdot i})^2],$$

$$\text{cov}\,(d_i, d_j) = -n[(p_{ij} + p_{ji}) + (p_{i\cdot} - p_{\cdot i})(p_{j\cdot} - p_{\cdot j})],$$

so that with $p_{i\cdot} - p_{\cdot i} = \Delta_i$, the matrix $\mathbf{\Sigma}$ in (7.1) is $\mathbf{\Sigma} = (\sigma_{ij})$, $\sigma_{ii} = p_{i\cdot} + p_{\cdot i} - 2p_{ii}$, $\sigma_{ij} = -(p_{ij} + p_{ji})$, $i, j = 1, 2, \cdots, c-1$.

The test statistic, the estimate of $2I(1:2)$, is

$$(7.7) \qquad 2\hat{I}(H_1:H_2) = \mathbf{d}'\mathbf{S}^{-1}\mathbf{d},$$

where $\mathbf{d}' = (d_1, d_2, \cdots, d_{c-1})$, the d's defined in (7.5), and $\mathbf{S} = (s_{ij})$, $s_{ii} = x_{i\cdot} + x_{\cdot i} - 2x_{ii}$, $s_{ij} = -(x_{ij} + x_{ji})$, $i, j = 1, 2, \cdots, c-1$. Under the null hypothesis H_2, $2\hat{I}(H_1:H_2)$ is asymptotically distributed as χ^2 with $(c-1)$ degrees of freedom. As in reparametrization, the conclusion is independent of which $c-1$ of the c d's are used.

8. PROBLEMS

8.1. Considering (3.18), what can be said about the range of values of $|\mathbf{R}|$?

8.2. What is the formal relation between (3.26) and the value in (3.5) of chapter 7 for $r = 1$?

8.3. Develop section 4.1 when the null hypothesis (insofar as the means are concerned) is changed to $H_2: \boldsymbol{\mu}_1 = \boldsymbol{\mu} + \boldsymbol{\delta}$, $\boldsymbol{\mu}_2 = \boldsymbol{\mu}$, with $\boldsymbol{\delta}$ specified, that is, the null hypothesis specifies that the difference of the means is $\boldsymbol{\delta}$.

8.4. What is the asymptotic distribution of $2\hat{I}(H_1:H_2(\cdot))$ in table 4.1 if the null hypothesis is not satisfied?

8.5. Show that $2\hat{I}(H_1:H_2(\cdot))$ in (4.12) yields $2\hat{I}(H_1:H_2)$ in (4.8) for $r = 2$.

8.6. Test the first and third samples in example 4.2 for homogeneity of the population means.

8.7. If you were to compute a linear discriminant function for the second and third samples in example 4.2 by the procedure of section 4.2 and by the procedure of section 8.1 of chapter 11, would you get different results?

8.8. What is the asymptotic distribution of $2\hat{I}(H_1:H_2(\cdot))$ in (5.14) if the null hypothesis is not satisfied?

8.9. Test the three covariance matrices in example 4.2 for homogeneity.

8.10. Develop the analysis of the data in example 5.1 according to table 5.2 and confirm Smith's (1947) remark that the correlations are not significant.

8.11. Complete the analysis of the data in examples 4.1 and 5.2 in accordance with table 5.2.

8.12. Discuss the similarities and differences of the test for the independence of two sets of variates in section 3.6 and the test in section 7 of chapter 11.

8.13. Write the probability densities in (6.2) for $k = 1$.

8.14. Verify the "algebraic manipulation" for (6.5).

8.15. Write the probability density in (6.19) for $p = 1$, that in (6.20) for $k_2 = 1$, and that in (6.23) for $k = 1$.

8.16. Wilks (1935b, p. 325) considered the following correlation matrix, given by Kelley (1928, p. 114), for a sample of 109 seventh-grade school children, in which the five variables are respectively arithmetic speed, arithmetic power, intellectual interest, social interest, activity interest:*

$$\mathbf{R} = \begin{pmatrix} 1 & 0.4249 & -0.0552 & -0.0031 & 0.1927 \\ 0.4249 & 1 & -0.0416 & 0.0495 & 0.0687 \\ -0.0552 & -0.0416 & 1 & 0.7474 & 0.1691 \\ -0.0031 & 0.0495 & 0.7474 & 1 & 0.2653 \\ 0.1927 & 0.0687 & 0.1691 & 0.2653 & 1 \end{pmatrix}.$$

Would you accept a null hypothesis that the set of the first two variables is independent of the set of the last three variables?

8.17. Bartlett and Rajalakshman (1953, p. 119) concluded that the observed correlation matrix $\mathbf{R}$, with $N = 29$, is significantly different from the hypothetical correlation matrix $\mathbf{P}_2$, where

$$\mathbf{P}_2 = \begin{pmatrix} 1 & 0.7071 & 0.7071 & 0.5000 \\ 0.7071 & 1 & 0.5000 & 0.7071 \\ 0.7071 & 0.5000 & 1 & 0.7071 \\ 0.5000 & 0.7071 & 0.7071 & 1 \end{pmatrix},$$

$$\mathbf{R} = \begin{pmatrix} 1 & 0.2676 & 0.5931 & 0.1269 \\ 0.2676 & 1 & 0.3753 & 0.5941 \\ 0.5931 & 0.3753 & 1 & 0.6796 \\ 0.1269 & 0.5941 & 0.6796 & 1 \end{pmatrix}.$$

Verify this conclusion.

* Reprinted from *Crossroads in the Mind of Man* by Truman L. Kelley with the permission of the publishers, Stanford University Press. Copyright 1928 by the Board of Trustees of Leland Stanford Junior University.

8.18. Box (1950, p. 387) gives the following covariance matrices for three treatment groups on growth data for rats:

$$9\mathbf{S}_1 = \begin{pmatrix} 210.5 & 13.5 & -7.5 & -13.5 \\ 13.5 & 202.5 & 224.5 & 110.5 \\ -7.5 & 224.5 & 310.9 & 117.5 \\ -13.5 & 110.5 & 117.5 & 258.5 \end{pmatrix},$$

$$6\mathbf{S}_2 = \begin{pmatrix} 111.4 & 83.0 & 78.4 & 39.7 \\ 83.0 & 246.0 & 292.0 & 157.0 \\ 78.4 & 292.0 & 473.4 & 264.7 \\ 39.7 & 157.0 & 264.7 & 174.9 \end{pmatrix},$$

$$9\mathbf{S}_3 = \begin{pmatrix} 260.4 & -54.0 & -126.4 & -100.8 \\ -54.0 & 160.5 & 110.0 & 77.0 \\ -126.4 & 110.0 & 262.4 & 76.8 \\ -100.8 & 77.0 & 76.8 & 419.6 \end{pmatrix}.$$

Box concludes that there is no reason to doubt the homogeneity of the covariance matrices. Verify this conclusion.

8.19. Suppose that in the analysis in table 3.1 there are only two sets, with $k_1 = 1$, $k_2 = k - 1$. Show that $2\hat{I}(H_1:H_2(\cdot)) = -N\log(1 - r^2_{1\cdot 23\cdots k})$, with $k - 1$ degrees of freedom and $\beta^2 = (k^2 - 1)/2N$, where $r_{1\cdot 23\cdots k}$ is the observed multiple correlation of x_1 with $x_2, x_3, \cdots, x_k$. [See (7.18) in chapter 9.]

8.20. Show that problem 10.12 in chapter 9 is equivalent to

$$H_2' \rightleftharpoons H_2'(\rho^2_{1\cdot 23\cdots k} = 0) \cap H_2'(\rho^2_{2\cdot 3\cdots k} = 0) \cap \cdots \cap H_2'(\rho^2_{k-1\cdot k} = 0),$$

where H_2' is the hypothesis of independence in (3.18) and $H_2'(\rho^2_{j\cdot j+1,\cdots,k} = 0)$ is the hypothesis that the multiple correlation of x_j with $x_{j+1}, \cdots, x_k$ is zero, $j = 1, 2, \cdots, k - 1$.

8.21. Show that $-N\log|\mathbf{R}| = -N\log(1 - r^2_{1\cdot 23\cdots k}) - N\log(1 - r^2_{2\cdot 3\cdots k}) - \cdots - N\log(1 - r^2_{k-1\cdot k})$, that is, $2\hat{I}(H_1:H_2') = 2\hat{I}(H_1:H_2'(\rho^2_{1\cdot 23\cdots k} = 0)) + \cdots + 2\hat{I}(H_1:H_2'(\rho^2_{k-1\cdot k} = 0))$, where $2\hat{I}(H_1:H_2')$ is given in (3.18) and $2\hat{I}(H_1:H_2'(\rho^2_{j\cdot j+1,\cdots,k} = 0)) = -N\log(1 - r^2_{j\cdot j+1,\cdots,k})$.

8.22. Show that table 8.1 is an analysis of $2\hat{I}(H_1:H_2')$ in (3.18).

8.23. Show that $-N\log(1 - r^2_{1\cdot 2\cdots k}) = -N\log(1 - r^2_{1k\cdot 23\cdots k-1}) - N\log(1 - r^2_{1k-1\cdot 23\cdots k-2}) - \cdots - N\log(1 - r^2_{13\cdot 2}) - N\log(1 - r^2_{12})$, where $r_{1j\cdot 23\cdots j-1}$, $j = 2, \cdots, k$, is a partial correlation coefficient.

8.24. Show that table 8.2 is an analysis of $2\hat{I}(H_1:H_2'(\rho^2_{1\cdot 23\cdots k} = 0))$.

8.25. Show that problem 10.15 in chapter 9 is equivalent to $H_2(\mathbf{\Sigma}_{ii}) \rightleftharpoons H_2(|\mathbf{\Sigma}_{22}| = |\mathbf{\Sigma}_{22\cdot 1}|) \cap H_2(|\mathbf{\Sigma}_{33}| = |\mathbf{\Sigma}_{33\cdot 12}|) \cap \cdots \cap H_2(|\mathbf{\Sigma}_{mm}| = |\mathbf{\Sigma}_{mm\cdot 12\cdots m-1}|)$, where $H_2(\mathbf{\Sigma}_{ii})$ is the hypothesis in (3.28) and $H_2(|\mathbf{\Sigma}_{jj}| = |\mathbf{\Sigma}_{jj\cdot 12\cdots j-1}|)$ is the hypothesis that $|\mathbf{\Sigma}_{jj}| = |\mathbf{\Sigma}_{jj\cdot 12\cdots j-1}|$, $j = 2, \cdots, m$.

8.26. Show that table 8.3 is an analysis of $2\hat{I}(H_1:H_2(\cdot))$ of table 3.1.

TABLE 8.1

Component due to	Information	D.F.	β^2
$\rho^2_{k-1\cdot k} = 0$	$-N \log (1 - r^2_{k-1\cdot k})$	1	$\frac{3}{2N}$
. . .	. . .	. . .	. . .
$\rho^2_{j\cdot j+1,\cdots,k} = 0$	$-N \log (1 - r^2_{j\cdot j+1,\cdots,k})$	$k - j$	$\frac{(k-j)^2 + 2(k-j)}{2N}$
. . .	. . .	. . .	. . .
$\rho^2_{1\cdot 2\cdots k} = 0$	$-N \log (1 - r^2_{1\cdot 2\cdots k})$	$k - 1$	$\frac{k^2 - 1}{2N}$
H_2'	$-N \log \lvert\mathbf{R}\rvert$	$\frac{k(k-1)}{2}$	$\frac{k(k-1)(2k+5)}{12N}$

TABLE 8.2

Component due to	Information	D.F.	β^2
$\rho^2_{12} = 0$	$-N \log (1 - r^2_{12})$	1	$\frac{3}{2N}$
$\rho^2_{13\cdot 2} = 0$	$-N \log (1 - r^2_{13\cdot 2})$	1	$\frac{5}{2N}$
. . .	. . .	. . .	. . .
$\rho^2_{1j\cdot 23\cdots j-1} = 0$	$-N \log (1 - r^2_{1j\cdot 23\cdots j-1})$	1	$\frac{2j - 1}{2N}$
. . .	. . .	. . .	. . .
$\rho^2_{1k\cdot 23\cdots k-1} = 0$	$-N \log (1 - r^2_{1k\cdot 23\cdots k-1})$	1	$\frac{2k - 1}{2N}$
$\rho^2_{1\cdot 2\cdots k} = 0$	$-N \log (1 - r^2_{1\cdot 2\cdots k})$	$k - 1$	$\frac{k^2 - 1}{2N}$

8.27. Show that the analysis in table 8.3 (problem 8.26) for $k_1 = k_2 = \cdots = k_m = 1$ is similar to that in table 8.1 (problem 8.22).

8.28. Show that an analysis of the information component due to $|\mathbf{\Sigma}_{jj\cdot 12\cdots j-1}|$ in table 8.3 (problem 8.26) is given by table 8.4, with $l = k_1 + k_2 + \cdots + k_j$.

8.29. Show that for $k_j = k_{j-1} = 1$ the partial independence component in table 8.4 (problem 8.28) reduces to that for the hypothesis $\rho^2_{jj-1\cdot 12\cdots j-2} = 0$, as would be given by a result similar to that in table 8.2 (problem 8.24). (Cf. problem 10.20 of chapter 9.)

8.30. Relate the analysis in table 8.1 (problem 8.22) for $k = 3$ with that in table 3.3 of chapter 8.

8.31. Relate the analysis in table 8.2 (problem 8.24) for $k = 3$ with that in table 3.5 of chapter 8.

8.32. Let the random vector $\mathbf{x}$ be subjected to the nonsingular linear transformation $\mathbf{y} = \mathbf{A}\mathbf{x}$. Show that (see section 3 in chapter 9):

(*a*) $\boldsymbol{\mu}_y = \mathbf{A}\boldsymbol{\mu}_x$.

(*b*) $\mathbf{\Sigma}_y = \mathbf{A}\mathbf{\Sigma}_x\mathbf{A}'$.

(*c*) $\bar{\mathbf{y}} = \mathbf{A}\bar{\mathbf{x}}$.

(*d*) $\mathbf{S}_y = \mathbf{A}\mathbf{S}_x\mathbf{A}'$.

(*e*) $\hat{I}(*:2)$ in (3.14) is equal to

$$\frac{n}{2}(\bar{\mathbf{y}} - \boldsymbol{\mu}_y)'\mathbf{\Sigma}_{y2}^{-1}(\bar{\mathbf{y}} - \boldsymbol{\mu}_y) + \frac{N}{2}\left(\log\frac{|\mathbf{\Sigma}_{y2}|}{|\mathbf{S}_y|} - k + \operatorname{tr}\mathbf{S}_y\mathbf{\Sigma}_{y2}^{-1}\right).$$

8.33. In problem 8.32 let

$$\mathbf{\Sigma}_x = \begin{pmatrix}\mathbf{\Sigma}_{11} & \mathbf{\Sigma}_{12}\\ \mathbf{\Sigma}_{21} & \mathbf{\Sigma}_{22}\end{pmatrix},\qquad \mathbf{S}_x = \begin{pmatrix}\mathbf{S}_{11} & \mathbf{S}_{12}\\ \mathbf{S}_{21} & \mathbf{S}_{22}\end{pmatrix},\qquad \mathbf{A} = \begin{pmatrix}\mathbf{I}_{k_1} & \mathbf{0}\\ -\mathbf{\Sigma}_{21}\mathbf{\Sigma}_{11}^{-1} & \mathbf{I}_{k_2}\end{pmatrix}.$$

Show that (see section 7 in chapter 9):

(*a*) $\mathbf{A}^{-1} = \begin{pmatrix}\mathbf{I}_{k_1} & \mathbf{0}\\ \mathbf{\Sigma}_{21}\mathbf{\Sigma}_{11}^{-1} & \mathbf{I}_{k_2}\end{pmatrix}.$

(*b*) $\mathbf{\Sigma}_y = \begin{pmatrix}\mathbf{\Sigma}_{y11} & \mathbf{\Sigma}_{y12}\\ \mathbf{\Sigma}_{y21} & \mathbf{\Sigma}_{y22}\end{pmatrix}$, $\mathbf{\Sigma}_{y11} = \mathbf{\Sigma}_{11}$, $\mathbf{\Sigma}_{y12} = \mathbf{0} = \mathbf{\Sigma}'_{y21}$, $\mathbf{\Sigma}_{y22} = \mathbf{\Sigma}_{22} - \mathbf{\Sigma}_{21}\mathbf{\Sigma}_{11}^{-1}\mathbf{\Sigma}_{12} = \mathbf{\Sigma}_{22\cdot 1}$.

(*c*) $\mathbf{S}_y = \begin{pmatrix}\mathbf{S}_{y11} & \mathbf{S}_{y12}\\ \mathbf{S}_{y21} & \mathbf{S}_{y22}\end{pmatrix}$, $\mathbf{S}_{y11} = \mathbf{S}_{11}$, $\mathbf{S}_{y12} = \mathbf{S}_{12} - \mathbf{S}_{11}\mathbf{\Sigma}_{11}^{-1}\mathbf{\Sigma}_{12} = \mathbf{S}'_{y21}$, $\mathbf{S}_{y22} = \mathbf{S}_{22} - \mathbf{\Sigma}_{21}\mathbf{\Sigma}_{11}^{-1}\mathbf{S}_{12} - \mathbf{S}_{21}\mathbf{\Sigma}_{11}^{-1}\mathbf{\Sigma}_{12} + \mathbf{\Sigma}_{21}\mathbf{\Sigma}_{11}^{-1}\mathbf{S}_{11}\mathbf{\Sigma}_{11}^{-1}\mathbf{\Sigma}_{12}$.

(*d*) $\mathbf{S}_{y22\cdot 1} = \mathbf{S}_{y22} - \mathbf{S}_{y21}\mathbf{S}_{y11}^{-1}\mathbf{S}_{y12} = \mathbf{S}_{22\cdot 1} = \mathbf{S}_{22} - \mathbf{S}_{21}\mathbf{S}_{11}^{-1}\mathbf{S}_{12}$.

(*e*) $2\hat{I}(H_1:H_2)$ in (3.15) is equal to

$$N\left(\log\frac{|\mathbf{\Sigma}_{y11}|}{|\mathbf{S}_{y11}|} - k_1 + \operatorname{tr}\mathbf{S}_{y11}\mathbf{\Sigma}_{y11}^{-1} + \log\frac{|\mathbf{\Sigma}_{y22}|}{|\mathbf{S}_{y22}|} - k_2 + \operatorname{tr}\mathbf{S}_{y22}\mathbf{\Sigma}_{y22}^{-1} + \log\frac{|\mathbf{S}_{y11}||\mathbf{S}_{y22}|}{|\mathbf{S}_y|}\right).$$

TABLE 8.3

Component due to	Information	D.F.	β^2
$\lvert\mathbf{\Sigma}_{22\cdot1}\rvert$	$N\log\dfrac{\lvert\mathbf{S}_{22}\rvert}{\lvert\mathbf{S}_{22\cdot1}\rvert}=N\log\dfrac{\lvert\mathbf{R}_{22}\rvert}{\lvert\mathbf{R}_{22\cdot1}\rvert}$	k_1k_2	$\dfrac{k_1k_2(k_1+k_2+1)}{2N}$
. . .	. . .	. . .	. . .
$\lvert\mathbf{\Sigma}_{jj\cdot12\cdots j-1}\rvert$	$N\log\dfrac{\lvert\mathbf{S}_{jj}\rvert}{\lvert\mathbf{S}_{jj\cdot12\cdots j-1}\rvert}=N\log\dfrac{\lvert\mathbf{R}_{jj}\rvert}{\lvert\mathbf{R}_{jj\cdot12\cdots j-1}\rvert}$	$(k_1+\cdots+k_{j-1})k_j$	$\dfrac{(k_1+\cdots+k_{j-1})k_j(k_1+\cdots+k_j+1)}{2N}$
. . .	. . .	. . .	. . .
$\lvert\mathbf{\Sigma}_{mm\cdot12\cdots m-1}\rvert$	$N\log\dfrac{\lvert\mathbf{S}_{mm}\rvert}{\lvert\mathbf{S}_{mm\cdot12\cdots m-1}\rvert}=N\log\dfrac{\lvert\mathbf{R}_{mm}\rvert}{\lvert\mathbf{R}_{mm\cdot12\cdots m-1}\rvert}$	$(k_1+\cdots+k_{m-1})k_m$	$\dfrac{(k_1+\cdots+k_{m-1})k_m(k_1+\cdots+k_m+1)}{2N}$
$2\hat{I}(H_1:H_2(\cdot))$	$N\log\dfrac{\lvert\mathbf{S}_{11}\rvert\cdots\lvert\mathbf{S}_{mm}\rvert}{\lvert\mathbf{S}\rvert}=N\log\dfrac{\lvert\mathbf{R}_{11}\rvert\cdots\lvert\mathbf{R}_{mm}\rvert}{\lvert\mathbf{R}\rvert}$	$\sum_{i<j}k_ik_j$	$\dfrac{2k^3+3k^2-k-\sum_{i=1}^{m}(2k_i^3+3k_i^2-k_i)}{12N}$

TABLE 8.4

Component due to	Information	D.F.	β^2
$\lvert\boldsymbol{\Sigma}_{jj\cdot 12\cdots j-2}\rvert$	$N \log \dfrac{\lvert\mathbf{S}_{jj}\rvert}{\lvert\mathbf{S}_{jj\cdot 12\cdots j-2}\rvert} = N \log \dfrac{\lvert\mathbf{R}_{jj}\rvert}{\lvert\mathbf{R}_{jj\cdot 12\cdots j-2}\rvert}$	$(l - k_j - k_{j-1})k_j$	$\dfrac{(l - k_j - k_{j-1})k_j(l - k_{j-1} + 1)}{2N}$
Partial independence	$N \log \dfrac{\lvert\mathbf{S}_{jj\cdot 12\cdots j-2}\rvert}{\lvert\mathbf{S}_{jj\cdot 12\cdots j-1}\rvert} = N \log \dfrac{\lvert\mathbf{R}_{jj\cdot 12\cdots j-2}\rvert}{\lvert\mathbf{R}_{jj\cdot 12\cdots j-1}\rvert}$	$k_{j-1}k_j$	$\dfrac{k_j k_{j-1}(2l - k_j - k_{j-1} + 1)}{2N}$
$\lvert\boldsymbol{\Sigma}_{jj\cdot 12\cdots j-1}\rvert$	$N \log \dfrac{\lvert\mathbf{S}_{jj}\rvert}{\lvert\mathbf{S}_{jj\cdot 12\cdots j-1}\rvert} = N \log \dfrac{\lvert\mathbf{R}_{jj}\rvert}{\lvert\mathbf{R}_{jj\cdot 12\cdots j-1}\rvert}$	$(l - k_j)k_j$	$\dfrac{(l - k_j)k_j(l + 1)}{2N}$

8.34. Show that table 8.5 is an analysis of $2\hat{I}(H_1:H_2)$ in (3.15) (see problems 8.32 and 8.33, and table 3.1, $k = k_1 + k_2$).

TABLE 8.5

Component due to	Information	D.F.	β^2
$\mathbf{\Sigma}_{11}$	$N\left(\log \frac{\lvert\mathbf{\Sigma}_{11}\rvert}{\lvert\mathbf{S}_{11}\rvert} - k_1 + \text{tr } \mathbf{S}_{11}\mathbf{\Sigma}_{11}^{-1}\right)$	$\frac{k_1(k_1+1)}{2}$	$\frac{2k_1^3 + 3k_1^2 - k_1}{12N}$
$\mathbf{\Sigma}_{22\cdot1}$	$N\left(\log \frac{\lvert\mathbf{\Sigma}_{22\cdot1}\rvert}{\lvert\mathbf{S}_{y22}\rvert} - k_2 + \text{tr } \mathbf{S}_{y22}\mathbf{\Sigma}_{22\cdot1}^{-1}\right)$	$\frac{k_2(k_2+1)}{2}$	$\frac{2k_2^3 + 3k_2^2 - k_2}{12N}$
$\mathbf{\Sigma}_{21}\mathbf{\Sigma}_{11}^{-1}$	$N \log \frac{\lvert\mathbf{S}_{y22}\rvert}{\lvert\mathbf{S}_{22\cdot1}\rvert}$	k_1k_2	$\frac{k_1k_2(k+1)}{2N}$
$2\hat{I}(H_1:H_2)$ (3.15)	$N\left(\log \frac{\lvert\mathbf{\Sigma}_2\rvert}{\lvert\mathbf{S}\rvert} - k + \text{tr } \mathbf{S}\mathbf{\Sigma}_2^{-1}\right)$	$\frac{k(k+1)}{2}$	$\frac{2k^3 + 3k^2 - k}{12N}$

8.35. In (3.13) let $\mathbf{\Sigma}_2 = \sigma^2\mathbf{\Sigma}_3$ and denote the null hypothesis with σ^2 and $\mathbf{\Sigma}_3$ specified by $H_3(\sigma^2)$, and the null hypothesis with $\mathbf{\Sigma}_3$ specified, but σ^2 not specified, by $H_3(\cdot)$. Show that [cf. Anderson (1958, p. 262), Mauchly (1940)]:

(*a*) $2\hat{I}(H_1:H_3(\sigma^2)) = N\left(\log \frac{\lvert\mathbf{\Sigma}_3\rvert}{\lvert\mathbf{S}\rvert} + k \log \sigma^2 - k + \frac{1}{\sigma^2} \text{tr } \mathbf{S}\mathbf{\Sigma}_3^{-1}\right)$.

(*b*) $\min_{\sigma^2} 2\hat{I}(H_1:H_3(\sigma^2))$ is given for $\hat{\sigma}^2 = \frac{1}{k} \text{tr } \mathbf{S}\mathbf{\Sigma}_3^{-1}$.

(*c*) $2\hat{I}(H_1:H_3(\cdot)) = \min_{\sigma^2} 2\hat{I}(H_1:H_3(\sigma^2)) = N \log \frac{\lvert\mathbf{\Sigma}_3\rvert}{\lvert\mathbf{S}_3\rvert}$, where $\mathbf{S} = \hat{\sigma}^2\mathbf{S}_3$.

CHAPTER 13

Linear Discriminant Functions

1. INTRODUCTION

In this chapter we shall continue the discussion initiated in section 9 of chapter 9. We have already studied linear discriminant functions, with assumptions of equality about the means or covariance matrices, in section 8 of chapter 11, section 3.5 of chapter 12, section 4.2 of chapter 12, and section 5.2 of chapter 12. For these linear discriminant functions, we obtained the same coefficient matrix (vector) $\boldsymbol{\alpha}$ of $y = \boldsymbol{\alpha}'\mathbf{x}$ whether we determined $\boldsymbol{\alpha}$ to maximize $I(1:2; y)$ or $J(1, 2; y)$. However, in section 9 of chapter 9 we saw that different linear discriminant functions arise according as we maximize $I(1:2; y)$, $I(2:1; y)$, or $J(1, 2; y)$.

2. ITERATION

In section 9 of chapter 9 we formulated the equations to be solved for the coefficients of the linear discriminant function as (9.5) of chapter 9, that is,

$$\boldsymbol{\Sigma}_1\boldsymbol{\alpha} - \lambda\boldsymbol{\Sigma}_2\boldsymbol{\alpha} = \gamma\boldsymbol{\delta}, \tag{2.1}$$

where λ and γ are defined in section 9 of chapter 9 according as it is $I(1:2; y)$, $I(2:1; y)$, or $J(1, 2; y)$ which is to be maximized. We remark that in the derivation of (9.5) in chapter 9, dividing by an appropriate factor, we might also have formulated the equations as

$$\boldsymbol{\Sigma}_2\boldsymbol{\alpha} - \lambda'\boldsymbol{\Sigma}_1\boldsymbol{\alpha} = \gamma'\boldsymbol{\delta}, \tag{2.2}$$

where for maximizing $I(1:2; y)$,

$$\lambda' = \frac{\boldsymbol{\alpha}'\boldsymbol{\Sigma}_2\boldsymbol{\alpha}(\boldsymbol{\alpha}'\boldsymbol{\Sigma}_2\boldsymbol{\alpha} - \boldsymbol{\alpha}'\boldsymbol{\Sigma}_1\boldsymbol{\alpha})}{\boldsymbol{\alpha}'\boldsymbol{\Sigma}_1\boldsymbol{\alpha}(\boldsymbol{\alpha}'\boldsymbol{\Sigma}_2\boldsymbol{\alpha} - \boldsymbol{\alpha}'\boldsymbol{\Sigma}_1\boldsymbol{\alpha} - (\boldsymbol{\alpha}'\boldsymbol{\delta})^2)}, \tag{2.3}$$

$$\gamma' = \frac{(\boldsymbol{\alpha}'\boldsymbol{\Sigma}_2\boldsymbol{\alpha})(\boldsymbol{\alpha}'\boldsymbol{\delta})}{\boldsymbol{\alpha}'\boldsymbol{\Sigma}_1\boldsymbol{\alpha} - \boldsymbol{\alpha}'\boldsymbol{\Sigma}_2\boldsymbol{\alpha} + (\boldsymbol{\alpha}'\boldsymbol{\delta})^2};$$

for maximizing $I(2:1; y)$,

$$(2.4)\quad \lambda' = \frac{\boldsymbol{\alpha}'\boldsymbol{\Sigma}_2\boldsymbol{\alpha}(\boldsymbol{\alpha}'\boldsymbol{\Sigma}_1\boldsymbol{\alpha} - \boldsymbol{\alpha}'\boldsymbol{\Sigma}_2\boldsymbol{\alpha} - (\boldsymbol{\alpha}'\boldsymbol{\delta})^2)}{\boldsymbol{\alpha}'\boldsymbol{\Sigma}_1\boldsymbol{\alpha}(\boldsymbol{\alpha}'\boldsymbol{\Sigma}_1\boldsymbol{\alpha} - \boldsymbol{\alpha}'\boldsymbol{\Sigma}_2\boldsymbol{\alpha})}, \qquad \gamma' = \frac{(\boldsymbol{\alpha}'\boldsymbol{\Sigma}_2\boldsymbol{\alpha})(\boldsymbol{\alpha}'\boldsymbol{\delta})}{\boldsymbol{\alpha}'\boldsymbol{\Sigma}_1\boldsymbol{\alpha} - \boldsymbol{\alpha}'\boldsymbol{\Sigma}_2\boldsymbol{\alpha}};$$

and for maximizing $J(1, 2; y)$,

$$(2.5)\quad \lambda' = \frac{\boldsymbol{\alpha}'\boldsymbol{\Sigma}_2\boldsymbol{\alpha}((\boldsymbol{\alpha}'\boldsymbol{\Sigma}_2\boldsymbol{\alpha})^2 - (\boldsymbol{\alpha}'\boldsymbol{\Sigma}_1\boldsymbol{\alpha})^2 + (\boldsymbol{\alpha}'\boldsymbol{\delta})^2(\boldsymbol{\alpha}'\boldsymbol{\Sigma}_2\boldsymbol{\alpha}))}{\boldsymbol{\alpha}'\boldsymbol{\Sigma}_1\boldsymbol{\alpha}((\boldsymbol{\alpha}'\boldsymbol{\Sigma}_2\boldsymbol{\alpha})^2 - (\boldsymbol{\alpha}'\boldsymbol{\Sigma}_1\boldsymbol{\alpha})^2 - (\boldsymbol{\alpha}'\boldsymbol{\delta})^2(\boldsymbol{\alpha}'\boldsymbol{\Sigma}_1\boldsymbol{\alpha}))},$$

$$\gamma' = \frac{(\boldsymbol{\alpha}'\boldsymbol{\delta})(\boldsymbol{\alpha}'\boldsymbol{\Sigma}_2\boldsymbol{\alpha})(\boldsymbol{\alpha}'\boldsymbol{\Sigma}_1\boldsymbol{\alpha} + \boldsymbol{\alpha}'\boldsymbol{\Sigma}_2\boldsymbol{\alpha})}{(\boldsymbol{\alpha}'\boldsymbol{\Sigma}_1\boldsymbol{\alpha})^2 - (\boldsymbol{\alpha}'\boldsymbol{\Sigma}_2\boldsymbol{\alpha})^2 + (\boldsymbol{\alpha}'\boldsymbol{\delta})^2(\boldsymbol{\alpha}'\boldsymbol{\Sigma}_1\boldsymbol{\alpha})}.$$

For convenience, setting the proportionality factors γ and γ' equal to 1, (2.1) and (2.2) are

$$(2.6)\qquad \boldsymbol{\Sigma}_1\boldsymbol{\alpha} - \lambda\boldsymbol{\Sigma}_2\boldsymbol{\alpha} = \boldsymbol{\delta}, \qquad \boldsymbol{\Sigma}_2\boldsymbol{\alpha} - \lambda'\boldsymbol{\Sigma}_1\boldsymbol{\alpha} = \boldsymbol{\delta},$$

where $\lambda = 1/\lambda'$ for each case. When λ, λ' are not solutions of $|\boldsymbol{\Sigma}_1 - \lambda\boldsymbol{\Sigma}_2| = 0$, $|\boldsymbol{\Sigma}_2 - \lambda'\boldsymbol{\Sigma}_1| = 0$ respectively, (2.6) yields the following implicit solution for $\boldsymbol{\alpha}$:

$$(2.7)\qquad \boldsymbol{\alpha} = (\boldsymbol{\Sigma}_1 - \lambda\boldsymbol{\Sigma}_2)^{-1}\boldsymbol{\delta}, \qquad \boldsymbol{\alpha} = (\boldsymbol{\Sigma}_2 - \lambda'\boldsymbol{\Sigma}_1)^{-1}\boldsymbol{\delta}.$$

If λ is a known number, (2.7) yields directly the value of $\boldsymbol{\alpha}$. However λ, λ' in all instances are functions of $\boldsymbol{\alpha}$. Initial or entering values of $\boldsymbol{\alpha}$ are therefore required to begin an iterative procedure.

The entering value for $\boldsymbol{\alpha}$ is taken to be, respectively, as

$$(2.8)\qquad \boldsymbol{\alpha}_0 = \boldsymbol{\Sigma}_1^{-1}\boldsymbol{\delta}, \qquad \boldsymbol{\alpha}_0 = \boldsymbol{\Sigma}_2^{-1}\boldsymbol{\delta}.$$

It should be clear that the same initial value of $\boldsymbol{\alpha}$ will serve each of the iterations necessary to maximize either $I(1:2; y)$, $I(2:1; y)$, or $J(1, 2; y)$.

With $\boldsymbol{\alpha}_0$ determined, values for $\boldsymbol{\alpha}_0'\boldsymbol{\delta}$, $\boldsymbol{\alpha}_0'\boldsymbol{\Sigma}_1\boldsymbol{\alpha}_0$, $\boldsymbol{\alpha}_0'\boldsymbol{\Sigma}_2\boldsymbol{\alpha}_0$ are found and then λ_0 or λ_0'. Cycle 1 is begun by entering with λ_0 or λ_0' to find a new set of α's from

$$(2.9)\qquad \boldsymbol{\alpha}_1 = (\boldsymbol{\Sigma}_1 - \lambda_0\boldsymbol{\Sigma}_2)^{-1}\boldsymbol{\delta}, \qquad \boldsymbol{\alpha}_1 = (\boldsymbol{\Sigma}_2 - \lambda_0'\boldsymbol{\Sigma}_1)^{-1}\boldsymbol{\delta},$$

and then determining $\boldsymbol{\alpha}_1'\boldsymbol{\delta}$, $\boldsymbol{\alpha}_1'\boldsymbol{\Sigma}_1\boldsymbol{\alpha}_1$, $\boldsymbol{\alpha}_1'\boldsymbol{\Sigma}_2\boldsymbol{\alpha}_1$, and then λ_1 or λ_1', thus completing the first cycle. This procedure is continued until the difference in successive $\boldsymbol{\alpha}$'s, or more appropriately in successive α_i/α_1, is as small as desired.

We shall replace population parameters by the best unbiased sample estimates.

3. EXAMPLE

We shall illustrate the procedures described with data from Smith (1947) (see example 5.1 of chapter 12). The computations were performed by S. W. Greenhouse. The pertinent values are:

$$\bar{\mathbf{x}}_1 = \begin{pmatrix} 20.80 \\ 12.32 \end{pmatrix}, \quad \bar{\mathbf{x}}_2 = \begin{pmatrix} 12.80 \\ 36.40 \end{pmatrix}, \quad \mathbf{d} = \begin{pmatrix} 8.00 \\ -24.08 \end{pmatrix},$$

$$\mathbf{S}_1 = \begin{pmatrix} 6.92 & -5.27 \\ -5.27 & 40.89 \end{pmatrix}, \quad \mathbf{S}_2 = \begin{pmatrix} 36.75 & 13.92 \\ 13.92 & 287.92 \end{pmatrix}, \quad |\mathbf{S}_1| = 255.1859,$$

$$|\mathbf{S}_2| = 10387.2936, \qquad \mathbf{S}_1^{-1} = \begin{pmatrix} 0.16023613 & 0.02065161 \\ 0.02065161 & 0.02711749 \end{pmatrix},$$

$$\mathbf{S}_2^{-1} = \begin{pmatrix} 0.02771848 & -0.00134010 \\ -0.00134010 & 0.00353798 \end{pmatrix},$$

$$\hat{I}(1:2) = \frac{1}{2}\left(\log \frac{|\mathbf{S}_2|}{|\mathbf{S}_1|} - 2 + \operatorname{tr} \mathbf{S}_1\mathbf{S}_2^{-1}\right) + \frac{1}{2}\mathbf{d}'\mathbf{S}_2^{-1}\mathbf{d}$$

$$= 1.028432 + 2.170861 = 3.199293,$$

$$\hat{I}(2:1) = 4.282444 + 9.010994 = 13.293438,$$

$$\hat{J}(1, 2) = 5.310876 + 11.181855 = 16.492731.$$

We shall find the linear discriminant function $y = a_1x_1 + a_2x_2$, $a_1 = 1$, $a_2 = \alpha_2/\alpha_1$, maximizing $\hat{I}(1:2; y)$; similar steps occur for the procedure leading to the linear discriminant function maximizing $\hat{I}(2:1; y)$ and $\hat{J}(1, 2; y)$.

We obtain the initial value from (2.8), that is,

$$\boldsymbol{\alpha}_0 = \mathbf{S}_2^{-1}\mathbf{d} = \begin{pmatrix} 0.02771848 & -0.00134010 \\ -0.00134010 & 0.00353798 \end{pmatrix}\begin{pmatrix} 8.00 \\ -24.08 \end{pmatrix} \tag{3.1}$$

$$= \begin{pmatrix} 0.25401745 \\ -0.09591536 \end{pmatrix},$$

so that $a_{01} = 1.000000$, $a_{02} = -0.377594$. From these we get,

$$\mathbf{a}_0'\mathbf{d} = (1, -0.377594)\begin{pmatrix} 8.00 \\ -24.08 \end{pmatrix} = 17.092464,$$

$$\mathbf{a}_0'\mathbf{S}_1\mathbf{a}_0 = (1, -0.377594)\begin{pmatrix} 6.92 & -5.27 \\ -5.27 & 40.89 \end{pmatrix}\begin{pmatrix} 1 \\ -0.377594 \end{pmatrix} = 16.729814,$$

$$\mathbf{a}_0'\mathbf{S}_2\mathbf{a}_0 = (1, -0.377594)\begin{pmatrix} 36.75 & 13.92 \\ 13.92 & 287.92 \end{pmatrix}\begin{pmatrix} 1 \\ -0.377594 \end{pmatrix} = 67.288553,$$

and from (2.3),

$$\lambda_0' = \frac{(67.288553)(67.288553 - 16.729814)}{16.729814(67.288553 - 16.729814 - (17.092464)^2)} = -0.8417.$$

Cycle 1

$$(\mathbf{S}_2 + 0.8417\mathbf{S}_1) = \begin{pmatrix} 42.574564 & 9.484241 \\ 9.484241 & 322.337113 \end{pmatrix},$$

$$|\mathbf{S}_2 + 0.8417\mathbf{S}_1| = 13633.4112,$$

$$\boldsymbol{\alpha}_1 = (\mathbf{S}_2 + 0.8417\mathbf{S}_1)^{-1}\mathbf{d} = \begin{pmatrix} 0.02364317 & -0.00069566 \\ -0.00069566 & 0.00312281 \end{pmatrix}\begin{pmatrix} 8.00 \\ -24.08 \end{pmatrix}$$

$$= \begin{pmatrix} 0.20589685 \\ -0.08076254 \end{pmatrix},$$

$$a_{11} = 1, \qquad a_{12} = -0.392248,$$

$$\mathbf{a}_1'\mathbf{d} = (1, -0.392248)\begin{pmatrix} 8.00 \\ -24.08 \end{pmatrix} = 17.445332,$$

$$\mathbf{a}_1'\mathbf{S}_1\mathbf{a}_1 = (1, -0.392248)\begin{pmatrix} 6.92 & -5.27 \\ -5.27 & 40.89 \end{pmatrix}\begin{pmatrix} 1 \\ -0.392248 \end{pmatrix} = 17.345548,$$

$$\mathbf{a}_1'\mathbf{S}_2\mathbf{a}_1 = (1, -0.392248)\begin{pmatrix} 36.75 & 13.92 \\ 13.92 & 287.92 \end{pmatrix}\begin{pmatrix} 1 \\ -0.392248 \end{pmatrix} = 70.128611,$$

and from (2.3),

$$\lambda_1' = \frac{70.128611(70.128611 - 17.345548)}{17.345548(70.128611 - 17.345548 - (17.445332)^2)} = -0.848333.$$

Cycle 2

$$(\mathbf{S}_2 + 0.8483\mathbf{S}_1) = \begin{pmatrix} 42.620236 & 9.449459 \\ 9.449459 & 322.606987 \end{pmatrix},$$

$$|\mathbf{S}_2 + 0.8483\mathbf{S}_1| = 13660.2936,$$

$$\boldsymbol{\alpha}_2 = (\mathbf{S}_2 + 0.8483\mathbf{S}_1)^{-1}\mathbf{d} = \begin{pmatrix} 0.02361640 & -0.00069175 \\ -0.00069175 & 0.00312001 \end{pmatrix}\begin{pmatrix} 8.00 \\ -24.08 \end{pmatrix}$$

$$= \begin{pmatrix} 0.20558854 \\ -0.08066384 \end{pmatrix},$$

$$a_{21} = 1, \qquad a_{22} = -0.392356,$$

$$\mathbf{a}_2'\mathbf{d} = (1, -0.392356)\begin{pmatrix} 8.00 \\ -24.08 \end{pmatrix} = 17.447932,$$

$$\mathbf{a}_2'\mathbf{S}_1\mathbf{a}_2 = (1, -0.392356)\begin{pmatrix} 6.92 & -5.27 \\ -5.27 & 40.89 \end{pmatrix}\begin{pmatrix} 1 \\ -0.392356 \end{pmatrix} = 17.350162,$$

$$\mathbf{a}_2'\mathbf{S}_2\mathbf{a}_2 = (1, -0.392356)\begin{pmatrix} 36.75 & 13.92 \\ 13.92 & 287.92 \end{pmatrix}\begin{pmatrix} 1 \\ -0.392356 \end{pmatrix} = 70.150078,$$

and from (2.3),

$$\lambda_2' = \frac{70.150078(70.150078 - 17.350162)}{17.350162(70.150078 - 17.350162 - (17{\cdot}447932)^2)} = -\ 0.848388.$$

A third cycle was computed, although two cycles would seem to be sufficient in view of the negligible change in λ'. The value of

$$\hat{I}(1:2;y) = \frac{1}{2}\left(\log\frac{\boldsymbol{\alpha}'\mathbf{S}_2\boldsymbol{\alpha}}{\boldsymbol{\alpha}'\mathbf{S}_1\boldsymbol{\alpha}} - 1 + \frac{\boldsymbol{\alpha}'\mathbf{S}_1\boldsymbol{\alpha}}{\boldsymbol{\alpha}'\mathbf{S}_2\boldsymbol{\alpha}} + \frac{(\boldsymbol{\alpha}'\mathbf{d})^2}{\boldsymbol{\alpha}'\mathbf{S}_2\boldsymbol{\alpha}}\right)$$

was also computed for the initial value and cycle 3. The various values are summarized in table 3.1.

TABLE 3.1

i	0	1	2	3
λ_i	−0.8417	−0.848333	−0.848388	−0.8483901
a_{i2}	−0.377594	−0.392248	−0.392356	−0.392357
$\mathbf{a}_i'\mathbf{S}_1\mathbf{a}_i$	16.729814	17.345548	17.350162	17.350213
$\mathbf{a}_i'\mathbf{S}_2\mathbf{a}_i$	67.288553	70.128611	70.150078	70.150338
$\mathbf{a}_i'\mathbf{d}$	17.092464	17.445332	17.447932	17.447957
$\hat{I}(1:2;y)$	2.4911	2.492030	2.492031	2.492031

When the basis for the iteration is $\boldsymbol{\alpha} = (\boldsymbol{\Sigma}_1 - \lambda\boldsymbol{\Sigma}_2)^{-1}\boldsymbol{\delta}$, the corresponding values are summarized in table 3.2.

TABLE 3.2

i	0	1	2	3
λ_i	−0.9409	−1.1763	−1.1787	−1.178703
a_{i2}	−0.621689	−0.395810	−0.392385	−0.392357
$\mathbf{a}_i'\mathbf{S}_1\mathbf{a}_i$	29.276464	17.497910	17.351408	17.350213
$\mathbf{a}_i'\mathbf{S}_2\mathbf{a}_i$	130.722394	70.837924	70.155892	70.150338

For this example, note that both procedures yield the same a_{32} and exactly reciprocal λ's after only 3 cycles. This number of cycles need not be 3 in general. We remark that the values across the rows of each table are monotonic.

The values for the linear discriminant function maximizing $\hat{I}(2:1;y)$ are summarized in table 3.3.

TABLE 3.3

i	0	1	2	3
λ_i	0.0361	0.036582	0.036596	0.036597
a_{i2}	−0.621689	−0.843193	−0.848904	−0.849072
$\mathbf{a}_i'\mathbf{S}_1\mathbf{a}_i$	29.276464	44.878981	45.334336	45.347760
$\mathbf{a}_i'\mathbf{S}_2\mathbf{a}_i$	130.722394	217.979141	220.602606	220.679986
$\hat{I}(2:1;y)$	9.9956	10.063654	10.063671	10.063678

The values for the linear discriminant function maximizing $\hat{J}(1, 2; y)$ are summarized in table 3.4.

TABLE 3.4

i	0	1	2	3
λ_i	0.00206	0.00231	0.0023397	0.0023435
a_{i2}	−0.621689	−0.628501	−0.629353	−0.629456
$\mathbf{a}_i'\mathbf{S}_1\mathbf{a}_i$	29.276464	29.696523	29.749296	29.755698
$\mathbf{a}_i'\mathbf{S}_2\mathbf{a}_i$	130.722394	132.984963	133.269606	133.304168
$\hat{J}(1, 2; y)$	12.3739	12.37405	12.37406	12.37406

We thus have the three linear discriminant functions:

$$\begin{aligned} &\max \hat{I}(1:2;y): \quad y = x_1 - 0.3924x_2, \\ &\max \hat{I}(2:1;y): \quad y = x_1 - 0.8491x_2, \\ &\max \hat{J}(1,2;y): \quad y = x_1 - 0.6295x_2. \end{aligned} \tag{3.2}$$

4. REMARK

Although it is clear that the procedure, including that for obtaining the initial values, did converge, we have no general proof that this procedure converges, or that a solution yielded by this procedure is the only one satisfying (2.6). For any two-variate problem however, $\hat{I}(1:2;y)$, $\hat{I}(2:1;y)$, and $\hat{J}(1, 2;y)$ are essentially functions of one unknown, the ratio α_2/α_1. The maximizing condition is a polynomial in this ratio and the properties of the roots can be studied. For $\hat{I}(1:2;y)$ and $\hat{I}(2:1;y)$ the polynomial is quartic, and for $\hat{J}(1,2;y)$ it is of the sixth degree. In each instance in section 3, there were only two real roots, a negative root yielding maximum $\hat{I}(1:2;y)$, $\hat{I}(2:1;y)$, and $\hat{J}(1,2;y)$ and a positive root yielding the minimum value in each case. The equations were solved by Newton's method and the negative roots maximizing $\hat{I}(1:2;y)$, $\hat{I}(2:1;y)$, and $J(1,2;y)$ respectively were -0.392357, -0.849083, and -0.629468. Reference to tables 3.2, 3.3, and 3.4 clearly indicates that the iteration is converging to these values and that the values obtained at the end of 2 cycles are correct to 4 decimals.

5. OTHER LINEAR DISCRIMINANT FUNCTIONS

Smith (1947) computed a linear discriminant function for these data by assuming the covariance matrix to be the same in both populations. The solution for $\boldsymbol{\alpha}$ is then

$$\boldsymbol{\alpha} = \mathbf{S}^{-1}\mathbf{d}, \tag{5.1}$$

where $N\mathbf{S} = N_1\mathbf{S}_1 + N_2\mathbf{S}_2$, $N = N_1 + N_2$. (See the last part of example 4.1 in chapter 12 and section 8.1 of chapter 11.) Smith's values, reduced to a basis comparable to (3.2), that is, so that $\alpha_1 = 1$, yield the discriminant function

$$y = x_1 - 0.3947x_2. \tag{5.2}$$

Since the two samples are of equal size, the linear discriminant function computed in accordance with section 4.2 of chapter 12, that is, the value of $\boldsymbol{\alpha}$ satisfying

$$\boldsymbol{\alpha} = \left(\frac{1}{n_1}\mathbf{S}_1 + \frac{1}{n_2}\mathbf{S}_2\right)^{-1}\mathbf{d}, \tag{5.3}$$

yields the linear discriminant function in (5.2).

Note that the linear discriminant function in (5.2) is almost the same here as the one in (3.2) resulting from maximizing $\hat{I}(1:2;y)$.

The discriminant function is often used to classify an individual on the basis of the observational vector $(x_1, x_2, \cdots, x_k)$ and a given linear compound $y = \gamma_1 x_1 + \cdots + \gamma_k x_k$ into one of two populations. [We use the matrix $\boldsymbol{\gamma}' = (\gamma_1, \gamma_2, \cdots, \gamma_k)$ to avoid confusion with the error probability α below.] The classification usually proceeds according to some rule such as: if y falls into the region A^*, classify into population π_1, say, and if y does not fall into A^*, classify into population π_2. It is clear that associated with this or any other classification scheme are two kinds of errors, namely, assigning y to population π_1 when it is in fact from population π_2, and assigning y to population π_2 when it is in fact from population π_1. Denote the probability of the first error by α and that of the second error by β. We can then form a minimum error criterion for finding a linear discriminant function, namely, for a given β, what linear function of the x's will minimize α? Since α and β are monotone functions of the normal deviates t_α and t_β respectively, it is simpler to work with the latter.

It may be shown that, for a given β, α will be minimized by maximizing

$$t_\alpha = \frac{\boldsymbol{\gamma}'\boldsymbol{\delta} - t_\beta(\boldsymbol{\gamma}'\boldsymbol{\Sigma}_1\boldsymbol{\gamma})^{1/2}}{(\boldsymbol{\gamma}'\boldsymbol{\Sigma}_2\boldsymbol{\gamma})^{1/2}}.$$

The usual calculus procedures lead to the equation

$$[t_\beta(\boldsymbol{\gamma}'\boldsymbol{\Sigma}_2\boldsymbol{\gamma})^{1/2}\boldsymbol{\Sigma}_1 + t_\alpha(\boldsymbol{\gamma}'\boldsymbol{\Sigma}_1\boldsymbol{\gamma})^{1/2}\boldsymbol{\Sigma}_2]\boldsymbol{\gamma} = (\boldsymbol{\gamma}'\boldsymbol{\Sigma}_1\boldsymbol{\gamma})^{1/2}(\boldsymbol{\gamma}'\boldsymbol{\Sigma}_2\boldsymbol{\gamma})^{1/2}\boldsymbol{\delta}, \tag{5.4}$$

which is nonlinear in the γ's. (The same equation is obtained if α is given and β minimized.) The solution here is best carried out, as in sections 2 and 3, by an iterative procedure on an equation of the form

$$(\boldsymbol{\Sigma}_1 + \lambda\boldsymbol{\Sigma}_2)\boldsymbol{\gamma} = \boldsymbol{\delta}, \tag{5.5}$$

where $\lambda = t_\alpha(\boldsymbol{\gamma}'\boldsymbol{\Sigma}_1\boldsymbol{\gamma})^{1/2}/t_\beta(\boldsymbol{\gamma}'\boldsymbol{\Sigma}_2\boldsymbol{\gamma})^{1/2}$. The iteration follows the identical steps given in section 2. The initial values of $\boldsymbol{\gamma}$ are obtained from $\boldsymbol{\gamma} = \boldsymbol{\Sigma}_1^{-1}\boldsymbol{\delta}$, which in turn determine $\boldsymbol{\gamma}'\boldsymbol{\Sigma}_1\boldsymbol{\gamma}$, $\boldsymbol{\gamma}'\boldsymbol{\Sigma}_2\boldsymbol{\gamma}$, and for a fixed t_β, t_α, these determine λ so that (5.5) becomes an explicit equation in $\boldsymbol{\gamma}$. The cycles can be continued until changes in t_α become as small as desired.

In this manner, two functions were found; one for $\beta = 0.05$ ($t_\beta = 1.645$) and the other for $\beta = 0.16$ ($t_\beta = 1.000$),

$$\max t_\alpha(t_\beta = 1.645):\quad y = x_1 - 0.4173x_2, \tag{5.6}$$

$$\max t_\alpha(t_\beta = 1.000):\quad y = x_1 - 0.3990x_2. \tag{5.7}$$

Although the linear discriminant function derived from the minimum error criterion is of interest in its own right, our interest in it at this point is to provide a base line for errors of classification with which the corresponding errors of the other linear discriminant functions may be compared. Note that the minimum error criterion does not provide a unique function, but yields a different discriminant for each t_β. Furthermore, the criterion used here gives only an approximation (although a very good one) to the actual linear function minimizing α for a fixed β. This is because the procedure assumes that the region for assigning to π_1, say $y > y_0$ (or $y < y_0$), is optimal when $\boldsymbol{\Sigma}_1 \neq \boldsymbol{\Sigma}_2$ as it is when $\boldsymbol{\Sigma}_1 = \boldsymbol{\Sigma}_2$. It is known that this is not so [see, for example, Penrose (1947) and section 2 of chapter 5].

6. COMPARISON OF THE VARIOUS LINEAR DISCRIMINANT FUNCTIONS

Before comparing the different linear discriminant functions obtained in sections 3 and 5, we present in table 6.1 the discrimination information values for the original x variables for x_1 and x_2 separately and jointly. Note that the values for x_2 are larger than the values for x_1 in all three measures; that is, an observation on the x_2 characteristic from either population has greater discrimination information in distinguishing between the two populations than does an observation on the x_1 characteristic. Reference to the lower portion of table 6.1, which presents the error made in classifying an observation from $\pi_2(\alpha)$ for a given error in classifying an observation from $\pi_1(\beta)$, indicates that x_2 also does better under an error criterion than does x_1.

TABLE 6.1

Information Measures	x_1	x_2	x_1 and x_2 Jointly
$\hat{I}(1:2)$	1.2997	1.5539	3.1993
$\hat{I}(2:1)$	5.9448	9.1351	13.2934
$\hat{J}(1, 2)$	7.2445	10.6890	16.4927
Errors			
α for $\beta = 0.01$	0.3782	0.2937	
α for $\beta = 0.05$	0.2723	0.2123	
α for $\beta = 0.16$	0.1879	0.1486	
min $(\alpha + \beta)$	0.3154	0.2580	
β	0.0738	0.0553	
α	0.2416	0.2027	

The last column in table 6.1 gives $\hat{I}(1:2)$, $\hat{I}(2:1)$, and $\hat{J}(1, 2)$ for x_1 and x_2 assumed to have a bivariate normal distribution in each of the two populations. To compute the efficiencies of the linear discriminant functions of x_1 and x_2, we note that the maximum $\hat{I}(1:2; y)$, $\hat{I}(2:1; y)$, and $\hat{J}(1, 2; y)$ each can attain is 3.1993, 13.2934, and 16.4927 respectively.

One last point of interest in table 6.1 is that in this example x_1 and x_2 jointly yield a value of $\hat{I}(1:2)$ which exceeds the sum of the value of $\hat{I}(1:2)$ for x_1 and for x_2. This is not true for $\hat{I}(2:1)$ and $\hat{J}(1, 2)$.

In table 6.2 are the data on six linear functions of x_1 and x_2, three obtained by maximizing the information measures, two obtained under an error principle, and one found by pooling variances and covariances between the two samples and proceeding as if the covariance matrices were the same. The upper portion of table 6.2 relates to the information measures in the linear compounds and the lower portion presents various error combinations in classifying observations, including the minimum total error that could be made with each function.

TABLE 6.2

Linear Discriminant Function Obtained by

	max $\hat{I}(1:2;y)$	max $\hat{I}(2:1;y)$	max $\hat{J}(1,2;y)$	Pooling Covariance Matrices	min α for $\beta = 0.05$	min α for $\beta = 0.16$
y	$x_1 - 0.3924x_2$	$x_1 - 0.8491x_2$	$x_1 - 0.6295x_2$	$x_1 - 0.3947x_2$	$x_1 - 0.4173x_2$	$x_1 - 0.3990x_2$
$\hat{I}(1:2;y)$	2.4920	2.2272	2.3728	2.4920	2.4897	2.4918
$\hat{I}(1:2;y)/\hat{I}(1:2)$	0.779	0.696	0.742	0.779	0.778	0.779
$\hat{I}(2:1;y)$	9.5962	10.0637	10.0012	9.6040	9.6711	9.6172
$\hat{I}(2:1;y)/\hat{I}(2:1)$	0.722	0.757	0.752	0.722	0.728	0.723
$\hat{J}(1,2;y)$	12.0882	12.2909	12.3741	12.0960	12.1608	12.1090
$\hat{J}(1,2;y)/\hat{J}(1,2)$	0.733	0.745	0.750	0.733	0.737	0.734
Errors						
α for $\beta = 0.01$	0.1771	0.1948	0.1823	0.1770	0.1764	0.1769
α for $\beta = 0.05$	0.1029	0.1212	0.1096	0.1029	0.1027	0.1028
α for $\beta = 0.16$	0.0564	0.0719	0.0626	0.0564	0.0564	0.0564
min $(\alpha + \beta)$	0.1525	0.1708	0.1591	0.1525	0.1522	0.1523
β	0.0438	0.0446	0.0434	0.0438	0.0435	0.0437
α	0.1087	0.1262	0.1157	0.1087	0.1087	0.1086

It is clear that the four linear discriminant functions obtained by (*a*) maximizing $\hat{I}(1:2;y)$, (*b*) pooling variances and covariances, (*c*) minimizing α for $\beta = 0.05$, and (*d*) minimizing α for $\beta = 0.16$, are very much alike with regard to discrimination information, divergence, and errors of classification. Maximizing $\hat{I}(2:1;y)$ and $\hat{J}(1,2;y)$ yields linear discriminant functions which have greater efficiencies than the other four with regard to $\hat{I}(2:1,y)$ and $\hat{J}(1,2;y)$, but have smaller efficiencies with regard to $\hat{I}(1:2;y)$, and have larger errors of classification than the other four.

From the point of view of information theory, the most interesting feature when the covariance matrices are not equal is the fact that $\hat{I}(1:2;y) \neq \hat{I}(2:1;y)$ and therefore maximizing these two measures, and the divergence measure $\hat{J}(1,2;y)$, yields three different linear functions. The example does suggest that at least one of the discriminant functions so obtained, in addition to having optimum properties associated with the information measure leading to it, will also possess optimum properties associated with an error criterion for finding a linear discriminant function.

An interesting problem that arises is the investigation of the properties of max $\hat{I}(1:2; y)$, max $\hat{I}(2:1; y)$, and max $\hat{J}(1, 2; y)$ to determine the conditions that will make one of them the best from the error point of view in numerical applications. It is conjectured that if π_1 is always taken as the population with the smaller covariance matrix (see the remark following lemma 5.1 in chapter 3), the linear discriminant function resulting from maximizing $\hat{I}(1:2; y)$ will always give smaller errors than the other two.

Note also that although max $\hat{I}(2:1; y)$ and max $\hat{J}(1, 2; y)$ do not do as well as the other functions on an error basis, they differ most from the linear discriminants derived from a basis other than the information measures. Further study of these two linear discriminants may elicit important properties within the information theory approach.

Of general interest is the fact that the linear discriminant function obtained by pooling the covariance matrices does so well. Whether this would continue to be true in other examples, or is peculiar to this one, remains to be investigated.

7. PROBLEMS

7.1. Derive (2.2), (2.3), (2.4), (2.5).

7.2. Derive the values in table 3.2.

7.3. Derive the values in table 3.3.

7.4. Derive the values in table 3.4.

7.5. Derive the two quartic and the sixth-degree polynomials mentioned in section 4.

7.6. Derive (5.4) and (5.5).

7.7. Derive the values in (5.7).

7.8. Derive (5.4) by minimizing β for a given α.

References

S. H. Abdel-Aty (1954), "Approximate formulae for the percentage points and the probability integral of the non-central χ^2 distribution," *Biometrika*, Vol. 41, pp. 538–540.

B. P. Adhikari and D. D. Joshi (1956), "Distance-Discrimination et résumé exhaustif," *Publs. inst. statist. univ. Paris*, Vol. 5, Fasc. 2, pp. 57–74.

A. C. Aitken and H. Silverstone (1941–43), "On the estimation of statistical parameters," *Proc. Roy. Soc. Edinburgh*, Vol. 61, pp. 186–194. (Issued separately Apr. 2, 1942.)

R. L. Anderson and T. A. Bancroft (1952), *Statistical Theory in Research*, McGraw-Hill Book Co., New York.

T. W. Anderson (1951), "The asymptotic distribution of certain characteristic roots and vectors," *Proceedings of the Second Berkeley Symposium on Mathematical Statistics and Probability*, Univ. Calif. Press, pp. 103–130.

——— (1958), *An Introduction to Multivariate Statistical Analysis*, John Wiley & Sons, New York.

W. R. Ashby (1956), *An Introduction to Cybernetics*, John Wiley & Sons, New York.

R. R. Bahadur (1954), "Sufficiency and statistical decision functions," *Ann. Math. Statist.*, Vol. 25, pp. 423–462.

E. W. Barankin (1949), "Locally best unbiased estimates," *Ann. Math. Statist.*, Vol. 20, pp. 477–501.

——— (1951), "Concerning some inequalities in the theory of statistical estimation," *Skand. Aktuar. Tidskr.*, Vol. 34, pp. 35–40.

——— and J. Gurland (1951), "On asymptotically normal, efficient estimators: I," *Univ. Calif. Publ. Statist.*, Vol. 1, No. 6, pp. 89–130.

Y. Bar-Hillel (1955), "An examination of information theory," *Philos. Sci.*, Vol. 22, pp. 86–105.

——— and R. Carnap (1953), "Semantic information," *Brit. J. Phil. Sci.*, Vol. 4, pp. 147–157; also appears with a discussion in *Communication Theory*, W. Jackson (ed.), Academic Press, New York, 1953, pp. 503–512.

G. A. Barnard (1949), "Statistical inference," *J. Roy. Statist. Soc., Ser. B*, Vol. 11, pp. 115–149.

——— (1951), "The theory of information," *J. Roy. Statist. Soc., Ser. B*, Vol. 13, pp. 46–64.

M. S. Bartlett (1935), "Contingency table interactions," *J. Roy. Statist. Soc., Suppl.*, Vol. 2, pp. 248–252.

——— (1936), "Statistical information and properties of sufficiency," *Proc. Roy. Soc., Ser. A*, Vol. 154, pp. 124–137.

M. S. Bartlett (1937), "Properties of sufficiency and statistical tests," *Proc. Roy. Soc., Ser. A*, Vol. 160, pp. 268–282.

——— (1947), "Multivariate analysis," *J. Roy. Statist. Soc., Suppl.*, Vol. 9, pp. 176–197.

——— (1948), "Internal and external factor analysis," *Brit. J. Psychol.*, Vol. 1, pp. 73–81.

——— (1950), "Tests of significance in factor analysis," *Brit. J. Psychol., Stat. Sec.*, Vol. 3, pp. 77–85.

——— (1951a), "An inverse matrix adjustment arising in discriminant analysis," *Ann. Math. Statist.*, Vol. 22, pp. 107–111.

——— (1951b), "The effect of standardization on a χ^2 approximation in factor analysis," *Biometrika*, Vol. 38, pp. 337–344.

——— (1952), "The statistical significance of odd bits of information," *Biometrika*, Vol. 39, pp. 228–237.

——— (1954), "A note on the multiplying factors for various χ^2 approximations," *J. Roy. Statist. Soc., Ser. B*, Vol. 16, pp. 296–298.

——— (1955), *An Introduction to Stochastic Processes*, Cambridge Univ. Press.

——— and D. V. Rajalakshman (1953), "Goodness of fit tests for simultaneous autoregressive series," *J. Roy. Statist. Soc., Ser. B*, Vol. 15, pp. 107–124.

D. E. Barton (1956), "A class of distributions for which the maximum-likelihood estimator is unbiased and of minimum variance for all sample sizes," *Biometrika*, Vol. 43, pp. 200–202.

G. I. Bateman (1949), "The characteristic function of a weighted sum of non-central squares of normal variates subject to *s* linear restraints," *Biometrika*, Vol. 36, pp. 460–462.

D. A. Bell (1953), *Information Theory and its Engineering Applications* (1st ed.), Sir Isaac Pitman & Sons, London; 2nd ed., 1956.

A. Bhattacharyya (1943), "On a measure of divergence between two statistical populations defined by their probability distributions," *Bull. Calcutta Math. Soc.*, Vol. 35, pp. 99–109.

——— (1946a), "On a measure of divergence between two multinomial populations," *Sankhyā*, Vol. 7, pp. 401–406.

——— (1946b, 1947, 1948), "On some analogues of the amount of information and their use in statistical estimation," *Sankhyā*, Vol. 8, pp. 1–14; pp. 201–218; pp. 315–328.

F. E. Binet and G. S. Watson (1956), "Algebraic theory of the computing routine for tests of significance on the dimensionality of normal multivariate systems," *J. Roy. Statist. Soc., Ser. B*, Vol. 18, pp. 70–78.

D. Blackwell and M. A. Girshick (1954), *Theory of Games and Statistical Decisions*, John Wiley & Sons, New York.

A. Blanc-Lapierre and A. Tortrat (1956), "Statistical mechanics and probability theory," *Proceedings of the Third Berkeley Symposium on Mathematical Statistics and Probability*, Univ. Calif. Press, Vol. III, pp. 145–170.

M. Bôcher (1924), *Introduction to Higher Algebra*, The Macmillan Co., New York.

A. H. Bowker (1948), "A test for symmetry in contingency tables," *J. Am. Statist. Assoc.*, Vol. 43, pp. 572–574.

G. E. P. Box (1949), "A general distribution theory for a class of likelihood criteria," *Biometrika*, Vol. 36, pp. 317–346.

——— (1950), "Problems in the analysis of growth and wear curves," *Biometrics*, Vol. 6, pp. 362–389.

R. N. Bradt and S. Karlin (1956), "On the design and comparison of certain dichotomous experiments," *Ann. Math. Statist.*, Vol. 27, pp. 390–409.

L. Brillouin (1956), *Science and Information Theory*, Academic Press, New York.

L. de Broglie (chairman) (1951), *La Cybernétique*, Éditions de la Revue d'Optique Théorique et Instrumentale, Paris.

H. D. Brunk (1958), "On the estimation of parameters restricted by inequalities," *Ann. Math. Statist.*, Vol. 29, pp. 437–453.

M. G. Bulmer (1957), "Confirming statistical hypotheses," *J. Roy. Statist. Soc., Ser. B*, Vol. 19, pp. 125–132.

F. L. Campbell, G. W. Snedecor, and W. A. Simanton (1939), "Biostatistical problems involved in the standardization of liquid household insecticides," *J. Am. Statist. Assoc.*, Vol. 34, pp. 62–70.

A. H. Carter (1949), "The estimation and comparison of residual regressions where there are two or more related sets of observations," *Biometrika*, Vol. 36, pp. 26–46.

M. Castañs Camargo (1955), "Una teoria de la certidumbre," *Anales real soc. españ. fís. y quím., Ser. A*, Vol. 51, pp. 215–232.

——— and M. Medina e Isabel (1956), "The logarithmic correlation," *Anales real soc. españ. fís. y quím., Ser. A*, Vol. 52, pp. 117–136.

D. G. Chapman and H. Robbins (1951), "Minimum variance estimation without regularity assumptions," *Ann. Math. Statist.*, Vol. 22, pp. 581–586.

H. Chernoff (1952), "A measure of asymptotic efficiency for tests of a hypothesis based on the sum of observations," *Ann. Math. Statist.*, Vol. 23, pp. 493–507.

——— (1954), "On the distribution of the likelihood ratio," *Ann. Math. Statist.*, Vol. 25, pp. 573–578.

——— (1956), "Large-sample theory: parametric case," *Ann. Math. Statist.*, Vol. 27, pp. 1–22.

C. Cherry (ed.) (1955), *Information Theory, Papers Read at a Symposium on 'Information Theory,'* Royal Institution, London, Sept. 1955; Academic Press, New York, 1956.

——— (1957), *On Human Communication*, John Wiley & Sons, New York.

E. C. Cherry (1950), "An history of the theory of information," *Proceedings of a Symposium on Information Theory*, W. Jackson (ed.), Royal Society, London, 1950, published by Ministry of Supply, and subsequently by the IRE, Feb. 1953, pp. 161–168.

——— (1951), "An history of the theory of information," *Proc. I.E.E. (London)*, Vol. 98, Part III, pp. 383–393.

——— (1952), "The communication of information," *Am. Scientist*, Vol. 40, pp. 640–664.

W. G. Cochran (1952), "The χ^2 test of goodness of fit," *Ann. Math. Statist.*, Vol. 23, pp. 315–345.

——— (1954), "Some methods for strengthening the common χ^2 tests," *Biometrics*, Vol. 10, pp. 417–451.

——— and C. I. Bliss (1948), "Discriminant functions with covariance," *Ann. Math. Statist.*, Vol. 19, pp. 151–176.

E. A. Cornish (1957), "An application of the Kronecker product of matrices in multiple regression," *Biometrics*, Vol. 13, pp. 19–27.

H. Cramér (1937), *Random Variables and Probability Distributions*, Cambridge Tracts in Mathematics, No. 36, Cambridge.

——— (1938), "Sur un nouveau théorème-limite de la théorie des probabilités," *Actualités sci. et ind.*, No. 736.

H. Cramér (1946a), *Mathematical Methods of Statistics*, Princeton Univ. Press.

——— (1946b), "Contributions to the theory of statistical estimation," *Skand. Aktuar. Tidskr.*, Vol. 29, pp. 85–94.

——— (1955), *The Elements of Probability Theory and Some of its Applications*, John Wiley & Sons, New York.

G. Darmois (1936), *Méthodes d'Estimation, Actualités sci. et ind.* No. 356.

——— (1945), "Sur les limites de la dispersion de certaines estimations," *Rev. Inst. intern. Statist.*, Vol. 13, pp. 9–15.

H. Davis (chairman) (1954), Symposium on statistical methods in communication engineering, Berkeley, California, August 1953, *Trans. IRE, PGIT*-3, Mar.

W. L. Deemer and I. Olkin (1951), "The Jacobians of certain matrix transformations useful in multivariate analysis," *Biometrika*, Vol. 38, pp. 345–367.

L. Dolanský and M. P. Dolanský (1952), "Table of $\log_2 \frac{1}{p}$, $p \log_2 \frac{1}{p}$ and $p \log_2 \frac{1}{p} + (1-p) \log_2 \frac{1}{1-p}$," *Tech. Rept. No.* 227, R.L.E., M.I.T., Jan. 2.

J. L. Doob (1934), "Probability and statistics," *Trans. Am. Math. Soc.*, Vol. 36, pp. 759–775.

——— (1936), "Statistical estimation," *Trans. Am. Math. Soc.*, Vol. 39, pp. 410–421.

D. Dugué (1936a), "Sur le maximum de précision des lois limites d'estimation," *Compt. Rend.*, Vol. 202, p. 452.

——— (1936b), "Sur le maximum de précision des estimations gaussiennes à la limite," *Compt. Rend.*, Vol. 202, p. 193.

J. Durbin and M. G. Kendall (1951), "The geometry of estimation," *Biometrika*, Vol. 38, pp. 150–158.

P. S. Dwyer and M. S. MacPhail (1948), "Symbolic matrix derivatives," *Ann. Math. Statist.*, Vol. 19, pp. 517–534.

L. P. Eisenhart (1926), *Riemannian Geometry*, Princeton Univ. Press.

P. Elias (chairman) (1956), 1956 Symposium on Information Theory, M.I.T., September 1956, *IRE Trans. on Inform. Theory*, Vol. IT-2, No. 3.

R. M. Fano (chairman) (1954), 1954 Symposium on Information Theory, M.I.T., September 1954, *Trans. IRE, PGIT-4*.

W. T. Federer (1955), *Experimental Design*, The Macmillan Co., New York.

A. Feinstein (1958), *Foundations of Information Theory*, McGraw-Hill Book Co., New York.

W. Feller (1950), *An Introduction to Probability Theory and its Applications* (1st ed.), John Wiley & Sons, New York.

R. Féron (1952a), "Information et corrélation," *Compt. Rend.*, Vol. 234, pp. 1343–1345.

——— (1952b), "Convexité et information," *Compt. Rend.*, Vol. 234, pp. 1840–1841.

——— and C. Fourgeaud (1951), "Information et régression," *Compt. Rend.*, Vol. 232, pp. 1636–1638.

W. L. Ferrar (1941), *Algebra*, Oxford Univ. Press.

R. A. Fisher (1921), "On the 'probable error' of a coefficient of correlation deduced from a small sample," *Metron*, Vol. 1, pp. 3–32.

——— (1922a), "On the interpretation of χ^2 from contingency tables, and the calculation of *P*," *J. Roy. Statist. Soc.*, Vol. 85, pp. 87–94; *Contributions to Mathematical Statistics*, John Wiley & Sons, New York, 1950, paper 5.

——— (1922b), "On the mathematical foundations of theoretical statistics," *Phil. Trans. Roy. Soc. London, Ser. A*, Vol. 222, pp. 309–368; *Contributions to Mathematical Statistics*, John Wiley & Sons, New York, 1950, paper 10.

R. A. Fisher (1924), "The conditions under which χ^2 measures the discrepancy between observation and hypothesis," *J. Roy. Statist. Soc.*, Vol. 87, pp. 442–450; *Contributions to Mathematical Statistics*, John Wiley & Sons, New York, 1950, paper 8.

——— (1925a), *Statistical Methods for Research Workers* (1st ed.), Oliver & Boyd, London; 10th ed., 1948.

——— (1925b), "Theory of statistical estimation," *Proc. Camb. Phil. Soc.*, Vol. 22, pp. 700–725; *Contributions to Mathematical Statistics*, John Wiley & Sons, New York, 1950, paper 11.

——— (1928), "The general sampling distribution of the multiple correlation coefficient," *Proc. Royal Soc., Ser. A*, Vol. 121, pp. 654–673; *Contributions to Mathematical Statistics*, John Wiley & Sons, New York, 1950, paper 14.

——— (1935), "The logic of inductive inference," *J. Roy. Statist. Soc.*, Vol. 98, pp. 39–54; *Contributions to Mathematical Statistics*, John Wiley & Sons, New York, 1950, paper 26.

——— (1936), "The use of multiple measurements in taxonomic problems," *Ann. Eugenics*, Vol. 7, pp. 179–188; *Contributions to Mathematical Statistics*, John Wiley & Sons, New York, 1950, paper 32.

——— (1938), "The statistical utilization of multiple measurements," *Ann. Eugenics*, Vol. 8, pp. 376–386; *Contributions to Mathematical Statistics*, John Wiley & Sons, New York, 1950, paper 33.

——— (1939a), "The comparison of samples with possibly unequal variances," *Ann. Eugenics*, Vol. 9, pp. 174–180; *Contributions to Mathematical Statistics*, John Wiley & Sons, New York, 1950, paper 35.

——— (1939b), "The sampling distribution of some statistics obtained from non-linear equations," *Ann. Eugenics*, Vol. 9, pp. 238–249; *Contributions to Mathematical Statistics*, John Wiley & Sons, New York, 1950, paper 36.

——— (1950), "The significance of deviations from expectation in a Poisson series," *Biometrics*, Vol. 6, pp. 17–24.

——— (1956), *Statistical Methods and Scientific Inference*, Oliver & Boyd, London.

E. Fix (1949), "Tables of noncentral χ^2," *Univ. Calif. Publ. Statist.*, Vol. 1, No. 2, pp. 15–19.

F. G. Foster and D. H. Rees (1957), "Upper percentage points of the generalized Beta distribution. I," *Biometrika*, Vol. 44, pp. 237–247.

D. A. S. Fraser (1957), *Nonparametric Methods in Statistics*, John Wiley & Sons, New York.

——— and I. Guttman (1952), "Bhattacharyya bounds without regularity assumptions," *Ann. Math. Statist.*, Vol. 23, pp. 629–632.

R. A. Frazer, W. J. Duncan, and A. R. Collar (1938), *Elementary Matrices*, Cambridge Univ. Press.

M. Fréchet (1943), "Sur l'extension de certaines évaluations statistiques au cas de petits échantillons," *Rev. Inst. intern. Statist.*, Vol. 11, pp. 183–205.

W. R. Garner and W. J. McGill (1954), "Relation between uncertainty, variance, and correlation analyses," *Rep. No. 166-I-192, ONR Contract N5ori-166*, Johns Hopkins Univ.

——— (1956), "The relation between information and variance analyses," *Psychometrika*, Vol. 21, pp. 219–228.

I. M. Gel'fand, A. N. Kolmogorov, and A. M. Iaglom (1956), "On the general definition of the quantity of information," *Doklady Akad. Nauk S.S.S.R.*, Vol. 111, No. 4, pp. 745–748. (Translation by E. Kelly, Lincoln Laboratory.)

E. N. Gilbert (1958), "An outline of information theory," *Am. Statistician*, Vol. 12, pp. 13–19.

M. A. Girshick (1936), "Principal components," *J. Am. Statist. Assoc.*, Vol. 31, pp. 519–528.

——— (1939), "On the sampling theory of the roots of determinantal equations," *Ann. Math. Statist.*, Vol. 10, pp. 203–224.

——— (1946), "Contributions to the theory of sequential analysis, I, II, III," *Ann. Math. Statist.*, Vol. 17, pp. 123–143; 282–298.

——— and L. J. Savage (1951), "Bayes and minimax estimates for quadratic loss functions," *Proceedings of the Second Berkeley Symposium on Mathematical Statistics and Probability*, Univ. of Calif. Press, pp. 53–73.

S. Goldman (1953), *Information Theory*, Prentice-Hall, New York.

I. J. Good (1950), *Probability and the Weighing of Evidence*, Charles Griffin, London.

——— (1952), "Rational decisions," *J. Roy. Statist. Soc., Ser. B*, Vol. 14, pp. 107–114.

——— (1953), "The population frequencies of species and the estimation of population parameters," *Biometrika*, Vol. 40, pp. 237–264.

——— (1956), "Some terminology and notation in information theory," *Proc. I.E.E.*, Part C, Vol. 103, pp. 200–204.

——— (1957), "Saddle-point methods for the multinomial distribution," *Ann. Math. Statist.*, Vol. 28, pp. 861–881.

P. E. Green, Jr. (1956), "A bibliography of Soviet literature on noise, correlation, and information theory," *IRE Trans. on Inform. Theory*, Vol. IT-2, pp. 91–94.

——— (1957), "Information theory in the U.S.S.R." *IRE WESCON Convention Record*, Part 2, pp. 67–83.

S. W. Greenhouse (1954), "On the problem of discrimination between statistical populations," *M.A. Thesis*, George Washington Univ.

H. Grell (ed.) (1957), *Arbeiten zur Informationstheorie I*, Deutscher Verlag der Wissenschaften, Berlin. (Translations from Russian and Hungarian.)

D. G. C. Gronow (1951), "Test for the significance of the difference between means in two normal populations having unequal variances," *Biometrika*, Vol. 38, pp. 252–256.

P. M. Grundy (1951), "A general technique for the analysis of experiments with incorrectly treated plots," *J. Roy. Statist. Soc., Ser. B*, Vol. 13, pp. 272–283.

J. Gurland (1954), "On regularity conditions for maximum likelihood estimators," *Skand. Aktuar. Tidskr.*, Vol. 37, pp. 71–76.

J. B. S. Haldane (1955), "Substitutes for χ^2," *Biometrika*, Vol. 42, pp. 265–266.

P. R. Halmos (1950), *Measure Theory*, D. Van Nostrand Co., New York.

——— and L. J. Savage (1949), "Applications of the Radon-Nikodym theorem to the theory of sufficient statistics," *Ann. Math. Statist.*, Vol. 20, pp. 225–241.

G. H. Hardy, J. E. Littlewood, and G. Pólya (1934), *Inequalities* (1st ed.), Cambridge Univ. Press; 2nd ed., 1952.

R. V. L. Hartley (1928), "Transmission of information," *Bell System Tech. J.*, Vol. 7, pp. 535–563.

P. G. Hoel (1947), *Introduction to Mathematical Statistics* (1st ed.), John Wiley & Sons, New York; 2nd ed., 1954.

H. Hotelling (1933), "Analysis of a complex of statistical variables into principal components," *J. Educ. Psych.*, Vol. 24, pp. 417–441; 498–520.

——— (1936), "Relations between two sets of variates," *Biometrika*, Vol. 28, pp. 321–377.

——— (1947), "Multivariate quality control, illustrated by the air testing of sample bombsights," *Techniques of Statistical Analysis*, McGraw-Hill Book Co., New York, pp. 111–184.

H. Hotelling (1951), "A generalized T test and measure of multivariate dispersion," *Proceedings of the Second Berkeley Symposium on Mathematical Statistics and Probability*, Univ. of Calif. Press, pp. 23–41.

J. P. Hoyt (1953), "Estimates and asymptotic distributions of certain statistics in information theory," *Dissertation*, Graduate Council of George Washington Univ.

P. L. Hsu (1938), "Notes on Hotelling's generalized T," *Ann. Math. Statist.*, Vol. 9, pp. 231–243.

——— (1939), "On the distribution of roots of certain determinantal equations," *Ann. Eugenics*, Vol. 9, pp. 250–258.

——— (1941a), "On the problem of rank and the limiting distribution of Fisher's test function," *Ann. Eugenics*, Vol. 11, pp. 39–41.

——— (1941b), "On the limiting distribution of roots of a determinantal equation," *J. London Math. Soc.*, Vol. 16, pp. 183–194.

——— (1941–42), "On the limiting distribution of the canonical correlations," *Biometrika*, Vol. 32, pp. 38–45.

——— (1949), "The limiting distribution of functions of sample means and application to testing hypotheses," *Proceedings of the Berkeley Symposium on Mathematical Statistics and Probability*, Univ. of Calif. Press, pp. 359–402.

V. S. Huzurbazar (1949), "On a property of distributions admitting sufficient statistics," *Biometrika*, Vol. 36, pp. 71–74.

——— (1955), "Exact forms of some invariants for distributions admitting sufficient statistics," *Biometrika*, Vol. 42, pp. 533–537.

J. O. Irwin (1949), "A note on the subdivision of χ^2 into components," *Biometrika*, Vol. 36, pp. 130–134.

K. Ito (1956), "Asymptotic formulae for the distribution of Hotelling's generalized T_0^2 statistic," *Ann. Math. Statist.*, Vol. 27, pp. 1091–1105.

W. Jackson (ed.) (1950), *Proceedings of a Symposium on Information Theory*, Royal Society, London, 1950, published by Ministry of Supply, and subsequently by the IRE, Feb. 1953.

——— (ed.) (1952), *Communication Theory, Papers Read at a Symposium on "Applications of Communication Theory,"* IEE, London, Sept. 1952; Academic Press, New York, 1953.

G. S. James (1954), "Tests of linear hypotheses in univariate and multivariate analysis when the ratios of the population variances are unknown," *Biometrika*, Vol. 41, pp. 19–43.

E. T. Jaynes (1957), "Information theory and statistical mechanics," *Phys. Rev.*, Vol. 106, pp. 620–630.

H. Jeffreys (1946), "An invariant form for the prior probability in estimation problems," *Proc. Roy. Soc. (London), Ser. A*, Vol. 186, pp. 453–461.

——— (1948), *Theory of Probability* (2nd ed.), Oxford Univ. Press.

J. L. W. V. Jensen (1906), "Sur les fonctions convexes et les inégalités entre les valeurs moyennes," *Acta Math.*, Vol. 30, pp. 175–193.

D. D. Joshi (1957), "L'information en statistique mathématique et dans la théorie des communications," *Thèse*, Faculté des Sciences de l'Université de Paris, June.

T. L. Kelley (1928), *Crossroads in the Mind of Man*, Stanford Univ. Press.

J. L. Kelley, Jr. (1956), "A new interpretation of information rate," *Bell System Tech. J.*, Vol. 35, pp. 917–926.

O. Kempthorne (1952), *The Design and Analysis of Experiments*, John Wiley & Sons, New York.

M. G. Kendall (1943, 1946), *The Advanced Theory of Statistics*, Charles Griffin, London, Vol. I, 1943; Vol. II, 1946.

A. I. Khinchin (1949), *Mathematical Foundations of Statistical Mechanics*, Dover Publications, New York.

——— (1953), "The entropy concept in probability theory," *Uspekhi Matematicheskikh Nauk*, Vol. 8, No. 3 (55), pp. 3–20 (Russian).

——— (1956), "On the fundamental theorems of information theory," *Uspekhi Matematicheskikh Nauk*, Vol. 11, No. 1 (67), pp. 17–75 (Russian).

——— (1957), *Mathematical Foundations of Information Theory*, Dover Publications, New York. (English translation of the preceding two papers.)

J. Kiefer (1952), "On minimum variance estimators," *Ann. Math. Statist.*, Vol. 23, pp. 627–629.

A. W. Kimball (1954), "Short-cut formulas for the exact partition of χ^2 in contingency tables," *Biometrics*, Vol. 10, pp. 452–458.

A. N. Kolmogorov (1950), *Foundations of the Theory of Probability*, Chelsea Publishing Co., New York.

——— (1956), "On the Shannon theory of information transmission in the case of continuous signals," *IRE Trans. on Inform. Theory*, Vol. IT-2, pp. 102–108.

S. Kolodziejczyk (1935), "On an important class of statistical hypotheses," *Biometrika*, Vol. 27, pp. 161–190.

B. O. Koopman (1936), "On distributions admitting a sufficient statistic," *Trans. Am. Math. Soc.*, Vol. 39, pp. 399–409.

C. F. Kossack (1945), "On the mechanics of classification," *Ann. Math. Statist.*, Vol. 16, pp. 95–98.

S. Kullback (1952), "An application of information theory to multivariate analysis," *Ann. Math. Statist.*, Vol. 23, pp. 88–102.

——— (1953), "A note on information theory," *J. Appl. Phys.*, Vol. 24, pp. 106–107.

——— (1954), "Certain inequalities in information theory and the Cramér-Rao inequality," *Ann. Math. Statist.*, Vol. 25, pp. 745–751.

——— (1956), "An application of information theory to multivariate analysis, II," *Ann. Math. Statist.*, Vol. 27, pp. 122–145; correction p. 860.

——— and R. A. Leibler (1951), "On information and sufficiency," *Ann. Math. Statist.*, Vol. 22, pp. 79–86.

S. Kullback and H. M. Rosenblatt (1957), "On the analysis of multiple regression in k categories," *Biometrika*, Vol. 44, pp. 67–83.

M. Kupperman (1957), "Further applications of information theory to multivariate analysis and statistical inference," *Dissertation*, Graduate Council of George Washington Univ.

——— (1958), "Probabilities of hypotheses and information-statistics in sampling from exponential-class populations," *Ann. Math. Statist.*, Vol. 29, pp. 571–574.

R. G. Laha (1954), "On some properties of the Bessel function distributions," *Bull. Calcutta Math. Soc.*, Vol. 46, pp. 59–72.

H. O. Lancaster (1949), "The derivation and partition of χ^2 in certain discrete distributions," *Biometrika*, Vol. 36, pp. 117–129.

——— (1957), "Some properties of the bivariate normal distribution considered in the form of a contingency table," *Biometrika*, Vol. 44, pp. 289–292.

D. N. Lawley (1938), "A generalization of Fisher's z test," *Biometrika*, Vol. 30, pp. 180–187; correction, pp. 467–469.

D. N. Lawley (1940), "The estimation of factor loadings by the method of maximum likelihood," *Proc. Roy. Soc. Edinburgh*, Vol. 9, p. 64.

——— (1956), "A general method for approximating to the distribution of likelihood ratio criteria," *Biometrika*, Vol. 43, pp. 295–303.

J. L. Lawson and G. E. Uhlenbeck (1950), *Threshold Signals*, McGraw-Hill Book Co., New York.

L. Le Cam (1956), "On the asymptotic theory of estimation and testing hypotheses," *Proceedings of the Third Berkeley Symposium on Mathematical Statistics and Probability*, Univ. of Calif. Press, Vol. I, pp. 129–156.

E. L. Lehmann (1949), *Theory of Testing Hypotheses*, Notes recorded by Colin Blyth, Associated Students Store, Univ. of Calif., Berkeley, Calif.

———(1950a), *Notes on the Theory of Estimation*, Notes recorded by Colin Blyth, Associated Students Store, Univ. of Calif., Berkeley, Calif., Sept.

———(1950b), "Some principles of the theory of testing hypotheses," *Ann. Math. Statist.*, Vol. 21, pp. 1–26.

——— and H. Scheffé (1950), "Completeness, similar regions and unbiased estimation, Part I," *Sankhyā*, Vol. 10, pp. 305–340.

D. V. Lindley (1956), "On a measure of the information provided by an experiment," *Ann. Math. Statist.*, Vol. 27, pp. 986–1005.

——— (1957), "Binomial sampling schemes and the concept of information," *Biometrika*, Vol. 44, pp. 179–186.

E. H. Linfoot (1957), "An informational measure of correlation," *Information and Control*, Vol. 1, pp. 85–89.

M. Loève (1955), *Probability Theory*, D. Van Nostrand Co., New York.

C. H. McCall, Jr. (1957), "The linear hypothesis, information, and the analysis of variance," *Dissertation*, Graduate Council of George Washington Univ.

J. McCarthy (1956), "Measures of the value of information," *Proc. Nat. Acad. Sci., U.S.*, Vol. 42, pp. 654–655.

D. K. C. MacDonald (1952), "Information theory and its application to taxonomy," *J. Appl. Physics*, Vol. 23, pp. 529–531.

C. C. MacDuffee (1946), *The Theory of Matrices*, Chelsea Publishing Co., New York.

W. J. McGill (1954), "Multivariate information transmission," *Psychometrika*, Vol. 19, pp. 97–116.

D. M. MacKay (1950), "Quantal aspects of scientific information," *Phil. Mag.*, Vol. 41, Seventh Series, No. 314, pp. 289–311.

N. W. McLachlan (1939), *Complex Variable and Operational Calculus with Technical Applications*, Cambridge Univ. Press.

B. McMillan (1953), "The basic theorems of information theory," *Ann. Math. Statist.*, Vol. 24, pp. 196–219.

———, D. A. Grant, P. M. Fitts, F. C. Frick, W. S. McCulloch, G. A. Miller, and H. W. Brosin (1953), *Current Trends in Information Theory*, Univ. of Pittsburgh Press.

P. C. Mahalanobis (1936), "On the generalized distance in statistics," *Proc. Nat. Inst. Sci. India*, Vol. 12, pp. 49–55.

B. Mandelbrot (1953), "Contribution a la théorie mathématique des jeux de communication," *Publs. Inst. statist. univ. Paris*, Vol. 2, Fasc. 1 et 2, pp. 3–124.

——— (1956), "An outline of a purely phenomenological theory of statistical thermodynamics: I. Canonical ensembles," *IRE Trans. on Inform. Theory*, Vol. IT-2, pp. 190–203.

H. B. Mann and A. Wald (1943), "On stochastic limit and order relationships," *Ann. Math. Statist.*, Vol. 14, pp. 217–226.

F. H. C. Marriott (1952), "Tests of significance in canonical analysis," *Biometrika*, Vol. 39, pp. 58–64.

J. W. Mauchly (1940), "Significance test for sphericity of a normal *n*-variate distribution," *Ann. Math. Statist.*, Vol. 11, pp. 204–209.

G. A. Miller and W. G. Madow (1954), "On the maximum likelihood estimate of the Shannon-Wiener measure of information," *AFCRC-TR-54-75*, Air Force Cambridge Research Center, Air Research and Development Command, Bolling Air Force Base, Washington D.C., Aug.

G. A. Miller and P. M. Ross (1954), "Tables of $n \log_2 n$ and $n \log_{10} n$ for n from 1 to 1000," *Tech. Rep. No. 60*, Lincoln Laboratory, M.I.T., Feb. 10.

S. K. Mitra (1955), "Contributions to the statistical analysis of categorical data," N. C. Inst. of Statist. Mimeo Series No. 142, Dec.

E. C. Molina (1942), *Tables of Poisson's Exponential Limit*, D. Van Nostrand Co., New York.

A. M. Mood (1951), "On the distribution of the characteristic roots of normal second-moment matrices," *Ann. Math. Statist.*, Vol. 22, pp. 266–273.

E. Mourier (1946), "Étude du choix entre deux lois de probabilité," *Compt. Rend.*, Vol. 223, pp. 712–714.

——— (1951), "Tests de choix entre diverses lois de probabilité," *Trabajos Estadística*, Vol. 2, pp. 233–260.

J. Neyman (1929), "Contribution to theory of certain test criteria," *XVIII Session de l'Institut International de Statistique*, Varsovie, pp. 1–48.

——— (1935), "Su un teorema concernente le cosiddette statistiche sufficienti," *Giorn. Ist. ital. Attuari*, Vol. 6, p. 320–334.

——— (1949), "Contribution to the theory of the χ^2 test," *Proceedings of the Berkeley Symposium on Mathematical Statistics and Probability*, Univ. of Calif. Press, pp. 239–273.

——— (1950), *First Course in Probability and Statistics*, Henry Holt and Co., New York.

——— and E. S. Pearson (1928), "On the use and interpretation of certain test criteria for purposes of statistical inference," *Biometrika*, Vol. 20A, pp. 175–240; 263–294.

——— (1933), "On the problem of the most efficient tests of statistical hypotheses," *Phil. Trans. Roy. Soc. London, Ser. A*, Vol. 231, pp. 289–337.

H. W. Norton (1945), "Calculation of chi-square for complex contingency tables," *J. Am. Statist. Assoc.*, Vol. 40, pp. 251–258.

P. B. Patnaik (1949), "The non-central χ^2 and F distributions and their applications," *Biometrika*, Vol. 36, pp. 202–232.

E. S. Pearson and H. O. Hartley (1951), "Charts of the power function for analysis of variance tests, derived from the non-central F-distribution," *Biometrika*, Vol. 38, pp. 112–130.

E. S. Pearson and S. S. Wilks (1933), "Methods of statistical analysis appropriate for k samples of two variables," *Biometrika*, Vol. 25, pp. 353–378.

K. Pearson (1904), "Mathematical contributions to the theory of evolution, XIII, on the theory of contingency and its relation to association and normal correlation," *Drap. Co. Mem. Biom. Ser.*, No. 1.

——— (1911), "On the probability that two independent distributions of frequency are really samples from the same population," *Biometrika*, Vol. 8, pp. 250–253.

L. S. Penrose (1947), "Some notes on discrimination," *Ann. Eugenics*, Vol. 13, pp. 228–237.

J. R. Pierce (1956), *Electrons, Waves and Messages*, Hanover House, New York.

K. C. S. Pillai (1955), "Some new test criteria in multivariate analysis," *Ann. Math. Statist.*, Vol. 26, pp. 117–121.

E. J. G. Pitman (1936), "Sufficient statistics and intrinsic accuracy," *Proc. Camb. Phil. Soc.*, Vol. 32, pp. 567–579.

R. L. Plackett (1949), "A historical note on the method of least squares," *Biometrika*, Vol. 36, pp. 458–460.

K. H. Powers (1956), "A unified theory of information," *Tech. Rept. No. 311*, R.L.E., M.I.T., Feb. 1.

H. Quastler (ed.) (1953), *Information Theory in Biology*, Univ. of Illinois Press, Urbana.

——— (ed.) (1955), *Information Theory in Psychology*, The Free Press, Glencoe, Ill.

——— (1956), "A Primer on Information Theory," *Tech. Memo. 56-1*, Office of Ordnance Research, Box CM, Duke Station, Durham, N.C., Jan.

C. R. Rao (1945), "Information and the accuracy attainable in the estimation of statistical parameters," *Bull. Calcutta Math. Soc.*, Vol. 37, pp. 81–91.

——— (1952), *Advanced Statistical Methods in Biometric Research*, John Wiley & Sons, New York.

——— (1957), "Maximum likelihood estimation for the multinomial distribution," *Sankhyā*, Vol. 18, pp. 139–148.

——— and I. M. Chakravarti (1956), "Some small sample tests of significance for a Poisson distribution," *Biometrics*, Vol. 12, pp. 264–282.

E. Reich (1951), "On the definition of information," *J. Math. and Phys.*, Vol. 30, pp. 156–161.

D. D. Rippe (1951), "Statistical rank and sampling variation of the results of factorization of covariance matrices," *Doctoral Thesis*, on file at the Univ. of Michigan.

H. R. Roberts (1957), "On estimation and information," *M.S. Thesis*, George Washington Univ.

H. M. Rosenblatt (1953), "On a k sample multivariate regression problem," *Master's Thesis*, George Washington Univ.

J. Rothstein (1951), "Information, measurement, and quantum mechanics," *Science*, Vol. 114, pp. 171–175.

S. N. Roy (1939), "p-statistics, or some generalizations in analysis of variance appropriate to multivariate problems," *Sankhyā*, Vol. 4, pp. 381–396.

——— (1957), *Some Aspects of Multivariate Analysis*, John Wiley & Sons, New York.

——— and R. C. Bose (1953), "Simultaneous confidence interval estimation," *Ann. Math. Statist.*, Vol. 24, pp. 513–536.

S. N. Roy and M. A. Kastenbaum (1955), "A generalization of analysis of variance and multivariate analysis to data based on frequencies in qualitative categories or class intervals," N. C. Inst. of Statist. Mimeo Series No. 131, June 1.

——— (1956), "On the hypothesis of no 'interaction' in a multi-way contingency table," *Ann. Math. Statist.*, Vol. 27, pp. 749–757.

S. N. Roy and S. K. Mitra (1956), "An introduction to some non-parametric generalizations of analysis of variance and multivariate analysis," *Biometrika*, Vol. 43, pp. 361–376.

M. Sakaguchi (1952, 1955, 1957a), "Notes on statistical applications of information theory," *Repts. Statist. Appli. Research Union Japan. Scientists and Engineers*, Vol. 1, No. 4, pp. 27–31; "II," Vol. 4, No. 2, pp. 21–68; "III," Vol. 5, No. 1, pp. 9–16.

——— (1957b), "Notes on information transmission in multivariate probability distributions," *Rep. Univ. of Electro-Communications*, No. 9, Dec., pp. 25–31.

I. N. Sanov (1957), "On the probability of large deviations of random variables," *Mat. Sbornik* (*Moscow*), Vol. 42, No. 1 (84), pp. 11–44 (Russian). (Translation, N.C. Inst. of Statist. Mimeo Series No. 192, Mar. 1958.)

L. J. Savage (1954), *The Foundations of Statistics*, John Wiley & Sons, New York.

M. P. Schützenberger (1954), "Contribution aux applications statistiques de la théorie de l'information," *Publs. inst. statist. univ. Paris*, Vol. 3, Fasc. 1–2, pp. 3–117.

G. R. Seth (1949), "On the variance of estimates," *Ann. Math. Statist.*, Vol. 20, pp. 1–27.

C. E. Shannon (1948), "A mathematical theory of communication," *Bell System Tech. J.*, Vol. 27, pp. 379–423; 623–656.

——— (1949), "Communication in the presence of noise," *Proc. IRE*, Vol. 37, pp. 10–21.

——— (1956), "The bandwagon," *IRE Trans. on Inform. Theory*, Vol. IT-2, p. 3.

——— and W. Weaver (1949), *The Mathematical Theory of Communication*, Univ. of Illinois Press, Urbana.

W. A. Shewhart (1931), *Economic Control of Manufactured Product*, The Macmillan Co., New York.

J. B. Simaika (1941), "On an optimum property of two important statistical tests," *Biometrika*, Vol. 32, pp. 70–80.

C. A. B. Smith (1947), "Some examples of discrimination," *Ann. Eugenics*, Vol. 13, pp. 272–282.

H. F. Smith (1957), "Interpretation of adjusted treatment means and regressions in analysis of covariance," *Biometrics*, Vol. 13, pp. 282–308.

G. W. Snedecor (1946), *Statistical Methods* (4th ed.), Collegiate Press of Iowa State College, Ames.

A. Stuart (1953), "The estimation and comparison of strengths of association in contingency tables," *Biometrika*, Vol. 40, pp. 105–110.

——— (1955a), "A test for homogeneity of the marginal distributions in a two-way classification," *Biometrika*, Vol. 42, pp. 412–416.

——— (1955b), "A paradox in statistical estimation," *Biometrika*, Vol. 42, pp. 527–529.

F. L. H. M. Stumpers (1953), "A bibliography of information theory; communication theory–cybernetics" (R.L.E., M.I.T., Feb. 2, 1953); *IRE Trans.*, PGIT-2, Nov. 1953; First suppl., IT-1, Sept. 1955, pp. 31–47; Second suppl., IT-3, June 1957, pp. 150–166.

K. Suzuki (1956), "On 'amount of information'," *Proc. Japan Acad.*, Vol. 32, pp. 726–730.

——— (1957), "On the écart between two 'amounts of information'," *Proc. Japan Acad.*, Vol. 33, pp. 25–28.

Tables of the Binomial Probability Distribution (1949), Nat. Bur. Standards (U.S.), Applied Math. Series 6, Washington.

P. C. Tang (1938), "The power function of the analysis of variance tests with tables and illustrations of their use," *Statistical Research Memoirs*, Vol. 2, pp. 126–149.

G. Thomson (1947), "The maximum correlation of two weighted batteries," *Brit. J. Psychol., Stat. Sec.*, Vol. 1, pp. 27–34.

K. D. Tocher (1952), "The design and analysis of block experiments," *J. Roy. Statist. Soc., Ser. B*, Vol. 14, pp. 45–100.

J. W. Tukey (1949), "Sufficiency, truncation and selection," *Ann. Math. Statist.*, Vol. 20, pp. 309–311.

——— (1957), "Approximations to the upper 5% points of Fisher's B distribution and non-central χ^2," *Biometrika*, Vol. 44, pp. 528–530.

W. G. Tuller (1950), "Information theory applied to system design," *Trans. AIEE*, Vol. 69, Part II, pp. 1612–1614.

A. Wald (1943), "Tests of statistical hypotheses concerning several parameters when the number of observations is large," *Trans. Am. Math. Soc.*, Vol. 54, pp. 426–482.

——— (1945a), "Sequential tests of statistical hypotheses," *Ann. Math. Statist.*, Vol. 16, pp. 117–186.

A. Wald (1945b), "Sequential method of sampling for deciding between two courses of action," *J. Am. Statist. Assoc.*, Vol. 40, pp. 277–306.

——— (1947), *Sequential Analysis*, John Wiley & Sons, New York.

——— and R. J. Brookner (1941), "On the distribution of Wilks' statistic for testing the independence of several groups of variates," *Ann. Math. Statist.*, Vol. 12, pp. 137–152.

G. N. Watson (1944), *Bessel Functions* (2nd ed.), The Macmillan Co., New York.

M. Weibull (1953), "The distributions of t- and F-statistics and of correlation and regression coefficients in stratified samples from normal populations with different means," *Skand. Aktuar. Tidskr.*, Vol. 36, 1–2 Suppl., pp. 1–106.

B. L. Welch (1935), "Problems in the analysis of regression among k samples," *Biometrika*, Vol. 27, pp. 145–160.

——— (1938), "The significance of the difference between two means when the population variances are unequal," *Biometrika*, Vol. 29, pp. 350–362.

——— (1939), "Note on discriminant functions," *Biometrika*, Vol. 31, pp. 218–219.

E. T. Whittaker (1915), "On the functions which are represented by the expansions of the interpolatory theory," *Proc. Roy. Soc. Edinburgh*, Vol. 35, pp. 181–194.

N. Wiener (1948), *Cybernetics*, John Wiley & Sons, New York.

——— (1950), *The Human Use of Human Beings*, Houghton Mifflin Co., Boston.

——— (1956), "What is information theory?" *IRE Trans. on Inform. Theory*, Vol. IT-2, p. 48.

R. A. Wijsman (1957), "Random orthogonal transformations and their use in some classical distribution problems in multivariate analysis," *Ann. Math. Statist.*, Vol. 28, pp. 415–423.

S. S. Wilks (1932), "Certain generalizations in the analysis of variance," *Biometrika*, Vol. 24, pp. 471–494.

——— (1935a), "The likelihood test of independence in contingency tables," *Ann. Math. Statist.*, Vol. 6, pp. 190–196.

——— (1935b), "On the independence of k sets of normally distributed statistical variables," *Econometrica*, Vol. 3, pp. 309–326.

——— (1938a), "The large-sample distribution of the likelihood ratio for testing composite hypotheses," *Ann. Math. Statist.*, Vol. 9, pp. 60–62.

——— (1938b), "The analysis of variance and covariance in non-orthogonal data," *Metron*, Vol. 13, pp. 141–158.

——— (1943), *Mathematical Statistics*, Princeton Univ. Press.

E. J. Williams (1952), "Some exact tests in multivariate analysis," *Biometrika*, Vol. 39, pp. 17–31.

——— (1955), "Significance tests for discriminant functions and linear functional relationships," *Biometrika*, Vol. 42, pp. 360–381.

J. Wolfowitz (1947), "The efficiency of sequential estimates and Wald's equation for sequential processes," *Ann. Math. Statist.*, Vol. 18, pp. 215–230.

P. M. Woodward (1953), *Probability and Information Theory, with Applications to Radar*, McGraw-Hill Book Co., New York.

——— and I. L. Davies (1952), "Information theory and inverse probability in telecommunications," *Proc. I.E.E.*, Part III, Vol. 99, pp. 37–44.

G. U. Yule and M. G. Kendall (1937), *An Introduction to the Theory of Statistics* (11th ed.), Charles Griffin, London.

M. Zelen (1957), "The analysis of covariance for incomplete block designs," *Biometrics*, Vol. 13, pp. 309–332.

TABLE I. $\log_e n$ and $n \log_e n$ for Values of n from 1 through 1000

n	$\log_e n$	$n \log_e n$	n	$\log_e n$	$n \log_e n$
01	0.0000000000	0000.0000000000	47	3.8501476017	0180.9569372804
02	0.6931471805	0001.3862943611	48	3.8712010109	0185.8176485236
03	1.0986122886	0003.2958368660	49	3.8918202981	0190.6991946074
04	1.3862943611	0005.5451774445	50	3.9120230054	0195.6011502714
05	1.6094379124	0008.0471895622	51	3.9318256327	0200.5231072689
06	1.7917594692	0010.7505568154	52	3.9512437185	0205.4646733662
07	1.9459101490	0013.6213710434	53	3.9702919135	0210.4254714183
08	2.0794415416	0016.6355323334	54	3.9889840465	0215.4051385145
09	2.1972245773	0019.7750211960	55	4.0073331852	0220.4033251878
10	2.3025850929	0023.0258509299	56	4.0253516907	0225.4196946812
11	2.3978952727	0026.3768480008	57	4.0430512678	0230.4539222666
12	2.4849066497	0029.8188797975	58	4.0604430105	0235.5056946117
13	2.5649493574	0033.3443416470	59	4.0775374439	0240.5747091904
14	2.6390573296	0036.9468026146	60	4.0943445622	0245.6606737333
15	2.7080502011	0040.6207530165	61	4.1108738641	0250.7633057146
16	2.7725887222	0044.3614195558	62	4.1271343850	0255.8823318728
17	2.8332133440	0048.1646268490	63	4.1431347263	0261.0174877627
18	2.8903717578	0052.0266916421	64	4.1588830833	0266.1685173350
19	2.9444389791	0055.9443406042	65	4.1743872698	0271.3351725432
20	2.9957322735	0059.9146454711	66	4.1896547420	0276.5172129737
21	3.0445224377	0063.9349711922	67	4.2046926193	0281.7144054992
22	3.0910424533	0068.0029339739	68	4.2195077051	0286.9265239520
23	3.1354942159	0072.1163669664	69	4.2341065045	0292.1533488172
24	3.1780538303	0076.2732919284	70	4.2484952420	0297.3946669435
25	3.2188758248	0080.4718956217	71	4.2626798770	0302.6502712699
26	3.2580965380	0084.7105099886	72	4.2766661190	0307.9199605692
27	3.2958368660	0088.9875953821	73	4.2904594411	0313.2035392038
28	3.3322045101	0093.3017262849	74	4.3040650932	0318.5008168971
29	3.3672958299	0097.6515790696	75	4.3174881135	0323.8116085152
30	3.4011973816	0102.0359214499	76	4.3307333402	0329.1357338618
31	3.4339872044	0106.4536033390	77	4.3438054218	0334.4730174827
32	3.4657359027	0110.9035488896	78	4.3567088266	0339.8232884818
33	3.4965075614	0115.3847495284	79	4.3694478524	0345.1863803449
34	3.5263605246	0119.8962578369	80	4.3820266346	0350.5621307739
35	3.5553480614	0124.4371821521	81	4.3944491546	0355.9503815285
36	3.5835189384	0129.0066817844	82	4.4067192472	0361.3509782757
37	3.6109179126	0133.6039627678	83	4.4188406077	0366.7637704471
38	3.6375861597	0138.2282740696	84	4.4308167988	0372.1886111028
39	3.6635616461	0142.8789041991	85	4.4426512564	0377.6253568017
40	3.6888794541	0147.5551781646	86	4.4543472962	0383.0738674778
41	3.7135720667	0152.2564547349	87	4.4659081186	0388.5340063229
42	3.7376696182	0156.9821239679	88	4.4773368144	0394.0056396741
43	3.7612001156	0161.7316049748	89	4.4886363697	0399.4886369062
44	3.7841896339	0166.5043438924	90	4.4998096703	0404.9828703297
45	3.8066624897	0171.2998120397	91	4.5108595065	0410.4882150930
46	3.8286413964	0176.1175042385	92	4.5217885770	0416.0045490885

TABLE I (continued)

n	$\log_e n$	$n \log_e n$	n	$\log_e n$	$n \log_e n$
93	4.5325994931	0421.5317528633	139	4.9344739331	0685.8918767052
94	4.5432947822	0427.0697095334	140	4.9416424226	0691.8299391653
95	4.5538768916	0432.6183047021	141	4.9487598903	0697.7751445433
96	4.5643481914	0438.1774263809	142	4.9558270576	0703.7274421794
97	4.5747109785	0443.7469649148	143	4.9628446302	0709.6867821272
98	4.5849674786	0449.3268129097	144	4.9698132995	0715.6531151389
99	4.5951198501	0454.9168651633	145	4.9767337424	0721.6263926510
100	4.6051701859	0460.5170185988	146	4.9836066217	0727.6065667694
101	4.6151205168	0466.1271722010	147	4.9904325867	0733.5935902565
102	4.6249728132	0471.7472269550	148	4.9972122737	0739.5874165171
103	4.6347289882	0477.3770857877	149	5.0039463059	0745.5879995859
104	4.6443908991	0483.0166535107	150	5.0106352940	0751.5952941144
105	4.6539603501	0488.6658367665	151	5.0172798368	0757.6092553591
106	4.6634390941	0494.3245439759	152	5.0238805208	0763.6298391686
107	4.6728288344	0499.9926852874	153	5.0304379213	0769.6570019730
108	4.6821312271	0505.6701725294	154	5.0369526024	0775.6907007717
109	4.6913478822	0511.3569191630	155	5.0434251169	0781.7308931225
110	4.7004803657	0517.0528402372	156	5.0498560072	0787.7775371309
111	4.7095302013	0522.7578523457	157	5.0562458053	0793.8305914397
112	4.7184988712	0528.4718735851	158	5.0625950330	0799.8900152183
113	4.7273878187	0534.1948235145	159	5.0689042022	0805.9557681530
114	4.7361984483	0539.9266231170	160	5.0751738152	0812.0278104374
115	4.7449321283	0545.6671947618	161	5.0814043649	0818.1061027625
116	4.7535901911	0551.4164621683	162	5.0875963352	0824.1906063076
117	4.7621739347	0557.1743503713	163	5.0937502008	0830.2812827315
118	4.7706846244	0562.9407856869	164	5.0998664278	0836.3780941632
119	4.7791234931	0568.7156956803	165	5.1059454739	0842.4810031936
120	4.7874917427	0574.4990091338	166	5.1119877883	0848.5899728672
121	4.7957905455	0580.2906560172	167	5.1179938124	0854.7049666736
122	4.8040210447	0586.0905674575	168	5.1239639794	0860.8259485397
123	4.8121843553	0591.8986757108	169	5.1298987149	0866.9528828220
124	4.8202815656	0597.7149141350	170	5.1357984370	0873.0857342985
125	4.8283137373	0603.5392171628	171	5.1416635565	0879.2244681620
126	4.8362819069	0609.3715202759	172	5.1474944768	0885.3690500119
127	4.8441870864	0615.2117599802	173	5.1532915944	0891.5194458481
128	4.8520302639	0621.0598737817	174	5.1590552992	0897.6756220633
129	4.8598124043	0626.9158001627	175	5.1647859739	0903.8375454366
130	4.8675344504	0632.7794785592	176	5.1704839950	0910.0051831267
131	4.8751973232	0638.6508493394	177	5.1761497325	0916.1785026656
132	4.8828019225	0644.5298537814	178	5.1817835502	0922.3574719520
133	4.8903491282	0650.4164340535	179	5.1873858058	0928.5420592455
134	4.8978397999	0656.3105331934	180	5.1929568508	0934.7322331602
135	4.9052747784	0662.2120950892	181	5.1984970312	0940.9279626591
136	4.9126548857	0668.1210644601	182	5.2040066870	0947.1292170480
137	4.9199809258	0674.0373868385	183	5.2094861528	0953.3359659700
138	4.9272536851	0679.9610085517	184	5.2149357576	0959.5481794001

TABLE I (continued)

n	$\log_e n$	$n \log_e n$	n	$\log_e n$	$n \log_e n$
185	5.2203558250	0965.7658276395	231	5.4424177105	1257.1984911305
186	5.2257466737	0971.9888813107	232	5.4467373716	1263.6430702266
187	5.2311086168	0978.2173113518	233	5.4510384535	1270.0919596808
188	5.2364419628	0984.4510890120	234	5.4553211153	1276.5451409937
189	5.2417470150	0990.6901858463	235	5.4595855141	1283.0025958239
190	5.2470240721	0996.9345737105	236	5.4638318050	1289.4643059860
191	5.2522734280	1003.1842247569	237	5.4680601411	1295.9302534490
192	5.2574953720	1009.4391114293	238	5.4722706736	1302.4004203338
193	5.2626901889	1015.6992064586	239	5.4764635519	1308.8747889116
194	5.2678581590	1021.9644828583	240	5.4806389233	1315.3533416021
195	5.2729995585	1028.2349139199	241	5.4847969334	1321.8360609712
196	5.2781146592	1034.5104732092	242	5.4889377261	1328.3229297299
197	5.2832037287	1040.7911345614	243	5.4930614433	1334.8139307318
198	5.2882670306	1047.0768720775	244	5.4971682252	1341.3090469715
199	5.2933048247	1053.3676601202	245	5.5012582105	1347.8082615835
200	5.2983173665	1059.6634733096	246	5.5053315359	1354.3115578394
201	5.3033049080	1065.9642865199	247	5.5093883366	1360.8189191471
202	5.3082676974	1072.2700748750	248	5.5134287461	1367.3303290489
203	5.3132059790	1078.5808137455	249	5.5174528964	1373.8457712197
204	5.3181199938	1084.8964787442	250	5.5214609178	1380.3652294656
205	5.3230099791	1091.2170457234	251	5.5254529391	1386.8886877221
206	5.3278761687	1097.5424907707	252	5.5294290875	1393.4161300529
207	5.3327187932	1103.8727902059	253	5.5333894887	1399.9475406481
208	5.3375380797	1110.2079205779	254	5.5373342670	1406.4829038227
209	5.3423342519	1116.5478586606	255	5.5412635451	1413.0222040154
210	5.3471075307	1122.8925814507	256	5.5451774444	1419.5654257868
211	5.3518581334	1129.2420661635	257	5.5490760848	1426.1125538181
212	5.3565862746	1135.5962902305	258	5.5529595849	1432.6635729098
213	5.3612921657	1141.9552312961	259	5.5568280616	1439.2184679802
214	5.3659760150	1148.3188672147	260	5.5606816310	1445.7772240640
215	5.3706380281	1154.6871760474	261	5.5645204073	1452.3398263112
216	5.3752784076	1161.0601360598	262	5.5683445037	1458.9062599854
217	5.3798973535	1167.4377257183	263	5.5721540321	1465.4765104628
218	5.3844950627	1173.8199236880	264	5.5759491031	1472.0505632306
219	5.3890717298	1180.2067088298	265	5.5797298259	1478.6284038863
220	5.3936275463	1186.5980601975	266	5.5834963087	1485.2100181359
221	5.3981627015	1192.9939570354	267	5.5872486584	1491.7953917929
222	5.4026773818	1199.3943787756	268	5.5909869805	1498.3845107769
223	5.4071717714	1205.7993050356	269	5.5947113796	1504.9773611129
224	5.4116460518	1212.2087156155	270	5.5984219589	1511.5739289296
225	5.4161004022	1218.6225904960	271	5.6021188208	1518.1742004584
226	5.4205349992	1225.0409098355	272	5.6058020662	1524.7781620325
227	5.4249500174	1231.4636539683	273	5.6094717951	1531.3858000855
228	5.4293456289	1237.8908034016	274	5.6131281063	1537.9971011503
229	5.4337220035	1244.3223388139	275	5.6167710976	1544.6120518583
230	5.4380793089	1250.7582410523	276	5.6204008657	1551.2306389379

TABLE I (continued)

n	$\log_e n$	$n \log_e n$	n	$\log_e n$	$n \log_e n$
277	5.6240175061	1557.8528492139	323	5.7776523232	1866.1817004009
278	5.6276211136	1564.4786696060	324	5.7807435157	1872.9608991167
279	5.6312117818	1571.1080871282	325	5.7838251823	1879.7431842572
280	5.6347896031	1577.7410888874	326	5.7868973813	1886.5285463255
281	5.6383546693	1584.3776620828	327	5.7899601708	1893.3169758834
282	5.6419070709	1591.0177940045	328	5.7930136083	1900.1084635500
283	5.6454468976	1597.6614720330	329	5.7960577507	1906.9030000018
284	5.6489742381	1604.3086836378	330	5.7990926544	1913.7005759720
285	5.6524891802	1610.9594163766	331	5.8021183753	1920.5011822498
286	5.6559918108	1617.6136578945	332	5.8051349689	1927.3048096803
287	5.6594822157	1624.2713959230	333	5.8081424899	1934.1114491635
288	5.6629604801	1630.9326182792	334	5.8111409929	1940.9210916542
289	5.6664266881	1637.5973128645	335	5.8141305318	1947.7337281614
290	5.6698809229	1644.2654676644	336	5.8171111599	1954.5493497476
291	5.6733232671	1650.9370707469	337	5.8200829303	1961.3679475287
292	5.6767538022	1657.6121102623	338	5.8230458954	1968.1895126733
293	5.6801726090	1664.2905744420	339	5.8260001073	1975.0140364020
294	5.6835797673	1670.9724515976	340	5.8289456176	1981.8415099875
295	5.6869753563	1677.6577301202	341	5.8318824772	1988.6719247537
296	5.6903594543	1684.3463984799	342	5.8348107370	1995.5052720754
297	5.6937321388	1691.0384452244	343	5.8377304471	2002.3415433779
298	5.6970934865	1697.7338589786	344	5.8406416573	2009.1807301364
299	5.7004435733	1704.4326284438	345	5.8435444170	2016.0228238758
300	5.7037824746	1711.1347423969	346	5.8464387750	2022.8678161700
301	5.7071102647	1717.8401896894	347	5.8493247799	2029.7156986416
302	5.7104270173	1724.5489592472	348	5.8522024797	2036.5664629615
303	5.7137328055	1731.2610400693	349	5.8550719222	2043.4201008486
304	5.7170277014	1737.9764212275	350	5.8579331544	2050.2766040692
305	5.7203117766	1744.6950918653	351	5.8607862234	2057.1359644365
306	5.7235851019	1751.4170411974	352	5.8636311755	2063.9981738105
307	5.7268477475	1758.1422585093	353	5.8664680569	2070.8632240975
308	5.7300997829	1764.8707331559	354	5.8692969131	2077.7311072494
309	5.7333412768	1771.6024545614	355	5.8721177894	2084.6018152638
310	5.7365722974	1778.3374122185	356	5.8749307308	2091.4753401833
311	5.7397929121	1785.0755956877	357	5.8777357817	2098.3516740953
312	5.7430031878	1791.8169945966	358	5.8805329864	2105.2308091315
313	5.7462031905	1798.5615986391	359	5.8833223884	2112.1127374673
314	5.7493929859	1805.3093975752	360	5.8861040314	2118.9974513221
315	5.7525726388	1812.0603812301	361	5.8888779583	2125.8849429582
316	5.7557422135	1818.8145394935	362	5.8916442118	2132.7752046809
317	5.7589017738	1825.5718623191	363	5.8944028342	2139.6682288381
318	5.7620513827	1832.3323397241	364	5.8971538676	2146.5640078198
319	5.7651911027	1839.0959617884	365	5.8998973535	2153.4625340576
320	5.7683209957	1845.8627186540	366	5.9026333334	2160.3638000249
321	5.7714411231	1852.6326005247	367	5.9053618480	2167.2677982360
322	5.7745515455	1859.4055976653	368	5.9080829381	2174.1745212462

TABLE I (continued)

n	$\log_e n$	$n \log_e n$	n	$\log_e n$	$n \log_e n$
369	5.9107966440	2181.0839616510	415	6.0282785202	2501.7355858957
370	5.9135030056	2187.9961120862	416	6.0306852602	2508.7650682687
371	5.9162020626	2194.9109652274	417	6.0330862217	2515.7969544901
372	5.9188938542	2201.8285137896	418	6.0354814325	2522.8312387953
373	5.9215784196	2208.7487505271	419	6.0378709199	2529.8679154474
374	5.9242557974	2215.6716682330	420	6.0402547112	2536.9069787365
375	5.9269260259	2222.5972597389	421	6.0426328336	2543.9484229803
376	5.9295891433	2229.5255179146	422	6.0450053140	2550.9922425232
377	5.9322451874	2236.4564356679	423	6.0473721790	2558.0384317366
378	5.9348941956	2243.3900059442	424	6.0497334552	2565.0869850184
379	5.9375362050	2250.3262217262	425	6.0520891689	2572.1378967929
380	5.9401712527	2257.2650760338	426	6.0544393462	2579.1911615108
381	5.9427993751	2264.2065619233	427	6.0567840132	2586.2467736486
382	5.9454206086	2271.1506724877	428	6.0591231955	2593.3047277090
383	5.9480349891	2278.0974008562	429	6.0614569189	2600.3650182201
384	5.9506425525	2285.0467401937	430	6.0637852086	2607.4276397357
385	5.9532433342	2291.9986837008	431	6.0661080901	2614.4925868347
386	5.9558373694	2298.9532246134	432	6.0684255882	2621.5598541215
387	5.9584246930	2305.9103562025	433	6.0707377280	2628.6294362251
388	5.9610053396	2312.8700717738	434	6.0730445341	2635.7013277996
389	5.9635793436	2319.8323646676	435	6.0753460310	2642.7755235236
390	5.9661467391	2326.7972282582	436	6.0776422433	2649.8520181002
391	5.9687075599	2333.7646559543	437	6.0799331950	2656.9308062568
392	5.9712618397	2340.7346411979	438	6.0822189103	2664.0118827449
393	5.9738096118	2347.7071774646	439	6.0844994130	2671.0952423400
394	5.9763509092	2354.6822582634	440	6.0867747269	2678.1808798414
395	5.9788857649	2361.6598771359	441	6.0890448754	2685.2687900721
396	5.9814142112	2368.6400276568	442	6.0913098820	2692.3589678783
397	5.9839362806	2375.6227034328	443	6.0935697700	2699.4514081300
398	5.9864520052	2382.6078981032	444	6.0958245624	2706.5461057199
399	5.9889614168	2389.5956053391	445	6.0980742821	2713.6430555640
400	5.9914645471	2396.5858188432	446	6.1003189520	2720.7422526009
401	5.9939614273	2403.5785323499	447	6.1025585946	2727.8436917923
402	5.9964520886	2410.5737396248	448	6.1047932324	2734.9473681219
403	5.9989365619	2417.5714344645	449	6.1070228877	2742.0532765963
404	6.0014148779	2424.5716106963	450	6.1092475827	2749.1614122440
405	6.0038870671	2431.5742621781	451	6.1114673395	2756.2717701157
406	6.0063531596	2438.5793827983	452	6.1136821798	2763.3843452842
407	6.0088131854	2445.5869664751	453	6.1158921254	2770.4991328438
408	6.0112671744	2452.5970071569	454	6.1180971980	2777.6161279108
409	6.0137151560	2459.6094988215	455	6.1202974189	2784.7353256227
410	6.0161571596	2466.6244354763	456	6.1224928095	2791.8567211386
411	6.0185932144	2473.6418111580	457	6.1246833908	2798.9803096387
412	6.0210233493	2480.6616199320	458	6.1268691841	2806.1060863243
413	6.0234475929	2487.6838558929	459	6.1290502100	2813.2340464178
414	6.0258659738	2494.7085131637	460	6.1312264894	2820.3641851622

TABLE I (continued)

n	$\log_e n$	$n \log_e n$	n	$\log_e n$	$n \log_e n$
645	6.4692503167	4172.6664543333	691	6.5381398237	4517.8546182235
646	6.4707995037	4180.1364794436	692	6.5395859556	4525.3934812874
647	6.4723462945	4187.6080525421	693	6.5410299991	4532.9337894386
648	6.4738906963	4195.0811712363	694	6.5424719605	4540.4755405917
649	6.4754327167	4202.5558331410	695	6.5439118455	4548.0187326675
650	6.4769723628	4210.0320358783	696	6.5453496603	4555.5633635928
651	6.4785096422	4217.5097770778	697	6.5467854107	4563.1094313001
652	6.4800445619	4224.9890543762	698	6.5482191027	4570.6569337281
653	6.4815771292	4232.4698654175	699	6.5496507422	4578.2058688214
654	6.4831073514	4239.9522078530	700	6.5510803350	4585.7562345304
655	6.4846352356	4247.4360793411	701	6.5525078870	4593.3080288112
656	6.4861607889	4254.9214775473	702	6.5539334040	4600.8612496261
657	6.4876840184	4262.4084001444	703	6.5553568918	4608.4158949429
658	6.4892049313	4269.8968448121	704	6.5567783561	4615.9719627353
659	6.4907235345	4277.3868092372	705	6.5581978028	4623.5294509826
660	6.4922398350	4284.8782911135	706	6.5596152374	4631.0883576702
661	6.4937538398	4292.3712881420	707	6.5610306658	4638.6486807889
662	6.4952655559	4299.8657980303	708	6.5624440936	4646.2104183352
663	6.4967749901	4307.3618184932	709	6.5638555265	4653.7735683113
664	6.4982821494	4314.8593472524	710	6.5652649700	4661.3381287251
665	6.4997870406	4322.3583820361	711	6.5666724298	4668.9040975901
666	6.5012896705	4329.8589205799	712	6.5680779114	4676.4714729253
667	6.5027900459	4337.3609606257	713	6.5694814204	4684.0402527554
668	6.5042881735	4344.8644999225	714	6.5708829623	4691.6104351105
669	6.5057840601	4352.3695362258	715	6.5722825426	4699.1820180262
670	6.5072777123	4359.8760672980	716	6.5736801669	4706.7549995438
671	6.5087691369	4367.3840909080	717	6.5750758405	4714.3293777099
672	6.5102583405	4374.8936048316	718	6.5764695690	4721.9051505765
673	6.5117453296	4382.4046068509	719	6.5778613577	4729.4823162014
674	6.5132301109	4389.9170947549	720	6.5792512120	4737.0608726473
675	6.5147126908	4397.4310663390	721	6.5806391372	4744.6408179824
676	6.5161930760	4404.9465194050	722	6.5820251388	4752.2221502806
677	6.5176712729	4412.4634517616	723	6.5834092221	4759.8048676208
678	6.5191472879	4419.9818612236	724	6.5847913923	4767.3889680873
679	6.5206211275	4427.5017456124	725	6.5861716548	4774.9744497696
680	6.5220927981	4435.0231027557	726	6.5875500148	4782.5613107628
681	6.5235623061	4442.5459304878	727	6.5889264775	4790.1495491669
682	6.5250296578	4450.0702266492	728	6.5903010481	4797.7391630872
683	6.5264948595	4457.5959890868	729	6.5916737320	4805.3301506343
684	6.5279579176	4465.1232156538	730	6.5930445341	4812.9225099240
685	6.5294188382	4472.6519042096	731	6.5944134597	4820.5162390771
686	6.5308776277	4480.1820526200	732	6.5957805139	4828.1113362197
687	6.5323342922	4487.7136587568	733	6.5971457018	4835.7077994829
688	6.5337888379	4495.2467204981	734	6.5985090286	4843.3056270031
689	6.5352412710	4502.7812357284	735	6.5998704992	4850.9048169214
690	6.5366915975	4510.3172023380	736	6.6012301187	4858.5053673845

TABLE I (continued)

n	$\log_e n$	$n \log_e n$	n	$\log_e n$	$n \log_e n$
737	6.6025878921	4866.1072765435	783	6.6631326959	5217.2329009608
738	6.6039438246	4873.7105425551	784	6.6644090203	5224.8966719547
739	6.6052979209	4881.3151635807	785	6.6656837177	5232.5617184592
740	6.6066501861	4888.9211377867	786	6.6669567924	5240.2280388494
741	6.6080006252	4896.5284633444	787	6.6682282484	5247.8956315045
742	6.6093492431	4904.1371384302	788	6.6694980898	5255.5644948080
743	6.6106960447	4911.7471612253	789	6.6707663208	5263.2346271474
744	6.6120410348	4919.3585299158	790	6.6720329454	5270.9060269142
745	6.6133842183	4926.9712426928	791	6.6732979677	5278.5786925042
746	6.6147256002	4934.5852977520	792	6.6745613918	5286.2526223170
747	6.6160651851	4942.2006932942	793	6.6758232216	5293.9278147564
748	6.6174029779	4949.8174275249	794	6.6770834612	5301.6042682302
749	6.6187389835	4957.4354986544	795	6.6783421146	5309.2819811502
750	6.6200732065	4965.0549048978	796	6.6795991858	5316.9609519321
751	6.6214056517	4972.6756444749	797	6.6808546787	5324.6411789958
752	6.6227363239	4980.2977156103	798	6.6821085974	5332.3226607649
753	6.6240652277	4987.9211165333	799	6.6833609457	5340.0053956673
754	6.6253923680	4995.5458454780	800	6.6846117276	5347.6893821343
755	6.6267177492	5003.1719006830	801	6.6858609470	5355.3746186018
756	6.6280413761	5010.7992803917	802	6.6871086078	5363.0611035089
757	6.6293632534	5018.4279828521	803	6.6883547139	5370.7488352992
758	6.6306833856	5026.0580063169	804	6.6895992691	5378.4378124199
759	6.6320017773	5033.6893490433	805	6.6908422774	5386.1280333219
760	6.6333184332	5041.3220092931	806	6.6920837425	5393.8194964603
761	6.6346333578	5048.9559853327	807	6.6933236682	5401.5122002938
762	6.6359465556	5056.5912754332	808	6.6945620585	5409.2061432850
763	6.6372580312	5064.2278778700	809	6.6957989170	5416.9013239003
764	6.6385677891	5071.8657909232	810	6.6970342476	5424.5977406099
765	6.6398758338	5079.5050128773	811	6.6982680541	5432.2953918876
766	6.6411821697	5087.1455420213	812	6.6995003401	5439.9942762113
767	6.6424868013	5094.7873766487	813	6.7007311095	5447.6943920624
768	6.6437897331	5102.4305150574	814	6.7019603660	5455.3957379261
769	6.6450909695	5110.0749555498	815	6.7031881132	5463.0983122913
770	6.6463905148	5117.7206964328	816	6.7044143549	5470.8021136507
771	6.6476883735	5125.3677360173	817	6.7056390948	5478.5071405006
772	6.6489845500	5133.0160726191	818	6.7068623366	5486.2133913410
773	6.6502790485	5140.6657045581	819	6.7080840838	5493.9208646757
774	6.6515718735	5148.3166301584	820	6.7093043402	5501.6295590118
775	6.6528630293	5155.9688477488	821	6.7105231094	5509.3394728604
776	6.6541525201	5163.6223556622	822	6.7117403950	5517.0506047362
777	6.6554403503	5171.2771522357	823	6.7129562006	5524.7629531572
778	6.6567265241	5178.9332358108	824	6.7141705299	5532.4765166454
779	6.6580110458	5186.5906047333	825	6.7153833863	5540.1912937261
780	6.6592939196	5194.2492573532	826	6.7165947735	5547.9072829283
781	6.6605751498	5201.9091920248	827	6.7178046950	5555.6244827846
782	6.6618547405	5209.5704071064	828	6.7190131543	5563.3428918310

TABLE I (continued)

n	$\log_e n$	$n \log_e n$	n	$\log_e n$	$n \log_e n$
829	6.7202201551	5571.0625086072	875	6.7742238863	5927.4459005629
830	6.7214257007	5578.7833316562	876	6.7753660909	5935.2206956603
831	6.7226297948	5586.5053595249	877	6.7765069923	5942.9966323104
832	6.7238324408	5594.2285907632	878	6.7776465936	5950.7737092116
833	6.7250336421	5601.9530239250	879	6.7787848976	5958.5519250653
834	6.7262334023	5609.6786575672	880	6.7799219074	5966.3312785756
835	6.7274317248	5617.4054902505	881	6.7810576259	5974.1117684498
836	6.7286286130	5625.1335205388	882	6.7821920560	5981.8933933980
837	6.7298240704	5632.8627469997	883	6.7833252006	5989.6761521333
838	6.7310181004	5640.5931682040	884	6.7844570626	5997.4600433717
839	6.7322107064	5648.3247827260	885	6.7855876450	6005.2450658320
840	6.7334018918	5656.0575891434	886	6.7867169506	6013.0312182361
841	6.7345916599	5663.7915860372	887	6.7878449823	6020.8184993086
842	6.7357800142	5671.5267719920	888	6.7889717429	6028.6069077770
843	6.7369669580	5679.2631455956	889	6.7900972355	6036.3964423719
844	6.7381524945	5687.0007054390	890	6.7912214627	6044.1871018263
845	6.7393366273	5694.7394501168	891	6.7923444274	6051.9788848765
846	6.7405193596	5702.4793782269	892	6.7934661325	6059.7717902614
847	6.7417006946	5710.2204883703	893	6.7945865808	6067.5658167227
848	6.7428806357	5717.9627791515	894	6.7957057751	6075.3609630051
849	6.7440591863	5725.7062491783	895	6.7968237182	6083.1572278560
850	6.7452363494	5733.4508970617	896	6.7979404129	6090.9546100255
851	6.7464121285	5741.1967214159	897	6.7990558620	6098.7531082667
852	6.7475865268	5748.9437208586	898	6.8001700683	6106.5527213354
853	6.7487595474	5756.6918940104	899	6.8012830344	6114.3534479900
854	6.7499311937	5764.4412394954	900	6.8023947633	6122.1552869919
855	6.7511014689	5772.1917559409	901	6.8035052576	6129.9582371051
856	6.7522703761	5779.9434419773	902	6.8046145200	6137.7622970965
857	6.7534379185	5787.6962962383	903	6.8057225534	6145.5674657355
858	6.7546040994	5795.4503173607	904	6.8068293603	6153.3737417945
859	6.7577689219	5803.2055039845	905	6.8079349436	6161.1811240484
860	6.7569323892	5810.9618547529	906	6.8090393060	6168.9896112749
861	6.7580945044	5818.7193683123	907	6.8101424501	6176.7992022544
862	6.7592552706	5826.4780433121	908	6.8112443786	6184.6098957700
863	6.7604146910	5834.2378784050	909	6.8123450941	6192.4216906073
864	6.7615727688	5841.9988722467	910	6.8134445995	6200.2345855549
865	6.7627295069	5849.7610234961	911	6.8145428972	6208.0485794038
866	6.7638849085	5857.5243308151	912	6.8156399900	6215.8636709478
867	6.7650389767	5865.2887928687	913	6.8167358805	6223.6798589832
868	6.7661917146	5873.0544083252	914	6.8178305714	6231.4971423091
869	6.7673431252	5880.8211758556	915	6.8189240652	6239.3155197271
870	6.7684932116	5888.5890941343	916	6.8200163646	6247.1349900415
871	6.7696419768	5896.3581618385	917	6.8211074722	6254.9555520592
872	6.7707894239	5904.1283776486	918	6.8221973906	6262.7772045896
873	6.7719355558	5911.8997402480	919	6.8232861223	6270.5999464449
874	6.7730803756	5919.6722483229	920	6.8243736700	6278.4237764396

TABLE I (continued)

n	$\log_e n$	$n \log_e n$	n	$\log_e n$	$n \log_e n$
921	6.8254600362	6286.2486933911	961	6.8679744089	6600.1234070205
922	6.8265452235	6294.0746961192	962	6.8690144506	6607.9919015404
923	6.8276292345	6301.9017834461	963	6.8700534117	6615.8614355616
924	6.8287120716	6309.7299541969	964	6.8710912946	6623.7320080046
925	6.8297937375	6317.5592071990	965	6.8721281013	6631.6036177921
926	6.8308742346	6325.3895412824	966	6.8731638342	6639.4762638493
927	6.8319535655	6333.2209552795	967	6.8741984954	6647.3499451033
928	6.8330317327	6341.0534480256	968	6.8752320872	6655.2246604837
929	6.8341087388	6348.8870183581	969	6.8762646118	6663.1004089222
930	6.8351845861	6356.7216651170	970	6.8772960714	6670.9771893525
931	6.8362592772	6364.5573871449	971	6.8783264682	6678.8550007109
932	6.8373328146	6372.3941832870	972	6.8793558044	6686.7338419355
933	6.8384052008	6380.2320523906	973	6.8803840821	6694.6137119670
934	6.8394764382	6388.0709933057	974	6.8814113036	6702.4946097478
935	6.8405465292	6395.9110048849	975	6.8824374709	6710.3765342229
936	6.8416154764	6403.7520859830	976	6.8834625864	6718.2594843392
937	6.8426832822	6411.5942354574	977	6.8844866520	6726.1434590458
938	6.8437499490	6419.4374521678	978	6.8855096700	6734.0284572941
939	6.8448154792	6427.2817349766	979	6.8865316425	6741.9144780374
940	6.8458798752	6435.1270827482	980	6.8875525716	6749.8015202313
941	6.8469431395	6442.9734943498	981	6.8885724595	6757.6895828336
942	6.8480052745	6450.8209686509	982	6.8895913083	6765.5786648041
943	6.8490662826	6458.6695045234	983	6.8906091201	6773.4687651047
944	6.8501261661	6466.5191008414	984	6.8916258970	6781.3598826994
945	6.8511849274	6474.3697564816	985	6.8926416411	6789.2520165545
946	6.8522425690	6482.2214703231	986	6.8936563546	6797.1451656382
947	6.8532990931	6490.0742412472	987	6.8946700394	6805.0393289208
948	6.8543545022	6497.9280681378	988	6.8956826977	6812.9345053749
949	6.8554087986	6505.7829498808	989	6.8966943316	6820.8306939749
950	6.8564619845	6513.6388853649	990	6.8977049431	6828.7278936973
951	6.8575140625	6521.4958734807	991	6.8987145343	6836.6261035210
952	6.8585650347	6529.3539131214	992	6.8997231072	6844.5253224266
953	6.8596149036	6537.2130031825	993	6.9007306640	6852.4255493969
954	6.8606636714	6545.0731425617	994	6.9017372066	6860.3267834166
955	6.8617113404	6552.9343301591	995	6.9027427371	6868.2290234728
956	6.8627579130	6560.7965648771	996	6.9037472575	6876.1322685543
957	6.8638033914	6568.6598456205	997	6.9047507699	6884.0365176520
958	6.8648477779	6576.5241712961	998	6.9057532763	6891.9417697588
959	6.8658910748	6584.3895408132	999	6.9067547786	6899.8480238699
960	6.8669332844	6592.2559530834	1000	6.9077552789	6907.7552789821

TABLE II.* $F(p_1, p_2) = p_1 \log \frac{p_1}{p_2} + q_1 \log \frac{q_1}{q_2}$, $p_1 + q_1 = 1 = p_2 + q_2$

p_2

p_1 \ p_2	0.01	0.02	0.03	0.04	0.05	0.10	0.15
0.01	0.0000000	0.0031170	0.0092198	0.0165994	0.0247332	0.0713311	0.1238648
0.02	0.0039160	0.0000000	0.0019456	0.0063448	0.0121424	0.0512682	0.0991756
0.03	0.0131606	0.0022116	0.0000000	0.0014188	0.0048802	0.0365339	0.0798150
0.04	0.0259124	0.0079304	0.0015616	0.0000000	0.0011252	0.0253068	0.0639616
0.05	0.0412940	0.0162790	0.0057530	0.0012110	0.0000000	0.0167095	0.0507380
0.10	0.1444790	0.0842990	0.0529870	0.0335430	0.0206510	0.0000000	0.0108970
0.15	0.2766080	0.1812630	0.1291650	0.0948190	0.0702460	0.0122345	0.0000000
0.20	0.4286740	0.2981640	0.2252800	0.1760320	0.1397780	0.0444060	0.0090400
0.25	0.5964975	0.4308225	0.3371525	0.2730025	0.2250675	0.0923350	0.0338375
0.30	0.7777260	0.5768860	0.4624300	0.3833780	0.3237620	0.1536690	0.0720400
0.35	0.9708980	0.7348930	0.5996510	0.5056970	0.4344000	0.2269465	0.1221860
0.40	1.1750840	0.9039140	0.7478860	0.6390300	0.5560520	0.3112380	0.1833460
0.45	1.3897125	1.0833775	0.9065635	0.7828055	0.6881465	0.4059720	0.2549485
p_1 0.50	1.6144600	1.2729600	1.0753600	0.9367000	0.8303600	0.5108250	0.3366700
0.55	1.8492245	1.4725595	1.2541735	1.1006115	0.9825905	0.6256950	0.4284085
0.60	2.0941080	1.6822780	1.4431060	1.2746420	1.1449400	0.7506840	0.5302660
0.65	2.3494340	1.9024390	1.6424810	1.4591150	1.3177320	0.8861155	0.6425660
0.70	2.6157740	2.1336140	1.8528700	1.6546020	1.5015380	1.0325610	0.7658800
0.75	2.8940575	2.3767325	2.0752025	1.8620325	1.6972875	1.1909500	0.9011375
0.80	3.1857460	2.6332560	2.3109400	2.0828680	1.9064420	1.3627440	1.0498000
0.85	3.4931920	2.9055370	2.5624350	2.3194610	2.1313640	1.5502955	1.2142200
0.90	3.8205750	3.1977550	2.8338670	2.5759910	2.3762030	1.7577840	1.3985770
0.95	4.1769020	3.5189170	3.1342430	2.8614650	2.6499960	1.9942165	1.6118780
0.96	4.2534228	3.5884048	3.1995736	2.9238152	2.7100100	2.0467584	1.6597936
0.97	4.3325734	3.6605224	3.2675340	2.9887952	2.7726538	2.1019301	1.7103390
0.98	4.4152312	3.7361472	3.3390016	3.0572824	2.8388048	2.1606090	1.7643916
0.99	4.5032176	3.8171006	3.4157978	3.1310982	2.9102844	2.2246165	1.8237728
	0.01	0.02	0.03	0.04	0.05	0.10	0.15

p_2

* For values of $p_1 \log \frac{p_1}{p_2} + q_1 \log \frac{q_1}{q_2}$ for $p_2 > 0.50$, enter the table using (q_1, q_2) as though they were (p_1, p_2).

TABLE II (continued)

p_1 \ p_2	0.20	0.25	0.30	0.35	0.40	0.45	0.50
0.01	0.1810018	0.2426649	0.3091418	0.3809692	0.4588834	0.5438455	0.6371488
0.02	0.1528296	0.2116158	0.2755796	0.3451244	0.4209028	0.5038170	0.5951136
0.03	0.1299860	0.1858953	0.2473460	0.3146082	0.3882508	0.4691171	0.5584070
0.04	0.1106496	0.1636820	0.2226196	0.2875992	0.3591060	0.4379244	0.5252076
0.05	0.0939430	0.1440985	0.2005230	0.2632200	0.3325910	0.4093615	0.4946380
0.10	0.0366870	0.0724580	0.1163170	0.1676010	0.2262930	0.2928240	0.3680670
0.15	0.0083750	0.0297615	0.0610550	0.1009260	0.1489390	0.2052305	0.2704400
0.20	0.0000000	0.0070020	0.0257300	0.0541880	0.0915220	0.1375740	0.1927500
0.25	0.0073825	0.0000000	0.0061625	0.0232075	0.0498625	0.0856750	0.1308175
0.30	0.0281700	0.0064030	0.0000000	0.0056320	0.0216080	0.0471810	0.0822900
0.35	0.0609010	0.0247495	0.0057810	0.0000000	0.0052970	0.0206305	0.0457060
0.40	0.1046460	0.0541100	0.0225760	0.0053820	0.0000000	0.0050940	0.0201360
0.45	0.1588335	0.0939130	0.0498135	0.0212065	0.0051455	0.0000000	0.0050085
0.50	0.2231400	0.1438350	0.0871700	0.0471500	0.0204100	0.0050250	0.0000000
0.55	0.2974635	0.2037740	0.1345435	0.0831105	0.0456915	0.0200670	0.0050085
0.60	0.3819060	0.2738320	0.1920360	0.1291900	0.0810920	0.0452280	0.0201360
0.65	0.4767910	0.3543325	0.2599710	0.1857120	0.1269350	0.0808315	0.0457060
0.70	0.5826900	0.4458470	0.3389200	0.2532480	0.1837920	0.1274490	0.0822900
0.75	0.7005325	0.5493050	0.4298125	0.3327275	0.2525925	0.1860100	0.1308175
0.80	0.8317800	0.6661680	0.5341100	0.4256120	0.3347980	0.2579760	0.1927500
0.85	0.9787850	0.7987885	0.6541650	0.5342540	0.4327610	0.3456995	0.2704400
0.90	1.1457270	0.9513460	0.7941570	0.6628330	0.5506610	0.4533600	0.3680670
0.95	1.3416130	1.1328475	0.9630930	0.8203560	0.6975050	0.5899645	0.4946380
0.96	1.3860456	1.1744032	1.0021356	0.8571160	0.7321292	0.6225408	0.5252076
0.97	1.4331080	1.2185887	1.0438080	0.8965058	0.7693832	0.6577469	0.5584070
0.98	1.4836776	1.2662814	1.0889876	0.9394028	0.8101444	0.6964602	0.5951136
0.99	1.5395758	1.3193027	1.1394958	0.9876284	0.8562342	0.7405021	0.6371488
	0.20	0.25	0.30	0.35	0.40	0.45	0.50

p_2

TABLE III. Noncentral χ^2

Table of 5% points of the distribution of Fisher's B^2

Values of β^2	Value of n, Degrees of Freedom 1	2	3	4	5	6	7
0	3.8415	5.9915	7.8147	9.4877	11.0705	12.5916	14.0671
0.04	3.9940	6.1108	7.9186	9.5821	11.1589	12.6750	14.1474
0.16	4.4394	6.4613	8.2254	9.8627	11.4217	12.9247	14.3868
0.36	5.1320	7.0209	8.7220	10.3202	11.8515	13.3349	14.7802
0.64	6.0050	7.7590	9.3881	10.9402	12.4383	13.8965	15.3225
1.00	7.0018	8.6424	10.2023	11.7073	13.1704	14.6000	16.0040
1.44	8.0946	9.6466	11.1462	12.6061	14.0340	15.4363	16.8166
1.96	9.2714	10.7558	12.2045	13.6242	15.0203	16.3952	17.7527
2.56	10.5294	11.9605	13.3671	14.7517	16.1186	17.4691	18.8035
3.24	11.8673	13.2569	14.6276	15.9824	17.3222	18.6486	19.9639
4.00	13.2853	14.6406	15.9808	17.3089	18.6261	19.9318	21.2281
4.84	14.7833	16.1098	17.4248	18.7299	20.0256	21.3130	22.5920
5.76	16.3612	17.6627	18.9564	20.2410	21.5185	22.7892	24.0522
6.76	18.0192	19.3002	20.5744	21.8416	23.1024	24.3572	25.6066
7.84	19.7571	21.0195	22.2775	23.5283	24.7745	26.0161	27.2526
9.00	21.5751	22.8216	24.0639	25.3019	26.5349	27.7634	28.9875
10.24	23.4731	24.7059	25.9346	27.1597	28.3801	29.5980	30.8114
11.56	25.4510	26.6710	27.8879	29.1017	30.3116	31.5192	32.7230
12.96	27.5090	28.7178	29.9242	31.1275	32.3272	33.5253	34.7204
14.44	29.6469	30.8458	32.0424	33.2352	34.4276	35.6158	36.8024
16.00	31.8649	33.0545	34.2412	35.4275	36.6098	37.7918	38.9701
17.64	34.1629	35.3442	36.5227	37.7008	38.8765	40.0499	41.2215
19.36	36.5408	37.7143	38.8864	40.0562	41.2241	42.3918	43.5574
21.16	38.9988	40.1652	41.3295	42.4934	43.6551	44.8163	45.9752
23.04	41.5367	42.6958	43.8549	45.0120	46.1679	47.3234	48.4764
25.00	44.1547	45.3077	46.4606	47.6128	48.7637	49.9128	51.0610

Entries in this table are the squares of the values of B and the values of β^2 are the squares of β_1 in the table on p. 665 of R. A. Fisher (1928).

Glossary

$\Rightarrow$	Implies
$\rightarrow$	Approaches
$\rightleftharpoons$	If and only if
$\cup$	Union
$\cap$	Intersection
$\subset$	Is contained in
$\supset$	Contains
$\sim$	Is asymptotically equal to
$\approx$	Is approximately equal to
:	Such that
$\{x: C\}$	Set of x's satisfying the condition C
$\in$	Belongs to
$[\lambda]$	Modulo λ, or except for sets of λ-measure 0
cov	Covariance
$E_i(\ )$	Expectation with respect to the probability measure μ_i
$O(n)$	Is at most of order n
$o(n)$	Is of smaller order than n
g.l.b.	Greatest lower bound, infimum, inf
l.u.b.	Least upper bound, supremum, sup
$\overline{\lim}$	Limit superior, lim sup
$\underline{\lim}$	Limit inferior, lim inf
lim	Limit
tr	Trace
var	Variance

Absolute Continuity: A measure μ is said to be absolutely continuous with respect to a second measure ν if for every set E for which $\nu(E) = 0$ it is true that $\mu(E) = 0$. For μ absolutely continuous with respect to ν, we write $\mu \ll \nu$. [μ and ν are defined on the same measurable space $(\mathscr{X}, \mathscr{S})$.]

Additive Class of Sets—a Field: Sometimes called "simply" additive to distinguish it from a "completely" additive class which is a Borel field. In other words, additive or simply additive refer to properties essentially dealing with a finite number of terms whereas completely additive refers to a denumerable number (finite or infinite).

Admissible: That which is regarded as a priori possible. Generally the property of belonging to a particular subset. For example, a parameter point is called an admissible point if it belongs to a set of the parameter space corresponding to a given hypothesis.

Asymptotic Confidence Interval: A confidence interval whose limits are statistics based on arbitrarily large samples.

Asymptotic Distribution Function: If the distribution function $F(c; n)$ of a random variable x depends upon a parameter n, then the distribution function (if any) to which $F(c; n)$ tends as $n \to \infty$ is called the asymptotic distribution function of the random variable.

Axiomatic Development: That development of a science which begins with the creation of a clearly defined set of axioms from which all theorems are deduced. The theorems are then applied to explain and predict the results of experiments—the "facts." Inductive development, by contrast, proceeds from a body of observed "facts" from which the theorems are obtained by a process of generalization. If then a set of axioms can be found which enables the theorems to be "proved," the two approaches produce equivalent results.

Basis: A set of linearly independent vectors such that every other vector of the space is a linear combination of the vectors of the set.

Best Estimate: That estimate of a parameter having minimum attainable variance.

Bias: The difference between the expected value of an estimate and the estimated parameter.

Biased Estimate: An estimate whose expected value is not the estimated parameter.

Binary Digit: A digit of a binary system of numbers.

Bit: Abbreviation for binary digit.

Borel Field: A field $\mathcal{S}$ such that the union of any denumerable number of sets of $\mathcal{S}$ is a set of $\mathcal{S}$.

Borel Set: A set of a Borel field. In an n-dimensional Euclidean space R^n a Borel set is one that is obtained by taking a finite or a denumerable number of unions, differences, and intersections of half-open intervals $(a_i < x_i \leq b_i)$, $i = 1, 2, \cdots, n$.

Characteristic Equation (of a square matrix $\mathbf{A}$): The determinantal equation in λ, $|\mathbf{A} - \lambda\mathbf{I}| = 0$, where $\mathbf{I}$ is the identity matrix (same order as $\mathbf{A}$).

Characteristic Function of a Set: The point function which is equal to 1 for any point of the set and which is 0 elsewhere.

Characteristic Vector (corresponding to a characteristic root of a characteristic equation for a square matrix $\mathbf{A}$): The vector $\mathbf{x}$ which satisfies the matrix equation $\mathbf{A}\mathbf{x} = \lambda\mathbf{x}$ for the particular characteristic root λ of the characteristic equation.

Class: Set of sets.

Communication Theory: Mathematics applied to communication processes.

Complement of One Set with Respect to Another: Set of all points of the second set which are not in the first. The complement of a set E with respect to the space $\mathcal{X}$ in which it is contained is the set of all points of $\mathcal{X}$ not in E.

Confidence Coefficient: The probability associated with a confidence interval.

Confidence Interval: An interval limited by two statistics such that the probability of a parameter value being covered by the interval is known.

Confidence Limits: The upper and lower limits of a confidence interval.

Consistent Estimate: One that converges in probability to the estimated parameter.

Converge Stochastically or Converge in Probability: Let $f(x), f_1(x), f_2(x), \cdots$ be random variables on an x-space. The sequence $f_n(x)$ is said to converge stochastically, or in probability, to $f(x)$ if $\lim_{n\to\infty} \text{Prob}\{|f_n(x) - f(x)| \geq \epsilon\} = 0$.

Converge with Probability 1: Let $f(x), f_1(x), f_2(x), \cdots$ be random variables on an x-space. If $\lim_{n\to\infty} f_n(x) = f(x)$ for almost all x, we say that $f_n(x)$ converges to $f(x)$ with probability 1.

Convex Set: A set such that the entire line segment connecting any two points of the set is contained in the set.

Cramér-Rao Inequality: See Information Inequality.

Denumerable: The property of being able to be placed in one-to-one correspondence with the set of positive integers.

Diagonal Element (of a square matrix): An element in the same row as column.

Disjoint Sets: Two sets having no common elements.

Distance (Function): A real-valued function d of points x, y, z such that $d(x, y) \geq 0$, $d(x, y) = 0$, if and only if $x = y$, $d(x, y) = d(y, x)$, and $d(x, y) \leq d(x, z) + d(z, y)$. The last relation is called the triangular inequality.

Dominated Set of Measures: A set M of measures μ_i defined on the measurable space $(\mathscr{X}, \mathscr{S})$ for which there exists a finite measure ν such that μ_i is absolutely continuous with respect to $\nu(\mu_i \ll \nu)$ for every μ_i belonging to M. ν need not be a member of M.

Efficient Estimate: An estimate of minimum possible variance.

Equivalent Measures: Two measures μ and ν such that μ is absolutely continuous with respect to ν (written $\mu \ll \nu$) and such that ν is absolutely continuous with respect to μ $(\nu \ll \mu)$. To indicate that two measures are equivalent we write $\mu \equiv \nu$.

Equivalent Set of Measures: A set of measures μ_i defined on the measurable space $(\mathscr{X}, \mathscr{S})$ for which there exists a measure ν such that each measure μ_i is equivalent to ν (written $\mu_i \equiv \nu$). This means that each μ_i is absolutely continuous with respect to ν and vice versa.

Estimator: A statistic selected to approximate (or to estimate) a given parameter (or function of such parameter).

Euclidean Space R^n of n Dimensions: A metric space made up of points (vectors) $x = (x_1, x_2, \cdots, x_n)$, where the x_i for $i = 1, 2, \cdots, n$ are real numbers and where for two points $x = (x_1, x_2, \cdots, x_n)$ and $y = (y_1, y_2, \cdots, y_n)$, the "distance" between x and y is defined as $\left(\sum_{i=1}^{n} (x_i - y_i)^2\right)^{1/2}$

Event: A set of the probability space $(\mathscr{X}, \mathscr{S}, \mu)$ belonging to $\mathscr{S}$.

Field: A class $\mathscr{S}$ of sets of a space $\mathscr{X}$ such that the union of any two sets in $\mathscr{S}$ is in $\mathscr{S}$, the intersection of any two sets in $\mathscr{S}$ is in $\mathscr{S}$, and the complement of any set in $\mathscr{S}$ with respect to $\mathscr{X}$ is in $\mathscr{S}$.

Finite Measure: A measure μ such that $\mu(\mathscr{X}) < \infty$ for a measurable space $(\mathscr{X}, \mathscr{S})$.

Fisher's Information Matrix: The $k \times k$ matrix whose element in the ith row and jth column is $\int f(x, \theta) \left[\frac{\partial}{\partial \theta_i} \log f(x, \theta)\right]\left[\frac{\partial}{\partial \theta_j} \log(f x, \theta)\right] d\lambda(x)$, where $\lambda(x)$ is a probability measure and $f(x, \theta)$, the generalized density, is a function of x and a k-dimensional parameter θ.

Generalized Probability Density: Let μ be a probability measure which is absolutely continuous with respect to λ on a probability space $(\mathscr{X}, \mathscr{S}, \lambda)$. Then the generalized probability density function corresponding to μ is that function $f(x)$, unique, positive, and finite except for sets of λ-measure zero, such that $\mu(E) = \int_E f(x)\, d\lambda(x)$ for all E belonging to $\mathscr{S}$.

Greatest Lower Bound (Abbreviated g.l.b., or called "the" lower bound): The largest of the lower bounds of a set (of real numbers).

Homogeneous Samples: Samples from populations with the same parameter values. If only some of the parameters are alike, the samples are said to be homogeneous with respect to these parameters only.

Homogeneous Set of Measures: A set of measures such that any two members of the set are absolutely continuous with respect to each other.

Hyperplane (of n dimensions): The set of all points in R^n which satisfy a single linear function $l(x_1, x_2, \cdots, x_n) = 0$. (See Linear Set.)

Hypothesis: A statement that a point of the parameter space belongs to a specified set of the parameter space.

Identity Matrix ($n \times n$): The matrix with all n diagonal elements equal to 1 and all other elements equal to 0.

Indicator of a Set: Same as characteristic function of a set.

Infimum (inf): Greatest lower bound.

Information Inequality: Consider $f(x, \theta)$ a density function corresponding to an absolutely continuous distribution function with parameter θ for a random variable X. Let $T(X)$ be any unbiased estimate of $\phi(\theta)$, a function of θ. Then the inequality

$$\text{variance of } T \geq \frac{(d\phi/d\theta)^2}{I},$$

where I is the variance of $\frac{1}{f}\frac{df}{d\theta}$, is called the information inequality.

Note: The range of X must be independent of θ and f must be differentiable with respect to θ under the integral sign. As defined by R. A. Fisher, I is the information on θ supplied by a sample of n observations.

Intersection of Two Sets: The set of points belonging to both sets. The intersection of sets A and B is written $A \cap B$.

Inverse Image of a Set: If a set G belongs to a space $\mathscr{Y}$ corresponding to a space $\mathscr{X}$ under a transformation $T(x)$, then the set of all points x of $\mathscr{X}$ whose transforms under $T(x)$ are in G is called the inverse image of G. It is denoted by $T^{-1}(G) = \{x: T(x) \in G\}$.

Inverse of a Matrix: An $n \times n$ square nonsingular matrix $\mathbf{A}$ is said to have an inverse $\mathbf{A}^{-1}$ if $\mathbf{AA}^{-1} = \mathbf{A}^{-1}\mathbf{A} = \mathbf{I}$, where $\mathbf{I}$ is the $n \times n$ identity matrix.

Jacobian of a Transformation: If $y_i = f_i(x_1, \cdots, x_k)$ for $i = 1, 2, \cdots, k$ is a transformation, then the determinant whose element in the ith row and jth column is $\partial f_i/\partial x_j$ is called the Jacobian of the transformation.

Khintchine's Theorem: Let $X_1, X_2, \cdots$ be identically distributed independent random variables with finite mean m. Then $\bar{X} = \frac{1}{n}\sum_{i=1}^{n} X_i$ converges in probability to m.

Least Upper Bound (Abbreviated l.u.b., or called "the" upper bound): The smallest of the upper bounds of a set (of real numbers).

Likelihood Ratio (at $X = x$): The ratio of $f_1(x)$ to $f_2(x)$, where $f_i(x)$ for $i = 1, 2$ is the generalized probability density for the observation $X = x$ under the hypothesis that the random variable X is from the population having the generalized probability density $f_i(X)$.

Limit Inferior (lim inf): The smallest limit point of a sequence (of real numbers bounded below). ($\underline{\lim}\, x = a$ if $x > a - \epsilon$ but never ultimately $> a + \epsilon$.) ($\underline{\lim}_{n\to\infty} x_n = \lim_{n\to\infty} m_n$, where m_1 is the lower bound of $x_1, x_2, x_3, \cdots$, m_2 is the lower bound of $x_2, x_3, \cdots$, m_3 is the lower bound of $x_3, x_4, \cdots$, etc.) ($\underline{\lim}_{n\to\infty} x_n = \sup_k \inf_{m \geq k} x_m$.)

Limit Point (of a sequence of real numbers): A point every neighborhood of which contains infinitely many points of the sequence.

Limit Superior (lim sup): The greatest limit point of a sequence (of real numbers bounded above). ($\overline{\lim}\, x = A$ if $x < A + \epsilon$ but never ultimately $< A - \epsilon$.) ($\overline{\lim}_{n\to\infty} x_n = \lim_{n\to\infty} M_n$, where M_1 is the upper bound of $x_1, x_2, \cdots$, M_2 is the upper bound of $x_2, x_3, \cdots$, M_3 is the upper bound of $x_3, x_4, \cdots$, etc.) ($\overline{\lim}_{n\to\infty} x_n = \inf_k \sup_{m \geqq k} x_m$.)

Linear Set of $n - p$ Dimensions: A set of points in a space R^n each of whose coordinates can be expressed as a linear function of $n - p$ arbitrary parameters. For $p = 1$ this set is a hyperplane and for $p = n - 1$, a straight line. Also: the set of points in R^n common to p linearly independent hyperplanes is a set of $n - p$ dimensions.

Linear Transformation: $\mathbf{y} = \mathbf{Ax}$ with $\mathbf{y}' = (y_1, y_2, \cdots, y_m)$, $\mathbf{x}' = (x_1, x_2, \cdots, x_n)$, $\mathbf{A} = (a_{ij})$, $i = 1, 2, \cdots, m$, $j = 1, 2, \cdots, n$.

Linearly Independent Functions (on R^n): A set of functions $f_i(x)$ defined on R^n such that no one of them can be expressed as a linear combination of the others with real numbers not all zero for coefficients.

Linearly Independent Vectors: A set of vectors is said to be linearly independent if none of them can be expressed as a linear combination of the rest.

Lower Bound (of a set E of real numbers): A real point c such that for every point x of E, $x \geqq c$.

Matrix ($m \times n$): A set of numbers arranged rectangularly in m rows and n columns.

Measurable Function: A real-valued function $f(x)$ of the points x of the measurable space $(\mathscr{X}, \mathscr{S})$ such that for every real number c, the set $\{x: f(x) < c\}$ belongs to $\mathscr{S}$. Such a function is called an $\mathscr{S}$-measurable function.

Measurable Set: Any subset of a measurable space $(\mathscr{X}, \mathscr{S})$ belonging to the Borel field $\mathscr{S}$ defined on the space $\mathscr{X}$.

Measurable Space: A space $\mathscr{X}$ on which is defined a Borel field $\mathscr{S}$ of subsets of $\mathscr{X}$. We denote this type of space by $(\mathscr{X}, \mathscr{S})$.

Measurable Transformation: A transformation $T(x)$ of the elements of a measurable space $(\mathscr{X}, \mathscr{S})$ into those of another measurable space $(\mathscr{Y}, \mathscr{T})$ such that for every set G belonging to the Borel field $\mathscr{T}$, the inverse image of G, $T^{-1}(G)$, belongs to the Borel field $\mathscr{S}$, where $T^{-1}(G) = \{x: T(x) \in G\}$.

Measure: A nonnegative, completely additive set function defined on a Borel field $\mathscr{S}$ of a measurable space $(\mathscr{X}, \mathscr{S})$.

Minor (of a matrix $\mathbf{A}$): The determinant of any square submatrix of $\mathbf{A}$.

Moment Generating Function (of a random variable X): A function of a real variable t equal to the expected value of e^{tX} with respect to the distribution function of X.

Most Powerful Test: That test among all tests of a given size giving the largest possible value to the probability of rejecting the null hypothesis when an alternative hypothesis is true.

Neighborhood of a Point: The neighborhood of a point $\mathbf{a}$ is the set of points $\mathbf{x}$ which satisfy an inequality of the form $|\mathbf{x} - \mathbf{a}| < \epsilon$, where $\epsilon > 0$ and $|\mathbf{x} - \mathbf{a}|$ means the distance between $\mathbf{x}$ and $\mathbf{a}$. (See Euclidean Space.)

Nonsingular Linear Transformation: A linear transformation with a nonsingular matrix.

Nonsingular Matrix: A square matrix $\mathbf{A}$ such that its determinant $|\mathbf{A}| \neq 0$. If $|\mathbf{A}| = 0$ the matrix is said to be singular.

Nonsingular Transformation: A one-to-one transformation which has an inverse.

One-sided Hypothesis: A hypothesis which places the value of a parameter as always greater than, or as always less than, some fixed constant.

One-to-One Transformation T: A transformation such that $T(x_1) = T(x_2)$ when and only when $x_1 = x_2$.

Open Set of R^n: A set all of whose points are interior points, that is, points such that a neighborhood of the point belongs entirely to the set.

Orthogonal Matrix: A matrix **C** such that $\mathbf{CC}' = \mathbf{I}$, where $\mathbf{C}'$ is the transpose of **C** and **I** is the identity matrix.

Parameter Space: The space of all admissible parameter points.

Point: Any element of a space $\mathscr{X}$. Generally a vector $(x_1, x_2, \cdots, x_k)$ for a vector space of k dimensions.

Point Function: A function defined (having a value) for every point of a space. Contrast is usually with set function.

Positive Definite Matrix: The matrix of a positive definite quadratic form.

Positive Definite Quadratic Form: A quadratic form which is never negative for real values of the variables and is zero only for all values of the variables equal to zero.

Positive Matrix: The matrix of a positive quadratic form.

Positive Quadratic Form: A quadratic form which is nonnegative and which may be zero for real values of the variables not all zero.

Power (of a test): The power of a test (of a given size) is the probability of rejecting the null hypothesis when an alternative hypothesis is true.

Principal Minor (of a square matrix): A minor whose diagonal elements are diagonal elements of the matrix.

Probability Measure: A measure μ such that $\mu(\mathscr{X}) = 1$ for the space $\mathscr{X}$ [which is a measurable space $(\mathscr{X}, \mathscr{S})$].

Probability Measure Space: Same as probability space.

Probability Space: A measurable space $(\mathscr{X}, \mathscr{S})$ on which a probability measure μ is defined. Designated as $(\mathscr{X}, \mathscr{S}, \mu)$.

Quadratic Form: An expression of the form $\mathbf{x}'\mathbf{A}\mathbf{x} = \sum_{i=1}^{n} \sum_{j=1}^{n} a_{ij}x_ix_j$, with $\mathbf{x}' = (x_1, x_2, \cdots, x_n)$ and the matrix $\mathbf{A} = (a_{ij})$ of the quadratic form symmetric.

R^n: Symbol for the Euclidean space of n dimensions.

Radon-Nikodym Theorem: If μ and ν are two σ-finite measures on the measurable space $(\mathscr{X}, \mathscr{S})$ such that ν is absolutely continuous with respect to μ, then there exists an $\mathscr{S}$-measurable function $f(x)$ such that $0 < f(x) < +\infty$, and for every set $E \in \mathscr{S}$ $\nu(E) = \int_E f(x)\, d\mu(x)$. The function $f(x)$ is unique in the sense that if there exists another function $g(x)$ with the same properties as $f(x)$, then $\mu(x{:}f(x) \neq g(x)) = 0$.

Random Variable: Any $\mathscr{S}$-measurable function $f(x)$ defined on a measurable space $(\mathscr{X}, \mathscr{S})$.

Rank of a Matrix: A matrix is of rank r if r is the largest integer such that at least one minor of the matrix of order r is not zero.

Region of Acceptance (Rejection): A set of the sample space such that if a sample point (or function thereof) falls inside (outside) the set we accept (reject) a given hypothesis.

Set: Any subset of a given set (space) $\mathscr{X}$. (The words set, subset, and space are among the "undefined elements" of the science, theory, or geometry of measure.)

Set Function: A function whose domain of definition is a class of sets.

σ-Algebra: Same as Borel field. A nonempty class of sets closed under (that is, contains the result of) the formation of complements and denumerable unions.

σ-Finite Measure: A measure μ for which a finite or denumerable sequence of measurable sets E_i can be found such that the union $\cup E_i = \mathscr{X}$ (the whole space) and $\mu(E_i) < \infty$ for every i.

Size (of a test): The probability of rejecting the null hypothesis when it is true.

Space: Any collection or set of elements x of any nature. Denoted by $\mathscr{X}$. (An "undefined" element of our science.)

Statistic: Any function of a sample not depending on any parameter. Itself a random variable.

Supremum (sup): Least upper bound.

Theory: A set of axioms and all logical deductions (theorems) therefrom. Synonyms: Science, geometry.

Trace (of a square matrix): The sum of the diagonal elements.

Transformation: A function $T(x) = y$ of the elements x of a space $\mathscr{X}$ which establishes a correspondence between those elements and the elements of a space $\mathscr{Y}$.

Transformation Matrix: The matrix $\mathbf{A} = (a_{ij})$ of a linear transformation.

Transpose (of a matrix $\mathbf{A}$): The matrix $\mathbf{A}'$ with the rows and columns of $\mathbf{A}$ interchanged.

Truncation: A process by which all observations outside a given interval are discarded. The remaining cases then yield a truncated distribution with the distribution function

$$F(x|a < \xi \leq b) = \begin{cases} 0 & \text{for } x \leq a \\ \dfrac{F(x) - F(a)}{F(b) - F(a)} & \text{for } a < x \leq b \\ 1 & \text{for } x > b, \end{cases}$$

where $F(x|a < \xi \leq b)$ is the conditional distribution function of the random variable ξ on the assumption that ξ lies on the interval $(a, b]$ (half open) and where $F(x)$ is the original distribution function of ξ on the whole x-space of ξ.

Type I *Error:* The error made in rejecting the null hypothesis when it is true.

Type II *Error:* The error made in accepting the null hypothesis when it is false.

Unbiased Estimate: An estimate whose expected value is the estimated parameter.

Uniformly Most Powerful Test: The test among all tests of a given size that is most powerful for *all* admissible alternative hypotheses.

Union of Two or More Sets: The set of all those points of a space $\mathscr{X}$ which belong to at least one of the sets. If E_i denotes the sets, for $i = 1, 2, \cdots, n$, then $\bigcup_{i=1}^{n} E_i$ denotes the union.

Upper Bound (of a set E of real numbers): A real point d such that for every point x of E, $x \leq d$.

Vector: A matrix consisting of a single row or of a single column.

Appendix

Note to page 38

Anticipating lemma 4.9 in section 4 we also state:

THEOREM 2.2. *If $f_1(x)$ and $f_2(x)$ are generalized densities of a dominated set of probability measures, $Y = T(x)$ is a measurable statistic such that $\int T(x)f_1(x)d\lambda(x)$ exists, and $M_2(\tau) = \int f_2(x)e^{\tau T(x)}\, d\lambda(x)$ exists for τ in some interval; then*

$$I(1:2) \geqq \theta\tau - \log M_2(\tau) = I(*:2),\ \theta = \frac{d}{d\tau}\log M_2(\tau),$$

for $f_1(x)$ ranging over the generalized densities for which $\int T(x)f_1(x)d\lambda(x) \geqq \theta$ and $\int T(x)f_2(x)\, d\lambda(x) < \theta$, with equality if and only if

$$f_1(x) = f^*(x) = e^{\tau T(x)}f_2(x)/M_2(\tau)\ [\lambda].$$

Note to page 70

THEOREM 2.1a. *Suppose that the probability measures in theorem 2.1 are such that*

$$\nu_i^{(N)}(G) = \int_G g_i^{(N)}(y)\, d\gamma(y),\ \nu_i(G) = \int_G g_i(y)\, d\gamma(y),\quad i = 1, 2,\ G \in \mathscr{T}.$$

If $\lim\limits_{N\to\infty} (g_1^{(N)}/g_1) = 1\ [\gamma]$, *uniformly and* $\lim\limits_{N\to\infty} \log (g_1^{(N)}/g_2^{(N)}) = \log (g_1/g_2)[\gamma]$, *uniformly, then* $\lim\limits_{N\to\infty} I(1^{(N)}:2^{(N)};\mathscr{Y}) = I(1:2;\mathscr{Y})$, *if* $I(1:2;\mathscr{Y})$ *is finite.*

Proof. $$|I(1:2;\mathscr{Y}) - I(1^{(N)}:2^{(N)};\mathscr{Y})| = \left|\int\left(g_1 \log\frac{g_1}{g_2} - g_1^{(N)}\log\frac{g_1^{(N)}}{g_2^{(N)}}\right) d\gamma\right|$$

$$\leqq \int\left|g_1\log\frac{g_1}{g_2} - g_1^{(N)}\log\frac{g_1}{g_2}\right| d\gamma + \int\left|g_1^{(N)}\log\frac{g_1}{g_2} - g_1^{(N)}\log\frac{g_1^{(N)}}{g_2^{(N)}}\right| d\gamma$$

$$\leqq \int\left|1 - \frac{g_1^{(N)}}{g_1}\right| g_1 \left|\log\frac{g_1}{g_2}\right| d\gamma + \int\left|\log\frac{g_1}{g_2} - \log\frac{g_1^{(N)}}{g_2^{(N)}}\right|\frac{g_1^{(N)}}{g_1}\, g_1\, d\gamma.$$

For sufficiently large N, $\left|1 - \frac{g_1^{(N)}}{g_1}\right| < \epsilon_1$

$$\left|\log \frac{g_1}{g_2} - \log \frac{g_1^{(N)}}{g_2^{(N)}}\right| < \epsilon_2, \frac{g_1^{(N)}}{g_1} < 1 + \epsilon_1$$

so that

$$|I(1:2; \mathscr{Y}) - I(1^{(N)}:2^{(N)}:\mathscr{Y})|$$

$$\leqq \epsilon_1 \int g_1 \left|\log \frac{g_1}{g_2}\right| d\gamma + \epsilon_2(1 + \epsilon_1)$$

and since ϵ_1 and ϵ_2 are arbitrarily small the assertion is proven. (See S. Ikeda (1960), "A remark on the convergence of Kullback-Leibler's mean information", *Annals of the Institute of Statistical Mathematics*, Vol. 12, No. 1, pp. 81–88.)

Note that if $I(1^{(N)}:2^{(N)}; \mathscr{Y})$ in theorem 2.1, page 70, is a monotonically increasing function of N and $I(1^{(N)}:2^{(N)}; \mathscr{Y}) \leqq I(1:2; \mathscr{Y})$, then $\lim_{N\to\infty} I(1^{(N)}:2^{(N)}; \mathscr{Y}) = I(1:2; \mathscr{Y})$. Since $\liminf_{N\to\infty} I(1^{(N)}:2^{(N)}; \mathscr{Y}) \geqq I(1:2; \mathscr{Y})$, if $I(1:2; \mathscr{Y}) = \infty$, then $\lim_{N\to\infty} I(1^{(N)}:2^{(N)}; \mathscr{Y}) = \infty$.

Note to page 72

The following is the proof of lemma 2.2, page 72. In problems 7.29, 7.31, 7.32 on page 69 it is shown that

$$I(1:2) \geqq -2 \log \int (f_1(x)f_2(x))^{1/2}\, d\lambda(x) \geqq 2(1 - \int (f_1(x)f_2(x))^{1/2}\, d\lambda(x)$$

$$= \int ((f_1(x))^{1/2} - (f_2(x))^{1/2})^2\, d\lambda(x) \geqq \tfrac{1}{4} \int |f_1(x) - f_2(x)|\, d\lambda(x))^2.$$

(see S. Kullback (1966), "An information-theoretic derivation of certain limit relations for a stationary Markov Chain," *SIAM Journal on Control*, Vol. 4, No. 3.)

Accordingly $\frac{1}{4}(\int |f_1^{(N)}(x) - f_1(x)|d\lambda(x))^2 \leqq I(1^{(N)}:1) < \epsilon$ for sufficiently large N, that is $\int |f_1^{(N)}(x) - f_1(x)|d\lambda(x) \to 0$ as $N \to \infty$. The last assertions follow from Loève (1955, p. 140, problem 16 and page 158).

Note to page 306

In problem 8.34, page 341, table 8.5, the component due to $\Sigma_{21}\, \Sigma_{11}^{-1}$ reduces to the independence test for $\Sigma_{21} = \mathbf{0}$. The component in table 8.5 is then also a test for specified $\Sigma_{21}\, \Sigma_{11}^{-1}$ other than $\mathbf{0}$. The analysis for independence may also be set up as in table 8.6.

TABLE 8.6

Component due to	Information	D.F.	β^2
Independence within set 1	$-N \log \lvert R_{11}\rvert$	$\frac{k_1(k_1-1)}{2}$	$\frac{k_1(k_1-1)(2k_1+5)}{12N}$
Independence within set 2	$-N \log \lvert R_{22}\rvert$	$\frac{k_2(k_2-1)}{2}$	$\frac{k_2(k_2-1)(2k_2+5)}{12N}$
...	...	...	...
Independence within set m	$-N \log \lvert R_{mm}\rvert$	$\frac{k_m(k_m-1)}{2}$	$\frac{k_m(k_m-1)(2k_m+5)}{12N}$
Independence between sets	$\frac{-N \log \lvert R\rvert}{\lvert R_{11}\rvert \cdots \lvert R_{mm}\rvert}$	$\sum_{i<j} k_i k_j$	$\frac{(2k^3+3k^2-k-\sum_i(2k_i^3+3k_i^2-k_i))}{12N}$
Total independence	$-N \log \lvert R\rvert$	$\frac{k(k-1)}{2}$	$\frac{k(k-1)(2k+5)}{12N}$

Note that when R_{11} contains all but the last variable

$$-N \log |R| = -N \log |R_{11}| - N \log \frac{|R|}{|R_{11}|}$$

and

$$\log \frac{|R|}{|R_{11}|} = \log (1 - r^2_{m.12\cdots m-1}).$$

(See table 8.1, page 337.)

Index

A CATALOG OF SELECTED

DOVER BOOKS

IN SCIENCE AND MATHEMATICS

HANDBOOK OF MATHEMATICAL FUNCTIONS WITH FORMULAS, GRAPHS, AND MATHEMATICAL TABLES, edited by Milton Abramowitz and Irene A. Stegun. Vast compendium: 29 sets of tables, some to as high as 20 places. 1,046pp. 8 × 10½. 61272-4 Pa. $24.95

MATHEMATICAL METHODS IN PHYSICS AND ENGINEERING, John W. Dettman. Algebraically based approach to vectors, mapping, diffraction, other topics in applied math. Also generalized functions, analytic function theory, more. Exercises. 448pp. 5⅜ × 8¼. 65649-7 Pa. $10.95

A SURVEY OF NUMERICAL MATHEMATICS, David M. Young and Robert Todd Gregory. Broad self-contained coverage of computer-oriented numerical algorithms for solving various types of mathematical problems in linear algebra, ordinary and partial, differential equations, much more. Exercises. Total of 1,248pp. 5⅜ × 8½. Two volumes. Vol. I 65691-8 Pa. $14.95
Vol. II 65692-6 Pa. $14.95

TENSOR ANALYSIS FOR PHYSICISTS, J.A. Schouten. Concise exposition of the mathematical basis of tensor analysis, integrated with well-chosen physical examples of the theory. Exercises. Index. Bibliography. 289pp. 5⅜ × 8½.
65582-2 Pa. $8.95

INTRODUCTION TO NUMERICAL ANALYSIS (2nd Edition), F.B. Hildebrand. Classic, fundamental treatment covers computation, approximation, interpolation, numerical differentiation and integration, other topics. 150 new problems. 669pp. 5⅜ × 8½. 65363-3 Pa. $15.95

INVESTIGATIONS ON THE THEORY OF THE BROWNIAN MOVEMENT, Albert Einstein. Five papers (1905–8) investigating dynamics of Brownian motion and evolving elementary theory. Notes by R. Fürth. 122pp. 5⅜ × 8½.
60304-0 Pa. $4.95

CATASTROPHE THEORY FOR SCIENTISTS AND ENGINEERS, Robert Gilmore. Advanced-level treatment describes mathematics of theory grounded in the work of Poincaré, R. Thom, other mathematicians. Also important applications to problems in mathematics, physics, chemistry and engineering. 1981 edition. References. 28 tables. 397 black-and-white illustrations. xvii + 666pp. 6⅛ × 9¼.
67539-4 Pa. $17.95

AN INTRODUCTION TO STATISTICAL THERMODYNAMICS, Terrell L. Hill. Excellent basic text offers wide-ranging coverage of quantum statistical mechanics, systems of interacting molecules, quantum statistics, more. 523pp. 5⅜ × 8½. 65242-4 Pa. $12.95

STATISTICAL PHYSICS, Gregory H. Wannier. Classic text combines thermodynamics, statistical mechanics and kinetic theory in one unified presentation of thermal physics. Problems with solutions. Bibliography. 532pp. 5⅜ × 8½.
65401-X Pa. $12.95

THE FOUR-COLOR PROBLEM: Assaults and Conquest, Thomas L. Saaty and Paul G. Kainen. Engrossing, comprehensive account of the century-old combinatorial topological problem, its history and solution. Bibliographies. Index. 110 figures. 228pp. 5⅜ × 8½. 65092-8 Pa. $6.95

CATALYSIS IN CHEMISTRY AND ENZYMOLOGY, William P. Jencks. Exceptionally clear coverage of mechanisms for catalysis, forces in aqueous solution, carbonyl- and acyl-group reactions, practical kinetics, more. 864pp. 5⅜ × 8½. 65460-5 Pa. $19.95

PROBABILITY: An Introduction, Samuel Goldberg. Excellent basic text covers set theory, probability theory for finite sample spaces, binomial theorem, much more. 360 problems. Bibliographies. 322pp. 5⅜ × 8½. 65252-1 Pa. $9.95

LIGHTNING, Martin A. Uman. Revised, updated edition of classic work on the physics of lightning. Phenomena, terminology, measurement, photography, spectroscopy, thunder, more. Reviews recent research. Bibliography. Indices. 320pp. 5⅜ × 8¼. 64575-4 Pa. $8.95

PROBABILITY THEORY: A Concise Course, Y.A. Rozanov. Highly readable, self-contained introduction covers combination of events, dependent events, Bernoulli trials, etc. Translation by Richard Silverman. 148pp. 5⅜ × 8¼. 63544-9 Pa. $6.95

AN INTRODUCTION TO HAMILTONIAN OPTICS, H. A. Buchdahl. Detailed account of the Hamiltonian treatment of aberration theory in geometrical optics. Many classes of optical systems defined in terms of the symmetries they possess. Problems with detailed solutions. 1970 edition. xv + 360pp. 5⅜ × 8½. 67597-1 Pa. $10.95

STATISTICS MANUAL, Edwin L. Crow, et al. Comprehensive, practical collection of classical and modern methods prepared by U.S. Naval Ordnance Test Station. Stress on use. Basics of statistics assumed. 288pp. 5⅜ × 8½. 60599-X Pa. $7.95

DICTIONARY/OUTLINE OF BASIC STATISTICS, John E. Freund and Frank J. Williams. A clear concise dictionary of over 1,000 statistical terms and an outline of statistical formulas covering probability, nonparametric tests, much more. 208pp. 5⅜ × 8½. 66796-0 Pa. $7.95

STATISTICAL METHOD FROM THE VIEWPOINT OF QUALITY CONTROL, Walter A. Shewhart. Important text explains regulation of variables, uses of statistical control to achieve quality control in industry, agriculture, other areas. 192pp. 5⅜ × 8½. 65232-7 Pa. $7.95

THE INTERPRETATION OF GEOLOGICAL PHASE DIAGRAMS, Ernest G. Ehlers. Clear, concise text emphasizes diagrams of systems under fluid or containing pressure; also coverage of complex binary systems, hydrothermal melting, more. 288pp. 6½ × 9¼. 65389-7 Pa. $10.95

STATISTICAL ADJUSTMENT OF DATA, W. Edwards Deming. Introduction to basic concepts of statistics, curve fitting, least squares solution, conditions without parameter, conditions containing parameters. 26 exercises worked out. 271pp. 5⅜ × 8½. 64685-8 Pa. $9.95